TOXIGENIC *FUSARIUM* SPECIES
Identity and Mycotoxicology

W.F.O. Marasas,
Paul E. Nelson,
and
T.A. Toussoun

The Pennsylvania State University Press
University Park and London

The authors gratefully acknowledge the contribution of the South African Medical Research Council, whose financial support and encouragement to W.F.O. Marasas made possible his continuous survey of the vast number of publications dealing with the mycotoxicology of *Fusarium* species and resulted in the writing of this book. The contribution of The Pennsylvania State University, which established the Fusarium Research Center and has supported the Center's research projects for the past fourteen years, is also gratefully acknowledged.

Preparation and publication of this book was supported in part by Contract No. 223-78-2018 from the Department of Health and Human Services, Public Health Service, Food and Drug Administration, Washington, D.C.

Library of Congress Cataloging in Publication Data

Marasas, W. F. O.
  Toxigenic Fusarium species, identity and mycotoxicology.

  Includes bibliography and index.
  1. Fusarium—Classification. 2. Fusarium—Toxicology.
3. Fusarium—Catalogs and collections—Pennsylvania—
University Park. 4. Toxigenic fungi—Classification.
5. Toxigenic fungi—Catalogs and collections—Pennsylvania—University Park. 6. Mycotoxicoses. 7. Mycotoxins.
8. Fungi—Classification. 9. Fungi—Catalogs and collections—Pennsylvania—University Park. 10. International Toxigenic Fusarium Reference Collection—Catalogs.
I. Nelson, Paul E., 1927–   II.. Toussoun, T.A., 1925–   III. Title.
QK625.T8M37 1984   589.2'4   82-42779
ISBN 0-271-00348-0

Copyright © 1984 The Pennsylvania State University
All rights reserved
Designed by Dolly Carr
Printed in the United States of America

Contents

PREFACE	vii
INTRODUCTION	ix
CONTRIBUTORS OF *FUSARIUM* CULTURES	xiii
LIST OF ABBREVIATIONS	xvii
CHEMICAL NAMES OF *FUSARIUM* MYCOTOXINS	xix
SECTION EUPIONNOTES	1
F. episphaeria	
F. merismoides	
SECTION SPICARIOIDES	3
F. decemcellulare	
SECTION ARACHNITES	5
F. nivale	
F. larvarum	
SECTION SPOROTRICHIELLA	12
F. sporotrichioides	
F. chlamydosporum	
F. poae	
F. tricinctum	

Contents

SECTION ROSEUM	92
F. avenaceum	
SECTION ARTHROSPORIELLA	98
F. semitectum	
SECTION GIBBOSUM	111
F. equiseti	
F. acuminatum	
SECTION DISCOLOR	133
F. heterosporum	
F. sambucinum	
F. culmorum	
F. graminearum	
SECTION LATERITIUM	212
F. lateritium	
SECTION LISEOLA	216
F. moniliforme	
F. proliferatum	
F. subglutinans	
F. anthophilum	
SECTION ELEGANS	253
F. oxysporum	
SECTION MARTIELLA AND VENTRICOSUM	263
F. solani	
LITERATURE CITED	273
FUSARIUM STRAINS CROSS-REFERENCE INDEX	311
INDEX	321

Preface

The genus *Fusarium* contains important mycotoxin-producing species that have been implicated in human diseases, such as alimentary toxic aleukia, Urov or Kashin-Beck disease, Akakabi-byo or scabby grain intoxication, and esophageal cancer. Many of these mycotoxin-producing species have also been implicated in several animal diseases, including hemorrhagic, estrogenic, emetic, and feed refusal syndromes, fescue foot, degnala disease, moldy sweet potato toxicosis, bean hulls poisoning, and equine leukoencephalomalacia. The interest in toxigenic *Fusarium* species is increasing world-wide due to the discovery of a growing number of naturally occurring *Fusarium* mycotoxins that have practical importance as threats to human and animal health.

Taxonomy of the genus *Fusarium* is complex and difficult to apply, partly because of the use of different taxonomic systems in different countries. The major taxonomic systems currently in use are discussed in a companion volume, "*Fusarium* Species: An Illustrated Manual for Identification" by Nelson et al. (1983). The taxonomy of the genus is further complicated by the extreme variability of *Fusarium* species in culture and the fact that they mutate and degenerate rapidly, particularly under conditions of repeated subculturing on common laboratory media. This situation has led to great confusion in the extensive literature on *Fusarium* mycotoxicology because the same fungus is known under a variety of different names, because different fungi are lumped together under names such as *F. tricinctum* Corda emend. Snyd. & Hans. and *F. roseum* Lk. emend.

Snyd. & Hans., because several *Fusarium* toxins have been named for mis-identified producing species, because elaborate chemical and pathological studies have been reported in the literature and attributed to incorrectly named species, and because many toxigenic *Fusarium* strains have become degenerate and lost their toxigenic ability due to maltreatment in laboratories that do not specialize in the maintenance of *Fusarium* cultures. Consequently it has become impossible to relate toxicological studies done in different laboratories to each other because of the existing confusion with regard to the taxonomy and nomenclature of toxigenic *Fusarium* species (Smalley et al., 1977).

In this volume we discuss the toxigenic *Fusarium* species in relation to their i) incidence and distribution, ii) association with human and/or animal diseases, iii) toxicity to experimental animals, and iv) mycotoxins produced. This information is followed by a listing of the toxigenic strains of *Fusarium* species deposited in the International Toxic *Fusarium* Reference Collection. This volume provides all the available information on these strains through December 1981.

The preparation of the manuscript for this volume was made possible by the dedicated cooperation and assistance of Miss Rhoda Klass and Mrs. Carleen Schreuder, National Research Institute for Nutritional Disease, South African Medical Research Council; and Mrs. Jo Barnes, Mrs. Hendra de Villiers, Mrs. Laetitia van der Merwe, and Miss Elize Birch, Institute for Medical Literature, South African Medical Research Council. Mrs. Lois Klotz and Mrs. Nancy Fisher Gregory, *Fusarium* Research Center, The Pennsylvania State University, prepared cultures for examination and storage in the International Toxic *Fusarium* Reference Collection and also assisted in many other ways. Our special thanks to all of the individuals involved in research on toxigenic *Fusarium* species who supplied us with strains and information to use in establishing the International Toxic *Fusarium* Reference Collection. They are too numerous to name here but we express our appreciation to each of them.

Finally we express our appreciation to the Director of the Pennsylvania State University Press, Chris W. Kentera, and his fine staff for their patience and understanding during the preparation of this volume. Our association with the Press staff has been a most rewarding one and their assistance always makes our task easier.

Introduction

Complete descriptions and illustrations of *Fusarium* species are given in the companion volume (Nelson et al., 1983). These descriptions and illustrations should be consulted for any information concerning the taxonomy and nomenclature of the *Fusarium* species discussed in this book.

The numbering of sections in this volume is the same as the numbering system used in Nelson et al. (1983). None of the *Fusarium* species in the sections Macroconia, Submicrocera, and Pseudomicrocera have been reported to be toxigenic and these sections are therefore not discussed.

The discussion of each toxigenic *Fusarium* species follows the standard outline given below.

- Incidence and Distribution
- Association with Human and/or Animal Diseases
- Toxicity to Experimental Animals
- Mycotoxins Produced
- Toxigenic Strains in the International Toxic *Fusarium* Reference Collection (ITFRC)

The main emphasis of this book is on the identity and toxigenicity of *Fusarium* strains. Thus more attention is given to the toxicity of cultures to animals and to the mycotoxins produced in culture by *Fusarium* strains than to the toxicology of pure *Fusarium* myco-

Table A. Toxigenic *Fusarium* Strains in the International *Fusarium* Reference Collection (ITFRC)

Section	Species	Number of Toxigenic Strains
Eupionnotes	*F. merismoides*	1
Arachnites	*F. nivale*	1
	F. larvarum	1
Sporotrichiella	*F. sporotrichioides*	43
	F. chlamydosporum	3
	F. poae	8
	F. tricinctum	8
Roseum	*F. avenaceum*	5
Arthrosporiella	*F. semitectum*	8
Gibbosum	*F. equiseti*	15
	F. acuminatum	4
Discolor	*F. sambucinum*	6
	F. culmorum	12
	F. graminearum	34
Liseola	*F. moniliforme*	32
	F. proliferatum	1
	F. subglutinans	4
	F. anthophilum	1
Elegans	*F. oxysporum*	9
Martiella	*F. solani*	7
	Total	203

toxins. In the section titled "Association with Human and/or Animal Diseases," we are not implying that the *Fusarium* species discussed under these diseases are the direct causes of these diseases. In cases where a species of *Fusarium* has been proven to be the cause of a particular disease, this is clearly stated. In many cases *Fusarium* species have been implicated as part of a complex of factors associated with a human and/or animal disease. In other cases, *Fusarium* species have merely been isolated from foods or feeds associated with a disease without any additional epidemiological or experimental evidence of a causative relationship. In the section titled "Mycotoxins Produced," the main emphasis is on the four major groups of *Fusarium* mycotoxins—i.e., butenolide, moniliformin, trichothecenes, and zearalenone—rather than the large number of antibiotics, phytotoxins, pigments, and other metabolites (Cole & Cox, 1981; Palti, 1978; Turner, 1971; Vesonder & Hesseltine, 1981) known to be produced by *Fusarium*. The literature review for all sections was completed on 31 December 1981.

Strains of *Fusarium* species reported to be toxigenic in the literature were solicited from the investigators who published on them. A total of 203 toxigenic strains of *Fusarium* were sent to us. These strains have been deposited in the ITFRC and have been identified according to Nelson et al. (1983) to belong in the 20 species in 10 sections of the genus *Fusarium* as shown in Table A.

Many *Fusarium* strains that have been reported in the literature to be toxigenic could not be examined by us for a variety of reasons, most often because we were unable to obtain the strains in question. In discussing strains not represented in the ITFRC, we have used the names given in the literature, but the reader should realize that some of these strains may have been mis-identified and some of the names used in the literature are incorrect.

Strains deposited in the ITFRC are treated in great detail. As far as possible, all of the references in which each strain has been cited are given. In addition, all of the names that have been used for each strain in the literature are given. This information is as complete as possible for literature published to December 1981.

In the text and tables the reader will note that each culture has several numbers. There is a strain number used as a means of identifying the strains described in the section of the text titled "Toxigenic Strains in the ITFRC." These numbers consist of the section number and the number of the strain in the order in which they are discussed in the text. For example, strain 4.1. is the first strain discussed in section 4, Sporotrichiella, and the numbers and species name for this culture are:

Strain No.	ITFRC No.	MRC No.	
4.1	T-345	1704	*F. tricinctum* NRRL 3510

ITFRC No. T-345 is the accession number of the culture as deposited in the International Toxic *Fusarium* Reference Collection; MRC 1704 is the accession number of the culture as deposited in the Medical Research Council Collection. The species name is the name used in published reports about this isolate followed by any number or numbers that were used in publications. In this way information is brought together that gives the reader the published name and numbers of the species, and the accession numbers for this strain in the two collections that have these cultures available for distribution. The address for each culture collection is given below.

International Toxic Fusarium Reference Collection
Fusarium Research Center
211 Buckhout Laboratory
The Pennsylvania State University
University Park, PA 16802
USA

Medical Research Council Collection
National Research Institute for Nutritional Diseases
P O Box 70
Tygerberg 7505
Republic of South Africa

A list of names and addresses of contributors of *Fusarium* cultures deposited in the ITFRC follows. Lists of abbreviations used in the text and of chemical names of *Fusarium* mycotoxins are also included.

Contributors of *Fusarium* Cultures

1. ATCC, American Type Culture Collection, 12301 Parklawn Drive, Rockville, Maryland 20852, USA.
2. BAF, Bundesanstalt für Fleischforschung, Oskar-von-Miller Strasse 20, Kulmbach 8650, Federal Republic of Germany.
3. Bhat, R.V., National Institute of Nutrition, Indian Council of Medical Research, Jamai-Osmania P.O., Hyderabad 500 007, India.
4. Blanchfield, B.J., Health Protection Branch, Health and Welfare Canada, Tunney's Pasture, Ottawa, Ontario K1A, 0L2, Canada.
5. Bottalico, A., Instituto Tossine e Micotossine da Parassiti Vegetali, Instituto Di Patologia Vegetale Dell'Universita, Via G Amendola 165/A, Bari 70126, Italy.
6. Claridge, C.A., Research Division, Bristol Laboratories, Bristol-Meyers Company, Syracuse, New York 13201, USA.
7. CMI, Commonwealth Mycological Institute, Kew, Surrey, England.
8. Cole, R.J., National Peanut Research Laboratory, USDA, P.O. Box 637, Dawson, Georgia 31742, USA.
9. Cullen, D., Department of Plant Pathology, University of Wisconsin, Madison, Wisconsin 53706, USA.

10. DAOM, Canadian National Collection of Fungus Cultures, Agriculture Canada, Biosystematics Research Institute, Wm. Saunders Building, C.E.F., Ottawa, Ontario K1A 0C6, Canada.
11. Diener, U.L., Department of Botany and Microbiology, Auburn University, Auburn, Alabama 36380, USA.
12. Di Menna, M.E., Ruakura Research Station, Private Bag, Hamilton, New Zealand.
13. El-Gholl, N.E., Division of Plant Industry, Florida Department of Agriculture and Consumer Services, P.O. Box 1269, Gainesville, Florida 32601, USA.
14. Eppley, R.M., Division of Chemistry and Physics, Food and Drug Administration, 200 C Street, Washington D.C. 20204, USA.
15. FRC, Fusarium Research Center, Department of Plant Pathology, The Pennsylvania State University, University Park, Pennsylvania 16802, USA.
16. Gerlach, W., Biologische Bundesanstalt für Land- und Forstwirtschaft, Königin-Luise-Strasse 19, 1000 Berlin 33 Dahlem, Federal Republic of Germany.
17. Hart, L.P., Department of Botany and Plant Pathology, Michigan State University, East Lansing, Michigan 48824, USA.
18. Ichinoe, M., National Institute of Hygienic Sciences, Kami-yoga 1-Chome, Setagaya-Ku, Tokyo 158, Japan.
19. Kurata, H., National Institute of Hygienic Sciences, Kami-yoga 1-Chome, Setagaya-Ku, Tokyo 158, Japan.
20. Langerfeld, E., Biologische Bundesanstalt für Land- und Forstwirtschaft, Messeweg 11/12, Braunschweig 3300, Federal Republic of Germany.
21. Marasas, W.F.O., National Research Institute for Nutritional Diseases, P.O. Box 70, Tygerberg 7505, South Africa.
22. Matuo, T., Faculty of Textile Science, Shinshu University, Ueda, Nagano 386, Japan.
23. Mirocha, C.J., Department of Plant Pathology, University of Minnesota, St. Paul, Minnesota 55101, USA.
24. Möller, J.M., Staatliche Landwirtschaftliche Untersuchungs- und Forschungsanstalt Augustenberg, Postfach 410943, 7500 Karlsruhe 41, Federal Republic of Germany.
25. MRC, South African Medical Research Council, P.O. Box 70, Tygerberg 7505, South Africa.
26. NRRL, Northern Regional Research Laboratory, USDA, 1815 North University Street, Peoria, Illinois 61604.
27. Palyusik, M., Veterinary Medical Research Institute, Hungarian Academy of Sciences, P.O. Box 18, H-1581 Budapest, Hungary.
28. Paul Gupta, R.K., Department of Veterinary Pathology, Haryana Agricultural University, Hissar 125004, India.
29. Rabie, C.J., National Research Institute for Nutritional Diseases, P.O. Box 70, Tygerberg 7505, South Africa.
30. Richard, J.L., National Animal Disease Center, USDA, P.O. Box 70, Ames, Iowa 50010, USA.
31. Smalley, E.B., Department of Plant Pathology, University of Wisconsin, Madison, Wisconsin 53706, USA.

32. Sutton, J.C., Department of Environmental Biology, University of Guelph, Guelph, Ontario N1G 2W1, Canada.
33. Ueno, Y., Department of Toxicology and Microbial Chemistry, Tokyo University of Science, 12, Ichigaya Funagawara-Machi, Shinjuki-ku, Tokyo 162, Japan.
34. Vora, V.C., Central Drug Research Institute, Chatter Manzil, Post Box 173, Lucknow 226001, India.
35. Wilson, B.J., Department of Biochemistry, School of Medicine, Vanderbilt University, Nashville, Tennessee 37232, USA.
36. Ylimäki, A., Institute of Plant Pathology, Agricultural Research Centre, PL 18, SF-01301 Vantaa 30, Finland.
37. Yoshizawa, T., Department of Food Science, Faculty of Agriculture, Kagawa University, Ikenobe, Miki-cho, Kita-gun, Kagawa, Japan.

MRC	South African Medical Research Council, Tygerberg, South Africa
NMR	Nuclear magnetic resonance
NRRL	Northern Regional Research Laboratory, Peoria, Illinois, USA
NUV	Near ultraviolet
PDA	Potato dextrose agar
PSC	Peptone-supplemented Czapek-Dox medium
TLC	Thin-layer chromatography
UV	Ultraviolet
YES	Yeast extract sucrose medium

Chemical Names of *Fusarium* Mycotoxins

4-Acetoxyscirpenediol = 4β-acetoxy-3α,15-dihydroxy-12,13-epoxytrichothec-9-ene. A similar compound, monodeacetylanguidin = 4- or 15-acetylscirpentriol.

3-Acetyldeoxynivalenol (= Deoxynivalenol monoacetate) = 3α-acetoxy-7α,15-dihydroxy-12,13-epoxytrichothec-9-en-8-one.

8-Acetylneosolaniol (= Neosolaniol monoacetate) = 4β,8α,15-triacetoxy-3α-hydroxy-12,13-epoxytrichothec-9-ene.

4- or 15-Acetylscirpentriol. See 4-Acetoxyscirpenediol.

Acetyl T-2 toxin = 3α,4β,15-triacetoxy-8α-(3-methylbutyryloxy)-12,13-epoxytrichothec-9-ene.

Anguidin. *See* Diacetoxyscirpenol.

Butenolide = 4-acetamido-4-hydroxy-2-butenoic-acid γ-lactone.

Calonectrin = 3α,15-diacetoxy-12,13-epoxytrichothec-9-ene.

15-Deacetylcalonectrin (= 15-De-0-acetylcalonectrin) = 3α-acetoxy-15-hydroxy-12,13-epoxytrichothec-9-ene.

Deoxynivalenol (= Rd toxin, = Vomitoxin) = 3α,7α,15-trihydroxy-12,13-epoxytrichothec-9-en-8-one.

Deoxynivalenol diacetate. See Diacetyldeoxynivalenol.

Deoxynivalenol monoacetate. See 3-Acetyldeoxynivalenol.

Diacetoxyscirpendiol. See 7α-Hydroxydiacetoxyscirpenol.

Diacetoxyscirpenol (= Anguidin) = 4β,15-diacetoxy-3α-hydroxy-12,13-epoxytrichothec-9-ene.

Diacetoxyscirpentriol. See 7α,8α-Dihydroxydiacetoxyscirpenol.

Diacetyldeoxynivalenol (= Deoxynivalenol diacetate) = 3α,15-diacetoxy-7α-hydroxy-12,13-epoxytrichothec-9-en-8-one.

Diacetylnivalenol (= Nivalenol diacetate) = 4β,15-diacetoxy-3α,7α-dihydroxy-12,13-epoxytrichothec-9-en-8-one.

7α 8α-Dihydroxydiacetoxyscirpenol (= Diacetoxyscirpentriol) = 4β,15-diacetoxy-3α,7α,8α-trihydroxy-12,13-epoxytrichothec-9-ene.

Fusarenon. See Fusarenon-X.

Fusarenon-X (=Fusarenon, = Monoacetylnivalenol, = Nivalenol monoacetate) = 4β-acetoxy-3α,7α,15-trihydroxy-12,13-epoxytrichothec-9-en-8-one.

Fusaric acid (= Fusarinic acid) = 5-butylpicolinic acid.

Fusarinic acid. See Fusaric acid.

F-2. See Zearalenone

HT-2 toxin = 15-acetoxy-3α,4β-dihydroxy-8α-(3-methylbutyryloxy)-12,13-epoxytrichothec-9-ene.

7α-Hydroxydiacetoxyscirpenol (= Diacetoxyscirpendiol) = 4β,15-diacetoxy-3α,7α-dihydroxy-12,13-epoxytrichothec-9-ene.

8α-Hydroxydiacetoxyscirpenol. See Neosolaniol.

1,4-Ipomeadiol = 1-(3-furyl)-1,4-pentanediol.

Ipomeanine = 1-(3-furyl)-1,4-pendanetione.

1-Ipomeanol = 1-(3-furyl)-1-hydroxy-4-pentanone.

4-Ipomeanol = 1-(3-furyl)-4-hydroxy-4-pentanone.

Moniliformin = potassium or sodium salt of 1-hydroxycyclobut-1-ene-3,4-dione.

Monoacetoxyscirpenol = 15-acetoxy-3α,4β-dihydroxy-12,13-epoxytrichothec-9-ene.

Monoacetylnivalenol. See Fusarenon-X

Monodeacetylanguidin. See 4-Acetoxyscirpenediol.

Neosolaniol (= 8α-Hydroxydiacetoxyscirpenol) = 4β,15-diacetoxy-3α,8α-dihydroxy-12,13-epoxytrichothec-9-ene.

Neosolaniolacetate. See 8-Acetylneosolaniol.

Neosolaniol monoacetate. See 8-Acetylneosolaniol.

Nivalenol = 3α,4β,7α,15-tetrahydroxy-12,13-epoxytrichothec-9-en-8-one.

Nivalenol diacetate. See Diacetylnivalenol.

Nivalenol monoacetate. See Fusarenon-X.

NT-1 toxin (=T-1 toxin) = 4β, 8α-diacetoxy-3α,15-dihydroxy-12,13-epoxytrichothec-9-ene.

NT-2 toxin = 4β-acetoxy-3α,8α,15-trihydroxy-12,13-epoxytrichothec-9-ene.

Rd toxin. See Deoxynivalenol.

Scirpentriol = 3α,4β,15-trihydroxy-12,13-epoxytrichothec-9-ene.

Solaniol. See Neosolaniol.

T-1 toxin. *See* NT-1 toxin.

T-2 toxin = 4β,15-diacetoxy-3α-hydroxy-8α-(3-methylbutyryloxy)-12,13-epoxytrichothec-9-ene.

Triacetoxyscirpendiol = 4β,8α,15-triacetoxy-3α,7α-dihydroxy-12,13-epoxytrichothec-9-ene.

Triacetoxyscirpenol = 3α, 4β,15-triacetoxy-12,13-epoxytrichothec-9-ene.

Vomitoxin. *See* Deoxynivalenol.

Zearalenol = 2,4-dihydroxy-6-(6,10-dihydroxy-trans-1-undecenyl)-benzoic acid μ-lactone.

Zearalenone = 6-(10-hydroxy-6-oxo-trans-1-undecenyl)-β-resorcylic acid lactone.

1. Section Eupionnotes

Fusarium episphaeria (Tode) Snyd. & Hans.

The name *F. episphaeria* as used by Snyder & Hansen (1945) combines the characteristics of all the *Fusarium* species in the sections Eupionnotes and Macroconia as defined by Wollenweber & Reinking (1935). As used by Snyder & Hansen (1945), this name can be considered as a synonym of all the species in these two sections.

We have identified the toxigenic Japanese strain referred to as *F. episphaeria* Fn-M by Ishii et al. (1974) and Ueno et al. (1971c, 1972b, 1973a, 1975), and as *F. tricinctum* NRRL 6491 by Burmeister et al. (1981), as *F. sporotrichioides* (see Strain 4.10). Consequently this strain is excluded from the section Eupionnotes.

Fusarium merismoides Corda

    Perfect State: Unknown

Incidence and Distribution

Fusarium merismoides has been isolated from slimy exudates of trees (Wollenweber & Reinking, 1935) and is common in soil and in polluted water or sludge in temperate and to a lesser extent also in tropical areas (Booth, 1971; Domsch et al., 1980; Meyer & Frank, 1979). This species has also been associated with stored potatoes in Finland (Jamalainen, 1955) and Germany (Siegfried & Langerfeld, 1978).

Association with Human and/or Animal Diseases

Fusarium merismoides has not been associated with any human or animal toxicoses.

Toxicity to Experimental Animals

According to Siegfried & Langerfeld (1978), extracts of potato tubers inoculated with a strain of *F. merismoides* (see Strain 1.1) isolated from potatoes in Germany were slightly toxic to brine shrimp.

Mycotoxins Produced

The metabolite(s) toxic to brine shrimp produced in inoculated potato tubers by the strain of *F. merismoides* (see Strain 1.1) reported by Siegfried & Langerfeld (1978) have not been identified.

Toxigenic Strains in the ITFRC

We have identified the following toxigenic strain as *F. merismoides* according to Nelson et al. (1983):

Strain 1.1 *F. merismoides* **DE 8** (=E-114; MRC 2196)

This strain was isolated from rotting potato tubers in northern Germany by Siegfried & Langerfeld (1978). Potato tubers were artificially inoculated with this strain, incubated at 20°C, and lyophilized. Extracts of the infected potato tissue proved to be slightly toxic to brine shrimp (20–50% mortality within 4 hours), but the chemical nature of the mycotoxin(s) has not yet been determined.

A culture of this strain was received on 20 January 1981 from E. Langerfeld as "*F. merismoides* DE 8."

2. Section Spicarioides

Fusarium decemcellulare Brick

Perfect State: *Calonectria rigidiuscula* (Berk. & Br.) Sacc.

Fusarium decemcellulare (= *F. rigidiusculum*) is best known as a pathogen associated with dieback and gall formation on cacao (*Theobroma cacao* L.), but also occurs as a wound parasite and saprophyte on a wide variety of hosts in tropical and sub-tropical regions of Africa, America, Asia, and Australasia (Booth, 1971; CMI Descriptions of Pathogenic Fungi and Bacteria No 21, 1964; Doidge, 1938; Gordon, 1956, 1960a; Meyer & Frank, 1979; Wollenweber & Reinking, 1935).

We have identified one of the three strains of *Fusarium* spp. (M-1-3, M-10-1, and M-13-3) isolated from moldy bean hulls associated with outbreaks of "bean hulls poisoning" of horses (see *F. sporotrichioides,* bean hulls poisoning, and *F. moniliforme,* equine leukoencephalomalacia) in Hokkaido, Japan, by Ueno et al. (1972a), and subsequently referred to as "*F. rigidiusculum* M-1-3" by Ueno et al. (1973a), as *F. graminearum* (Strain 8.29). Consequently this strain is excluded from *F. decemcellulare.*

Culture filtrates of one unnumbered isolate of "*F. rigidiusculum*" from foodstuffs in Uganda were reported to be toxic to mice upon intraperitoneal injection by Itakura & Kinosita (1975). The mycotoxins produced by this strain were not identified.

Spicarioides (2)

There are no toxigenic strains of *F. decemcellulare* represented in the ITFRC. We have identified the strain isolated from moldy bean hulls in Japan (as *Fusarium* sp. M-1-3) and found to be toxic to mice (Ueno et al., 1972a) and subsequently reported (as *F. rigidiusculum* M-1-3) to produce diacetoxyscirpenol, neosolaniol, and T-2 toxin (Ueno et al., 1973a) as *F. graminearum* (Strain 8.29)

3. Section Arachnites

Fusarium nivale (Fr.) Ces.
Perfect State: *Monographella nivalis* (Schaffnit) Müller

Incidence and Distribution

Fusarium nivale typically occurs as a pathogen of grasses and cereals, particularly under a snow cover (=snow mold), in areas with cold to temperate climates in Asia, Europe, North America, Australia, and New Zealand (Booth, 1971; CMI Descriptions No. 309, 1971; Domsch et al., 1980; Gams & Müller, 1980; Gordon, 1952; Joffe, 1960a, b; Meyer & Frank, 1979; Wollenweber & Reinking, 1935); it also occurs as a seed-borne pathogen in harvested cereal grains, e.g., barley and wheat in the United Kingdom (Hacking et al., 1976, 1977; Hewett, 1967).

Association with Human and/or Animal Diseases

Fescue Foot. We have identified *F. nivale* NRRL 3249, one of the strains originally referred to as *F. nivale* in connection with research on "fescue foot" of cattle in the United States, as *F. sporotrichioides* (Strain 4.8). Consequently this strain is excluded from *F. nivale* and fescue foot is discussed under section Sporotrichiella (see *F. sporotrichioides,* fescue foot).

Akakabi-byo (Scabby Grain Intoxication). We have identified *F. nivale* Fn-2B, one of the strains originally referred to as *F. nivale* in connection with research on "akakabi-byo" in Japan, as *F. sporotrichioides* (Strain 4.9). Consequently this strain is excluded from *F. nivale* and akakabi-byo is discussed under section Sporotrichiella (see *F. sporotrichioides, akakabi-byo*). Because of the fact that *F. graminearum* is probably also involved in this syndrome, akakabi-byo is also discussed under section Discolor (see *F. graminearum, akakabi-byo*).

Toxicity to Experimental Animals

Cattle. The strain originally identified as *F. nivale* and reported (as *F. nivale* NRRL 13,318, *F. nivale* NRRL 3249, or *F. tricinctum* NRRL 3249) to be toxic to cattle by Tookey et al. (1972) and Yates et al. (1969) is *F. sporotrichioides* (Strain 4.8).

Guinea Pigs. *Fusarium nivale* Fn-2 reported to be toxic to guinea pigs by Saito et al. (1969) is *F. sporotrichioides* (Strain 4.9).

Insects. According to G.R.F. Davis et al. (1975), cultures of two Canadian isolates of *F. nivale* (DE 7247 and 7249) on autoclaved rye incubated at 10°C for 54 days, caused a reduction in weight gain, but not mortality, in larvae of the yellow meal worm (*Tenebrio molitor*). The metabolite(s) responsible for the growth retardation have not been identified.

Mice. The strain originally identified as *F. nivale* and reported (as *F. nivale* NRRL 13,318, *F. nivale* NRRL 3249, or *F. tricinctum* NRRL 3249) to be toxic to mice by Ellis & Yates (1971), Keyl et al. (1967), Ueno et al. (1971a, 1973a) and Yates et al. (1968, 1969, 1970) is *F. sporotrichioides* (Strain 4.8).

Fusarium nivale Fn-2B (= *F. tricinctum* NRRL 3509, = *F. tricinctum* NRRL 6490) reported to be toxic to mice by Burmeister et al. (1981), Saito & Ohtsubo (1974), Saito & Okubo (1970), Saito et al. (1969), Tatsuno (1968, 1976), Ueno (1971), Ueno & Fukushima (1968), and Ueno et al. (1969a, 1970a, 1971a, b, 1973a) is *F. sporotrichioides* (Strain 4.9).

Pigeons. Extracts of cultures of *F. nivale* No. 3125 were reported to cause emesis in pigeons by Prentice & Dickson (1968) and Prentice et al. (1959), but Ellison & Kotsonis (1973) found this strain (as *F. nivale* NRRL 3289) to be negative for emetic activity in pigeons. We have identifed this strain as *F. nivale* (Strain 3.1). The emetic metabolites produced by this strain have not been identified, but the production of low levels of deoxynivalenol (1 μg/g) by *F. nivale* NRRL 3289 (Strain 3.1) has been reported by Vesonder et al. (1981a).

Pigs. Cultures on autoclaved corn of *F. nivale* NRRL 3289 were found to induce a feed refusal response in pigs by Vesonder et al. (1977b, 1981a). We have identified this strain as *F. nivale* (Strain 3.1). The low level (1 μg/g) of deoxynivalenol detected in this culture material by Vesonder et al. (1981a) could not explain the rejection level (42 to 48%) by pigs, and the material was negative for T-2 toxin, diacetoxyscirpenol, nivalenol, fusarenon, and monoacetoxyscirpenol. These findings led Vesonder et al. (1981a) to conclude that the factors responsible for the feed refusal by pigs of cultures of *F. nivale* NRRL 3289 had not been identified.

Rabbits. The strain originally identified as *F. nivale* and reported (as *F. nivale* NRRL 13,318, *F. nivale* NRRL 3249, or *F. tricinctum* NRRL 3249) to be dermotoxic to rabbit skin (Ellis & Yates, 1971; Keyl et al., 1967; Yates et al., 1968, 1969) is *F. sporotrichioides* (Strain 4.8).

Fusarium nivale Fn-2B reported to be dermotoxic to rabbit skin (Ueno, 1971; Ueno et al., 1970b, 1971a, b,) is *F. sporotrichioides* (Strain 4.9).

Joffe (1960a, b) reported that extracts of cultures of two out of 13 strains of *F. nivale* isolated from overwintered cereals associated with outbreaks of alimentary toxic aleukia (ATA) in the USSR (see *F. sporotrichioides*, ATA) were mildly dermotoxic to rabbit skin. The mycotoxin(s) responsible for the dermotoxic effect were not identified.

Rats. *Fusarium nivale* Fn-2B reported to be toxic to rats (Saito & Ohtsubo, 1974; Saito et al., 1969) is *F. sporotrichioides* (Strain 4.9).

Mycotoxins Produced

Butenolide. The strain originally identified as *F. nivale* NRRL 13,318 and reported (as *F. nivale* NRRL 3249 or *F. tricinctum* NRRL 3249) to produce a toxic butenolide (Burkhardt et al., 1968; Ellis & Yates, 1971; Yates et al., 1967, 1968, 1969) is *F. sporotrichioides* (Strain 4.8).

Calonectrin. The strain (as *Calonectria nivalis* CMI 14764, = *Micronectriella nivalis* ATCC 26559) reported to produce calonectrin (Gardner et al., 1972; Yoshizawa et al., 1980) is *F. culmorum* (Strain 8.18).

Culmorin. The strain (as *Calonectria nivalis* CMI 14764) reported to produce culmorin (Gardner et al., 1972) is *F. culmorum* (Strain 8.18).

15-Deacetylcalonectrin. The strain (as *Calonectria nivalis* CMI 14764) reported to produce 15-deacetylcalonectrin (Gardner et al., 1972) is *F. culmorum* (Strain 8.18).

Deoxynivalenol (=Vomitoxin). Vesonder et al. (1977 b) could not detect deoxynivalenol in culture material on autoclaved corn of *F. nivale* NRRL 3289 by TLC, but subsequent analysis by GC-MS revealed the presence of 1μg/g of deoxynivalenol in culture material refused by swine (Vesonder et al., 1981a). We have identified this strain as *F. nivale* (Strain 3.1). The finding by Vesonder et al. (1981a) that NRRL 3289 produces deoxynivalenol, constitutes the only report of the production of a known *Fusarium* toxin by an authenticated strain of *F. nivale* (see Strain 3.1).

Fusarenon-X (=Fusarenon, = Nivalenol Monoacetate). The Japanese isolate *F. nivale* Fn-2B reported to produce fusarenon-X (Fujimoto et al., 1972; Morooka & Tatsuno, 1970; Morooka et al., 1971; Tatsuno, 1976; Tatsuno et al., 1970, 1973; Ueno, 1971; Ueno et al., 1969a, 1970a, c, 1971a, b, c, 1973a, 1975; Yoshizawa & Morooka, 1975b; Yoshizawa et al., 1979, 1980) is *F. sporotrichioides* (Strain 4.9).

Nivalenol. The Japanese strain *F. nivale* Fn-2B reported to produce nivalenol (Fujimoto et al., 1972; Morooka & Tatsuno, 1970; Tatsuno, 1968, 1976; Tatsuno et al., 1968, 1969, 1970, 1973; Ueno, 1971; Ueno et al., 1969a, 1970a, c, 1971a, 1973a; Yoshizawa et al., 1979) is *F. sporotrichioides* (Strain 4.9).

Nivalenol Diacetate (=Diacetylnivalenol). The Japanese strain *F. nivale* Fn-2B reported to produce nivalenol diacetate (Fujimoto et al., 1972; Tatsuno, 1976; Tatsuno et al., 1970, 1973; Yoshizawa et al., 1980) is *F. sporotrichioides* (Strain 4.9).

T-2 Toxin. The strain originally identified as *F. nivale* NRRL 13,318 (= NRRL 3249) and reported (as *F. tricinctum* NRRL 3249) to produce T-2 toxin (Burmeister et al., 1971; Ellis & Yates, 1971; Yates et al., 1968, 1969, 1970) is *F. sporotrichioides* (Strain 4.8).

The Japanese strain *F. nivale* Fn-2B reported (as *F. sporotrichioides* var. *tricinctum* NRRL 3509) to produce T-2 toxin (Joffe, 1978b; Joffe & Yagen, 1977) is *F. sporotrichioides* (Strain 4.9).

Bottalico (1977c) reported negative results for T-2 toxin production by three isolates of *F. nivale* from cereals in Italy cultured on autoclaved corn at 24°C for 15 days followed by 10°C for 60 days.

Zearalenone. One of two Japanese isolates of *F. nivale* was reported to produce zearalenone (7.3 µg/g of rice culture) by Ichinoe et al. (1977). This reported zearalenone-producing isolate was unfortunately not specified and it is consequently not known whether this isolate was identical or similar to the Japanese isolate *F. nivale* Fn-2B, which we have identified as *F. sporotrichioides* (Strain 4.9).

According to Hacking et al. (1976, 1977), one of 21 isolates of *F. nivale* from barley in the United Kingdom produced zearalenone at unspecified levels in cultures on autoclaved rice incubated for 4 weeks at 25°C followed by 6 weeks at 12°C. A culture of this reported zeralenone-producing isolate of *F. nivale* is no longer available (W.R. Rosser, personal communication, 3 April 1981).

Negative results for zearalenone production by *F. nivale* have been reported for seven isolates by Caldwell et al. (1970), two isolates by Ishii et al. (1974), and six isolates by Bottalico (1977b).

In conclusion, two isolates out of 38 tested have been reported to produce zearalenone. Unfortunately it has not been possible to authenticate either of these two reported zearalenone-producing isolates as *F. nivale* and the question as to whether *F. nivale* is capable of producing zearalenone remains unresolved at present.

Toxigenic Strains in the ITFRC

We have identified only one strain (=NRRL 3289) in the ITFRC as *F. nivale* (Table 3.1) according to Nelson et al. (1983). Toxigenic strains referred to as *F. nivale* in the literature but identified as other species by us are listed in Table 3.2.

Strain 3.1 *F. nivale* 3125, NRRL 3289 (=N-41; MRC 2229)

According to Ellison & Kotsonis (1973) and Vesonder et al. (1977b), *F. nivale* NRRL 3289 is *F. nivale* 3125 that was used by Prentice & Dickson (1968). This strain (as *F. nivale* 3125) was originally obtained from the Canadian Department of Agriculture, Winnipeg, Manitoba (Prentice & Dickson, 1968; Sloey & Prentice, 1962). This culture was presumably identified by W. L. Gordon, but the original source is not known.

This strain was shown to have a general inhibitory effect on the malting of barley (Sloey & Prentice, 1962). Subsequently diethyl-ether extracts of cultures of this strain in nutrient broth (it grew poorly in Richard's solution) incubated in shake culture at room temperature were shown to cause emesis in pigeons when injected into the wing vein (Prentice & Dickson, 1968; Prentice et al., 1959). The chemical nature of the emetic material produced in nutrient broth by this strain was not determined.

Ellison & Kotsonis (1973) re-examined the strain of *F. nivale* (as NRRL 3289) used by Prentice & Dickson (1968) for the production of emetic material. Ethyl acetate extracts of cultures on autoclaved corn incubated at 8° or 25°C or of cultures in Richard's solution incubated at 25°C were negative for emetic activity upon dosing to pigeons. These extracts did not contain T-2 toxin as determined by TLC. Extracts of cultures of *F. nivale* NRRL 3289 in PSC medium incubated in stationary culture at 26°C for 12 days were not acutely toxic to mice, did not inhibit protein synthesis in rabbit reticulocytes, and did not

Table 3.1 *Fusarium* Strains of the Section Arachnites in the International Toxic *Fusarium* Reference Collection (ITFRC)

Strain No.	ITFRC No.	MRC No.	Published Name(s) and Number(s)	Source	Toxigenicity
Fusarium nivale					
3.1	N-41	2229	*F. nivale* 3125 (= NRRL 3289)	NRRL	a, b
Fusarium larvarum					
3.2	N-40	1802	*F. larvarum* 27 (= IMB 62239)	W. Gerlach	a, c

a = Toxic to experimental animals
b = Produces deoxynivalenol
c = Produces dihydroisocoumarin derivatives

TABLE 3.2 *Fusarium* Strains Published as *F. nivale* but Identified as Other Species in the International Toxic *Fusarium* Reference Collection (ITFRC)

Strain No.	ITFRC No.	MRC No.	Published Name(s) and Number(s)	Source	Present Identification
8.18	R-6353	2434	*Calonectria nivalis* CMI 14764 (=*Micronectriella nivalis* ATCC 26559)	ATCC	*F. culmorum*
4.8	T-344	1705	*F. nivale* (=*F. tricinctum*) NRRL 3249	NRRL	*F. sporotrichioides*
4.9	T-436	1767	*F. nivale* Fn 2 (=*F. tricinctum*) NRRL 3509	NRRL	*F. sporotrichioides*
4.9	T-436	1962	= *F. nivale* Fn-2B	H. Kurata	*F. sporotrichioides*
4.9	T-563	2476	= *F. nivale* Fn-2B	T. Yoshizawa	*F. sporotrichioides*
4.9	T-564	2533	= *F. nivale* Fn-2B	Y. Ueno	*F. sporotrichioides*

contain chemically detectable trichothecenes (Ueno et al., 1972a, 1973a). Vesonder et al. (1977b) reported that cultures of this strain on autoclaved corn incubated at 28°C for 13 days induced a feed refusal response in swine, but did not contain deoxynivalenol, T-2 toxin, HT-2 toxin, acetyl-T-2 toxin, or fusarenon-X that could be detected by TLC. In a continuation of this study, Vesonder et al. (1981a) analyzed extracts of corn cultures of this strain that were refused by swine, using GC-MS to analyze for trichothecenes. Deoxynivalenol (= vomitoxin) was detected by this analytical method at a level of 1 μg/g, but no T-2 toxin, diacetoxyscirpenol, nivalenol, fusarenon-X or monoacetoxyscirpenol could be detected chemically. The rejection level of this culture material by swine was 42 to 48%, and the small amount of deoxynivalenol detected could not explain this level of rejection. These findings led Vesonder et al. (1981a) to conclude that the factor(s) responsible for feed refusal in the culture material were not chemically identified.

The report by Vesonder et al. (1981a) that *F. nivale* NRL 3289 produces deoxynivalenol, albeit at the low level of 1 μg/g, is extremely important because it constitutes the only report of the production of a known *Fusarium* toxin by an authenticated strain of *F. nivale*. This finding has important implications in establishing relationships between *F. nivale* and other *Fusarium* species, since *F. nivale* has recently been excluded from *Fusarium* by Gams & Müller (1980).

Vesonder et al. (1981a) stated that *F. nivale* NRRL 3289 was "a slow grower at 28°C because it grew in the vegetative state" and also that "it has been necessary to carry strain NRRL 3289 as refrigerated agar slant cultures with periodic renewal every 6 to 8 months." We conclude from these statements that this strain does not sporulate well and could consequently not be lyophilized at NRRL.

A slant culture was received on 9 March 1981 from NRRL as "*F. nivale* NRRL 3289." This culture produces macroconidia in orange sporodochia on CLA and we consider it to be a typical representative of *F. nivale*. All attempts to lyophilize this culture have, however, been unsuccessful and it is being stored in liquid nitrogen.

Fusarium larvarum Fuckel

Perfect State: *Nectria aurantiicola* Berk. & Br.

Incidence and Distribution

Fusarium larvarum is a parasite of scale insects and certain other insects such as woolly aphids on a number of hosts in temperate as well as tropical areas (Booth, 1971; Claydon et al., 1979; CMI Descriptions of Pathogenic Fungi and Bacteria No. 714, 1981; Gerlach, 1977; Gordon, 1956; Smirnoff, 1970; Wollenweber & Reinking, 1935).

Association with Human and/or Animal Diseases

Fusarium larvarum has not been associated with any human or animal toxicoses.

Toxicity to Experimental Animals

Insects. Solvent extracts of cultures in Raulin-Thom medium of two strains of *F. larvarum* (No. 26 and 27), showed insecticidal activity on injection in the blowfly, *Calliphora erythrocephala* (Claydon et al., 1979; Grove & Pople, 1979, 1981). We have confirmed the identification of Strain 27 as *F. larvarum* (Strain 3.2). The insecticidal activity of the extracts was shown to be due to a group of dihydroisocoumarin derivatives (see Mycotoxins produced, below).

Mice. The oral administration of aqueous suspensions of mycelium, chlamydospores and conidia of *F. larvarum* isolated from the woolly aphid *Adelges piceae* (Ratz.) in Canada, did not cause mortality or other changes in white mice (Smirnoff, 1970).

Mycotoxins Produced

The major insecticidal compounds isolated in yields of 5–15 mg/ℓ from two strains of *F. larvarum* (No. 26 and 27) were the dihydroisocoumarin derivatives monocerin, fusarentin 6-methyl ether, and fusarentin 6,7-dimethyl ether (Claydon et al., 1979; Grove & Pople, 1979, 1981). In addition, the following minor compounds were also isolated in yields of 1–2 mg/ℓ from *F. larvarum* Strain No. 27 (Strain 3.2): (+)-mellein, 7-0-demethylmonocerin, and fusarentin 6,8-dimethyl ether (Claydon et al., 1979; Grove & Pople, 1979).

Toxigenic Strains in the ITFRC

We have identified the following strain as *F. larvarum* (Table 3.1) according to Nelson et al. (1983):

Strain 3.2 *F. larvarum* **Strain 27, IMB 62239** (= N-40; MRC 1802).
This strain (as *F. larvarum* Stamm 62239) was originally isolated from a scale insect (*Quadraspidiotus perniciosus*) in Iran during 1968 (Gerlach, 1977; Gerlach & Ershad, 1970). We assume that this is the same strain (No. 27) "that had been isolated from a scale insect in Iran (CMI 141200)" according to Claydon et al. (1979). *F. larvarum* CMI 141200 is not listed in the Catalogue of the Culture Collection of the CMI, Seventh Edition, 1975. Since only one strain of *F. larvarum* (= No. 62239) was isolated from a scale insect in Iran according to Gerlach (1977) and Gerlach & Ershad (1970), we assume that it is this strain that was referred to as *F. larvarum* Strain 27 (= CMI 141200) by Claydon et al. (1979) and Grove & Pople (1979).

Solvent extracts of cultures of this strain (as *F. larvarum* No. 27) in Raulin-Thom medium incubated for 5 weeks or longer were insecticidal upon injection in the blowfly *Calliphora erythrocephala* (Claydon et al., 1979; Grove & Pople, 1979). The insecticidal activity of these extracts was due to a group of dihydroisocoumarin derivatives and the major components isolated in yields of 5–15 mg/ℓ from this strain were monocerin, fusarentin 6-methyl ether, and fusarentin 6,7-dimethyl ether (Claydon et al., 1979; Grove & Pople, 1979, 1981). In addition, the following minor compounds were also isolated in yields of 1–2 mg/ℓ: (+)-mellein, 7-0-demethylmonocerin, and fusarentin 6,8-dimethyl ether (Claydon et al., 1979; Grove & Pople, 1979). No epoxy-trichothecenes were isolated from this strain (Claydon et al., 1979).

A culture on a sterile barley ear of this strain was received on 18 April 1979 from W. Gerlach as "*F. larvarum,* IMB, Stamm 62239."

4. Section Sporotrichiella

Serious confusion has been introduced into the mycotoxicological literature by the use of the name *F. tricinctum* (Corda) Sacc. emend. Snyd. & Hans. in the sense of Snyder & Hansen (1945), which encompasses at least four distinct species. In the literature, this name has been used for a large number of *Fusarium* strains that have been implicated in several human and veterinary mycotoxicoses, shown experimentally to cause toxicoses in many animal species, and reported to produce a variety of mycotoxins. This is most unfortunate because in our experience, *F. tricinctum* (Corda) Sacc. sensu stricto has a very low degree of toxicity compared to other species in the section Sporotrichiella. In many cases it is impossible to deduce from the literature which species in the section is indicated in any specific instance or which toxins are produced by which species. This problem is further complicated by the fact that several other toxigenic strains of *Fusarium* that most probably belong in Sporotrichiella have been incorrectly identified and are referred to in the literature under a variety of other names. The only solution to this dilemma is a taxonomic study of all the *Fusarium* strains that have been reported to be toxic in the literature. Although we have attempted to examine as many published toxigenic strains of *Fusarium* as possible in this study, many toxicologically important strains, particularly strains of the section Sporotrichiella from the USSR, were not available. These complications have made the following discussion of the mycotoxicology of the four species in Sporotrichiella (Nelson et al., 1983) very difficult, and the remaining problems can only be resolved by an examination of the relevant cultures.

Fusarium sporotrichioides Sherb.

Perfect State: Unknown.

Incidence and Distribution

Fusarium sporotrichioides occurs in soil and on a wide variety of host plants, particularly cereals and grasses, in the temperate to cold areas of the world such as northern Europe, northern USA, Canada, Japan and the USSR (Bilai, 1970; Booth 1971; Domsch et al., 1980; Gordon, 1952; Joffe, 1960a, b, 1962, 1974a, b; Kvashnina, 1979; Meyer & Frank, 1979; Palti, 1978; Seemüller, 1968; Wollenweber & Reinking, 1935; Ylimäki, 1981; Ylimäki et al., 1979).

Fusarium sporotrichioides can grow at very low temperatures and has been isolated from cereals that overwintered under snow in Canada (Mills & Frydman, 1980; Scott et al., 1980) and the USSR (Bilai, 1960, 1970; Joffe 1960a, b, 1962, 1965, 1969, 1971, 1973a, 1974b, c, 1978b; Joffe & Palti, 1974, 1975; Joffe & Yagen, 1977; Kvashnina, 1979). Reports on the occurrence of *F. sporotrichioides* in subtropical and tropical areas of the world such as equatorial West Africa (Dominik & Ihnatowicz, 1975) and Swaziland (Martin et al., 1971) are questionable (see notes on *F. sporotrichioides*, Strain 4.9).

Association with Human and/or Animal Diseases

Alimentary Toxic Aleukia (ATA) (=Septic Angina, = Endemic Panmyelotoxicosis). Overwintered cereal grains colonized by *F. sporotrichioides* (see also *F. poae*, ATA) caused the deaths of hundreds of thousands of people in the USSR during the closing years of World War II. These deaths were due to a toxicosis known as alimentary toxic aleukia (ATA), characterized by stomatitis, dermatitis, hemorrhage, suppression of the hematopoietic system, and depletion of the bone marrow with resultant leukopenia, thrombocytopenia, and immunosuppression (Akhmeteli, 1977; Bamburg & Strong, 1971; Bilai, 1970; Forgacs & Carll, 1962; Gajdusek, 1953; Joffe, 1960a, b, 1962, 1963, 1965, 1971, 1974c, 1978b; Leonov, 1977; Mayer, 1953a, b; Palti, 1978; Rubinstein, 1960a, b; Smalley & Strong, 1974; Ueno, 1977a).

Cultures of *F. sporotrichioides* (see also *F. poae*, toxicity to experimental animals) isolated from the toxic overwintered grain associated with outbreaks of ATA proved to be highly toxic to a variety of experimental animals. These cultures caused some of the characteristic clinical signs and pathological changes of ATA in experimental animals including cats, cattle, dogs, guinea pigs, horses, monkeys, mice, pigeons, pigs, poultry, rabbits, rats, and sheep (Akhmeteli et al., 1972a, b, 1973; Bilai, 1970; Getsova, 1960; Joffe, 1960a, 1971, 1974c, 1978a, b; Joffe & Yagen, 1978; Kurmanov, 1978a, b; Kvashnina, 1976, 1978, 1979; Lutsky et al., 1978; Mayer, 1953b, Rubinstein, 1960a, b; Rubinstein & Lyass, 1948; Rubinstein et al., 1967; Sarkisov, 1960; Schoental & Joffe, 1974; Schoental et al., 1979; Yagen et al., 1977). We have identified two toxigenic *Fusarium* strains isolated from overwintered cereals associated with ATA in the USSR as *F. sporotrichioides*: ·*F. tricinctum* NRRL 3510 (Strain 4.1) and *F. tricinctum* NRRL 3511 (Strain 4.2).

According to Joffe (1978b), Lutsky et al. (1978), and Rubinstein (1960a, b), the typical lesions of ATA were reproduced successfully for the first time by Sarkisov and his co-workers during the 1940s in cats dosed with toxic overwintered grain infected by *F. sporotrichioides* and with pure cultures of the fungus. Several authors have subsequently confirmed that a satisfactory experimental model of ATA can be obtained in cats dosed with culture material of *F. sporotrichioides* (see also *F. poae*, toxicity to experimental animals) (Joffe, 1960a, 1971, 1974c, 1978b; Lutsky et al., 1978; Rubinstein, 1960a, b; Rubinstein & Lyass, 1948; Yagen et al., 1977) or with crystalline T-2 toxin produced by *F. sporotrichioides* (Lutsky et al., 1978; Lutsky & Mor, 1981a, b; Sato et al., 1975; Yagen et al., 1977).

According to Rubinstein (1960b), a "particularly convincing model" of ATA was obtained in monkeys dosed with toxic grain by Rubinstein and Lyass (1948). This publication was not available to us, but according to Mayer (1953b), monkeys dosed with toxic grain by these authors vomited, had diarrhea, and lost appetite. Within 20 to 30 days petechial hemorrhages developed on the skin and hematological examination revealed progressive leukopenia, relative lymphocytosis, anemia, and thrombocytopenia—all characteristic findings of ATA in humans. Some of these characteristic clinical signs and pathological changes of ATA have also been induced in monkeys with pure diacetoxyscirpenol (Stähelin et al., 1968) and semi-purified T-2 toxin (Rukmini et al., 1980).

Many of the characteristic features of ATA—including nausea, vomiting, diarrhea, fever and chills, headaches, skin erythema with burning sensation; neurological signs such as confusion, somnolence, disorientation, hallucinations, and psychomotor seizures; and hematological abnormalities such as leuko- and thrombocytopenia, and myelosuppression—have also been recorded in humans treated with diacetoxyscirpenol (=anguidine) as an antineoplastic agent (Belt et al., 1979; De Simone et al., 1979; Diggs et al., 1978; Goodwin et al., 1978; Yap et al., 1979). Massive hemorrhage reportedly occurs in humans exposed to trichothecenes in the form of a chemical warfare agent known as "Yellow Rain" according to Seagrave (1981) and numerous newspaper and magazine reports since September 1981 (e.g., *Time*, 14 September 1981, p. 24; *Newsweek*, 28 September 1981, p. 44; *Nature*, 1 October 1981, p. 327; *Nature*, 22 October 1981, p. 598; *Science*, 27 November 1981, p. 1008).

Russian scientists implicated the steroidal glycosides sporofusarin and poaefusarin and their aglycones sporofusariogenin and poaefusariogenin, isolated from *F sporotrichioides* and *F. poae*, respectively, as the fusariotoxins responsible for ATA (Olifson, 1960, 1965; Olifson et al., 1972, 1975, 1978). However, the striking pathological similarities between ATA and the lesions caused in experimental animals by a group of mycotoxins known as 12,13-epoxytrichothecenes led to suggestions that ATA was more likely caused by these trichothecenes than by the steroidal compounds implicated by the Russian scientists (Bamburg et al., 1969; Bamburg & Strong, 1971). Further evidence for this hypothesis was provided by the reports that *F. sportrichioides* NRRL 3510 (Strain 4.1) isolated from millet involved in ATA (Ueno et al., 1972b) and "authentic toxic isolates (*F. poae* 5627 and *F. sporotrichioides* 5328a) obtained from the USSR" (Szathmary et al., 1976) produced several trichothecenes, including T-2 toxin, in culture. Moreover, the major toxic component of an "authentic sample of poaefusarin" isolated from *F. poae* in the USSR was identified as T-2 toxin by modern analytical methods (Mirocha & Pathre, 1973; Mirocha et al., 1977b). Six sterols isolated from *F. sporotrichioides* No. 921, one of the most toxic strains and highest producers of T-2 toxin isolated from overwintered cereals associated with a fatal outbreak of ATA, have also been found to be non-toxic (Yagen et al., 1977, 1980). Recent findings that more than 95% of *F. sporotrichioides* and *F. poae* strains from cereals associated with clinical ATA produced T-2 toxin in culture, together with the reproduction of the characteristic pathological changes of ATA in cats with crystalline T-2 toxin, have led to the conclusion that T-2 toxin was primarily responsible for the outbreaks of ATA in the USSR (Joffe, 1978b, Joffe & Yagen, 1977; Lutsky & Mor, 1981a, b; Lutsky et al., 1978; Yagen & Joffe, 1976; Yagen et al., 1977, 1980). This conclusion would be strengthened considerably if it were possible to detect T-2 toxin in samples of toxic overwintered cereals that were associated with ATA in the USSR, assuming such samples still exist in the USSR.

Hemorrhagic Syndrome (Moldy Corn Toxicosis). Outbreaks of a hemorrhagic syndrome characterized by bloody diarrhea, necrotic oral lesions, hemorrhagic gastro-enteritis, and extensive hemorrhages in many organs occur sporadically in animals such as cattle, pigs, and poultry in the north central USA and elsewhere (Albright et al., 1964; Bailey & Groth, 1959; Bamburg & Strong, 1971; Bamburg et al., 1969; Burnside et al., 1957; Dahlgren & Williams, 1972; Forgacs, 1965; Forgacs & Carll, 1962; Forgacs et al., 1962; Hibbs et al., 1974; Hsu et al., 1972; Kurmanov, 1978a, b; Marasas & Smalley, 1972; Mirocha et al., 1977b; Petrie et al., 1977; Ribelin, 1978; Smalley, 1973; Smalley & Strong, 1974; Smalley et al., 1970; Wyatt et al., 1972b). The disease is associated with the ingestion of moldy cereals, particularly corn (= moldy corn toxicosis), and some of the most toxic fungi isolated from these feeds proved to be *Fusarium* species of the section Sporotrichiella that have frequently been referred to in the literature as *F. tricinctum*. We have identified several of these toxic *Fusarium* strains isolated from moldy cereals in various parts of the world as *F. sporotrichioides*, e.g., *F. tricinctum* strain T-2 (=NRRL 3299, Strain 4.3), *F. tricinctum* ATCC 24044 and 24045 (Strains 4.4 and 4.5), *F. tricinctum* 2061-C (Strain 4.6), *F. poae* 7452 (Strain 4.7) and many others.

Many of the characteristic pathological changes observed in field outbreaks of the hemorrhagic syndrome have been reproduced experimentally with cultures, crude extracts of cultures, and crystalline trichothecenes such as T-2 toxin and diacetoxyscirpenol isolated from cultures of toxic strains of *F. sporotrichioides*. These characteristic pathological changes that have been reproduced experimentally in animals include:

1. Extensive hemorrhages in many organs in cats (Lutsky & Mor, 1981a, b; Lutsky et al., 1978; Sato et al., 1975; Yagen et al., 1977); cattle (Grove et al., 1970; Kosuri et al., 1970; Kurmanov, 1978a; Pier et al., 1976; Ribelin, 1978; Tookey et al., 1972); monkeys (Stähelin et al., 1968; Rukmini et al., 1980); poultry (Forgacs et al., 1962; Joffe, 1978a; Joffe & Yagen, 1978; Kurmanov, 1978b; Palyusik & Koplik-Kovacs, 1975a, b; Palyusik et al., 1968; Pearson, 1978; Wyatt et al., 1973b); pigs (Weaver et al., 1977, 1978c); rodents (De Nicola et al., 1978; Hayes & Schiefer, 1980; Hayes et al., 1980; Joffe, 1960a; Korpinen & Uoti, 1974; Korpinen & Ylimäki, 1972; Kosuri et al., 1971); and sheep (Kurmanov, 1978a).

2. Dermatitis and/or necrotic oral lesions in cattle (Kosuri et al., 1970; Patterson et al., 1979); pigs (Weaver et al., 1981); poultry (Chi & Mirocha, 1978; Chi et al., 1977b, c; Christensen et al., 1972a; Coffin & Combs, 1981; Hamilton et al., 1971; Hoerr & Carlton, 1979; Joffe & Yagen, 1978; Palyusik & Koplic-Kovacs, 1975a, b; Palyusik et al., 1968; Richard et al., 1978; Speers et al., 1972, 1977; Weaver et al., 1977; Witlock et al., 1977; Wyatt et al., 1972a, 1973b, 1975a); rodents (Bamburg & Strong, 1971; Bamburg et al., 1968a, b; Chung et al., 1974; Gilgan et al., 1966; Hayes & Schiefer, 1979, 1980; Hayes et al., 1980; Joffe, 1960a, b, 1969, 1973a, 1974b; Joffe & Palti, 1974; Joffe & Yagen, 1977; Lansden et al., 1978; Marasas et al., 1969; Sato & Ueno, 1977; Ueno et al., 1970b; Wei et al., 1972; Yagen & Joffe, 1976); and sheep (Kurmanov, 1978a).

3. Erosions, hyperkeratosis and ulceration of the squamous epithelium of the esophagus and stomach in cats (Lutsky & Mor, 1981a, b,; Lutsky et al., 1978; Yagen et al., 1977); poultry (Greenway & Puls, 1976; Palyusik & Koplik-Kovacs, 1975a, b; Palyusik et al., 1968; Puls & Greenway, 1976); pigs (Weaver et al., 1978d); and rodents (De Nicola et al., 1978; Hayes & Schiefer, 1980; Hayes et al., 1980; Ohtsubo & Saito, 1977; Rubinstein et al., 1967; Schoental & Joffe, 1974; Schoental et al., 1979).

4. Radiomimetic cellular damage in the hematopoietic and lymphopoietic systems resulting in hematological abnormalities and immunosuppression in cats (Lutsky & Mor, 1981a, b; Lutsky et al., 1978; Sato et al., 1975; Yagen et al., 1977); pigs (Weaver et al., 1977, 1978a, c); poultry (Boonchuvit et al., 1975; Doerr et al., 1974, 1981; Hoerr et al., 1981a, b; Joffe & Yagen, 1978; Richard et al., 1978); rodents (De Nicola et al., 1978; Frayssinet & Lafont, 1976; Fromentin et al., 1980, 1981; Hayes & Schiefer, 1980; Lafarge-Frayssinet et al., 1979; Lafont et al., 1977; Masuko et al., 1977; Otokawa et al., 1979; Rosenstein et al., 1978, 1979, 1981; Salazar et al., 1980; Ueno et al., 1971c); and sheep (Kurmanov, 1978a).

The highly irritant properties of T-2 toxin and other trichothecenes suggest that they may be carcinogenic, particularly to the skin and the gastro-intestinal tract. However, no papillomas were induced in mice by the repeated topical application of either T-2 (Lindenfelser et al., 1974; Marasas et al., 1969) or diacetoxyscirpenol (Lindenfelser et al., 1974). Chronic feeding experiments with T-2 toxin in trout and rats (Marasas et al., 1969) and with diacetoxyscirpenol in rats (Stähelin et al., 1968) also failed to induce tumors. These findings that T-2 toxin and diacetoxyscirpenol are non-carcinogenic are supported by the fact that both of these, as well as several other trichothecenes, are non-mutagenic to *Salmonella* in the Ames test (Kuczuk et al., 1978; Nagao et al., 1976; Ueno, 1977b; Ueno et al., 1978; Wehner et al., 1978), although fusarenon-X was found to be mutagenic by Nagao et al. (1976). On the other hand, lung adenomas have been observed in mice dosed with extracts of *F. sporotrichioides* (Ahkmeteli et al., 1972a, b, 1973), and a low incidence of tumors in various organs in mice fed culture material of *F. sporotrichioides* (=*F. nivale*) Fn-2B (Strain 4.9) has also been reported (Ohtsubo & Saito, 1977; Saito & Ohtsubo, 1974). Low incidences of tumors in various organs have also been reported in mice administered pure T-2 toxin (Ohtsubo & Saito, 1977) and fusarenon-X (Ohtsubo & Saito, 1977; Saito & Ohtsubo, 1974) and in rats with T-2 toxin (Schoental et al., 1979) and fusarenon-X (Ohtsubo & Saito, 1977; Saito & Ohtsubo, 1974; Saito et al., 1980). The latter findings have led to suggestions that T-2 toxin and other trichothecenes produced by *Fusarium* may be involved in the development of tumors in the gastro-intestinal tract of man (Schoental, 1975, 1977a, b, 1978, 1979a, b, 1980a, b, c, 1981a, b, c, d, e; Schoental et al., 1976, 1978a, 1979).

Although many isolates of *F. sporotrichioides* have been shown to produce T-2 toxin, diacetoxyscirpenol, and several other trichothecenes (see mycotoxins produced, below) in culture, there are only a few reliable reports in the literature on the natural occurrence of these mycotoxins in moldy feeds associated with field outbreaks of the hemorrhagic syndrome. This is probably due to chemical problems in detecting trichothecenes in natural products. Moldy corn implicated in a lethal hemorrhagic toxicosis of cattle in Wisconsin, USA, showing the characteristic clinical and pathological features of "moldy corn toxicosis" has been shown to contain 2 ppm of T-2 toxin (Hsu et al., 1972). T-2 toxin has also been detected chemically in mixed feeds associated with bloody stools in cattle and in emesis and diarrhea in pigs, and diacetoxyscirpenol has been detected in mixed feeds implicated in the hemorrhagic bowel syndrome in pigs (Mirocha, 1979; Mirocha et al., 1976; Pathre & Mirocha, 1977, 1979). The increasing use of GLC and capillary GC-MS methods to analyze for T-2 toxin and other trichothecenes in corn and mixed feeds suspected of causing mycotoxicoses (Eppley, 1979; Jemmali et al., 1978; Kamimura et al., 1981; Kuroda et al., 1979; Lafont & Lafont, 1980; Renault et al., 1979; Szathmary et al., 1980) can be expected to result in more information on the natural occurrence of these compounds and their involvement in human and animal diseases.

In conclusion, there is little doubt that the hemorrhagic syndrome or moldy corn toxicosis in farm animals and alimentary toxic aleukia in humans (see ATA, above) are closely related if not identical syndromes and that both are caused by trichothecene mycotoxins such as T-2 toxin and diacetoxyscirpenol produced primarily by *F. sporotrichioides.*

Fescue Foot. Winter pastures of tall fescue (*Festuca arundinacea* Schreb.) in the United States, Australia, and New Zealand are associated with sporadic outbreaks of a disease in cattle, known as fescue foot, characterized by lameness, loss of weight, arched back, elevated body temperature, and dry gangrene involving the hind feet, tail tip, and ears, with sloughing of the most distal parts of these extremities (Ellis & Yates, 1971; Garner & Cornell, 1978; Yates, 1962, 1971; Yates et al., 1969). Although the clinical signs of fescue foot are very reminiscent of the gangrenous form of ergot poisoning, ergot sclerotia are apparently not involved in the disease (Garner & Cornell, 1978; Yates, 1971; Yates et al., 1969). The typical clinical signs of fescue foot have been reproduced experimentally in cattle with ethanol extracts of toxic fescue hay (Ellis & Yates, 1971; Yates et al., 1969).

The finding that extracts of toxic hay caused a necrotic reaction on rabbit skin led to the isolation of several dermotoxic *Fusarium* species from the toxic hay (Burmeister & Hesseltine, 1970; Ellis & Yates, 1971; Keyl et al., 1967; Yates et al., 1967, 1968, 1969, 1970). One of these dermatoxic *Fusarium* strains isolated from toxic fescue hay was originally referred to as *F. nivale* NRRL 13,318 (Keyl et al., 1967) or *F. nivale* NRRL 3249 (Burkhardt et al., 1968; Ellis & Yates, 1971; Yates, 1971; Yates et al., 1967, 1968), but was later considered to be an atypical strain of *F. tricinctum* (Burmeister et al., 1971; Ellis & Yates, 1971; Tookey et al., 1972; Ueno et al., 1972b, 1973a; Yates, 1971; Yates et al., 1969, 1970). *F. tricinctum* NRRL 3249 was eventually identified as *F. sporotrichioides* by Joffe & Palti (1975) and has also been referred to as *F. sporotrichioides* NRRL 3249 by Joffe (1978b) and Joffe & Yagen (1977). We have confirmed the identification of this strain as *F. sporotrichioides* (Strain 4.8).

Fusarium sporotrichioides NRRL 3249 (Strain 4.8) produces at least three mycotoxins in culture: butenolide, T-2 toxin, and an unidentified toxin (Burkhardt et al., 1968; Burmeister et al., 1971, 1981; Ellis & Yates 1971; Joffe, 1978b; Joffe & Yagen, 1977; Yates et al., 1967, 1968, 1969, 1970). Extracts of cultures of this strain are dermotoxic to rabbit skin and acutely toxic to mice, causing visceral hemorrhages (Ellis & Yates, 1971; Joffe, 1978b; Joffe & Palti, 1975; Joffe & Yagen, 1977; Keyl et al., 1967; Yates et al., 1968, 1969, 1970). According to Tookey et al. (1972), cultures of this strain on nutrient enriched non-toxic fescue hay caused pathological changes including edema and perivascular serous infiltration in the tails of four out of four heifers following administration by ruminal fistula for 69 to 74 days. No butenolide could be detected chemically in these cultures. Ethanol extracts of cultures of this strain on fescue hay were acutely toxic to cattle, but did not produce signs of toxicosis at lower dosage rates (Yates et al., 1969).

One of the characteristic pathological lesions of fescue foot, i.e., necrosis of the tip of tail, has been reproduced experimentally in cattle by prolonged daily administration of high doses of the butenolide known to be produced by *F. sporotrichioides* (= *F. tricinctum*) NRRL 3249 (Grove et al., 1970; Kosuri et al., 1970; Tookey et al., 1972). This finding tends to support the hypothesis that butenolide-producing *F. sporotrichioides* strains may be involved in the etiology of fescue foot. However, all of the clinical signs of the disease have not been reproduced with the butenolide and this compound has not yet been shown to occur naturally in toxic fescue hay (Garner & Cornell, 1978). Thus the causative role of *F. sporotrichioides* in fescue foot of cattle has not been proven.

Degnala disease (see *F. equiseti,* degnala disease) of buffaloes and cattle in the rice-growing areas of India and Pakistan is clinically and pathologically very similar to fescue foot (Kalra et al., 1973). The disease has been reproduced experimentally in buffaloes with rice straw naturally infected by *F. equiseti* (Kalra et al., 1977a; Kwatra & Singh, 1973) and also with cultures of this fungus on rice straw (Kalra et al., 1977a). Another similar disease known as "sore foot disease" of cattle (see *F. equiseti,* degnala disease) in China has reportedly also been reproduced experimentally with cultures of *F. equiseti* (and *F. semitectum*) by Qin et al. (1981). *F. equiseti* is a known producer of butenolide (Burmeister et al., 1971; White, 1967) and has also been found to be associated with toxic fescue hay in the USA (Burmeister et al., 1971). The possible role of *F. equiseti* in the etiology of fescue foot requires further investigation.

Akakabi-byo (Scabby Grain Intoxication). In Japan, sporadic epiphytotics of akakabi-byo (red mold disease or scab) of wheat, barley, oats, and rye can affect more than one-third of the national production (Kurata, 1978; Saito & Ohtsubo, 1974; Saito & Okubo, 1970; Tsunoda, 1970; Ueno, 1971, 1973a, b, 1977a, 1980; Ueno et al., 1971a). This disease of cereals is caused mainly by *F. graminearum* and the mycotoxicoses in humans and animals caused by the consumption of cereals infected by this fungus will be discussed under *F. graminearum* (see *F. graminearum,* association with human and/or animal diseases). However, a fungus isolated from 5 to 6% of scabby wheat grains in southern Japan during the severe epiphytotic of scab in 1963 and referred to as *F. nivale* proved to be extremely toxic to experimental animals and to produce several trichothecenes in culture (Morooka & Tatsuno, 1970; Morooka et al., 1971; Saito & Okubo, 1970; Saito & Ohtsubo, 1974; Saito et al., 1969; Tatsuno, 1968; Tatsuno et al., 1968; Tsunoda, 1970; Ueno, 1971, 1977a, 1980; Ueno et al., 1969a, 1970a, b, 1971a).

The highly toxic strain *F. nivale* Fn-2B (Strain 4.9) isolated from scabby wheat in southern Japan was identified as *F. nivale* because it was found to be identical to a referral strain, *F. nivale* NRRL 3249 (Tsunoda, 1970). However, *F. nivale* NRRL 3249 was subsequently classified as an "abnormal strain of *Fusarium tricinctum* by W.C. Snyder" (Tsunoda, 1970), and was eventually identified as *F. sporotrichioides* by Joffe & Palti (1975) and by us (Strain 4.8). We have also identified *F. nivale* Fn-2B (=*F. tricinctum* NRRL 3509) as *F. sporotrichioides* (Strain 4.9).

Fusarium sporotrichioides (= *F. nivale*) Fn-2B (Strain 4.9) has been reported to produce four trichothecenes in culture, i.e., fusarenon-X, nivalenol, diacetylnivalenol, and T-2 toxin (Fujimoto et al., 1972; Joffe, 1978b; Joffe & Yagen, 1977; Morooka & Tatsuno, 1970; Morooka et al., 1971; Ohta et al., 1977, 1978; Tatsuno, 1968, 1976; Tatsuno et al., 1968, 1969, 1970, 1973; Tsunoda, 1970; Ueno, 1971, 1977a; Ueno et al., 1969a, b, 1970a, b, c, 1971a, b, c, 1972b, 1975; Yoshizawa & Morooka, 1975; Yoshizawa et al., 1979, 1980). The only one of these trichothecenes that has been reported to occur naturally in moldy cereals in Japan is nivalenol, which was detected (together with deoxynivalenol) at unspecified levels in scabby barley infected by *F. graminearum* (=*F. roseum*) (Morooka et al., 1972). Nivalenol has also been reported to occur naturally at a level of 4 to 8 mg/kg (together with deoxynivalenol, T-2 toxin, and zearalenone) in corn in France (Jemmali et al., 1978). It is not clear which species of *Fusarium* produced the nivalenol in these cereals since *F. sporotrichioides* (=*F. nivale*) (Strain 4.9) as well as *F. graminearum* (see Strains 8.35 and 8.36) are known producers of nivalenol.

Extensive toxicological investigations, utilizing crude extracts of cultures of *F. sporotrichioides* (= *F. nivale*) Fn-2B (Strain 4.9) as well as the trichothecenes produced by it, and involving experimental animals including cats, ducklings, guinea pigs, mice, and rats, revealed that the mycotoxins produced by this fungus: (1) can cause clinical signs such as vomiting, diarrhea, and drowsiness; (2) are dermotoxic to exposed skin; and (3) cause hemorrhage and characteristic radiomimetic cytotoxic effects in the proliferating cells of the hematopoietic tissues in the bone marrow, spleen, thymus, and lymph nodes and also in the intestinal epithelium and testes (Ohtsubo & Saito, 1977; Saito & Ohtsubo, 1974; Saito & Okubo, 1970; Saito et al., 1969, 1980; Ueno, 1971, 1977a; Ueno et al., 1969a, 1970a, b, 1971a, b, c, 1973a). In long-term feeding studies in mice and rats with either culture material of *F. nivale* Fn-2B or pure fusarenon-X isolated from this strain, malignant tumors were observed in various organs, but their incidence was very low (Ohtsubo & Saito, 1977; Saito & Ohtsubo, 1974; Saito et al., 1980).

These toxicological and pathological findings suggest that the trichothecenes produced by *F. sporotrichioides* (referred to as *F. nivale* in the literature), possibly together with the same and/or other trichothecenes produced by *F. graminearum* (see *F. graminearum*, akakabi-byo), are probably responsible for the clinical signs of vomiting, diarrhea, headache, and drowsiness in humans and feed refusal and hemorrhagic gastro-enteritis in animals associated with the consumption of scabby cereals in Japan (Ueno, 1971, 1977a; Ueno et al., 1971a).

It is interesting that fusarenon-X, nivalenol, and diacetylnivalenol are so-called Group B trichothecenes (i.e., they have a carbonyl group at C-8) and that two of the *F. sporotrichioides* strains that are known to produce them—*F. nivale* Fn-2b (Strain 4.9) and *F. episphaeria* Fn-M (Strain 4.10)—have been isolated in the relatively warm southern part of Japan (Ueno, 1977a; Ueno et al., 1972a, 1973a). The geographical origin of a third strain of *F. sporotrichioides* that produces these trichothecenes, *F. oxysporum* "niveum" Melon-1 (Strain 4.11), is unknown to us. A search for *Fusarium* species that produce fusarenon-X, nivalenol, and diacetylnivalenol, and investigations on the natural occurrence of these trichothecenes in the warmer and subtropical areas of the world are needed.

Bean Hulls Poisoning. In Hokkaido, the northern island of Japan, mass outbreaks of a disease in horses known as "bean hulls poisoning," characterized by clinical signs such as marked perspiration, involuntary forward movement, cyclic movement, mental excitation, icterus, and convulsions, have occurred sporadically during late winter and early spring (Konishi & Ichijo, 1970a, b; Ueno, 1977c, 1980; Ueno et al., 1972a). The main pathological changes have been described as hepatic icterus and disturbances of the central nervous and circulatory systems characterized by hemorrhage in the leptomeninges and brain tissue, scattered degenerative changes of the nerve cells in the cerebral cortex, and degenerative changes in the renal tubular epithelium and myocardium (Konishi & Ichijo, 1970a, b; Ueda et al., 1967; Ueno et al., 1972a). The disease has been reproduced experimentally in horses with naturally infected bean hulls (Konishi & Ichijo, 1970b). In a mycotoxicological study of bean hulls in Hokkaido, Ueno et al. (1972a) found that they were extensively colonized by *Fusarium* spp. and that cultures of many of the *Fusarium* isolates were actually toxic to mice. One of the most toxic strains, designated as *F. solani* M-1-1, caused radiomimetic pathological changes in mice characteristic of trichothecene intoxication, such as hemorrhage, cellular degeneration, and karyorrhexis in the actively dividing cells of the hematopoietic tissues, small intestine, and testes (Ueno et al., 1972a). We have identified this strain as *F. sporotrichioides* (Strain 4.12). These findings resulted in the isolation of T-2 toxin, HT-2 toxin, diacetoxyscirpenol, neosolaniol, and two new trichothecenes, NT-1 and NT-2, from the culture filtrate of *F. sporotrichioides* (=*F. solani*) M-1-1 (Ishii & Ueno, 1981; Ishii et al., 1971; Tatsuno et al., 1973; Ueno et al., 1972a, b, 1973a, 1975).

Nervous disturbances have not been produced in mice with either the crude extract or any of the trichothecenes produced by F. sporotrichioides (=F. solani) M-1-1 (Ueno et al., 1972a). Culture material of this strain has apparently not been administered to horses and very little information is available on the effects of trichothecenes on horses. Saito & Ohtsubo (1974) stated that "1 to 2 g of fusarenon-X or T-2 toxin is sufficient to kill a horse weighing about 500 kg," but no details of the pathological changes were given. According to Ueno (1980), Okubo & Ueno (unpublished) "carried out an extensive toxicological investigation on the effect of trichothecenes on horses." One male horse weighing 550 kg was administered 500 mg of T-2 toxin orally and an additional 1000 mg after 12 hours. Marked bloody diarrhea resulted in 20 to 24 hours and the horse was sacrificed 25 hours later. Another male horse weighing 470 kg was administered 500 mg of fusarenon-X orally. Marked bloody diarrhea was observed and the horse died after 13 hours. Pathological changes included: karyorrhexis and cellular degeneration in the lymph nodes, intestinal epithelia, bone marrow, testes, and thymus; pseudorosette formation at the subependymal tissue of the third ventricle; and degeneration of striated muscle. No nervous disturbances were observed (Ueno, 1980).

It has to be concluded that there is no experimental evidence that "bean hulls poisoning" of horses is a form of trichothecene toxicosis, except for the fact that trichothecene-producing *Fusarium* species, including *F. sporotrichioides* (=F. solani M-1-1, Strain 4.12), have been isolated from bean hulls in Hokkaido. Consequently statements in the literature to the effect that T-2 toxin and neosolaniol have been determined to be the cause of "bean hulls poisoning" of horses are without foundation in fact.

The clinical signs and pathological changes associated with "bean hulls poisoning" of horses in Japan are very similar to those of equine leukoencephalomalacia (LEM) caused by *F. moniliforme* (see *F. moniliforme*, LEM), except that the characteristic leukoencephalomalacic lesions have not been specifically mentioned by the Japanese investigators (Pienaar et al., 1981). Consequently it is not certain at present whether "bean hulls poisoning" is identical to equine LEM or not. The possible causative role of *F. moniliforme* in "bean hulls poisoning" of horses in Hokkaido warrants further investigation.

Toxicity to Experimental Animals

Brine Shrimp (*Artemia salina*). Extracts of cultures of 20 of 38 *Fusarium* isolates from overwintered cereals in Canada were found to be toxic to brine shrimp by Scott et al. (1980). We have identified seven of these isolates from overwintered cereals toxic to brine shrimp as *F. sporotrichioides*: HPB 071178-10, -12, -13, -18, -20, -21, and -23 (Strains 4.13, 4.14, 4.15, 4.16, 4.17, 4.18, and 4.19). All of these toxic strains identified as *F. sporotrichioides* were reported to produce T-2 toxin by Scott et al. (1980), and some produced HT-2 toxin and neosolaniol as well. Brine shrimp larvae are known to be very sensitive to pure trichothecenes, including T-2 toxin and diacetoxyscirpenol (Durachova et al., 1977; Eppley, 1974; Harwig & Scott, 1971; Scott et al., 1980; Tanaka et al., 1975).

Cats. According to Joffe (1960a, 1971, 1974c, 1978b), Lutsky et al. (1978) and Rubinstein (1960b), the characteristic clinical signs and pathological lesions of alimentary toxic aleukia (ATA) in man (see ATA, above) were reproduced successfully for the first time by Sarkisov and his co-workers in the USSR during the 1940s in cats fed naturally infected overwintered cereals or cultures of *F. sporotrichioides* (referred to in the Russian literature as *F. sporotrichiella* var. *sporotrichioides*).

Cultures on agar or millet of unspecified isolates of *F. sporotrichioides* from toxic overwintered cereals associated with outbreaks of ATA in the USSR were dosed to cats by Joffe (1960a, 1971, 1974c, 1978b). The death of the cats followed a breakdown in blood production manifested by a decline in the number of erythrocytes, leukocytes, and neutrophils and an increase in the number of lymphocytes. In most cases the cats died within 6 to 12 days. Autopsies revealed marked hyperemia and hemorrhages of internal organs and histopathological changes in the hematopoietic system. Joffe (1974c) emphasized that "cats were found to be the best model on which the whole clinical picture of ATA could be reproduced."

A detailed toxicological study in cats with *F. sporotrichioides* strain 921 was reported by Lutsky et al. (1978) and Yagen et al. (1977). This strain was originally isolated in 1947 from overwintered rye grain from a household affected by ATA in the USSR (Joffe, 1960a; Yagen et al., 1977). Strain 921 was the most toxic among 106 isolates of *F. sporotrichioides* tested for dermotoxicity to rabbit skin by Yagen et al. (1977), and 4.1 g of T-2 toxin was isolated from 1 kg of millet inoculated with this strain and incubated at 12°C for 21 days. Clinical signs in cats dosed either with ethanolic extracts of cultures of Strain 921 containing T-2 toxin or with crystalline T-2 toxin included vomiting, ataxia, bloody feces, anorexia, paralysis, and death (Lutsky et al., 1978; Yagen et al., 1977). Consistent hematological findings were a generalized pancytopenia with marked leukopenia, anemia, and thrombocytopenia. Necropsy findings were severe emaciation, intestinal hemorrhage, and enlarged lymph nodes. Histopathological changes included necrosis of the intestinal epithelium, pulmonary edema, and hemorrhagic lymph nodes. Cats receiving the same crude extract with the T-2 toxin removed remained healthy. These results led Lutsky et al. (1978) and Yagen et al. (1977) to conclude that T-2 toxin was the active metabolite in the extract of *F. sporotrichioides* 921 that caused the syndrome in cats similar to ATA in humans.

The toxicity of crystalline T-2 toxin to cats has been confirmed by Lutsky & Mor (1981a, b,) and Sato et al. (1975). Pure fusarenon-X isolated from *F. sporotrichioides* (= *F. nivale*) Fn-2B (Strain 4.9) has also been shown to cause emesis (Ueno et al., 1974a) and radiomimetic cytotoxic changes in the acively dividing cells of the small intestine, thymus, lymph nodes, spleen, and bone marrow in cats (Ueno et al., 1971c).

Cattle. According to Kurmanov (1978a), clinical signs induced in cattle by cultures of *F. sporotrichioides* (as *F. sporotrichiella* var. *sporotrichioides*) isolated in the USSR included depression, loss of weight, loss of hair, hyperemia of the oral cavity, and necrotic foci on the lips. There was a marked increase in the leukocyte count; neutrophils were increased and lymphocytes were decreased. Many isolates of *F. sporotrichioides* from the USSR have subsequently been found to produce large amounts of T-2 toxin in culture (Joffe, 1978b; Joffe & Yagen, 1977; Yagen & Joffe 1976).

Cultures of *F. tricinctum* (=*F. nivale*) NRRL 3249 originally isolated from toxic fescue hay associated with outbreaks of fescue foot (see fescue foot, above) in cattle caused edema and perivascular serous infiltration of the tails of four heifers following administration by ruminal fistula for 69 to 74 days (Tookey et al., 1972). We have identified this strain as *F. sporotrichioides* (Strain 4.8). The surface temperatures of the tails of the treated animals were lower than those of the controls, but no evidence of vasoconstriction could be found (Tookey et al., 1972). Ethanol extracts of cultures of this strain were acutely toxic to cattle and high dosage rates caused rapid death, but lower dosage rates did not cause signs of toxicosis (Yates et al., 1969). This strain is known to produce T-2 toxin and butenolide (see Strain 4.8), and necrosis of the tail tip has been reproduced experimentally in cattle by prolonged administration of high doses of the pure butenolide (Grove et al., 1970; Kosuri et al., 1970; Tookey et al., 1972).

Patterson et al. (1979) dosed Friesian calves with cultures and extracts of cultures of *F. tricinctum* NRRL 3299 containing T-2 toxin. We have identified this strain as *F. sporotrichioides* (Strain 4.3). One of the calves dosed with extract died after 7 days, but none of the other animals exhibited abnormal clinical signs, except that hematological changes including blood clotting factor deficiencies, leukopenia, and neutrophilia were observed in two calves dosed with culture material. The extracts were also dermotoxic when applied to the shaved skin of a calf. Traces of T-2 toxin were found in the kidney tissue of one of the calves used in this experiment (Patterson et al., 1980).

Some of the characteristic features of field outbreaks of the hemorrhagic syndrome (see hemorrhagic syndrome, above) in cattle (Dahlgren & Williams, 1972; Hibbs et al., 1974; Hsu et al., 1972; Petrie et al., 1977; Ribelin, 1978), including extensive hemorrhages, epistaxis, and bloody feces, have been reproduced experimentally in cattle with crystalline T-2 toxin (Grove et al., 1970; Kosuri et al., 1970; Pier et al., 1976; Ribelin, 1978). Cattle are, however, relatively resistant to the effects of fusariotoxins (Kurmanov, 1978a; Patterson et al., 1979), and it has not always been possible to induce the characteristic hemorrhagic lesions experimentally in cattle with *Fusarium* cultures or pure T-2 toxin (Matthews et al., 1977; Patterson et al., 1979; Weaver et al., 1977, 1980).

Chickens. *Fusarium sporotrichioides* No. 921, originally isolated from overwintered grain associated with outbreaks of ATA (see ATA, above) in the USSR, was used in a detailed toxicological study in New Hampshire chicks by Joffe & Yagen (1978). This strain is known to produce large amounts of T-2 toxin (4.1g/kg) in culture on wheat (Joffe, 1978b; Joffe & Yagen, 1977). Culture material for use in the toxicity experiments of Joffe & Yagen (1978) was obtained by inoculating wheat with *F. sporotrichioides* No. 921 and incubating at 12°C for 21 days. Dry molded grain was added to diets to obtain T-2 toxin levels from 23 to 460 mg/kg feed. Crude ethanol extracts of this culture material were also dosed at levels from 0.35 to 0.90 mg T-2 toxin/kg body weight and pure T-2 toxin isolated from this strain was incorporated into diets at levels of 8 and 16 mg/kg feed. Clinical signs in chicks that received diets containing culture material or crude extracts included depression, inappetence, feed refusal, diarrhea, hyperkeratosis of the corner of the beak, and hemorrhages of the mucous membrane of the oral cavity. Hematological changes in these chicks were characterized by severe leuko- and thrombocytopenia and relative decreases of the erythrocyte counts and hemoglobin values. Pathological changes included multiple hemorrhages and necrosis in the intestinal tract and other organs and damage to the hematopoietic system. The only signs and pathological changes observed in the chicks fed diets containing pure T-2 toxin at levels of 8 and 16 mg/kg feed were oral lesions and leukopenia. Joffe & Yagen (1978) concluding from these results that culture material of *F. sporotrichioides* No. 921 containing T-2 toxin caused a lethal toxicosis in chicks with clinical and pathological changes similar to those observed in human ATA and that T-2 toxin played an essential role in the human disease. They attributed the absence of hemorrhage and marked hematological changes in the chicks that received pure T-2 toxin to the fact that the dietary levels of 8 and 16 mg/kg were too low, since chicks that received culture material containing 46 mg/kg or more of T-2 toxin or dosed with crude extract at levels of 0.35 mg or more of T-2 toxin/kg body weight developed these characteristic lesions. However, the fact that the culture material and crude extracts caused a more severe toxicosis than T-2 toxin alone may also have been due to the presence of other toxic metabolites.

A strain of "*F. tricinctum*" isolated from moldy corn meal in Wisconsin, USA, was reported to be acutely toxic to White Leghorn cockerels fed cultures on corn incubated for 21 days at either 10 or 25°C (Marasas & Smalley, 1972). Cultures of this strain no longer exist, but we believe that it was probably a strain of *F. sporotrichioides* similar to other isolates of this species (Strains 4.4 and 4.5) from corn in Wisconsin.

The strain of "*F. tricinctum*" used in toxicity experiments in laying hens by Speers et al. (1972, 1977) was the same isolate (i.e., 2061-C) used by Christensen et al. (1972a), according to Mirocha & Christensen (1974). We have identified this strain as *F. sporotrichioides* (Strain 4.6). Diets containing 1, 3, or 5% of corn inoculated with this strain caused marked decreases in feed intake, weight gain, and egg production in laying hens, and there was a high incidence of mouth lesions in these birds (Speers et al., 1972). In a continuation of this study, Speers et al. (1977) fed rations containing 2.5 and 5% of culture material of this strain to White Leghorn laying hens. The ration containing 5% culture material contained 16 ppm of T-2 and caused reductions in feed consumption, body weight gain, egg production, and egg quality. Mouth lesions developed on the medial surfaces of the mandibles, on the dorsal surfaces of the tongue, and on the skin at the corners of the mouth of these birds. The diet containing 2.5% of culture material (=8 ppm of T-2 toxin) had no significant effects on the performance of the hens, but caused the development of mouth lesions in some of the birds (Speers et al., 1977). The T-2 toxin produced by this strain (Joffe, 1978b; Joffe & Yagen, 1977; Speers et al., 1977) was presumably responsible for the toxicity to laying hens and for the development of the characteristic mouth lesions (Mirocha & Christensen, 1974).

Field outbreaks of necrotic oral lesions in broiler chicks have been reported in the USA (Wyatt et al., 1972b). Necrotic oral lesions of the circumscribed proliferative yellow caseous plaque type, identical to those seen in field outbreaks, have been reproduced in chickens with crystalline T-2 toxin at levels as low as 1 mg/kg feed, with a dose-related increase in the incidence and severity of lesions at higher dietary levels (Chi & Mirocha, 1978; Chi et al., 1977b, c; Coffin & Combs, 1981; Hamilton et al., 1971; Hoerr & Carlton, 1979; Joffe & Yagen 1978; Richard et al., 1978; Speers et al., 1977; Weaver et al., 1977; Witlock et al., 1977; Wyatt et al., 1972a, 1973b, 1975a). Oral lesions have also been induced in chickens with pure diacetoxyscirpenol, which appeared to be more caustic than T-2 toxin in broiler chicks with respect to oral lesion formation (Chi & Mirocha, 1978).

Other clinical signs and pathological lesions that have been induced in chickens with pure T-2 toxin include: asthenia, diarrhea, inappetence, and decreased weight (Chi et al., 1977a, b, 1978; Hoerr & Carlton, 1979); decreased egg production and decreases in egg quality such as hatchability and thinner egg shells in laying hens (Chi et al., 1977c; Hamilton et al., 1974; Speers et al., 1977; Wyatt et al., 1975a); neurotoxic signs such as abnormal positioning of wings, hysteroid seizures, and impaired righting reflex (Witlock et al., 1977; Wyatt et al., 1973a); abnormal feathering (Wyatt et al., 1975b); atrophy of the bursa of Fabricius (Boonchuvit et al., 1975; Hoerr & Carlton, 1979; Wyatt et al., 1973b); coagulopathy (Doerr et al., 1974, 1981); and increased susceptibility to bacterial infections (Boonchuvit et al., 1975). Recently Stuart et al. (1981) reported that the clinical signs and pathological changes seen in field outbreaks of the "malabsorption syndrome" in broiler chicks—i.e., poor feed conversion, hyperplasia of the proventriculus, gizzard erosion and dilation, decreased spleen and bursa size, and occasionally heart enlargement—had been reproduced experimentally with "multiple combinations of trichothecene mycotoxins, histamines and diamines."

Ducklings. According to Joffe (1978a), cultures of *F. sporotrichioides* No. 347 isolated from millet associated with outbreaks of ATA (see ATA, above) in the USSR, induced vomiting in ducklings. This culture material caused pronounced swelling and inflammation in many organs and tissues. Marked leukopenia and reduction in erythrocyte counts were also noted in these ducklings (Joffe, 1978a). *F. sporotrichioides* No. 347 was subsequently shown to produce large amounts of T-2 toxin (23.2 mg/10 g of grain) in cultures on wheat incubated at 12°C for 21 days (Joffe, 1978b; Joffe & Yagen, 1977). Emesis is a characteristic response of ducklings to pure trichothecenes such as T-2 toxin, diacetoxyscirpenol, neosolaniol, and fursarenon-X (Ueno et al., 1971c, 1974a).

Geese. In Hungary, cultures of *F. sporotrichioides* have been found to cause necrotic oral lesions in geese (Palyusik et al., 1968) and inhibition of spermatogenesis involving decreases in sperm volume and azoospermy in ganders (Palyusik et al., 1974). The toxicity of corn cultures of an unspecified Hungarian isolate of *F. sporotrichioides* to laying geese was reported on by Palyusik & Koplik-Kovacs (1975a, b). We have identified this strain as *F. sporotrichioides* (Strain 4.20). Culture material of this strain was incorporated in the feed, and diets containing 3 ppm of T-2 toxin fed to one-year-old Rhineland geese. Egg production stopped completely within 10 days. Clinical signs included feed refusal, loss of weight, and extensive necrotic oral lesions. All the birds died within 10 weeks and pathological changes included marked emaciation, cachexia, hypoplasia of the ovaries and oviducts, dilatation of the heart and myocardial degeneration, hemorrhagic enteritis, and necrotic and diphteroid lesions in the oral, glottal, and pharyngeal mucosa. The culture material of this strain of *F. sporotrichioides* contained three unidentified trichothecenes in addition to T-2 toxin and Palyusik & Koplik-Kovacs (1975b) concluded that "apart from T-2, other factors share the responsibility for the severe lesions found in this study." This strain (as *F. sporotrichioides* F-38; see Strain 4.20) was subsequently found to produce HT-2 toxin, T-2 tetraol, and neosolaniol in addition to T-2 toxin (Szathmary et al., 1976, 1977).

Barley that overwintered in the field in northern Canada caused field outbreaks in geese of a fatal toxicosis characterized by vomiting, depression, and severe necrosis of the esophagus, proventriculus, and gizzards (Greenway & Puls, 1976). Clinical signs observed in geese force fed the incriminated, *Fusarium*-infected barley were tremors of the head and legs prior to death (Puls & Greenway, 1976). The main pathological change was severe necrosis of the digestive tract similar to that observed in the field cases. Extracts of the barley were dermotoxic to the skin of a guniea pig and TLC analysis indicated the presence of 25 ppm of T-2 toxin. However, analysis of this barley by another laboratory failed to confirm the presence of T-2 toxin and indicated that other unidentified mycotoxins may have been involved (Puls & Greenway, 1976).

Goats. The crude toxin of *F. nivale* Fn-2B was administered orally to 25 goats by Kubota (1977). We have identified this strain, which is known to produce fusarenon-X, nivalenol, diacetylnivalenol, and T-2 toxin, as *F. sporotrichioides* (Strain 4.9). Clinical signs included depression, anorexia, poor coat condition, lassitude, and salivation. Ulceration was induced at the site of application on the skin or mucous membrane. No gross changes were seen at *post mortem* and no abnormalities were detected in the blood, serum, or urine. Histopathological examination revealed severe damage to the central nervous system, including degeneration and atrophy of nerve cells. Radiomimetic cellular damage also occurred in the spleen, bone marrow, and intestines. Another notable finding was inhibition of spermatogenesis leading to aspermy or reduced viability of spermatozoa (Kubota, 1977).

Guinea Pigs. Rubinstein (1960b) reported that cultures on millet of *F. sporotrichiella* var. *sporotrichioides* No. 319 isolated in the USSR were lethal to guinea pigs. According to Joffe (1960a, 1971, 1974c, 1978b), culture filtrates and powdered dry mycelium of unspecified isolates of *F. sporotrichioides* from overwintered cereals associated with ATA (see ATA, above) in the USSR were toxic to guinea pigs. The animals died within 5 to 21 days and autopsies revealed hemorrhages in many organs and tissues. Many isolates of *F. sporotrichioides* from the USSR have subsequently been shown to produce large amounts of T-2 toxin in culture (Joffe, 1978b; Joffe & Yagen, 1977; Yagen & Joffe, 1976).

Extracts containing T-2 toxin from cultures of *F. tricinctum* NRRL 3299 were reported to be dermotoxic to the shaved skin of a guinea pig by Patterson et al. (1979). We have identified this strain as *F. sporotrichioides* (Strain 4.3).

Guinea pigs dosed with crude methanol-extracted toxin of *F. nivale* Fn-2B showed neurological signs such as unusual jumping and *post mortem* examination revealed degeneration of the nerve cells of their brains together with radiomimetic pathological changes in the actively dividing tissues (Saito & Tatsuno, 1971; Saito et al., 1969). We have identified this strain, which is known to produce fusarenon-X, nivalenol, diacetylnivalenol, and T-2 toxin, as *F. sporotrichioides* (Strain 4.9).

Crystalline fusarenon-X and nivalenol isolated from *F. sporotrichioides* (=*F. nivale*) Fn-2B (Strain 4.9) and diacetoxyscirpenol have been shown to be dermotoxic to the shaved skin of guinea pigs (Ueno et al., 1970b, 1971c). Guinea pigs were more sensitive to the dermotoxic effects of these trichothecenes than rabbits or mice. Pure fusarenon-X has also been shown to induce radiomimetic pathological changes including necrosis and karyorrhexis of the mucosal epithelium of the stomach and small and large intestines, and cellular degeneration in the lymph nodes, spleen, and bone marrow (Ueno et al., 1971c). In a detailed study of the mycotoxicosis produced in guinea pigs by pure T-2 toxin, De Nicola et al. (1978) found gross lesions, hematological and pathological changes including gastric and caecal hyperemia and hemorrhage, edematous intestinal lymphoid tissue, adrenal gland hyperemia, decrease in erythrocyte numbers, leukopenia, absolute lymphopenia, necrosis of lymphoid tissue, bone marrow, and testes, and necrosis and ulceration of the gastro-intestinal tract.

Horses. According to Joffe (1960a, 1971, 1974c, 1978b), the feeding of a horse with 16.4 kg of cereals infected with *F. sporotrichioides* in the USSR caused the development of stomatitis and gingivitis. Very little information is available on the effects of trichothecenes produced by *F. sporotrichioides* on horses. According to Saito & Ohtsubo (1974), "1 to 2 g of fusarenon-X or T-2 toxin is sufficient to kill a horse weighing about 500 kg." Ueno (1980), summarizing the unpublished results of Okubo & Ueno on the toxicity of these two trichothecenes to two horses, reported that both T-2 toxin and fusarenon-X caused bloody diarrhea and radiomimetic pathological changes including karyorrhexis and cellular degeneration in the lymph nodes, intestinal epithelia, bone marrow, thymus, and testes (Ueno, 1980).

Insects. Cultures of a *Fusarium* isolate from horse feed associated with sickness and death of horses in Canada were reported, as *Fus.* sp. MCH 7452 by Davis & Smith (1977) and as *F. poae* 7452 by Davis and Smith (1981), to be toxic to the larvae of the yellow meal worm (*Tenebrio molitor*). We have identified this strain as *F. sporotrichioides* (Strain 4.7). The greatest mortality and/or reduction in weight gain of larvae occurred with cultures incubated at temperatures lower than 25°C (Davis & Smith, 1977) and with cultures incubated in the dark rather than in continuous light (Davis & Smith, 1981). The larval toxicity of culture material of this strain was attributed to fusariotoxins such as T-2 toxin because this strain was found to produce "a high concentration of a metabolite resembling T-2 toxin" by Davis & Smith (1981).

Mice. Rubinstein (1960b) reported that cultures on millet of F. sporotrichiella var sporotrichioides No. 319, isolated in the USSR, were lethal to white mice. According to Joffe (1960a, 1971, 1974c, 1978b), cultures and extracts of cultures of unspecified isolates of F. sporotrichioides from overwintered cereals associated with ATA (see ATA, above) in the USSR were lethal to white mice. Many isolates of F. sporotrichioides from the USSR have subsequently been shown to produce large amounts of T-2 toxin in culture (Joffe, 1978b; Joffe & Yagen, 1977; Yagen & Joffe, 1976). The toxicity to mice of F. tricinctum NRRL 3510, originally isolated from overwintered millet in the USSR, has been reported by Burmeister et al. (1981) and Ueno et al. (1972b, 1973a). We have identified this strain as F. sporotrichioides (Strain 4.1). Ueno et al. (1972b) reported that culture filtrates and extracts of this strain were lethal to mice and caused marked bleeding of the small intestine together with radiomimetic cellular damage characterized by mitotic injury and karyorrhexis in the actively dividing cells in the small intestine, spleen, thymus, testes, and bone marrow. This strain of F. sporotrichioides is known to produce T-2 toxin as well as other trichothecenes including HT-2 toxin, neosolaniol, and diacetoxyscirpenol (see Strain 4.1). Another strain that was originally isolated from overwintered millet in the USSR and that has been reported (as F. tricinctum NRRL 3511) to be lethal to mice (Burmeister et al., 1981) has also been identified as F. sporotrichioides (Strain 4.2). The mycotoxicosis caused in mice by an unspecified "authentic isolate from overwintered grain" of F. sporotrichioides has been described by Schoental & Joffe (1974). Ethanol extracts of cultures of this strain on wheat incubated for 40 days at temperatures that fluctuated from 7 to 24°C were administered to mice by local application, subcutaneous and intraperitoneal injections, and intragastric intubation. Small doses applied to the skin, esophagus, and stomach had local cytotoxic effects that were followed by regeneration and basal cell hyperplasia of the squamous epithelium. High doses caused death within one or more days regardless of the route of administration. Doses of the order LD_{50-70} caused depletion of the lymphoid tissues and subsequent widespread infections suggestive of an immunosuppressive action (Schoental & Joffe, 1974). Extracts of F. sporotrichioides No. 738, which was isolated from overwintered cereals in the USSR and is known to produce T-2 toxin, HT-2 toxin, diacetoxyscirpenol, and neosolaniol, have also been shown to cause pathological changes in the thymus and spleen resulting in immunosuppression in mice (Lafarge-Frayssinet et al., 1979; Lafont et al., 1977).

A mouse bioassay in which culture filtrates or extracts of cultures are injected intraperitoneally in mice has been used extensively in Japan for the detection of toxigenic Fusarium strains (Ueno et al., 1971a, 1972b, 1973a). The following strains that we have identified as F. sporotrichioides have been found to be lethal to mice in this assay: F. tricinctum NRRL 3510 (Strain 4.1), F. tricinctum NRRL 3511 (Strain 4.2), F. tricinctum NRRL 3299 (Strain 4.3), F. nivale NRRL 3249 (Strain 4.8), F. nivale Fn-2B (Strain 4.9), F. episphaeria Fn-M (Strain 4.10), F. oxysporum "niveum" Melon-2 (Strain 4.11), F. solani M-1-1 (Strain 4.12), and F. poae NRRL 3287 (Strain 4.25). All of these strains of F. sporotrichioides have been reported to cause the radiomimetic cytotoxic damage to the actively dividing cells of the hematopoietic and lymphopoietic systems in mice characteristic of trichothecenes (Saito & Okubo, 1970; Saito et al., 1969; Tatsuno, 1968, 1976; Ueno, 1971; Ueno & Fukushima, 1968; Ueno et al., 1969a, 1970a, 1971a, b, 1972b,

1973a). All of these strains have also been reported to produce trichothecenes such as diacetoxyscirpenol, fusarenon-X, neosolaniol, nivalenol, and T-2 toxin in culture.

Three isolates of *F. sporotrichioides* from hay associated with respiratory symptoms in a cow in Finland were lethal to mice (Korpinen & Ylimäki, 1972). Cultures grown on a mixture of wheat, barley, and oats were incubated for one month at three different temperatures (22, 8 and 0°C). Incubation temperature did not have much effect on the toxicity of culture material to mice. Another Finnish strain referred to as *F. tricinctum* has been reported to be toxic to mice by Nummi (1977) and Nummi et al. (1975a). We have identified this strain as *F. sporotrichioides* (Strain 4.23). This strain has subsequently been found to produce T-2 toxin, HT-2 toxin, neosolaniol, and a new trichothecene, T-1 toxin (see Strain 4.23).

Chronic feeding trials in mice have been conducted with *F. sporotrichioides* No. 63 isolated in the USSR (Akhmeteli et al., 1972a, b, 1973) and with the Japanese isolate *F. sporotrichioides* (=*F. nivale*) Fn-2B (Strain 4.9) (Saito & Ohtsubo, 1974). Aqueous extracts of cultures of *F. sporotrichioides* No. 63 on barley were dosed by means of gastric sonde to male and female mice for 12 months and the surviving animals were sacrificed 18 months after the commencement of the experiment (Akhmeteli et al., 1972a, b, 1973). The extract caused a significant increase in lung adenomas, particularly in the males, and one adenocarcinoma of the lungs in a female. No significant difference was found in the incidence of hepatomas, but the only malignant hepatoma occurred in the treated group. Akhmeteli et al. (1972a, b, 1973) concluded from these results that the aqueous extract of *F. sporotrichioides* No. 63 was "mildly blastogenic" in mice. This strain (as *F. sporotrichiella* var. *sporotrichioides* strain 63) has subsequently been shown to produce T-2 toxin in cultures on rice incubated at 25°C for two weeks (Ermakov et al., 1978). Cultures on rice of *F. sporotrichioides* (=*F. nivale*) Fn-2B (Strain 4.9), which is known to produce fusarenon-X, nivalenol, diacetylnivalenol, and T-2 toxin, have been fed to male mice on a life-long basis (Ohtsubo & Saito, 1977; Saito & Ohtsubo, 1974). Malignant tumors occurred in various organs, but their incidences were very low, i.e., one case each of hepatoma, myeloic and lymphocytic leukemia, subcutaneous sarcoma, and adenocarcinoma of the jejunum.

Some of the clinical signs and pathological changes caused in mice by culture material and extracts of cultures of *F. sporotrichioides* have also been induced in mice with pure trichothecenes such as diacetoxyscirpenol (Fromentin et al., 1980, 1981; Rosenstein et al., 1978, 1979, 1981; Salazar et al., 1980; Ueno et al., 1970b), fusarenon-X (Ito et al, 1980; Saito & Ohtsubo, 1974; Saito & Okubo, 1970; Saito et al., 1969; Sato et al., 1978; Shimizu et al., 1979; Tatsuno, 1976; Ueno, 1971; Ueno et al., 1969a, 1970b, 1971b, c,), neosolaniol (Sato et al., 1978; Ueno et al., 1972a), nivalenol (Saito & Okubo, 1970; Saito et al., 1969; Tatsuno, 1968; Ueno & Fukushima, 1968; Ueno et al., 1970b), and T-2 toxin (Hayes & Schiefer, 1980; Hayes et al., 1980; Ohtsubo & Saito, 1977; Otokawa et al., 1979; Rosenstein et al., 1978, 1979, 1981; Sato et al., 1978; Schoental et al., 1978b; Ueno et al., 1972a). Carcinogenicity tests in mice with diacetoxyscirpenol (Lindenfelser et al., 1974), fusarenon-X (Ohtsubo & Saito, 1977; Saito & Ohtsubo, 1974), and T-2 toxin (Lindenfelser et al., 1974; Marasas et al., 1969; Ohtsubo & Saito, 1977), have yielded negative results.

Monkeys. According to Joffe (1960a, 1971, 1974c, 1978b), Mayer (1953b), and Rubinstein (1960b), clinical signs and pathological changes similar to ATA in man (see ATA, above) have been reproduced experimentally in monkeys with culture material of *F. sporotrichioides* by Rubenstein & Lyass (1948). This publication was not available to us, but according to a summary of this work by Mayer (1953b), monkeys dosed with the toxic grain vomited, had diarrhea, and lost appetite. Within 20 to 30 days, petechial spots and hemorrhages developed on the skin. Hematological examination revealed progressive leukopenia, relative lymphocytosis, anemia, and thrombocytopenia. Some of the characteristic lesions of ATA have also been reproduced in monkeys with pure diacetoxyscirpenol (Stähelin et al., 1968) and semi-purified T-2 toxin (Rukmini et al., 1980). Both of these trichothecenes are known to be produced by F. sporotrichioides (see mycotoxins produced, below).

Pigeons. Rubinstein (1960b) reported that a single feeding of millet cultures of F. sporotrichiella var. sporotrichioides No. 319, isolated in the USSR, was lethal to pigeons. According to Joffe (1978a), cultures on prosomillet of unspecified strains of F. sporotrichioides (as F. sporotrichiella var. sporotrichioides) isolated in the USSR caused emesis upon feeding to pigeons.

Extracts of cultures of F. moniliforme 111 (=NRRL 3197) and F. poae 2518 (=NRRL 3287) in Richard's solution caused prolonged emesis upon injection in the wing vein of pigeons (Prentice & Dickson, 1968; Prentice et al., 1959). We have identified these strains as F. sporotrichioides (Strains 4.24 and 4.25). Two compounds that caused prolonged emesis in pigeons were partially purified from cultures of F. moniliforme 111, but not chemically characterized (Prentice & Dickson, 1968). In addition to the emetic activity, one of these compounds was also toxic and caused death of the pigeons within 12 hours after injection. The emetic activity of these strains was re-examined by Ellison & Kotsonis (1973). These authors found no emetic activity in pigeons with F. moniliforme NRRL 3197 (Strain 4.24) and concluded that it had lost the ability to produce emetic compounds. Subsequently this strain has been found to produce deoxynivalenol and T-2 toxin in cultures on corn (see Strain 4.24). Extracts of cultures of F. poae NRRL 3287 (Strain 4.25) grown either on corn or in Richard's solution caused emesis in pigeons at nonlethal concentrations under conditions of oral and intravenous administration. The emetic metabolite in these cultures was identified as T-2 toxin by Ellison & Kotsonis (1973). Pure T-2 toxin has been shown to cause emesis in pigeons at non-lethal concentrations (Ellison & Kotsonis, 1973).

Pigs. Although F. sporotrichioides (=F. tricinctum) has been implicated in the etiology of field outbreaks of the hemorrhagic syndrome in pigs known as moldy corn toxicosis (see hemorrhagic syndrome, above), very little information is available on the toxicity of whole cultures of this fungus to pigs. According to Joffe (1971), pigs fed large quantities of toxic overwintered cereals in the USSR did not develop signs of ATA. In a toxicity experiment in pigs with culture material of an unspecified isolate of F. sporotrichioides in the USSR referred to by Mirocha (1979), clinical signs of acute intoxication included emesis, loss of appetite, diarrhea, frequent urination, uncoordinated movement, periodic muscle tremmors, and paresis of the hind quarters before death. The animals died rapidly (6 to 7.5 hours after treatment) and a catarrhal-hemorrhagic gastroenteritis and marked signs of hemorrhagic diathesis were observed together with increases in erythrocytes, hemoglobin, and leukocytes, decreases in thrombocytes, and neutrophilia. In chronic feeding experiments, ulcerous-necrotic lesions of the snout, lips, and mucous membrane of the mouth were noted. Hematological changes were characterized by marked leukopenia and thrombocytopenia. Patterson et al. (1979) found no abnormal clinical signs or pathological lesions in a pig administered a crude extract of F. sporotrichioides (=F. tricinctum) NRRL 3299 (Strain 4.3) containing T-2 toxin at a dosage rate of 0.1 mg T-2 toxin/kg/day for 14 days, but higher levels of T-2 toxin caused almost immediate emesis.

A marked feed refusal response in pigs has been induced with cultures on corn of *F. moniliforme* (=*F. tricinctum*) NRRL 3197 and *F. poae* NRRL 3287 (Vesonder et al., 1977b). We have identified both of these strains as *F. sporotrichioides* (Strains 4.24 and 4.25). Cultures of both strains were subsequently shown to contain T-2 toxin, and *F. sporotrichioides* NRRL 3197 (Strain 4.24) produced deoxynivalenol as well (Vesonder et al., 1981a).

Some of the above clinical signs and pathological lesions have been reproduced experimentally in pigs with pure T-2 toxin (Weaver et al., 1977, 1978a) and diacetoxyscirpenol (Weaver et al., 1977, 1978c, 1981), and both compounds caused pronounced emesis and feed refusal. Hemorrhage, however, could not be induced in pigs with pure T-2 toxin (isolated from *F. tricinctum* NRRL 3299, Strain 4.3) at the dosage regimes used (Patterson et al., 1979; Weaver et al., 1977, 1978a). Diacetoxyscirpenol caused hemorrhagic bowel lesions as well as severe ulcerative oral lesions in pigs (Weaver et al., 1977, 1978c, 1981). Pure T-2 toxin has also been shown to cause reproductive problems in pigs, including abortion (Weaver et al., 1977, 1978b), infertility, small litters, and very small piglets (Weaver et al., 1977, 1978d).

Rabbits. According to Joffe (1960a, 1971, 1974c, 1978b), culture filtrates and powdered dry mycelium of unspecified isolates of *F. sporotrichioides* from overwintered cereals associated with outbreaks of ATA (see ATA, above) in the USSR were acutely toxic to rabbits. The rabbits died within 8 to 24 hours and autopsies revealed hemorrhages in many organs and tissues. Many isolates of *F. sporotrichioides* from the USSR have subsequently been shown to produce large amounts of T-2 toxin in culture (Joffe, 1978b; Joffe & Yagen, 1977; Yagen & Joffe, 1976).

A skin test involving the application of ether or alcohol extracts of fungal cultures on the shaved skin of rabbits and assessment of the dermotoxic reaction has been used extensively for the screening of *Fusarium* strains, including *F. sporotrichioides*, for toxicity (Joffe, 1960a, b, 1960, 1973a, 1974b; Joffe & Palti, 1974; Joffe & Yagen, 1977; Kvashnina, 1976, 1978, 1979; Yagen & Joffe, 1976). In a study of *F. sporotrichioides* isolates from soil and overwintered cereals associated with outbreaks of ATA in the USSR, 65 strains were grown on various media at temperatures ranging from −5 to 8°C for 25 to 70 days and subjected to freezing and thawing (Joffe, 1960a, b). Extracts of 44 of these strains (68%) were found to be toxic or highly toxic and to cause an edematous-hemorrhagic skin reaction; topical application of some of the extracts even caused death of the treated rabbits. The highest levels of dermotoxicity to rabbit skin in all isolates of *F. sporotrichioides* tested were found in extracts of cultures incubated at low temperatures (5 to 8°C), although the optimal temperature for vegetative growth was 18 to 24°C (Joffe, 1974b). At all temperatures, cultures incubated in the dark were more toxic than those incubated in the light (Joffe, 1974b). A correlation has been found between the degree of dermotoxicity to rabbit skin of *F. sporotrichioides* isolates and their phytotoxicity (Joffe, 1973a; Joffe & Palti, 1974). A total of 106 isolates of *F. sporotrichioides* from overwintered cereals in the USSR were tested for dermotoxicity to rabbit skin by Yagen & Joffe (1976), using ethyl alcohol extracts of cultures on wheat incubated at 12°C for 21 days. Of the 106 isolates, 71 strains (67%) produced a strong to very strong dermotoxic reaction. All the dermotoxic strains produced T-2 toxin as detected by TLC and GLC (Yagen & Joffe, 1976).

Seven of the above isolates of *F. sporotrichioides* (No 60/10, 347, 351, 738, 921, 1182, and 1183) that had previously been found to be dermotoxic (Joffe, 1960a) were re-tested for dermotoxicity to rabbit skin by Joffe (1978b) and Joffe & Yagen (1977). All of these isolates were found to be strongly dermotoxic and to produce from 4.6 to 24.0 mg of T-2 toxin/10 g of wheat grain on which they had been cultured at 12°C for 21 days. Pathological examination confirmed that the crude extracts caused the same skin lesions as pure T-2 toxin, i.e., edema, hemorrhage, and necrosis of the epidermis, dermis, and hair follicles (Yagen & Joffe, 1976). There was a positive relationship between the rabbit skin irritation factor and the amount of T-2 toxin detected by TLC and GLC in these extracts, and it could be concluded that T-2 toxin was the dermotoxic compound produced by these strains of *F. sporotrichioides* (Joffe, 1978b; Joffe & Yagen, 1977; Yagen & Joffe, 1976).

The following isolates from other sources were also found to be dermotoxic to rabbit skin and to produce T-2 toxin by Joffe (1978b) and Joffe & Yagen (1977): 1) NRRL 3287, as *F. poae* (*F. sporotrichioides,* Strain 4.25); 2) NRRL 3299, as *F. poae* (*F. sporotrichioides,* Strain 4.3); 3) T-2, as *F. poae* by Joffe (1978b) but as *F. sporotrichioides* by Joffe & Yagen (1977) (same strain as NRRL 3299 above, which we have identified as *F. sporotrichioides,* Strain 4.3); 4) NRRL 3249, as *F. sporotrichioides* (Strain 4.8); 5) NRRL 5908, as *F. sporotrichioides* (we have identified this strain as *F. graminearum,* Strain 8.41, and it is consequently excluded from the section Sporotrichiella); 6) No. 2061-C, as *F. sporotrichioides* (Strain 4.6); 7) No. YN-13, as *F. sporotrichioides* (Strain 4.26); and 8) NRRL 3509, as *F. sporotrichioides* var. *tricinctum* (Strain 4.9).

The rabbit skin test has also been used to screen *Fusarium* isolates from toxic fescue hay associated with outbreaks of fescue foot (see fescue foot, above) in cattle in the USA for dermotoxicity (Ellis & Yates, 1971; Keyl et al., 1967; Yates et al., 1967, 1968, 1969, 1970). Extracts of several of these isolates were found to be dermotoxic to rabbit skin. One of the most toxic of these isolates has been referred to in the literature as *F. nivale* NRRL 13,318 (Keyl et al., 1967), *F. nivale* NRRL 3249 (Ellis & Yates, 1971; Yates et al., 1968, 1969), *F. tricinctum* NRRL 3249 (Ellis & Yates, 1971; Yates et al., 1969), and *F. sporotrichioides* NRRL 3249 (Joffe, 1978b; Joffe & Yagen, 1977); we have identified this isolate, which is known to produce butenolide, T-2 toxin, and an unidentified toxin, as *F. sporotrichioides* (Strain 4.8).

Extracts of the Japanese isolate referred to as *F. nivale* Fn-2B (=*F. tricinctum* NRRL 3509 = *F. sporotrichioides* var. *trincinctum* NRRL 3509) have been found to be dermotoxic to rabbit skin and to cause necrosis of the epidermis, degeneration and necrosis of the hair follicles, edema in the dermis, and coagulation necrosis of the muscle fibers (Joffe, 1978b; Joffe & Palti, 1975; Joffe & Yagen, 1977; Ueno, 1971; Ueno et al., 1970b, 1971a, b). We have identified this strain as *F. sporotrichioides* (Strain 4.9). This strain is known to produce four trichothecenes, i.e., fusarenon-X, nivalenol, diacetylnivalenol, and T-2 toxin (see Strain 4.9). The Japanese isolate *F. episphaeria* Fn-M has also been reported to be dermotoxic to rabbit skin (Ueno et al., 1971a). We have identified this strain, which is known to produce fusarenon-X and nivalenol, as *F. sporotrichioides* (Strain 4.10).

Pure trichothecenes that have been demonstrated to be dermotoxic to rabbit skin include diacetoxyscirpenol (Hayes & Schiefer, 1979; Ueno et al., 1970b), fusarenon-X and nivalenol (Ueno et al., 1970b), and T-2 toxin (Chung et al., 1974; Eppley et al., 1974; Hayes & Schiefer, 1979; Yagen & Joffe, 1976). Pure T-2 toxin has also been reported to cause anorexia, diarrhea, oral lesions, leukopenia, decreases in hematocrit, and prolonged plasma clotting time in rabbits (Gentry & Cooper, 1981).

Rats. An "authentic isolate" of *F. sporotrichioides* from overwintered cereals associated with outbreaks of ATA (see ATA, above) in the USSR has been reported to be toxic to rats by Schoental & Joffe (1974). Ethanol extracts of cultures of this strain on wheat incubated for 40 days at temperatures that fluctuated from 7 to 24°C were administered to rats by local application, subcutaneous and intraperitoneal injections, and intragastric intubation. Skin application of non-lethal doses caused irritation, congestion, edema, ulceration, keratinization, and desquamation. Larger doses caused death within one or more days regardless of the route of administration. The main pathological lesions were hemorrhage and necrosis of the lymph glands, hemorrhage and depletion of lymphoid elements in the spleen, and edema, congestion, and hemorrhage of the mucosa of the digestive tract. In weanling rats dosed by stomach tube, the lining of the esophagus and of the squamous part of the stomach became ulcerated, keratinized, and desquamated. Animals dosed repeatedly with smaller doses did not appear to have gross lesions, but microscopically there were pockets of distension and invagination in the esophagus and thickened areas in the forestomach lined with multicellular squamous epithelium with basal cell hyperplasia. In rats that died after several weeks, widespread infections suggestive of an immunosuppressive action were noted. According to Schoental et al. (1976), unspecified "chronic lesions" were also found in various organs, including the digestive tract, brain, and sex organs, of rats dosed with these extracts of *F. sporotrichioides* on a long-term basis. Schoental et al. (1979) referred to unpublished results of Schoental & Joffe, who found brain tumors, heart lesions, and pancreatic neoplasms in rats treated chronically with *F. sporotrichioides* extracts. Many isolates of *F. sporotrichioides* from the USSR, presumably including the one used in the toxicity experiments in rats by Schoental & Joffe (1974) and Schoental et al. (1976), have subsequently been found to produce large amounts of T-2 toxin in culture (Joffe, 1978b; Joffe & Yagen, 1977; Yagen & Joffe, 1976).

The Japanese isolate *F. nivale* Fn-2B has been reported to be acutely toxic to Donryu rats and to cause radiomimetic cellular damage characteristic of trichothecenes and marked damage to the deep gland cells of the glandular stomach (Saito et al., 1969). We have identified this strain as *F. sporotrichioides* (Strain 4.9). In a long-term feeding study in male rats with rice cultures of this strain, the prominent pathological lesions were atrophy of the thymus, spleen, and testes, and atypical hyperplasia in the epithelium of the gastric and intestinal mucosa and intrahepatic bile ducts, but no tumors occurred (Ohtsubo & Saito, 1977; Saito & Ohtsubo, 1974). This strain is known to produce at least four trichothecenes, i.e., fusarenon-X, nivalenol, diacetylnivalenol, and T-2 toxin (see Strain 4.9).

A rat skin test was used as bioassay in the original isolation and characterization of T-2 toxin from *F. tricinctum* strain T-2 (Bamburg, 1968; Bamburg et al., 1968a, b); we have identified this strain as *F. sporotrichioides* (Strain 4.3). Ethyl acetate extracts of cultures were applied to the shaved back of rats, and the severity of the resulting dermotoxic reaction was estimated visually. The reaction ranged from a just-noticeable skin reaction with the formation of a light scab to inflammation, hemorrhage and heavy scab formation, or death at high concentrations. Strain T-2 is known to produce T-2 toxin as well as HT-2 toxin, diacetoxyscirpenol, neosolaniol, and butenolide (see Strain 4.3). We have also identified the following isolates of "*F. tricinctum*" reported to be dermotoxic to rats and to produce T-2 toxin by Bamburg (1968) and Marasas (1969) as *F. sporotrichioides:* Nos. 223, 238, 347, 209 and 348 (Strains 4.4, 4.27, 4.28, 4.29, and 4.30).

Lyophilized culture material of *F. sporotrichioides* (=*F. tricinctum*) T-2 (Strain 4.3) containing T-2 toxin was acutely toxic to rats (Kosuri et al., 1971). Clinical signs included listlessness, insensitivity, irritability, diarrhea, hemorrhages from the urinary tract, and death within 48 to 72 hours. Pathological changes included ecchymotic hemorrhages in the peritoneal surfaces of the small and large intestines, hemorrhage into the intestinal lumens, and necrosis of intestinal lymph nodes.

Two isolates of *F. tricinctum* (No. 15 and 16) from moldy corn associated with a field outbreak of a hemorrhagic syndrome (see hemorrhagic syndrome, above) in dairy cattle in Wisconsin, USA, were lethal to albino rats (Hsu et al., 1972). We have identified isolate No. 16 (=ATCC 24045) as *F. sporotrichioides* (Strain 4.5). Although the moldy corn from which this strain was isolated contained T-2 toxin, Hsu et al. (1972) did not state whether the ethyl acetate extracts of cultures of this strain toxic to rats contained T-2 toxin or not. Another isolate of "*F. tricinctum*" from moldy corn in Wisconsin, USA, was reported to be acutely toxic to female albino rats fed cultures on corn incubated for 21 days at either 10 or 25°C (Marasas & Smalley, 1972). Cultures of this strain no longer exist, but we have reason to believe that this strain was probably *F. sporotrichioides* similar to other isolates of this species (Strains 4.4 and 4.5) from corn in Wisconsin.

Pure trichothecenes that have been demonstrated to be dermotoxic to rat skin include diacetoxyscirpenol (Gilgan et al., 1966; Hayes & Schiefer, 1979) and T-2 toxin (Bamburg et al., 1968a, b; Hayes & Schiefer, 1979; Marasas et al., 1969; Wei et al., 1972). Many of the above clinical signs and pathological changes have also been reproduced experimentally in rats with pure diacetoxyscirpenol (Stähelin et al., 1968), fusarenon-X (Ohtsubo & Saito, 1977; Ueno et al., 1971b, c), nivalenol (Saito et al., 1969), and T-2 toxin (Bamburg et al., 1968b; Kosuri et al., 1971; Marasas et al., 1969; Ohtsubo & Saito, 1977; Schoental et al., 1979).

Carcinogenicity tests in rats with pure trichothecenes have yielded inconclusive results. Stähelin et al. (1968) reported that diacetoxyscirpenol did not induce tumors in rats and Marasas et al. (1969) found that T-2 toxin was not carcinogenic in rats. On the other hand, Schoental et al. (1979) reported that T-2 toxin caused tumors in various organs including the stomach, duodenum, pancreas, brain, and pituitary in rats. A low incidence of tumors in various organs has also been reported in rats fed fusarenon-X (Ohtsubo & Saito, 1977; Saito & Ohtsubo, 1974; Saito et al., 1980).

Sheep. The toxicity to sheep of cultures of *F. sporotrichioides* (as *F. sporotrichiella* var. *sporotrichioides*) strains isolated in the USSR was reported by Kurmanov (1978a). Culture material introduced directly into the abomasum of sheep and lambs caused some of the characteristic clinical signs and pathological changes of trichothecene toxicoses including depression, ataxia, paralysis of the extremeties, hyperemia of the conjunctiva, loss of appetite, necrotic mouth lesions, bloody diarrhea, hemorrhages in the gastrointestinal tract, lungs, and heart, increases in erythrocytes and leukocytes and decreases in lymphocytes followed by progressive leukopenia and thrombocytopenia, and atrophy of the bone marrow. Many isolates of *F. sporotrichioides* from the USSR have subsequently been shown to produce large amounts of T-2 toxin in culture (Joffe, 1978b; Joffe & Yagen, 1977; Yagen & Joffe, 1976). No reports on the effects of pure trichothecenes in sheep have been found in the literature.

Turkeys. Cultures of *F. tricinctum* 2061-C were found to be acutely toxic to turkey poults by Christensen et al. (1972a). We have identified this strain as *F. sporotrichioides* (Strain 4.6). Consumption of diets containing as little as 1% of culture material of this strain resulted in mortality, decreased feed efficiency and weight gain, and the development of bilateral necrotic lesions at the angles of the mouth. According to Mirocha & Christensen

(1974), this strain produces T-2 toxin which was responsible for the necrotic mouth lesions and mortality in turkey poults reported by Christensen et al. (1972a). The production of T-2 toxin by this strain (as *F. sporotrichioides* 2061-C) has been confirmed by Joffe (1978b) and Joffe & Yagen (1977). Pure T-2 toxin (isolated from *F. tricinctum* NRRL 3299, Strain 4.3) has been shown to cause reduced feed efficiency and weight gain, oral lesions, mortality, and atrophy of the thymus in turkey poults (Richard et al., 1978).

Mycotoxins Produced

Acetyl T-2 Toxin. This trichothecene has only been isolated from cultures of *F. poae* NRRL 3287 in Richard's solution incubated at room temperature for 4 weeks (Kotsonis et al., 1975b). We have identified this strain, which has been reported to have emetic activity in pigeons (Ellison & Kotsonis, 1973; Prentice & Dickson, 1968), as *F. sporotrichioides* (Strain 4.25). The yield of pure acetyl T-2 toxin obtained from this strain by Kotsonis et al. (1975b) was 7 mg/7 ℓ of culture material.

Butenolide. Butenolide was first isolated from a strain of "*F. nivale*" isolated from toxic fescue hay associated with outbreaks of fescue foot in cattle (see fescue foot, above) in the United States (Burkhardt et al., 1968; Yates et al., 1967). The butenolide-producing strain has been referred to as *F. nivale* NRRL 13,318 (Keyl et al., 1967), *F. nivale* NRRL 3249 (Yates et al., 1968), *F. tricinctum* NRRL 3249 (Burmeister et al., 1971; Ellis & Yates, 1971; Tookey et al., 1972; Ueno et al., 1972b, 1973a; Yates, 1971; Yates et al., 1969, 1970) and *F. sporotrichioides* NRRL 3249 (Joffe, 1978b; Joffe & Yagen, 1977). We have identified this strain as *F. sporotrichioides* (Strain 4.8). According to Yates et al. (1968), about 29 mg of butenolide/plate could be isolated from cultures on Sabouraud's agar incubated at 15°C for 3 to 4 weeks and the yield of crystalline butenolide from cultures in liquid Sabouraud's maltose medium incubated at 3°C for 23 weeks was 1.0 g/ℓ of culture filtrate. Although these large amounts of butenolide were originally isolated from this strain (Yates et al., 1968), subcultures of the parent strain are known to lose the ability to produce the butenolide through repeated transfers on agar (Yates et al., 1969), and no butenolide could be detected chemically in cultures of strain NRRL 3249 by Burmeister et al. (1971), Tookey et al. (1972) and Yates et al. (1970).

In subsequent studies on butenolide production by *Fusarium* species on Sabouraud's agar incubated at 15°C, Yates et al. (1970) reported that 23 of 34 unidentified strains of *Fusarium* produced butenolide, while Burmeister et al. (1971) found 8 out of 14 isolates of "*F. tricinctum*" to be positive. In both of these studies *F. tricinctum* NRRL 3299 was shown to produce butenolide. We have identified this strain as *F. sporotrichioides* (Strain 4.3), which was also found to be one of five butenolide-producing strains out of 30 isolates of "*F. tricinctum*" tested by Marasas (1969). We have identified two of the other strains reported to produce butenolide by Marasas (1969) as *F. sporotrichioides*, i.e., No. 238 (Strain 4.27) and No. 209 (Strain 4.29).

Two isolates of *F. sporotrichioides* from overwintered cereals associated with outbreaks of ATA (see ATA, above) in the USSR have been reported to produce butenolide in culture. Ueno et al. (1972b) isolated 9.5 g of crystalline butenolide from 40 ℓ of culture material of *F. sporotrichioides* NRRL 3510 (Strain 4.1) in PSC medium incubated at 26°C for 12 days. *F. sporotrichioides* No. 738 (Joffe, 1960a, 1974b, 1978b; Joffe & Yagen, 1977), was reported to produce butenolide in culture in modified Czapek medium incubated at 15°C for 9 days (Lafarge-Frayssinet et al., 1979; Lafont et al., 1977).

In a study of butenolide production by 106 *Fusarium* isolates from barley and wheat grain in Japan in PSC medium incubated at 25°C for 2 weeks, Yoshizawa et al. (1979)

detected butenolide by TLC in 37 (= 34.9%) of the strains. Although most of these isolates belonged to *F. graminearum*, no indication was given of the identity of the remaining isolates. The butenolide standard used in this work by Yoshizawa et al. (1979) was isolated from "*F. solani* M-1-1 (ATCC 26533)," but this mycotoxin was not detected by TLC in toxic extracts of this strain by Ueno et al. (1972a). We have identified this strain as *F. sporotrichioides* (Strain 4.12).

According to Suzuki et al. (1981a), 3 out of 6 isolates of "*F. tricinctum*" from wheat and barley in Japan produced from 0.09 to 1.18 µg/g of butenolide in cultures on rice. We have not seen these isolates and it is not clear to which species these three butenolide-producing strains belong.

Deoxynivalenol. Analysis by GLC of corn cultures of *F. moniliforme* (= *F. tricinctum*) NRRL 3197 refused by swine revealed the presence of 1.5 µg/g of deoxynivalenol together with 33 µg/g of T-2 toxin (Vesonder et al., 1981a). We have identified this strain as *F. sporotrichioides* (Strain 4.24). This is the only known report of deoxynivalenol production by *F. sporotrichioides*.

Diacetoxyscirpenol. The production of diacetoxyscirpenol by *F. tricinctum* strain T-2 (= NRRL 3299) has been reported by Bamburg & Strong (1969), Ueno et al. (1973a), and Schmidt et al. (1981a—as *F. tricinctum* Sp. 897), but this mycotoxin was not detected in toxic extracts of this strain by Bamburg et al. (1968a, b) and Ueno et al. (1972b). We have identified this strain as *F. sporotrichioides* (Strain 4.3).

At least two isolates of *F. sporotrichioides* from overwintered cereals associated with outbreaks of ATA (see ATA, above) in the USSR, have been reported to produce diacetoxyscirpenol in culture. *F. sporotrichioides* NRRL 3510 (Strain 4.1) was reported to produce diacetoxyscirpenol by Ueno et al. (1973a) but not by Ueno et al. (1972b). *F. sporotrichioides* No. 738 (Joffe, 1960a, 1974b, 1978b; Joffe & Yagen, 1977) was found to produce diacetoxyscirpenol in culture in modified Czapek medium incubated at 15°C for 9 days (Lafarge-Frayssinet et al., 1979; Lafont et al., 1977).

The Japanese isolate *F. solani* M-1-1 was reported to produce trace amounts of diacetoxyscirpenol in PSC medium in stationary culture at 25°C for 12 days by Ishii et al. (1971). We have identified this strain as *F. sporotrichioides* (Strain 4.12). The production of diacetoxyscirpenol by this strain has subsequently been confirmed by Ohta et al. (1977), Schmidt et al. (1981a—as *F. solani* Sp. 900), Ueno et al. (1972a, b, 1973a, 1975), and Tatsuno et al. (1973).

Diacetylnivalenol (= Nivalenol Diacetate). This trichothecene was first isolated from cultures of the Japanese isolate *F. nivale* Fn-2B in PSC medium incubated in stationary culture at 25°C for 4 days followed by 2 weeks at 15°C (Tatsuno et al., 1970). We have identified this strain as *F. sporotrichioides* (Strain 4.9). The yield of diacetylnivalenol obtained by Tatsuno et al. (1970) was 60 mg from 50 ℓ of culture material. The production of diacetylnivalenol by this strain has subsequently been confirmed by Fujimoto et al. (1972), Ohta et al. (1977, 1978), Tatsuno (1976), Tatsuno et al. (1973), and Yoshizawa et al. (1980).

In addition to *F. sporotrichioides* (= *F. nivale*) Fn-2B, only two other *Fusarium* strains, i.e., *F. oxysporum* "niveum" Melon-1 and Melon-2, have been reported to produce diacetylnivalenol (Ueno et al., 1973a). We have identified isolate Melon-1 as *F. sporotrichioides* (Strain 4.11). Diacetylnivalenol was detected by TLC in toxic extracts of cultures of this strain in PSC medium incubated at 25 to 27°C for 2 weeks (Ueno et al., 1973a).

Fusarenon-X (=Fusarenon, =Nivalenol Monoacetate, =Monoacetyl Nivalenol). This trichothecene was first isolated from the Japanese isolate *F. nivale* Fn-2B by Morooka & Tatsuno (1970) and Tatsuno et al. (1969, 1970), who referred to this compound as fusarenon, and by Ueno et al. (1969a, 1970c), who referred to it as fusarenon-X. We have identified this strain as *F. sporotrichioides* (Strain 4.9). Although fusarenon and fusarenon-X were originally thought to be slightly different chemically (Morooka & Tatsuno, 1970; Saito & Okubo, 1970; Tatsuno et al., 1969; Ueno et al., 1969a, 1970c, 1971b), they are now considered to be identical (Morooka et al., 1971) and the name fusarenon-X is apparently preferred. Yields of 20 mg of fusarenon-X from 3 kg of culture material of strain Fn-2B on rice incubated at 25 to 27°C for 20 days were obtained by Morooka & Tatsuno (1970) and Tatsuno et al. (1970), while Ueno et al. (1969a, 1970c) obtained 200 mg from 10 ℓ of culture material in PSC medium incubated at 25 to 27°C for 2 weeks. Maximal yields of fusarenon-X (approximately 200 mg/10 ℓ) have been obtained with strain Fn-2B in stationary culture in PSC medium after 6 to 8 days of incubation at 27°C, with the yield decreasing at lower temperatures (Ueno et al., 1975). *F. sporotrichioides* (= *F. nivale*) Fn-2B has been used extensively in Japan for the production of fusarenon-X for use in toxicological investigations (Fujimoto et al., 1972; Kato et al., 1979; Matsuoka & Kubota, 1981; Matsuoka et al., 1979; Ohta et al., 1977, 1978; Saito & Okubo, 1970; Sato et al., 1978; Tatsuno, 1976; Tatsuno et al., 1973; Ueno, 1971; Ueno et al., 1970a, 1971a, b, c, 1973a, 1975, 1978; Yoshizawa & Morooka, 1975b; Yoshizawa et al., 1979, 1980).

Two other Japanese isolates have also been reported to produce fusarenon-X: *F. episphaeria* Fn-M (Ueno et al., 1971a, c, 1972b, 1973a, 1975) and *F. oxysporum* "niveum" Melon-1 (Ueno et al., 1973a, 1975). We have identified both of these strains as *F. sporotrichioides* (Strains 4.10 and 4.11). High yields of fusarenon-X (0.8 g/10 ℓ of culture filtrate) have been isolated from cultures of *F. sporotrichioides* (= *F. episphaeria*) Fn-M (Strain 4.10) in PSC medium incubated at 27°C for 2 weeks (Ueno et al., 1971a).

HT-2 Toxin. This trichothecene was first isolated from cultures of *F. tricinctum* Strain T-2 in Gregory's medium incubated at 24°C for 2 to 3 weeks (Bamburg, 1968; Bamburg & Strong, 1969). We have identified this strain as *F. sporotrichioides* (Strain 4.3). The yield of crystalline HT-2 toxin obtained by Bamburg (1968) and Bamburg & Strong (1969) was 109 mg/6 ℓ of culture material. Ikediobi et al. (1971) detected 750 mg/kg of HT-2 toxin by GLC in corn-rice cultures of this strain incubated at 27°C for 21 days followed by 24 days at 8°C. The production of HT-2 toxin by this strain has been confirmed by Schmidt et al. (1981a—as *F. tricinctum* Sp. 897) and Ueno et al. (1972b, 1973a—as *F. tricinctum* NRRL 3299). HT-2 toxin for use in toxicological studies has also been isolated from cultures of this strain on rice incubated at 11°C for 3 months followed by 10 days at room temperature (Chi et al., 1978) and prepared chemically from T-2 toxin produced by this strain (Lee & Chu, 1981; Wei et al., 1971).

The Japanese isolate *F. sporotrichioides* (= *F. solani*) M-1-1 (Strain 4.12) has been reported to produce HT-2 toxin by Ueno et al. (1972b, 1973a). The yield of crystalline HT-2 toxin obtained from cultures of this strain in PSC medium incubated at 25°C for 20 days was 275 mg/87 ℓ of culture filtrate (Ishii & Ueno, 1981). The production of HT-2 toxin by this strain has also been confirmed by Schmidt et al. (1981a—as *F. solani* Sp. 900) and Yoshizawa et al. (1979, 1980—as *F. sporotrichioides* ATCC 26553).

At least four isolates of *F. sporotrichioides* from overwintered cereals associated with outbreaks of ATA (see ATA, above) in the USSR have been reported to produce HT-2 toxin in culture. *F. sporotrichioides* NRRL 3510 (Strain 4.1) was found to produce HT-2 toxin by Ueno et al. (1972b), but not by Ueno et al. (1973a). *F. sporotrichioides* 738 (Joffe, 1960a, 1974b, 1978b; Joffe & Yagen, 1977) was reported to produce HT-2 toxin by Lafont et al. (1977) and strain 5328 by Kotik et al. (1970) and Szathmary et al. (1976, 1977). According to Yagen et al. (1977), *F. sporotrichioides* 921 was the most toxic among 106 isolates of *F. sporotrichioides* from overwintered cereals in the USSR and 0.030 g of crystalline HT-2 toxin could be isolated from 1 kg of culture material on millet incubated at 12°C for 21 days.

The presence of HT-2 toxin in culture material of *F. poae* (= *F. tricinctum*) NRRL 3287 with emetic activity in pigeons was reported by Kotsonis & Ellison (1975). We have identified this strain as *F. sporotrichioides* (Strain 4.25). These authors detected HT-2 toxin by a combination of TLC and fluorodensitometry in shake cultures of this strain in Richard's solution incubated at room temperature. Kotsonis & Ellison (1975) also demonstrated that T-2 toxin is produced prior to HT-2 toxin and that it can be converted to HT-2 toxin by cultures of this strain.

Three Hungarian isolates of *F. sporotrichioides* (F-38, F-24, and F-26) have been reported to produce HT-2 toxin (Szathmary et al., 1976, 1977). We have confirmed the identification of these isolates as *F. sporotrichioides* (Strains 4.20, 4.21, and 4.22). The maximal yield of HT-2 toxin (1000 µg/g) was obtained with *F. sporotrichioides* F-38 (Strain 4.20) in cultures on millet containing 2% malt extract incubated at 23 to 27°C for one week, 5 to 8°C for 2 weeks and again at 23 to 27°C for one week (Szathmary et al., 1976, 1977).

Fusarium tricinctum 72187 isolated from cereal grain in Finland has been reported to produce HT-2 toxin in culture on a mixture of cereal grains incubated at 6°C for 4 months (Ilus et al., 1977; Niku-Paavola & Nummi, 1977; Niku-Paavola et al., 1977). We have identified this strain as *F. sporotrichioides* (Strain 4.23). This strain (as *F. sporotrichioides* 72187-12 = *F. tricinctum* 72187-10k) has also been used for the production of HT-2 toxin for use in toxicological studies (Ilus et al., 1981; Ylimäki et al., 1979).

In a study of the toxigenicity of *Fusarium* isolates from overwintered cereals in Canada, Scott et al. (1980) detected HT-2 toxin by TLC in two of 20 toxic strains and also found that a third strain could produce HT-2 toxin under certain conditions. We have confirmed that two of these strains (HPB 071178-13 and -18) are *F. sporotrichioides* (Strains 4.15 and 4.16). Maximal yields of HT-2 toxin (160 µg/g) were obtained with *F. sporotrichioides* HPB 071178-18 (Strain 4.16) cultured on frost-killed barley or wheat grain incubated at temperatures fluctuating between 0 and 14°C (Scott et al., 1980).

Neosolaniol (= 8-Hydroxydiacetoxyscirpenol). This trichothecene, originally called solaniol, was first isolated and characterized from the Japanese strain *F. solani* M-1-1 by Ishii et al. (1971). We have identified this strain as *F. sporotrichioides* (Strain 4.12). The name of this compound was changed from solaniol to neosolaniol by Ueno et al. (1972a). The production of neosolaniol by this strain has been confirmed by Ishii & Ueno (1981), Tatsuno et al. (1973), and Ueno et al. (1972a, b, 1973a, b, 1975) and this strain has also been used for the production of neosolaniol used in toxicological studies (Kato et al., 1979; Matsumoto et al., 1978; Ohta et al., 1977; Sato et al., 1978; Yoshizawa et al., 1979, 1980). The yield of neosolaniol obtained by Ishii et al. (1971) and Ueno et al. (1972a) from stationary cultures of this strain in PSC medium incubated at 25°C for 12 days was 200 mg/20 ℓ, while Ishii & Ueno (1981) isolated 4114 mg of crystalline neosolaniol from 87 ℓ of culture material in the same medium incubated at 25°C for 20 days.

Fusarium sporotrichioides (= *F. tricinctum*) NRRL 3299 (Strain 4.3) has been reported to produce neosolaniol by Ueno et al. (1972b, 1973a). Neosolaniol for use in toxicological studies has also been isolated from cultures of this strain on rice incubated at 11°C for 3 months followed by 10 days at room temperature (Chi et al., 1978).

At least four isolates of *F. sporotrichioides* from overwintered cereals associated with outbreaks of ATA (see ATA, above) in the USSR have been reported to produce neosolaniol in culture. *F. sporotrichioides* NRRL 3510 (Strain 4.1) was reported to produce neosolaniol by Ueno et al. (1972b, 1973a, 1975a), *F. sporotrichioides* 738 (Joffe, 1960a, 1974b, 1978b; Joffe & Yagen, 1977) by Lafont et al. (1977), and strain 5328a by Szathmary et al. (1976, 1977). According to Yagen et al. (1977), *F. sporotrichioides* 921 was the most toxic among 106 isolates of *F. sporotrichioides* from overwintered cereals in the USSR and 0.025 g of crystalline neosolaniol could be isolated from 1 kg of cultural material on millet incubated at 12°C for 21 days.

The production of neosolaniol by *F. poae* NRRL 3287 in PSC medium incubated at 26°C for 12 days has been reported by Ueno et al., (1972b, 1973a). We have identified this strain as *F. sporotrichioides* (Strain 4.25).

Three Hungarian isolates of *F. sporotrichioides* (F-38, F-24, and F-26) have been reported to produce neosolanol (Szathmary et al., 1976, 1977). We have confirmed the identification of these isolates as *F. sporotrichioides* (Strains 4.20, 4.21, and 4.22). The maximal yield of neosolaniol (1480 µg/g) was obtained with *F. sporotrichioides* F-14 (Strain 4.21) in cultures on millet containing 2% malt extract incubated at 23 to 27°C for one week, 5 to 8°C for 2 weeks and again at 23 to 27°C for one week (Szathmary et al., 1976, 1977).

Fusarium tricinctum 72187 isolated from cereal grain in Finland has been reported to produce neosolaniol in culture on a mixture of cereal grains incubated at 6°C for 4 months (Ilus et al., 1977; Niku-Paavola & Nummi, 1977; Niku-Paavola et al., 1977). We have identified this strain as *F. sporotrichioides* (Strain 4.23). This strain (as *F. sporotrichioides* 72187-12 = *F. tricinctum* 72187-10k) has also been used for the production of neosolaniol for use in toxicological studies (Ilus et al., 1981; Ylimäki et al., 1979).

In a study of the toxigenicity of *Fusarium* isolates from overwintered cereals in Canada, Scott et al. (1980) detected neosolaniol by TLC in 5 of 20 toxic strains. We have identified four of these neosolaniol-producing strains (HPB 071178-13, -18, -21, and -23) as *F. sporotrichioides* (Strains 4.15, 4.16, 4.18, and 4.19). Maximal yields of neosolaniol (220 µg/g) were obtained with *F. sporotrichioides* HPB 071178-18 (Strain 4.16) cultured on frost-killed wheat grain incubated at temperatures fluctuating between 0 and 14°C (Scott et al., 1980).

Nivalenol. This trichothecene was first isolated from the Japanese isolate *F. nivale* Fn-2B by Tatsuno (1968) and Tatsuno et al. (1968, 1969). We have identified this strain as *F. sporotrichioides* (Strain 4.9). Yields of 150 mg of nivalenol from 3 kg of rice incubated at 25 to 27°C for 20 days were obtained by Morooka & Tatsuno (1970). This strain also produces nivalenol in stationary culture in PSC medium incubated at 25 to 27°C for 2 weeks (Fujimoto et al., 1972; Tatsuno, 1976; Tatsuno et al., 1970; Ueno & Fukushima, 1968; Ueno et al., 1969a, 1970a, b, c, 1972a, 1973a). Yields of nivalenol in this medium have been reported to be 10 mg/10 ℓ by Ueno et al. (1969a) and 2 to 3 mg/ℓ by Ueno et al. (1970c). According to Ueno et al. (1970), maximal production of nivalenol by strain Fn-2B in PSC medium occurs after 6 to 8 days at 27°C and in cultures on rice incubated at 27°C, and lower yields are obtained at lower temperatures. Nivalenol is produced later and in lower yields than fusarenon-X in liquid media by strain Fn-2B (Ueno et al., 1969a, 1970a). Nivalenol lacks one acetyl group of fusarenon-X, and it appears that nivalenol may be produced from fusarenon-X by a chemical or enzymatic deacetylation reaction during the cultivation of strain Fn-2B (Ueno et al., 1970a, 1971b). However, in biological deacetylation experiments, the growing mycelium of this strain failed to convert fusarenon-X to nivalenol (Yoshizawa & Morooka, 1975b). *F. sporotrichioides* (= *F. nivale*) Fn-2B (Strain 4.9) has been used extensively in Japan for the production of nivalenol for use in toxicological investigations (Ueno & Fukushima, 1968; Ueno et al., 1970b, 1973b; Yoshizawa & Morooka, 1975b; Yoshizawa et al., 1979, 1980). Nivalenol has also been prepared chemically from fusarenon-X isolated from this strain (Ohta et al., 1977, 1978). Another Japanese strain, *F. episphaeria* Fn-M, has also been identified as *F. sporotrichioides* (Strain 4.10). This strain produces nivalenol in stationary cultures in PSC medium incubated at 25 to 27°C for 2 weeks (Ueno et al., 1971a, 1973a).

NT-1 and NT-2. These two new trichothecenes were isolated and characterized from cultures of *F. sporotrichioides* (= *F. solani*) M-1-1 (Strain 4.12) in PSC medium incubated at 25°C for 20 days by Ishii & Ueno (1981). The yields of crystalline NT-1 and NT-2 from 87 ℓ of culture filtrate were 320 mg and 172 mg, respectively. A compound identical to NT-1 was isolated from *F. sporotrichioides* 72187-12 (= *F. tricinctum* 72187-10k, strain 4.23) by Ilus et al. (1977) and referred to as T-1 toxin by Ylimäki et al. (1979).

T-1 Toxin. This new trichothecene was isolated and characterized from cultures of the Finnish isolate *F. tricinctum* 72187 on cereal grains incubated at 6°C for 4 months by Ilus et al. (1977). We have identified this strain as *F. sporotrichioides* (Strain 4.23). The new trichothecene was called T-1 toxin by Ylimäki et al. (1979), who referred to the producing strain as *F. sporotrichioides* 72187-12 (= *F. tricinctum* 72187-10k). The same compound was isolated from *F. sporotrichioides* (= *F. solani*) M-1-1 (Strain 4.12) and called NT-1 toxin by Ishii & Ueno (1981).

T-2 Toxin. This trichothecene was first isolated from *F. tricinctum* strain T-2 by Bamburg (1968) and Bamburg et al. (1968a). We have identified this strain as *F. sporotrichioides* (Strain 4.3). The yield of crystalline toxin from stationary cultures of this strain in Gregory's medium incubated at 8°C for 30 days was 1.5 g/15 ℓ (Bamburg, 1968; Bamburg et al., 1968a). The level of T-2 toxin produced by this strain (as *F. tricinctum* No. 203 by Bamburg, 1968, and as *F. tricinctum* T-2 by Marasas, 1969, and Marasas et al., 1971) in Gregory's medium incubated at 8°C for 14 days and determined by GC, was 103 mg/ℓ. Ikediobi et al. (1971) detected 1515 mg/kg of T-2 toxin by GLC in cultures on corn incubated at 8°C for 30 days, and 920 mg/kg in corn-rice cultures incubated at 27°C for 21 days followed by 24 days at 8°C. Burmeister (1971) reported that this strain (as *F. tricinctum* NRRL 3299) produced high yields (9.96 g/1200 g) of T-2 toxin in culture on corn grits incubated at 15°C for 3 weeks, and 2 to 3 g of crystalline product could be isolated from this culture material. In cultures on wheat incubated at 12°C for 21 days, Strain T-2 (as *F. poae* by Joffe, 1978b, and as *F. sporotrichioides* by Joffe & Yagen, 1977), was reported to produce 3.4 mg of T-2 toxin/10 g of culture material. The same strain (as *F. poae* NRRL 3299, see Strain 4.3) was found to produce 5.8 mg/10 g of T-2 toxin on this medium by Joffe (1978b) and Joffe & Yagen (1977). Maximal yields of T-2 toxin (339 mg/ℓ) by this strain (as *F. tricinctum* NRRL 3299) were obtained by Cullen & Smalley (1980) in stationary cultures in nutrient broth-moistened vermiculite incubated at 19°C for 12 days, compared to yields of 2.7 mg/g, 1.7 mg/g, 155 mg/ℓ, and 76 mg/ℓ on corn, rice, Gregory's medium, and PSC medium, respectively. Cullen (1981) reported a yield of 511 mg/ℓ of T-2 toxin in a vermiculite-supported nutritionally complex medium containing soya meal, corn steep liquor, and glucose after 24 days' incubation at 19°C. The production of T-2 toxin by this strain, usually referred to as *F. tricinctum* NRRL 3299 (see Strain 4.3), has also been reported on a variety of other media and it has been used extensively for the production of T-2 toxin used in toxicological investigations (Bamburg et al., 1968b; Boonchuvit et al., 1975; Burmeister et al., 1971, 1972, 1981; Chi & Mirocha, 1978; Chi et al., 1977a, b, c, 1978, 1981; Collins & Rosen, 1981; Coulter et al., 1977; Doerr et al., 1974, 1981; Grove et al., 1970; Hagler et al., 1981; Kosuri et al., 1970, 1971; Lee & Chu, 1981; Marasas et al., 1967, 1969; Müller & Thaler, 1981; Patterson et al., 1979; Richard et al., 1978; Schmidt et al., 1981a, b; Ueno et al., 1972b, 1973a; Weaver et al., 1978a, b, d, 1980; Wei et al., 1971; Witlock et al., 1977; Wyatt et al., 1973a, b, 1975a, b; Yates et al., 1969, 1970).

In a study of T-2 toxin production by 30 strains of "*F. tricinctum*" in stationary culture in Gregory's medium incubated at 8°C for 14 days, Bamburg (1968) and Marasas (1969) found that 13 isolates produced T-2 toxin that could be detected by TLC. We have identified the six T-2 toxin-producing strains available to us as *F. sporotrichioides*, i.e., Strain T-2 (= No. 203 of Bamburg, 1968) (Strain 4.3), and Nos. 223, 238, 347, 209, and 348 (Strains 4.4, 4.27, 4.28, 4.29, and 4.30). These isolates identified as *F. sporotrichioides* included the two highest producers of T-2 toxin in the study, No. 209 (Strain 4.29) and No. 348 (Strain 4.30), which produced 214 and 212 mg/ℓ of T-2 toxin, respectively, compared to 103 mg/ℓ produced by strain T-2 (Bamburg, 1968; Marasas, 1969).

During the course of investigations on the toxigenicity of *Fusarium* isolates from fescue hay associated with outbreaks of fescue foot (see fescue foot, above) of cattle in the United States, Yates et al. (1970) found that 21 of 34 unidentified strains of *Fusarium* produced T-2 toxin in culture on Sabouraud's agar incubated at 15°C. Burmeister et al. (1971) detected T-2 toxin by TLC in 10 of 14 isolates of "*F. tricinctum*" cultured on corn grits at 15°C for 21 days, while Burmeister et al. (1972) found that 5 of 9 toxic isolates of "*F. tricinctum*" produced T-2 toxin on the same medium. We have identified three of these T-2 toxin-producing isolates available to us as *F. sporotrichioides*, i.e., NRRL 3299 (Strain 4.3), NRRL 3249 (Strain 4.8), and NRRL 3510 (Strain 4.1). One of the most toxic T-2 toxin-producing isolates from fescue hay, *F. sporotrichioides* (= *F. nivale*, = *F. tricinctum*) NRRL 3249 (Strain 4.8) (Burmeister et al., 1971, 1981; Ellis & Yates, 1971; Yates et al., 1968, 1969, 1970), produced 14 mg/kg of T-2 toxin in cultures on autoclaved wheat incubated at 3°C for 10 to 12 weeks (Yates et al., 1968). This strain has also been reported to produce 4.0 mg of T-2 toxin/10 g of wheat culture material incubated at 12°C for 21 days (Joffe, 1978b; Joffe & Yagen, 1977).

The Japanese strain *F. solani* M-1-1, isolated during investigations of "bean hulls poisoning" of horses (see bean hulls poisoning, above) in Hokkaido, has been reported to produce T-2 toxin (Ishii & Ueno, 1981; Ishii et al., 1971; Tatsuno et al., 1973; Ueno et al., 1972a, b, 1973a, b, 1975). We have identified this strain as *F. sporotrichioides* (Strain 4.12). Reported yields of T-2 toxin from cultures of this strain include the following: 160 mg/20 ℓ in PSC medium in stationary culture at 25 to 27°C for 12 days (Ishii et al., 1971); 4114 mg/87 ℓ in PSC medium at 25°C for 20 days (Ishii & Ueno, 1981); and 1.54 g/11.2 ℓ in PSC medium at 25°C for 7 days (Sato & Amano, 1976). *F. sporotrichioides* (= *F. solani*) M-1-1 was also the highest producer among 12 isolates tested for T-2 toxin production in stationary, shake, and fermentor cultures (Ueno et al., 1975) and has been used for the production of T-2 toxin for use in toxicological investigations (Kato et al., 1979; Masuko et al., 1977; Matsumoto et al., 1978; Ohta et al., 1977; Sato et al., 1975, 1978; Ueno et al., 1978; Yoshizawa & Morooka, 1975a, b; Yoshizawa et al., 1979, 1980).

In a study of the toxigenicity of 106 isolates of *F. sporotrichioides* from overwintered cereals associated with outbreaks of ATA (see ATA, above) in the USSR, Yagen & Joffe (1976) found that 104 (= 98%) strains produced T-2 toxin in cultures on wheat incubated at 12°C for 21 days. The levels of T-2 toxin detected by GLC in purified ethanol extracts of these cultures varied from 0.1 to 21 mg/ml (Yagen & Joffe, 1976). According to Yagen et al. (1977), *F. sporotrichioides* No. 921 was the most toxic among the 106 isolates tested and produced 4.1 g of T-2 toxin/kg of culture material on wheat. This strain was also the highest T-2 toxin producer among seven isolates of *F. sporotrichioides* from the USSR tested by Joffe (1978b) and Joffe & Yagen (1977), and was reported to produce 24.0 mg of T-2 toxin/10 g of wheat. *F. sporotrichioides* No. 921 has been used for the production of T-2 toxin used in toxicological studies (Joffe & Yagen, 1978; Lutsy et al., 1978; Schoental et al., 1979; Yagen et al., 1977). Another isolate from the USSR that was shown to produce T-2 toxin by Joffe (1978b) and Joffe & Yagen (1977), *F. sporotrichioides* 738, has also been used for the production of T-2 toxin (Lafarge-Frayssinet et al., 1979; Lafont et al., 1977). In addition to the above isolates, at least three other strains of *F. sporotrichioides* from overwintered cereals in the USSR have also been reported to produce T-2 toxin, i.e., NRRL 3510 (Strain 4.1) (Burmeister et al., 1972, 1981; Ueno et al., 1972b, 1973a, 1975), strain 5328a (Kotik et al., 1979; Szathmary et al., 1976, 1977), and strain 63 (Ermakov et al., 1978).

The yields of T-2 toxin in cultures on wheat incubated at 12°C for 21 days by seven isolates of *F. sporotrichioides* from overwintered cereals in the USSR (Nos. 60/10, 347, 351, 738, 921, 1182 and 1183) were compared with those of the following eight isolates from other sources by Joffe (1978b) and Joffe & Yagen (1977): 1) NRRL 3287, as *F. poae* (*F. sporotrichioides*, Strain 4.25); 2) NRRL 3299, as *F. poae* (*F. sporotrichioides*, Strain 4.3); 3) T-2, as *F. poae* by Joffe (1978b) but as *F. sporotrichioides* by Joffe & Yagen (1977) (this is the same strain as NRRL 3299, i.e., *F. sporotrichioides*, Strain 4.3); 4) NRRL 3249, as *F. sporotrichioides* (*F. sporotrichioides*, Strain 4.8); 5) NRRL 5908, as *F. sporotrichioides* (we have identified this strain as *F. graminearum*, Strain 8.41, and it is consequently excluded from the section Sporotrichiella); 6) NRRL 2061-C, as *F. sporotrichioides* (*F. sporotrichioides*, Strain 4.6); 7) No. YN-13, as *F. sporotrichioides* (*F. sporotrichioides*, Strain 4.26); 8) NRRL 3509, as *F. sporotrichioides* var. *tricinctum* (*F. sporotrichioides*, Strain 4.9). The levels of T-2 toxin produced by the seven isolates of *F. sporotrichioides* from overwintered cereals in the USSR ranged from 4.6 to 24.0 mg/10 g wheat grain, while of the isolates listed above from other sources, the six identified as *F. sporotrichioides* produced 0.5 to 5.8 mg of T-2 toxin/10 g of wheat grain (Joffe, 1978b; Joffe & Yagen, 1977). The fact that the strains of *F. sporotrichioides* isolated from overwintered cereals associated with ATA in the USSR produced much higher levels of T-2 toxin than isolates from other sources led Joffe (1978b) and Joffe & Yagen (1977) to conclude that the extreme overwintering conditions of cereals from which these strains were isolated in the USSR during 1943/44 may have been conducive to the extraordinarily high levels of T-2 toxin that they produced in culture, even more than 30 years after their original isolation.

The emetic material in cultures of *F. poae* NRRL 3287, which was originally reported (as *F. poae* 2518) by Prentice & Dickson (1968) to cause emesis in pigeons, was identified as T-2 toxin by Ellison & Kotsonis (1973). We have identified this strain as *F. sporotrichioides* (Strain 4.25). The yields of T-2 toxin obtained by Ellison & Kotsonis (1973) from cultures of this strain, incubated on corn at 8°C and in Richard's solution at room temperature, were 4 mg/800 g and 59 mg/7 ℓ, respectively. The production of T-2 toxin by this strain has also been confirmed by Joffe (1978b), Joffe & Yagen (1977), Kotsonis & Ellison (1975), Ueno et al. (1972b, 1973a), and Vesonder et al., (1981a). Cultures on corn incubated at 28°C for 13 days were refused by pigs, but TLC analyses for T-2 toxin and other trichothecenes were negative (Vesonder et al., 1977b). However, subsequent analysis of the refused corn by GLC-MS revealed the presence of 30 µg/g of T-2 toxin (Vesonder et al., 1981a).

Three Hungarian isolates of *F. sporotrichioides* (F-38, F-14, and F-26) have been reported to produce T-2 toxin (Szathmary et al., 1976, 1977). We have confirmed the identification of these isolates as *F. sporotrichioides* (Strains 4.20, 4.21, and 4.22). The maximal yield of T-2 toxin (2300 µg/g) was obtained with *F. sporotrichioides* F-38 (Strain 4.20) in cultures on millet containing 2% malt extract incubated at 23 to 27°C for one week, 5 to 8°C for 2 weeks and again at 23 to 27°C for one week (Szathmary et al., 1976, 1977). *F. sporotrichioides* F-14 (= ATCC 34914) (Strain 4.21) has also been reported to produce T-2 toxin at levels up to 107 mg/kg in cultures on corn (Müller & Thaler, 1981).

Fusarium tricinctum 72187, isolated from cereal grain in Finland, has been reported to produce T-2 toxin in culture on a mixture of cereal grains incubated at 6°C for 4 months (Ilus et al., 1977, 1981; Niku-Paavola & Nummi, 1977; Niku-Paavola et al., 1977). We have identified this strain as *F. sporotrichioides* (Strain 4.23). This strain has also been used for the production of T-2 toxin used in toxicological investigations by Linnainmaa et al. (1979), Norppa et al. (1980), Sorsa et al. (1980), and Ylimäki et al. (1979), who referred to this strain as *F. sporotrichioides* 72187-12 (= *F. tricinctum* 72187-10k).

Among 47 isolates of *Fusarium* species from cereals in Italy, only one out of seven isolates of "*F. tricinctum*" produced T-2 toxin that could be detected by TLC (Bottalico, 1977c). This isolate produced 40 µg/g of T-2 toxin in cultures on corn kernels incubated at 24°C for 15 days followed by 60 days at 10°C. We have identified a culture of *F. tricinctum* ITH-194, isolated from corn ear rot in Lombardy and reportedly a "high T-2 toxin producer" (A. Bottalico, personal communication, 19 October 1981), as *F. sporotrichioides* (Strain 4.31). It is not certain whether this is the T-2 toxin-producing strain of *F. tricinctum* referred to by Bottalico (1977b).

In a study of the toxigenicity of *Fusarium* isolates from overwintered cereals in Canada, Scott et al. (1980) detected T-2 toxin by TLC in 10 of 20 toxic strains at levels ranging from 17 to 1350 µg/g. We have identified seven of these T-2 toxin-producing strains available to us (HPB 071178-10, -12, -13, -18, -20, -21, and -23) as *F. sporotrichioides* (Strains 4.13, 4.14, 4.15, 4.16, 4.17, 4,18, and 4.19). Maximal yields of T-2 toxin (1350 µg/g) were obtained with *F. sporotrichioides* HPB 071178-13 (Strain 4.15) in cultures on corn incubated at 14°C (Scott et al., 1980).

Cullen (1981) tested 16 isolates of "*F. tricinctum* sensu Booth" and seven isolates of *F. sporotrichioides* for T-2 toxin production. Cultures on corn incubated at 24°C for 2 weeks followed by 4 weeks at 8°C were analyzed for T-2 toxin by GLC. Nine of the 16 isolates of *F. tricinctum* produced T-2 levels ranging from 4 to 5950 µg/g, while all seven isolates of *F. sporotrichioides* produced levels of 20 to 1835 µg/g of T-2 toxin. We have identified seven of the nine T-2 toxin-producing strains of *F. tricinctum* available to us (T-340, T-285, T-288, M-350, F-696, F-14, and T-336) as *F. sporotrichioides* (Strains 4.32, 4.33, 4.34, 4.35, 4.36, 4.37, and 4.38). We have also confirmed the identity of all seven T-2 toxin-producing isolates of *F. sporotrichioides* (T-341, T-290, T-286, T-306, A-1, and 1973-33 = T-249) as *F. sporotrichioides* (Strains 4.20, 4.39, 4.40, 4.41, 4.42, and 4.43). Thus out of a total of 18 isolates reported to produce T-2 toxin at levels ranging from 4 to 5950 µg/g by Cullen (1981), 14 have been identified as *F. sporotrichioides*. Maximal yields of T-2 toxin (714 mg/ℓ) by the highest producing strain identified in this study, i.e., *F. sporotrichioides* (= *F. tricinctum*) T-340 (Strain 4.32), were obtained by incubating cultures in a vermiculite-supported, nutritionally complex medium containing soya meal, corn steep liquor, and glucose at 19°C for 24 days (Cullen, 1981).

The strain that was originally reported (as *F. moniliforme* 111) to produce emetic material (Prentice & Dickson, 1968; Prentice et al., 1959) has subsequently been shown (as *F. tricinctum* = *F. moniliforme* NRRL 3197) to produce 33 µg/g of T-2 toxin as detected by GLC-MS in cultures on corn incubated at 28°C for 13 days (Vesonder et al., 1981a). We have identified this strain as *F. sporotrichioides* (Strain 4.24).

In view of the voluminous literature on T-2 toxin production by strains of *F. sporotrichioides*, as well as the fact that at least 40 T-2 toxin-producing strains in the ITFRC have been identified as *F. sporotrichioides* (Table 4.1), we conclude that *F. sporotrichioides* is the most important T-2 toxin-producing *Fusarium* species. The strains of *F. sporotrichioides* that produce the most T-2 toxin have been isolated from overwintered cereals in the USSR, particularly strain 921, which has been reported to produce up to 4.1 g/kg of T-2 toxin (Yagen et al., 1977). Of the strains of *F. sporotrichioides* represented in the ITFRC that have been reported to give yields of T-2 toxin in the g/kg range, those that produce the most are probably strain T-2 (Strain 4.3), the Hungarian isolate F-38 (Strain 4.20), the isolate from overwintered cereals in Canada referred to as HPB 071178-13 (Strain 4.15), and three of the strains tested by Cullen (1981), i.e., T-340, T-285, and T-341 (Strains 4.32, 4.33, and 4.39).

T-2 Tetraol. This trichothecene has been detected by TLC, GLC, and GC-MS in cultures of two Hungarian isolates (F-38 and F-14) of *F. sporotrichioides* (Szathmary et al., 1976, 1977). We have confirmed the identification of both of these isolates as *F. sporotrichioides* (Strains 4.20 and 4.21). The level of T-2 tetraol produced by *F. sporotrichioides* F-14 (Strain 4.21) in cultures on millet incubated at 23 to 27°C for one week, 5 to 8°C for 2 weeks and again at 23 to 27°C for one week, was 50 µg/g. Although *F. sporotrichioides* F-38 (Strain 4.20), did not produce T-2 tetraol in cultures on millet, it produced 300 µg/g in cultures on corn (Szathmary et al., 1976, 1977).

Zearalenone. The production of zearalenone by 3 of 19 isolates of "*F. tricinctum*" in cultures on corn incubated at 24°C for 2 weeks followed by 8 weeks at 12°C was reported by Caldwell et al. (1970). Levels of 9.4, 14.4, and 14.7 µg/g of zearalenone were detected by TLC in cultures of "*F. tricinctum*" isolates FT 2, 3, and 12, respectively. The presence of zearalenone in the culture material of these three strains was confirmed in a uterotrophic bioassay in mice. The three zearalenone-producing isolates of "*F. tricinctum*" were described as having macroconidia with one to three septa, and microconidia that "varied from globose with a basal papilla to obclavate and were borne on bottle-shaped conidiophores" (Caldwell et al., 1970). It is not possible to determine from this description whether these three isolates were *F. sporotrichioides*, *F. poae*, or *F. tricinctum*. In a subsequent study, Caldwell & Tuite (1970) found that "*F. tricinctum* FT-3," which produced the highest level of zearalenone in culture, did not produce zearalenone in corn ears inoculated in the field.

According to Eugenio et al. (1970a), none of 12 isolates of "*F. tricinctum*" produced zearalenone in culture on rice incubated at 22 to 25°C for 2 weeks followed by 8 weeks at 8 to 12°C, and "*F. roseum* isolate No. 2" isolated from corn in Yankton, South Dakota, USA, was the second highest producer of zearalenone (1420 ppm) among 43 isolates of "*F. roseum*" tested. We have identified a culture of "*Fusarium* sp. YN-13," which was reported to cause uterine hypertrophy in rats by Christensen et al. (1965) and is the same strain as "*F. roseum* isolate No. 2" of Eugenio et al. (1970a) according to C.J. Mirocha (personal communication, 7 March 1980), as *F. sporotrichioides* (Strain 4.26). However, "*F. roseum* isolate No. 2" produced ascospores (Eugenio et al., 1970a) and we conclude that this strain was either a mixture of a perithecial isolate of *F. graminearum* and *F. sporotrichioides* YN-13 or the two strains are not the same. In either case there is no proof that *F. sporotrichioides* YN-13 (Strain 4.26) produces zearalenone.

Mirocha & Christensen (1974) listed *F. tricinctum* and *F. sporotrichioides* among the *Fusarium* species that had been found to produce zearalenone in their laboratory. According to Mirocha & Christensen (1974), the *F. tricinctum* isolate used in the experiments of Speers et al. (1971, 1972) is the same isolate that Christensen et al. (1972a) reported as highly toxic to turkey poults, and this strain produces "copious amounts" of zearalenone. The strain used by Christensen et al. (1972a) was *F. tricinctum* 2061-C and we have identified this strain as *F. sporotrichioides* (Strain 4.6). However, no zearalenone was detected in cultures of the unspecified isolate of *F. tricinctum* used in the experiments of Speers et al. (1977). The unnumbered zearalenone-producing isolate of *F. sporotrichioides* referred to by Mirocha & Christensen (1974) was "one of the original isolates reported by Russian scientists as causing alimentary toxic aleukia." Presumably this isolate, as well as the one referred to by Mirocha & Pathre (1973), was *F. sporotrichioides* 5328a, which was reported to produce zearalenone by Szathmary et al. (1976, 1977). The latter authors also found that one of three Hungarian isolates of *F. sporotrichioides* (F-38, Strain 4.20) produced zearalenone. This strain produced up to 1500 µg/g of zearalenone in cultures on millet incubated at 23 to 27°C for one week, 5 to 8°C for 2 weeks, and 23 to 27°C for one week, but only 200 µg/g in culture on corn (Szathmary et al., 1976, 1977).

Hungarian isolates of *F. tricinctum*, but not of "*F. sporotrichoella*," have also been reported to produce zearalenone in culture on corn incubated at 16°C for 3 weeks (Lasztity & Wöller, 1975a, b; Lasztity et al., 1977). However, no indication was given of the number of strains used or the levels produced.

Cullen (1981) tested 16 isolates of "*F. tricinctum* sensu Booth" and seven isolates of *F. sporotrichioides* for zearalenone production. Cultures on rice incubated at 24°C for 2 weeks followed by 4 weeks at 14°C were analyzed for zearalenone by GLC. Four of the 16 isolates of "*F. tricinctum*" produced zearalenone at levels ranging from 14 to 203 µg/g, while two out of the seven isolates of *F. sporotrichioides* produced levels of 100 to 123 µg/g of zearalenone. We have identified three of the four zearalenone-producing strains of "*F. tricinctum*" available to us (M-350, F-14, and T-336) as *F. sporotrichioides* (Strains, 4.35, 4.37, and 4.38). We have also confirmed the identity of both of the zearalenone-producing strains of *F. sporotrichioides* (T-341 and 1973-33) as *F. sporotrichioides* (Strains 4.39 and 4.20). Thus five out of six isolates reported to produce zearalenone at levels ranging from 14 to 203 µg/g by Cullen (1981) have been identified as *F. sporotrichioides*.

Traces of zearalenone (up to 3 µg/g) were detected in cultures of two unidentified *Fusarium* strains isolated from overwintered cereals in Canada (Scott et al., 1980). We have identified one of these strains (HPB 071178-19) as *F. sambucinum* (Strain 8.4). None of the seven Canadian isolates that we have identified as *F. sporotrichioides* (Strains 4.13 to 4.19) produced zearalenone in cultures on corn incubated at 14°C (Scott et al., 1980).

Negative results for the production of zearalenone by isolates of *F. sporotrichioides* (and *F. tricinctum*) have also been reported by several other authors: neither of 2 (= *F. tricinctum* No. 161 and *F. sporotrichioides* No. 214) by Bacon et al. (1977); none of 8 *F. tricinctum* by Bottalico (1977b, 1979); none of 20 *F. sporotrichioides* by Hacking et al., (1976, 1977); none of 4 *F. sporotrichioides* and none of 3 *F. tricinctum* by Ichinoe et al. (1977); none of 7 "*F. tricinctum* (Sporotrichiella)" but one of 4 *F. tricinctum* by Ishii et al. (1974); and none of 8 *F. sporotrichioides*, but one of 2 *F. tricinctum* by Vesela et al.

(1981). We have identified *F. tricinctum* A-B-2, which was reported to produce zearalenone by Ishii et al. (1974), as *F. graminearum* (Strain 8.33) and this strain is consequently excluded from Sporotrichiella.

In conclusion, only the following 10 isolates of *F. sporotrichioides*, out of approximately 110 tested, have definitely been shown to produce zearalenone: the three isolates FT 2, 3, and 12 of "*F. tricinctum*" (specific identity unknown) that produced from 9.4 to 14.4 µg/g (Caldwell et al., 1970); the five isolates reported by Cullen (1981) to produce from 14 to 203 µg/g (see Strains 4.20, 4.35, and 4.37 to 4.39); the Russian isolate *F. sporotrichioides* 5328a that produced 1300 µg/g (Szathmary et al., 1976, 1977); and the Hungarian isolate *F. sporotrichioides* F-38 (Strain 4.20) that produced up to 1500 µg/g (Szathmary et al., 1976, 1977). This last isolate, F-38 (Strain 4.20), is the highest known zearalenone-producing strain of *F. sporotrichioides* represented in the ITFRC.

Toxigenic Strains in the ITFRC

The *Fusarium* strains listed under *F. sporotrichioides* in Table 4.1 have been identified as *F. sporotrichioides* according to Nelson et al. (1983). Toxigenic strains referred to as species of the section Sporotrichiella in the literature but identified as other species by us are listed in Table 4.2.

Strain 4.1 *F. tricinctum* NRRL 3510 (= T-345; MRC 1704)

This strain was isolated from overwintered proso millet in the USSR by A.Z. Joffe (Burmeister et al., 1972, 1981; Ueno et al., 1972b). According to Burmeister et al. (1981), this strain had been identified as *F. sporotrichioides* when it was originally deposited at NRRL.

Culture filtrates and extracts of cultures in PSC medium incubated in stationary culture at 26°C for 12 days were lethal to mice upon intraperitoneal injection (Ueno et al., 1972b). Marked hemorrhage occurred in the small intestine together with radiomimetic damage characterized by mitotic injury and karyorrhexis in the actively dividing cells in the small intestine, spleen, thymus, testes, and bone marrow. The toxic extracts were analyzed by TLC and found to contain T-2 toxin, HT-2 toxin, neosolaniol, and butenolide; 9.5 g of crystalline butenolide was isolated from 40 ℓ of culture material (Ueno et al., 1972b). The toxicity to mice of this strain was confirmed by Ueno et al. (1973a), but the toxic extracts were reported to contain T-2 toxin, neosolaniol, and diacetoxyscirpenol. This strain did not produce either T-2 toxin or neosolaniol in shake culture in PSC medium, but either T-2 toxin alone or both mycotoxins were produced in shake culture in other liquid media containing glucose and yeast extract or corn steep liquor incubated at 27°C for 5 days (Ueno et al., 1975). Burmeister et al. (1972) reported that this strain produced T-2 toxin in cultures on white corn grits incubated at 15°C for 21 days. This was confirmed by Burmeister et al. (1981), who also found that crude extracts of the cultures on corn grits were lethal to mice.

A lyophilized culture of this strain was received on 5 September 1979 from NRRL as "*F. tricinctum* NRRL 3510."

Strain 4.2 *F. tricinctum* NRRL 3511 (= T-347; MRC 1692)

This strain was isolated from proso millet in the USSR according to Burmeister et al. (1972), but from wheat seed in the USSR according to Burmeister et al. (1981). This strain had been identified as *F. poae* when it was originally deposited at NRRL (Burmeister et al., 1981).

Extracts of cultures of this strain on white corn grits incubated at 15°C for 21 days

Table 4.1. *Fusarium* Strains of the Section Sporotrichiella in the International Toxic *Fusarium* Reference Collection (ITFRC)

Strain No.	ITFRC No.	MRC No.	Published Name(s) and Number(s)	Source	Toxigenicity
Fusarium sporotrichioides					
4.1	T-345	1704	*F. tricinctum* NRRL 3510	NRRL	a,b,c,d,e,f,l
4.2	T-347	1692	*F. tricinctum* NRRL 3511	NRRL	a,b
4.3	T-424	43	*F. tricinctum* Strain T-2 (= ATCC 24043)	E.B. Smalley	b,c,d,e,f,l
4.3	T-348	1768	= *F. tricinctum* NRRL 3299 (= ATCC 24631)	NRRL	
4.4	T-409	1661	*F. tricinctum* No. 223 (= ATCC 24044)	E.B. Smalley	b,c
4.5	T-602	2863	*F. tricinctum* Isolate No. 16 (= ATCC 24045)	ATCC	a,b,c?
4.6	T-405	1620	*F. tricinctum* 2061-C	C.J. Mirocha	a,b,c
4.7	T-592	2728	*F. poae* 7452 (= DAOM 165006)	DAOM	a,b,c?
4.8	T-344	1705	*F. tricinctum* (= *F. nivale*) NRRL 3249 (= ATCC 24529)	NRRL	a,b,c,l
4.9	T-346	1767	*F. nivale* Fn 2 (= NRRL 3509)	NRRL	a,b,c,g,h,i
4.9	T-436	1962	= *F. nivale* Fn-2B	H. Kurata	
4.9	T-563	2476	= *F. nivale* Fn-2B	T. Yoshizawa	
4.9	T-564	2533	= *F. nivale* Fn-2B (= ATCC 26532)	Y. Ueno	
4.10	T-565	2531	*F. episphaeria* Fn-M (= ATCC 26531, NRRL 6491)	Y. Ueno	b,g,h
4.11	T-566	2534	*F. oxysporum* "niveum" Melon-1	Y. Ueno	b,g,i
4.12	T-422	935	*F. solani* M-1-1 (= Sp. 900)	Y. Ueno (1973)	a,b,c,d,e,f,k,l
4.12	T-568	2557	= *F. solani* M-1-1	Y. Ueno (1981)	
4.12	T-492	2319	= *F. sporotrichioides* ATCC 26533	ATCC	
4.12	T-567	2478	= *F. sporotrichioides* ATCC 26553	T. Yoshizawa	
4.12	T-490	2034	= *F. solani* NHL-F-111 (= Sp. 934, Sp. 980)	BAF	
4.13	T-504	2182	*F. sporotrichioides* HPB 071178-10	B.J. Blanchfield	b,c
4.14	T-505	2183	*F. sporotrichioides* HPB 071178-12	B.J. Blanchfield	b,c
4.15	T-506	2184	*F. sporotrichioides* HPB 071178-13	B.J. Blanchfield	b,c,d,f
4.16	T-507	2186	*F. sporotrichioides* HPB 071178-18	B.J. Blanchfield	b,c,d,f
4.17	T-508	2188	*F. sporotrichioides* HPB 071178-20	B.J. Blanchfield	b,c
4.18	T-509	2189	*Fusarium* sp. HPB 071178-21	B.J. Blanchfield	b,c,f
4.19	T-510	2192	*F. sporotrichioides* HPB 071178-23	B.J. Blanchfield	b,c,f
4.20	T-249	2032	*F. sporotrichioides* F-38 (= ATCC 34911)	M. Palyusik	b,c,d,f,k,m
4.21	T-245	2031	*F. sporotrichioides* F-14 (= ATCC 34914)	M. Palyusik	c,d,f,k
4.22	T-248	2033	*F. sporotrichioides* F-26 (= ATCC 36318)	M. Palyusik	c,d,f
4.23	T-544	2625	*F. sporotrichioides* 72187-12 (= *F. tricinctum* 72187-10k)	A. Ylimäki	b,c,d,f,k
4.24	T-494	2323	*F. moniliforme* 111 (= *F. tricinctum* NRRL 3197)	NRRL	b,c,j
4.25	T-423	1786	*F. poae* 2518 (= *F. poae* = *F. tricinctum* NRRL 3287)	NRRL	b,c,d,f,k
4.26	T-432	1902	*Fusarium* sp. YN-13	C.J. Mirocha	b,c
4.27	T-406	1657	*F. tricinctum* No. 238	E.B. Smalley	b,c,l
4.28	T-407	1659	*F. tricinctum* No. 347	E.B. Smalley	b,c
4.29	T-408	1660	*F. tricinctum* No. 209	E.B. Smalley	b,c,l
4.30	T-411	1663	*F. tricinctum* No. 348	E.B. Smalley	b,c
4.31	T-570	2523	*F. tricinctum* ITH-194	A. Bottalico	c
4.32	T-340	2913	*F. tricinctum* CMI 175449	CMI	c
4.33	T-285	1709	*F. tricinctum* T-285	FRC	c
4.34	T-288	1708	*F. tricinctum* T-288	FRC	c
4.35	T-625	1606	*F. tricinctum* M-350	E.B. Smalley	c,m
4.36	T-522	2489	*F. tricinctum* F-696	D. Cullen	c
4.37	T-521	2488	*F. tricinctum* F-14	D. Cullen	c,m
4.38	T-336	2912	*F. tricinctum* T-336	FRC	c,m
4.39	T-341	2914	*F. sporotrichioides* T-341	CMI	c,m
4.40	T-290	2910	*F. sporotrichioides* T-290	FRC	c

Table 4.1. (Continued)

Strain No.	ITFRC No.	MRC No.	Published Name(s) and Number(s)	Source	Toxigenicity
4.41	T-286	2909	F. sporotrichioides T-286	FRC	c
4.42	T-306	2911	F. sporotrichioides T-306	FRC	c
4.43	T-520	2487	F. sporotrichioides A-1	D. Cullen	c
Fusarium chlamydosporum					
4.44	T-428	35	F. fusarioides MRC 35	C.J. Rabie	a,b,n
4.45	T-227	117	F. fusarioides T-227	W.F.O. Marasas	o
4.46	T-502	2486	F. tricinctum NRRL 6358 (= NRRL A-23377)	NRRL	b,k
Fusarium poae					
4.47	T-412	1658	F. tricinctum B-24 (= No. 215)	E.B. Smalley	a,b,e
4.48	T-626	1607	F. poae M-355	E.B. Smalley	c
4.49	T-403	1605	F. poae (= F. tricinctum) M-334	E.B. Smalley	c
4.50	T-410	1662	F. tricinctum No. 335	E.B. Smalley	o
4.51	T-503	2181	F. poae HPB 071178-7	B.J. Blanchfield	b,e
4.52	T-247	1893	F. poae Sp. 884	W.F.O. Marasas	b
4.53	T-339	1398	F. poae 72188	A. Ylimäki	o
4.54	T-280	2908	F. poae T-280	FRC	o
Fusarium tricinctum					
4.55	T-429	1894	F. tricinctum Sp. 879	W.F.O. Marasas	b
4.56	T-246	1895	F. tricinctum Sp. 880	W.F.O. Marasas	b
4.57	T-338	1399	F. tricinctum 72188	A. Ylimäki	b
4.58	T-624	1400	F. tricinctum 72183	A. Ylimäki	b
4.59	T-431	1799	F. tricinctum IMB 62448	W. Gerlach	o
4.60	T-399	1574	F. tricinctum R-P-1	N.E. El-Gholl	o
4.61	T-701	1575	F. tricinctum W-P-1	N.E. El-Gholl	o
4.62	T-402	1576	F. tricinctum B-P-1	N.E. El-Gholl	o

a = Associated with human and/or animal diseases
b = Culture toxic to experimental animals
c = Produces T-2 toxin
d = Produces HT-2 toxin
e = Produces diacetoxyscirpenol
f = Produces neosolaniol
g = Produces fusarenon-X
h = Produces nivalenol
i = Produces diacetylnivalenol
j = Produces deoxynivalenol
k = Produces other trichothecenes
l = Produces butenolide
m = Produces zearalenone
n = Produces moniliformin
o = Non-toxic

were lethal to mice upon intraperitoneal injection (Burmeister et al., 1981), but did not contain T-2 toxin that could be detected by TLC (Burmeister et al., 1972, 1981).

A lyophilized culture of this strain was received on 5 September 1979 from NRRL as "*F. tricinctum* NRRL 3511."

Strain 4.3 *F. tricinctum* **Strain T-2** (= T-424; MRC 43; ATCC 24043; Sp. 897)
=**NRRL 3299** (= T-348; MRC 1768; ATCC 24631)

This strain was originally isolated from corn in France (Bamburg, 1968; Marasas, 1969; Marasas et al., 1971). According to E.B. Smalley (personal communication, 1965), a culture of this strain was obtained by him from W.C. Snyder, who stated that it was originally isolated by him from corn in France. Consequently statements in the literature

Table 4.2 *Fusarium* Strains Published as Species of the Section Sporotrichiella but Identified as Other Species in the International Toxic *Fusarium* Reference Collection (ITFRC)

Strain No.	ITFRC No.	MRC No.	Published Name(s) and Number(s)	Source	Present Identification
8.41	R-5317	1781	*F. sporotrichioides* (= *F. tricinctum*) NRRL 5908	NRRL	*F. graminearum*
8.33	R-6777	2563	*F. tricinctum* A-B-2 (or Abashiri-2)	Y. Ueno	*F. graminearum*

that this strain was isolated from "toxic corn," or "corn kernels, Wisconsin, USA," or "South Africa," are in error.

Strain T-2 was deposited in NRRL as *F. tricinctum* NRRL 3299 (Burmeister, 1971; Burmeister et al., 1971, 1972, 1981; Grove et al., 1970; Yates et al., 1970). As *F. tricinctum* NRRL 3299 this strain has become one of the most widely used toxigenic strains of *Fusarium* in the world, and it has been cited as such in more publications than any other toxigenic *Fusarium* strain known to us.

In their "Taxonomic study of Fusaria of the Sporotrichiella Section used in recent toxicological work," Joffe & Palti (1975) identified "the T-2 isolate supplied from South Africa, originally isolated from *Zea mays* in France (Marasas, personal communication)" as *F. sporotrichioides*, while "isolate NRRL 3299, originally isolated from toxic corn," was identified as *F. poae*. Joffe & Yagen (1977) also referred to strain T-2 as *F. sporotrichioides* and to NRRL 3299 as *F. poae*, but Joffe (1978b) referred to both strains as *F. poae*. The reason for this discrepancy is unknown to us because NRRL 3299 is a subculture of strain T-2 and both are *F. sporotrichioides*.

T-2 toxin was first isolated and characterized from *F. tricinctum* strain T-2 by Bamburg (1968) and Bamburg et al. (1968a). Cultures were grown in still culture in Gregory's medium at 8°C for 30 days and a rat skin test was used as bioassay in the isolation of the toxin. Ethyl acetate extracts of macerated, lyophilized cultures were applied to the shaven back of rats. The reaction ranged from a just-noticeable skin reaction with the formation of a light scab, to inflammation, hemorrhage, and heavy scab formation. Application of too high a concentration led to death of the rat within 72 hours. In addition to rats, extracts of cultures of strain T-2 have also been shown to be dermotoxic to cattle (Patterson et al., 1979), guinea pigs (Patterson et al., 1979) and rabbits (Joffe, 1978b; Joffe & Yagen, 1977). During the work on the isolation of T-2 toxin from cultures of strain T-2, two laboratory investigators came into contact with T-2 toxin containing extracts and this resulted in a burning sensation, numbness and loss of sensitivity, callus formation, and peeling of the skin (Bamburg, 1968; Bamburg et al., 1968b).

The toxicity to cattle of cultures of *F. tricinctum* NRRL 3299 on cracked corn incubated at 25°C for one week followed by 2 weeks at 4°C was investigated by Patterson et al. (1979). The culture material was shown to contain T-2 toxin by TLC and GC and was dosed to two Friesian calves by means of a bolus gun to provide daily oral doses of 0.2 mg/kg of T-2 toxin for 78 days, followed by 1.0 mg/kg for 8 days. The whole cultures did not cause hemorrhage or other pathological lesions in the calf, except for slight blood clotting factor deficiencies, leukopenia, and neutrophilia. The crude ethyl acetate extract was administered to three calves at a daily oral dose rate of 0.2 mg/kg of T-2 toxin. One calf died after 7 days but the other two exhibited no abnormal clinical signs or pathological changes during a dosing period of 77 days. Administration of the crude extract to pigs at a dosage rate of 0.2 mg/kg of T-2 toxin caused almost immediate emesis, but no hemorrhage or other pathological changes were found in pigs dosed at a level of 0.1 mg/kg of T-2 toxin for 14 days (Patterson et al., 1979).

Extracts of cultures of *F. tricinctum* NRRL 3299 in PSC medium incubated at 25 to 27°C for 12 to 14 days were lethal to mice upon intraperitoneal injection and caused hemorrhage in the small intestine as well as the radiomimetic pathological changes characteristic of trichothecenes in the actively dividing tissues of the small intestine, spleen, thymus, testes, and bone marrow (Ueno et al., 1971a, 1972b, 1973a). Extracts of cultures of this strain on Sabouraud's agar (Yates et al., 1970) or corn grits incubated at 15°C for 3 weeks (Burmeister et al., 1971, 1981) were also lethal to mice.

Lyophilized culture material of *F. tricinctum* strain T-2 in Gregory's medium incubated at 8°C for 30 days was acutely toxic to rats (Kosuri et al., 1971). Clinical signs included listlessness, insensitivity, irritability, diarrhea, hemorrhages from the urinary tract, and death within 48 to 72 hours. Pathological changes included ecchymotic hemorrhages in the peritoneal surfaces of the small and large intestines, hemorrhage into the intestinal lumens, and necrosis of intestinal lymph nodes.

Mycotoxins. *Fusarium sporotrichioides* (= *F. tricinctum*) strain T-2 (= NRRL 3299) has been reported to produce the following five mycotoxins:

1. *Butenolide.* This strain produced the toxic butenolide in cultures on Sabouraud's agar incubated at 15°C for 21 days (Burmeister, 1971; Ellis & Yates, 1971; Marasas, 1969; Yates et al., 1970).

2. *Diacetoxyscirpenol.* There is some uncertainty about diacetoxyscirpenol production by strain T-2. Bamburg et al. (1968a) characterized the mycotoxin isolated from *F. tricinctum* strain B-24 as diacetoxyscirpenol, identified the "crystalline toxin related to that from strain B-24" isolated from *F. tricinctum* strain T-2 as T-2 toxin, but did not state that strain T-2 also produced diacetoxyscirpenol. However, Bamburg & Strong (1969) stated that "in addition to diacetoxyscirpenol (I) and T-2 toxin (II) the fungus *Fusarium tricinctum*, strain T-2, when grown at 24°C, produces a new trichothecene derivative, HT-2 toxin." Similarly Ueno et al. (1972b) did not detect diacetoxyscirpenol by TLC in toxic extracts of *F. tricinctum* NRRL 3299 cultured in PSC medium at 26°C for 12 days, but Ueno et al. (1973a) reported that this strain produces diacetoxyscirpenol in this medium. Recently Schmidt et al. (1981a) detected diacetoxyscirpenol in cultures of this strain (as *F. tricinctum* Sp. 897) on rice incubated at room temperature for 6 days.

3. *HT-2 toxin.* This trichothecene was first isolated from cultures of *F. tricinctum* Strain T-2 in Gregory's medium incubated at 24°C for 2 to 3 weeks (Bamburg, 1968; Bamburg & Strong, 1969). The yield of crystalline HT-2 toxin obtained by Bamburg (1968) and Bamburg and Strong (1969) was 109 mg/6 ℓ of culture material. Ikediobi et al. (1971) detected 750 mg/kg of HT-2 toxin by GLC in corn-rice cultures of this strain incubated at 27°C for 21 days followed by 24 days at 8°C. The production of HT-2 toxin by this strain has been confirmed by Schmidt et al. (1981a, as *F. tricinctum* Sp. 897) and Ueno et al. (1972b, 1973a, as *F. tricinctum* NRRL 3299). HT-2 toxin for use in toxicological studies has also been isolated from cultures of this strain on rice incubated at 11°C for 3 months followed by 10 days at room temperature (Chi et al., 1978) and prepared chemically from T-2 toxin produced by this strain (Lee & Chu, 1981; Wei et al., 1971).

4. *Neosolaniol. Fusarium tricinctum* NRRL 3299 was reported to produce neosolaniol by Ueno et al. (1972b, 1973a), who detected this trichothecene by TLC in toxic extracts of cultures in PSC medium incubated at 25 to 27°C for 12 to 14 days. Neosolaniol for use in toxicological studies has also been isolated from cultures of this strain on rice incubated at 11°C for 3 months followed by 10 days at room temperature (Chi et al., 1978).

5. *T-2 toxin.* This trichothecene was first isolated and characterized from *F. tricinctum* strain T-2 by Bamburg (1968) and Bamburg et al. (1968a). The yield of crystalline toxin from stationary cultures of this strain in Gregory's medium incubated at 8°C for 30 days was 1.5 g/15 ℓ (Bamburg, 1968; Bamburg et al., 1968a). The level of T-2 toxin produced by this strain (as *F. tricinctum* No. 203, by Bamburg, 1968, and as *F. tricinctum* T-2 by Marasas, 1969, and Marasas et al., 1971) in Gregory's medium incubated at 8°C for 14 days as determined by GC, was 103 mg/ℓ. Ikediobi et al. (1971) detected 1515 mg/kg of T-2 toxin by GLC in cultures on corn incubated at 8°C for 30 days and 920 mg/kg in corn-rice cultures incubated at 27°C for 21 days followed by 24 days at 8°C. Burmeister (1971) reported that this strain produced high yields (9.96 g/1200 g) of T-2 toxin in culture on corn grits incubated at 15°C for 3 weeks, and 2 to 3 g of crystalline product could be isolated from this cultural material. In cultures on wheat incubated at 12°C for 21 days, Strain T-2 (as *F. poae* by Joffe, 1978b, and as *F. sporotrichioides* by Joffe & Yagen, 1977) was reported to produce 3.4 mg T-2 toxin/10 g of culture material. The same strain (as *F. poae* NRRL 3299) was found to produce 5.8 mg/10 g of T-2 toxin on this medium by Joffe (1978b) and Joffe & Yagen (1977). Maximal yields of T-2 toxin (339 mg/ℓ) by this strain were obtained by Cullen & Smalley (1981) in stationary cultures in nutrient broth-moistened vermiculite incubated at 19°C for 12 days, compared to yields of 2.7 mg/g, 1.7 mg/g, 155 mg/ℓ, and 76 mg/ℓ on corn, rice, Gregory's medium, and PSC medium, respectively. Cullen (1981) reported a yield of 511 mg/ℓ in a vermiculite-supported, nutritionally complex medium containing soya meal, corn steep liquor and glucose after 24 days' incubation at 19°C. The production of T-2 toxin by this strain, usually referred to as *F. tricinctum* NRRL 3299, has also been reported on a variety of other media and it has been used extensively for the production of T-2 toxin used in toxicological investigations (Bamburg et al., 1968b; Boonchuvit et al., 1975; Burmeister et al., 1971, 1972, 1981; Chi & Mirocha, 1978; Chi et al., 1977a, b, c, 1978, 1981; Collins & Rosen, 1981; Coulter et al., 1977; Doerr et al., 1974, 1981; Grove et al., 1970; Hagler et al., 1981; Kosuri et al., 1970, 1971; Lee & Chu, 1981; Marasas et al., 1967, 1969; Müller & Thaler, 1981; Patterson et al., 1979; Richard et al., 1978; Schmidt et al., 1981a, b; Ueno et al., 1972b, 1973a; Weaver et al., 1978a, b, d, 1980; Wei et al., 1971; Witlock et al., 1977; Wyatt et al., 1973a, b, 1975a, b; Yates et al., 1969, 1970).

Cultures. We have examined the following cultures of this strain that have been deposited in the ITFRC:

Strain T-2. A slant culture of this strain was obtained during 1965 from E.B. Smalley as "*F. tricinctum* T-2, isolated from corn in France and originally obtained from W.C. Snyder." A subculture of strain T-2 was brought to South Africa by W.F.O. Marasas and has been maintained by lyophilization since 1970 as MRC 43. Subcultures of this lyophilized material of strain T-2 have also been lyophilized at BAF during 1978 as *F. tricinctum* Sp. 897 and in the ITFRC as *F. sporotrichioides* T-424. A subculture of strain T-2 obtained from E.B. Smalley has also been lyophilized at ATCC as *F. tricinctum* ATCC 24043.

NRRL 3299. A lyophilized culture of this strain was obtained from NRRL on 23 October 1979 as "*F. tricinctum* NRRL 3299." This is a subculture of strain T-2 obtained from E.B. Smalley which had been maintained by lyophilization since 1970 as *F. tricinctum* NRRL 3299 (Burmeister, 1971; Burmeister et al., 1971, 1972, 1981; Grove et al., 1970; Yates et al., 1970). Subcultures of this lyophilized material have also been lyophilized at ATCC as *F. tricinctum* ATCC 24631 and in the ITFRC as *F. sporotrichioides* T-348.

The different subcultures of the original *F. tricinctum* strain T-2 of E.B. Smalley that we have examined, i.e., MRC 43, T-424, Sp. 897, NRRL 3299, and T-348, show various stages of cultural degeneration. However, they all produce polyphialides and some carmine red pigment, spindle-shaped as well as pear-shaped microconidia are present, and we have identified all of them as *F. sporotrichioides* according to Nelson et al. (1983). At present it is not known whether these subcultures that differ in cultural appearance also differ in toxigenicity. The culture represented in the ITFRC that is in the best morphological condition is T-424 (= MRC 43).

Strain 4.4 *F. tricinctum* No. 223 (=T-409; MRC 1661; ATCC 24044)
This strain was isolated from field corn in Washington, USA (Bamburg, 1968; Marasas, 1969; Marasas et al., 1971). Extracts of cultures of this strain in Gregory's medium incubated at 8°C for 14 days were dermotoxic to rat skin and GLC analysis revealed the presence of 106 mg of T-2 toxin/ℓ of culture medium (Bamburg, 1968; Marasas, 1969; Marasas et al., 1971).

A slant culture of this strain was obtained during 1965 from E.B. Smalley as "*F. tricinctum* No. 223." A subculture has been maintained by lyophilization at MRC since 1970 as MRC 1661. Subcultures have also been lyophilized at ATCC as *F. tricinctum* ATCC 24044 and in the ITFRC as *F. sporotrichioides* T-409.

Strain 4.5 *F. tricinctum* Isolate No. 16 (= T-602; MRC 2863; ATCC 24045)
This strain was isolated from moldy corn that contained 2 mg/kg of T-2 toxin and was implicated in a field outbreak of a lethal hemorrhagic syndrome in dairy cattle in Wisconsin, USA, during 1970–71 (Hsu et al., 1972). Ethyl acetate extracts of corn cultures of this strain incubated at either 8 or 24°C for 30 days were lethal to albino rats (3 of 3 dead within 5 days) when force fed as a single dose (Hsu et al., 1972). The moldy corn from which this strain was isolated contained T-2 toxin, but Hsu et al. (1972) did not state whether the extracts of *F. tricinctum* Isolate No. 16 toxic to rats contained T-2 toxin and it is not known whether this strain produces T-2 toxin.

A lyophilized culture of this strain was received from ATCC on 25 October 1982 as "*F. tricinctum* ATCC 24045. E.B. Smalley 1971-3-16. Moldy corn. Produces T-2 toxin."

Strain 4.6 *F. tricinctum* 2061-C (= T-405; MRC 1620)
This strain was isolated from corn stored on the cob and suspected to have been involved in field outbreaks of illness in cattle in Minnesota, USA (Christensen et al., 1972a). Cultures on autoclaved corn incubated for 2 weeks at 22 to 25°C followed by 2 weeks at 14°C were fed to day-old turkey poults (Christensen et al., 1972a). Consumption of a ration containing 1% of culture material resulted in the death of 13% of the turkey poults within 35 days and in decreased feed efficiency and weight gain and moderate development of bilateral necrotic lesions at the angles of the mouth. Consumption of a ration containing 2% of culture material resulted in death of 60 to 83% of the birds, greatly reduced growth and feed efficiency in the survivors, and the development of severe beak lesions. Consumption of rations containing 5, 10, or 20% of culture material resulted in 100% mortality within 5 to 15 days.

The unspecified strain of *F. tricinctum* used in toxicity experiments in laying hens by Speers et al. (1972, 1977) was the same isolate (i.e., 2061-C) used by Christensen et al. (1972a), according to Mirocha & Christensen (1974). When corn invaded by this strain was used to make up 1, 3, or 5% of the diet, it caused a marked decrease in feed intake, subsequent body weight loss, and a reduction in egg production in laying hens; birds fed the 3% diet exhibited a high incidence of mouth lesions (Speers et al., 1972). In a continuation of this study, Speers et al. (1977) fed White Leghorn laying hens rations containing 2.5 and 5% of corn inoculated with this strain of *F. tricinctum* and incubated at 25°C for 2 weeks followed by 4 weeks at approximately 12°C. The ration containing 5% culture material, and 16 ppm of T-2 toxin as detected by GC, resulted in reductions in feed consumption, body weight gain, egg production, and egg quality. Mouth lesions developed on the medial surfaces of the mandibles, on the dorsal surface of the tongue, and on the skin at the corners of the mouth of some of these birds. The diet containing 2.5% of culture material, and 8 ppm of T-2 toxin, had no significant effects on performance of the hens, but did cause the development of mouth lesions in some of the birds (Speers et al., 1977). According to the chemical analyses of Speers et al. (1977), diets containing 5% of culture material of this strain contained 16 ppm T-2 toxin. Thus the level of T-2 toxin in the original corn culture can be calculated at 320 mg/kg. Speers et al. (1977) detected no zearalenone in this material. However, Mirocha & Christensen (1974) stated that this strain of *F. tricinctum* produces "copious amounts of zearalenone." Mirocha & Christensen (1974) also stated that this strain produces T-2 toxin, which was responsible for the bilateral necrotic mouth lesions and deaths observed in turkey poults by Christensen et al. (1972a) and in laying hens by Speers et al. (1972).

Joffe & Palti (1975) identified a culture of this strain received as "*F. tricinctum* 2061-C" from C.J. Mirocha as *F. sporotrichioides*. Joffe (1978b) and Joffe & Yagen (1977) cultured this strain (as *F. sporotrichioides* 2061-C) on wheat at 12°C for 21 days and reported that ethyl alcohol extracts were slightly dermotoxic to rabbit skin and that this strain produced 1.5 mg T-2 toxin/10 g of grain.

A soil culture of this strain was received on 6 March 1978 from C.J. Mirocha as "*F. tricinctum* 2061-C."

Strain 4.7 *F. poae* **7452** = DAOM 165006 (= T-592; MRC 2728)
This strain was isolated from horse feed associated with sickness and death of race horses in Saskatchewan, Canada, and was originally referred to as *Fusarium* sp. MCH 7452 (Davis & Smith, 1977) and later as *F. poae* 7452 (Davis & Smith, 1981).

Cultures of this strain (as *Fusarium* sp. MCH 7452) on autoclaved rye seed were incubated in the dark for 24 days at 10, 15, 20, or 25°C and the culture material fed to larvae of the yellow mealworm, *Tenebrio molitor* (Davis & Smith, 1977). The cultures incubated at all four temperatures were toxic to the larvae, but the highest mortality and greatest reduction in weight gain occurred within the cultures incubated at temperatures lower that 25°C. In a continuation of this study, Davis & Smith (1981) incubated cultures of this strain (as *F. poae* 7452, identified by R.A. Shoemaker) on rye in the dark and in light at 8, 14, 20, and 24°C for 14 days and fed the culture material to larvae of *T. molitor*. A greater depression of larval growth was observed with culture material of this strain grown in the dark than with those grown in continuous light at all incubation temperatures.

According to Davis & Smith (1981), this isolate produced "a high concentration of a metabolite resembling T-2 toxin" when grown on liquid and solid substrates. These authors concluded that fusariotoxins simlar to T-2 toxin were responsible for the toxicity of this isolate to *T. molitor* larvae.

A lyophilized culture of this strain was received on 20 May 1982 from DAOM as "*F. poae* DAOM 165006 = 7452," together with a note from the Director, Biosystematics Research Institute, Agriculture Canada, Ottawa, Canada, that this strain had originally been identified as *F. poae,* but was recently redetermined as *F. sporotrichioides* by G.A. Neish.

Strain 4.8 *F. nivale* (= *F. tricinctum*) **NRRL 3249** (= T-344; MRC 1705; ATCC 24529)
This strain (as *F. nivale* NRRL 13,318) was originally isolated from toxic fescue (*Festuca arundinacea* Schreb.) hay from a pasture where outbreaks of fescue foot in cattle (see

fescue foot, above) had occurred in Missouri, USA (Keyl et al., 1967). This strain has also been referred to as *F. nivale* NRRL 3249 (Burkhardt et al., 1968; Ellis & Yates, 1971; Yates et al., 1967, 1968, 1969). The Japanese isolate Fn-2 (Strain 4.9) was identified as *F. nivale* by Tsunoda (1970) because it was found to be "completely identical to the referral strain *F. nivale* NRRL 3249." Subsequently *F. nivale* NRRL 3249 was referred to as *F. tricinctum* NRRL 3249 (Burmeister et al., 1971; Ellis & Yates, 1971; Tookey et al., 1972; Ueno et al., 1972b, 1973a; Yates, 1971; Yates et al., 1969, 1970), and Ellis & Yates (1971) stated that "this strain was later identified by Professor W.C. Snyder, who found it to fit into the Snyder & Hansen concept of *F. tricinctum*. We now consider the strain atypical for the species *F. tricinctum*." Joffe & Palti (1975) identified a lyophilized culture of *F. tricinctum* NRRL 3249 obtained from NRRL as *F. sporotrichioides*. Joffe (1978b) and Joffe & Yagen (1977) subsequently also referred to this strain as *F. sporotrichioides* NRRL 3249.

Ethanol extracts of cultures of *F. tricinctum* (= *F. nivale*) NRRL 3249 on sterile, non-toxic fescue hay incubated at 7°C for 2 to 3 weeks were acutely toxic to cattle (Yates et al., 1969). Extract equivalent to 0.89 kg of moldy hay given orally killed one out of two animals within 24 hours. The second animal died after receiving another dose equivalent to 0.45 kg of moldy hay. In a subsequent experiment, lower daily dosage rates (equivalent to 0.15 kg of hay) did not cause any signs of toxicity. Cultures on enriched fescue or timothy hay incubated at 15°C for 18 to 21 days and administered by ruminal fistula to four heifers caused pathological changes in the tails of all the experimental animals (Tookey et al., 1972). The surface temperature of the tails of the treated animals were lower than those of the controls and the pathological changes included edema and perivascular serous infiltration of the tails, but no evidence of vasoconstriction was found. This culture material did not contain chemically detectable butenolide (Tookey et al., 1972). Butenolide was, however, originally isolated from this strain (Burkhardt et al., 1968; Ellis & Yates et al., 1967, 1968, 1969) and the pure compound has been reported to cause necrosis of the distal portion of the tail in cattle (Grove et al., 1970; Kosuri et al., 1970; Tookey et al., 1972).

Extracts of cultures of this strain on a variety of media incubated at low temperatures (5 to 15°C), but not at room temperature, have been reported to be highly toxic to rabbit skin and to cause erythema, edema, and necrosis (Ellis & Yates, 1971; Joffe, 1978b; Joffe & Palti, 1975; Joffe & Yagen, 1977; Keyl et al., 1967; Yates et al., 1968, 1969).

Extracts of cultures on Sabouraud's agar or corn grits incubated at 15°C for 21 days were acutely toxic to mice upon intraperitoneal injection and caused visceral hemorrhages (Burmeister et al., 1971, 1981; Ellis & Yates, 1971; Keyl et al., 1967; Yates et al., 1968, 1969, 1970). The radiomimetic pathological changes characteristic of trichothecenes, including necrosis and erosion in the mucosa of the small intestine and karyorrhexis and necrosis of the immature cells in the thymus, spleen, lymph nodes, and bone marrow, have also been induced in mice with extracts of this strain (as *F. nivale* NRRL 13,318) in PSC medium incubated at 25 to 27°C for 2 weeks (Ueno et al., 1971a).

Mycotoxins. *Fusarium sporotrichioides* (= *F. tricinctum* = *F. nivale*) NRRL 3249 has been reported to produce the following three mycotoxins:

1. *Butenolide*. The toxic butenolide was first isolated from this strain by Burkhardt et al. (1968) and Yates et al. (1967). According to Yates et al. (1968), about 29 mg of butenolide/plate could be isolated from cultures on Sabouraud's agar incubated at 15°C for 3 to 4 weeks and the yield of crystalline butenolide from cultures in liquid Sabouraud's maltose medium incubated at 3°C for 23 weeks was 1.0 g/ℓ of culture filtrate.

2. *T-2 toxin.* This strain has been reported to produce T-2 toxin in cultures on Sabouraud's agar incubated at 15°C for 3 weeks (Burmeister et al., 1971; Ellis & Yates, 1971; Yates et al., 1968, 1970), Sabouraud's maltose liquid medium incubated at 3°C for 20 to 30 weeks (Ellis & Yates, 1971; Yates et al., 1968, 1970), and white corn grits incubated at 15°C for 21 days (Burmeister et al., 1971, 1981). Yields of 14 mg T-2 toxin/kg of culture material have been reported in cultures of this strain on autoclaved wheat incubated at 3°C for 10–12 weeks (Yates et al., 1968) and 4.0 mg of T-2 toxin/10 g of wheat culture material incubated at 12°C for 21 days (Joffe, 1978b; Joffe & Yagen, 1977).

3. *Unidentified toxin.* This compound was extracted with ethyl acetate from cultures in Sabouraud's maltose liquid medium incubated at 3°C for 20 to 30 weeks and had about the same toxicity for mice as the butenolide (Ellis & Yates, 1971; Yates et al., 1968).

The ratio of the respective abundance of butenolide, the unknown toxin, and T-2 toxin in these extracts was 87:8:3 (Ellis & Yates, 1971; Yates et al., 1968).

Considerable variability has been reported in both the toxicity to experimental animals and the mycotoxins produced by NRRL 3249. For instance, Ueno et al. (1971a) reported that this isolate (as *F. nivale* NRRL 13,318 and NRRL 3249) was acutely toxic to and caused radiomimetic pathological changes in mice, and that it also markedly inhibited protein synthesis in rabbit reticulocytes. On the other hand, Ueno et al. (1972b, 1973a) found that cultures of *F. tricinctum* NRRL 3249 were not toxic to mice, were negative in the rabbit reticulocyte assay, and were negative for T-2 toxin. Although the toxin butenolide was first isolated from this strain (Burkhardt et al., 1968; Yates et al., 1967), which was originally reported to produce large amounts of the butenolide (Yates et al., 1968), cultures of this strain were subsequently found to be negative for butenolide (Burmeister et al., 1971; Tookey et al., 1972; Yates et al., 1970), but still positive for T-2 toxin (Burmeister et al., 1971, 1981; Yates et al., 1970). These conflicting results are probably the result of a progressive loss of toxigenicity of NRRL 3249 during maintenance in culture. Yates et al. (1968) stated that "certain *F. nivale* isolates from the parent strain appear to lose their ability to produce a toxic extract through repeated transfers on agar slants incubated at room temperature. This response sometimes occurs after only one transfer at room temperature." According to Yates et al. (1968), "such losses have been accompanied by a change in pigment production, i.e., in the cultures producing lesser amounts of toxins, the pigments are a dark reddish-brown and the colony appears tan"; Yates et al. (1969) also noted that a loss in toxicity was usually accompanied by a change in pigmentation in which yellow and red pigments were replaced by dark reddish-brown pigments.

A lyophilized culture of this strain was received on 5 September 1979 from NRRL as "*F. tricinctum* NRRL 3249." Subcultures of this lyophilized material are in relatively good condition: the colonies are red tinged with yellow, and the underside is carmine red; polyphialides are clearly visible; spindle-shaped as well as pear-shaped microconidia are present; and macroconidia are produced in sporodochia. We consider this to be a typical culture of *F. sporotrichioides*.

Strain 4.9 *F. nivale* **Fn 2, NRRL 3509** (= T-346; MRC 1767)
= *F. nivale* **Fn-2B** (= T-436; MRC 1962) (= T-563; MRC 2476) (= T-564; MRC 2533; ATCC 26532)

This strain was originally isolated (as *F. nivale* F. n-2) from a sample of wheat that was severely damaged by *Fusarium* blight (see akakabi-byo, above) during 1963 in Kyushu, Kumamoto Prefecture, Japan (Tsunoda, 1970).

The morphological characteristics of this isolate were found to be "completely identical to the referral strain, F. nivale NRRL 3249" (see Strain 4.8 = F. sporotrichioides) by Tsunoda (1970). The Japanese strain F. nivale Fn-2 has been deposited in NRRL as F. tricinctum NRRL 3509 (J.J. Ellis, personal communication, 2 June 1981). Joffe & Palti (1975) identified NRRL 3509 as F. sporotrichioides var. tricinctum and this name was also used for this strain by Joffe (1978b) and Joffe & Yagen (1977). F. nivale Fn-2B has also been deposited in NRRL as F. tricinctum NRRL 6490 (Burmeister et al., 1981) and in ATCC as Gerlachia nivalis ATCC 26532 (ATCC Catalogue of Strains, Fifteenth Ed., 1982).

In the Japanese mycotoxological literature on "F. nivale," one of the most toxic strains has been referred to either as Fn-2 (= F. n-2, Fn 2, F_n-2) (Saito & Okubo, 1970; Saito & Tatsuno, 1971; Saito et al., 1969; Tatsuno, 1968; Tatsuno et al., 1968, 1969; Tsunoda, 1970; Ueno & Fukushima, 1968; Ueno et al., 1970c), or Fn-B (Morooka et al., 1971), or Fn-2B (= Fn 2 B, Fn-2-B, Fn 2 b, Fn-2-L-A, Fn-2-L-B) (Fujimoto et al., 1972; Saito & Ohtsubo, 1974; Saito et al., 1980; Tatsuno, 1976; Tatsuno et al., 1970, 1973; Ueno, 1971; Ueno et al., 1969a, 1970a, 1971a, b, c, 1973a, 1975; Yoshizawa & Morooka, 1975a; Yoshizawa et al., 1979, 1980). We are assuming that all these different strain numbers are orthographical variants (or stock culture numbers, see Ueno et al., 1971a) and that they all refer to a single strain that was originally called F. nivale F. n-2 by Tsunoda (1970). In the original accounts of the isolation of toxic Fusarium species from a sample of wheat that was severely damaged by Fusarium blight during 1963 in Kyushu, Kumamoto Prefecture, Japan, it was stated that "F. nivale (F. n-2)" was toxic to mice (Tsunoda, 1970) and that "high toxicities were shown in one strain of F. nivale (Fn-2 strain)" (Saito & Okubo, 1970).

Extracts of cultures of F. nivale Fn-2B in PSC medium or on autoclaved rice incubated at 25 to 27°C for 2 weeks or longer inhibited protein synthesis in Ehrlich ascites tumor cells and in rabbit reticulocytes (Tatsuno, 1968; Ueno, 1971; Ueno & Fukushima, 1968; Ueno et al., 1969a, 1971a, 1973a). Such extracts were also dermotoxic to rabbit skin and caused necrosis of the epidermis, degeneration and necrosis of the hair follicles, edema in the dermis, and coagulation necrosis of the muscle fibers (Joffe, 1978b; Joffe & Palti, 1975; Joffe & Yagen, 1977; Ueno, 1971; Ueno et al., 1970b, 1971a, b).

Extracts of cultures of this strain have been found to be acutely toxic to mice and to cause hemorrhage and radiomimetic cytotoxic damage characterized by cellular degeneration and karyorrhexis in the actively dividing cells of the hematopoietic and lymphopoietic tissues in the bone marrow, spleen, thymus, and lymph nodes, and also of the intestinal mucosa and testes (Saito & Okubo, 1970; Saito et al., 1969; Tatsuno, 1968, 1976; Ueno, 1971; Ueno & Fukishima, 1968; Ueno et al., 1969a, 1970a, 1971a, b, 1973a). Cultures of F. nivale Fn-2B on rice have also been fed to male mice on a lifelong basis (Ohtsubo & Saito, 1977; Saito & Ohtsubo, 1974). The notable pathological lesions were atrophy of the thymus, spleen, and testes, and atypical hyperplasia in the gastric and intestinal mucosa as well as in the intrahepatic bile ducts. Malignant tumors occurred in various organs, but their incidences were very low, i.e., one case each of hepatoma, myeloic and lymphocytic leukemia, subcutaneous sarcoma, and adenocarcinoma of the jejunum (Ohtsubo & Saito, 1977; Saito & Ohtsubo, 1974).

Rats of the Donryu strain were found to be more than five times more sensitive than mice to the toxic metabolites produced by this strain (Saito et al., 1969). Histological findings in the rats were very similar to those in mice, but damage of the deep gland cells of the glandular stomach was more marked in rats than in mice. In a long-term feeding study in male rats with rice cultures of F. nivale Fn-2B, Ohtsubo & Saito (1977) and Saito & Ohtsubo (1974) reported that the prominent pathological lesions were hypoplasia of the intrahepatic bile ducts and bone marrow, atrophy of the thymus, spleen, and testes, and atypical hyperplasia in the intestinal mucosa; no tumors occurred, however. Guinea pigs dosed orally with extracts of cultures of this strain showed neurological disorders such as unusual jumping, and post-mortem examination disclosed degeneration of the nerve cells in the brain (Saito et al., 1969). Severe damage to the central nervous system characterized by degeneration and atrophy of nerve cells has also been reported in goats dosed with the crude toxin of this strain (Kubota, 1977). Radiomimetic cellular damage also occurred in the spleen, bone marrow, and intestines, and spermatogenesis was inhibited in these goats (Kubota, 1977).

Mycotoxins. *Fusarium sporotrichioides* (= *F. tricinctum* = *F. nivale*) Fn-2B has been reported to produce the following four trichothecenes:

1. *Diacetylnivalenol* (= *Nivalenol Diacetate*). This trichothecene was first isolated from cultures of *F. nivale* Fn-2B in PSC medium incubated in stationary culture at 25°C for 4 days followed by 2 weeks at 15°C (Tatsuno et al., 1970). The yield of diacetylnivalenol obtained by Tatsuno et al. (1970) was 60 mg from 50 ℓ of culture material. The production of diacetylnivalenol by this strain has subsequently been confirmed by Fujimoto et al. (1972), Ohta et al. (1977, 1978), Tatsuno (1976), Tatsuno et al. (1973), and Yoshizawa et al. (1980).

2. *Fusarenon-X* (= *Fusarenon,* = *Nivalenol Monoacetate,* = *Monoacetyl Nivalenol*). This trichothecene was first isolated from *F. nivale* Fn-2B by Morooka & Tatsuno (1970) and Tatsuno et al. (1969, 1970), who referred to this compound as fusarenon, and by Ueno et al. (1969a, 1970c), who referred to it as fusarenon-X. Although fusarenon and fusarenon-X were originally thought to be slightly different chemically (Morooka & Tatsuno, 1970; Saito & Okubo, 1970; Tatsuno et al., 1969; Ueno et al., 1969a, 1970c, 1971b), they are now considered to be identical (Morooka et al., 1971) and the name fusarenon-X is apparently preferred. Yields of 20 mg of fusarenon-X from 3 kg of culture material of strain Fn-2B on rice incubated at 25 to 27°C for 20 days were obtained by Morooka & Tatsuno (1970) and Tatsuno et al. (1970), while Ueno et al. (1969a, 1970c) obtained 200 mg from 10 ℓ of culture material in PSC medium incubated at 25 to 27°C for 2 weeks. Maximal yields of fusarenon-X (approximately 200 mg/10 ℓ) have been obtained with strain Fn-2B in stationary culture in PSC medium after 6 to 8 days of incubation at 27°C and the yield decreased at lower temperatures (Ueno et al., 1975). Fusarenon-X is produced prior to and in much higher yields (approximately 20:1) than nivalenol in liquid medium by *F. nivale* Fn-2B and is considered to be a major toxic metabolite of this strain (Ueno et al., 1969a, 1970a, 1971b). *Fusarium sporotrichioides* (= *F. nivale*) Fn-2B has been used extensively in Japan for the production of fusarenon-X for use in toxicological investigations (Fujimoto et al., 1972; Kato et al., 1979; Matsuoka & Kubota, 1981; Matsuoka et al., 1979; Ohta et al., 1977, 1978; Saito & Okubo, 1970; Sato et al., 1978; Tatsuno, 1976; Tatsuno et al., 1973; Ueno 1971d; Ueno et al., 1970a, 1971a, b, c, 1973a, 1975a, 1978; Yoshizawa & Morooka, 1975b, Yoshizawa et al., 1979, 1980).

3. *Nivalenol.* This trichothecene was first isolated from *F. nivale* Fn-2B by Tatsuno (1968) and Tatsuno et al. (1968, 1969). Yields of 150 mg of nivalenol from 3 kg of rice incubated at 25 to 27°C for 20 days were obtained by Morooka & Tatsuno (1970). This strain also produces nivalenol in stationary culture in PSC medium incubated at 25 to 27°C for 2 weeks (Fujimoto et al., 1972; Tatsuno, 1976; Tatsuno et al., 1970; Ueno & Fukushima, 1968; Ueno et al., 1969a, 1970a, b, c, 1971a, 1973a). Yields of nivalenol in this medium have been reported to be 10 mg/10 ℓ by Ueno et al. (1969a) and 2 to 3 mg/ℓ by Ueno et al. (1970c). According to Ueno et al. (1970a), maximal production of nivalenol

by strain Fn-2B in PSC medium occurs after 6 to 8 days at 27°C and in cultures on rice incubated at 27°C, and lower yields are obtained at lower temperatures. Nivalenol is produced later and in lower yields than fusarenon-X in liquid media by strain Fn-2B (Ueno et al., 1969a, 1970a). Nivalenol lacks one acetyl group of fusarenon-X, and it appears that nivalenol may be produced from fusarenon-X by a chemical or enzymatic deacetylation reaction during the cultivation of strain Fn-2B (Ueno et al., 1970a, 1971b). However, in biological deacetylation experiments, the growing mycelium of this strain failed to convert fusarenon-X to nivalenol (Yoshizawa & Morooka, 1975b). *Fusarium sporotrichioides* (= *F. nivale*) Fn-2B has been used extensively in Japan for the production of nivalenol for use in toxicological investigations (Ueno & Fukushima, 1968; Ueno et al., 1970b, 1973b; Yoshizawa & Morooka, 1975; Yoshizawa et al., 1979, 1980). Nivalenol has also been prepared chemically from fusarenon-X isolated from this strain (Ohta et al., 1977, 1978).

4. *T-2 toxin.* The production of low levels of T-2 toxin (0.5 mg/10 g of grain) by this strain (as *F. sporotrichioides* var. *tricinctum* NRRL 3509) on autoclaved wheat incubated at 12°C for 21 days has been reported by Joffe (1978b) and Joffe & Yagen (1977). On the other hand, Burmeister et al. (1981) found that extracts of cultures of this strain (as *F. tricinctum* NRRL 6490) on corn incubated at 15°C for 21 days were toxic to mice, but did not contain T-2 toxin. Negative results for T-2 toxin production by this strain (as *F. tricinctum* T-346) have also been reported by Cullen (1981).

Cultures. We have obtained from different sources four subcultures of this strain that have been deposited in the ITFRC: 1) A lyophilized culture that was received on 23 October 1979 from NRRL as "*F. tricinctum* NRRL 3509" (= T-346, MRC 1767)—according to J.J. Ellis (personal communication, 2 June 1981), "Strain NRRL 3509 *Fusarium tricinctum* was received as Fn 2 from H. Tsunoda, Food Research Institute, Tokyo, Japan"; 2) a slant culture that was received on 29 April 1980 from H. Kurata as "*F. nivale* Fn-2B from Dr. Tsunoda" (= T-436; MRC 1962); 3) a slant culture that was received on 30 September 1981 from T. Yoshizawa as "*F. nivale* Fn-2B" (= T-563; MRC 2476); and 4) a slant culture that was received in November 1981 from Y. Ueno as "*F. nivale* Fn-2B" (= T-564; MRC 2533).

The culture that had been maintained by lyophilization as *F. tricinctum* NRRL 3509 (= T-346; MRC 1767) is in better condition than the other three, which were maintained by subculturing on agar. Subcultures of NRRL 3509 grow rapidly on PDA and are floccose-powdery, red tinged with yellow in the center, and carmine red on the underside. Microconidia, produced on monophialides and on conspicuous polyphialides in the aerial mycelium, are of two types, i.e., spindle-shaped and pear-shaped. Many microconidia become 1–3 separate and fusoid. Macroconidia are produced in sporodochia and are typically 3–5 septate, straight to slightly curved, and either apedicellate or possessing a weakly developed foot cell. Some micro- as well as macroconidia appear to be dividing by a process similar to "budding" in yeast cells. The other three cultures are basically similar, but show various degrees of degeneration, such as a loss of the carmine red pigment, increase in the number of micro- and macroconidia that exhibit "budding," and degeneration of phialides so that they are misshapen, swollen with constrictions, and difficult to differentiate as mono- or polyphialides.

In our opinion, NRRL 3509 is a typical representative of *F. sporotrichioides* because of the presence of polyphialides and two types of microconidia. The other three cultures are much more difficult to identify because of the abnormal phialides and the prevalence of "budding" conidia. We have observed similar characteristics in two other Japanese strains, i.e., *F. episphaeria* Fn-M (Strain 4.10) and *F. oxysporum* "niveum" Melon-1

(Strain 4.11). The two latter isolates are so similar to each other and to the above three cultures of "*F. nivale*" that one can almost consider them to be subcultures of the same parent isolate.

Following our detailed comparative study of the above cultures of *F. nivale*, *F. episphaeria*, and *F. oxysporum* "niveum," we can only conclude that they represent degenerate cultures of *F. sporotrichioides*. We fully realize, however, that all these "atypical" Japanese strains of *F. sporotrichioides* produce fusarenon-X and nivalenol, two mycotoxins that have not yet been detected in *F. sporotrichioides* strains from other geographical areas. It is also remarkable that *F. nivale* Fn-2B was isolated in Kyushu in the relatively warm southern part of Japan (Tsunoda, 1970) and *F. episphaeria* Fn-M (Strain 4.10) was isolated in Shikoku, a southern island of Japan where "the weather in winter and spring is rather warm" (Ueno et al., 1971a). The origin of *F. oxysporum* "niveum" Melon-1 (Strain 4.11) is unknown to us. The occurrence of *F. sporotrichioides* in these relatively warm geographical areas is unexpected.

It is possible that the "atypical" characteristics of "degenerate" phialides and conidial "budding" may be correlated with the ability to produce fusarenon-X and nivalenol. We may also be dealing here with a specific, geographically restricted strain of *F. sporotrichioides*. However, until such time as more isolates exhibiting these characteristics can be examined, we are concluding that *F. nivale* Fn-2B is *F. sporotrichioides*.

Strain 4.10   *F. episphaeria* Fn-M (= T-565; MRC 2531)

This strain was isolated from "a house covered with vinyl plates (so-called vinyl house) in a farm of Shikoku district, a southern island of Japan" (Ueno et al., 1971a). Cultures of this strain have been deposited in ATCC as *F. aquaeductuum* var. *medium* ATCC 26531 (ATCC Catalogue of Strains, Fifteenth Ed., 1982) and in NRRL as *F. tricinctum* NRRL 6491 (Burmeister et al., 1981).

Extracts of cultures of *F. episphaeria* Fn-M in PSC medium incubated in stationary culture at 25 to 27°C for 2 weeks inhibited protein synthesis in rabbit reticulocytes, were dermotoxic to rabbit skin, and caused the radiomimetic pathological changes characteristic of trichothecenes in mice (Ueno et al., 1971a, 1973a). The presence of fusarenon-X (Ueno et al., 1971a, c, 1972b, 1973a, 1975) and nivalenol (Ueno et al., 1971a, 1973a) was detected by TLC in these toxic extracts. Crystalline fusarenon-X has been isolated in high yields (0.8 g/10 ℓ) from cultures of this strain in PSC medium incubated at 27°C for 2 weeks (Ueno et al., 1971a). This strain has also been used for the production of fusarenon-X for use in toxicological studies (Ueno et al., 1971c, 1972b, 1975). According to Burmeister et al. (1981), extracts of cultures of this strain (as *F. tricinctum* NRRL 6491) on corn grits incubated at 15°C for 3 weeks were lethal to mice, but did not contain T-2 toxin.

A slant culture of this strain was received in November 1981 from Y. Ueno as "*F. episphaeria* Fn-M." This culture has abnormal phialides and "budding" conidia identical to those of the degenerate cultures of *F. nivale* Fn-2B (see Strain 4.9) and of *F. oxysporum* "niveum" Melon-1 (Strain 4.11). We have identified this strain as *F. sporotrichioides*, but with the same reservations as discussed under *F. nivale* Fn-2B (Strain 4.9).

Strain 4.11   *F. oxysporum* "niveum" Melon-1 (= T-566; MRC 2534)

The source of this strain was not given by Ueno et al. (1973a), who reported that extracts of cultures of this strain, as well as of *F. oxysporum* "niveum" Melon-2, in PSC medium incubated at 25 to 27°C for 2 weeks were lethal to mice upon intraperitoneal injection. These extracts also inhibited protein synthesis in rabbit reticulocytes, and diacetylnivalenol and fusarenon-X were detected by TLC. According to Ueno et al. (1975), this strain (as *F. oxysporum* T-M-1) produced small amounts of fusarenon-X in shake culture in a medium containing glucose, corn steep liquor, and yeast extract, but not in PSC medium, incubated at 27°C for 5 days.

A slant culture of this strain was received in November 1981 from Y. Ueno as "*F. oxysporum* Melon-1." This culture has abnormal phialides and "budding" conidia identical to those of the degenerate cultures of *F. nivale* Fn-2B (see Strain 4.9) and of *F. episphaeria* Fn-M (Strain 4.10). We have identified this strain as *F. sporotrichioides*, but with the same reservations as discussed under *F. nivale* Fn-2B (Strain 4.9).

Strain 4.12 *F. solani* M-1-1 (= T-422; MRC 935; Sp. 900), (= T-567; MRC 2478), (= T-568; MRC 2557)
 = *F. solani* **NHL-F-111** (= T-490; MRC 2034; Sp. 934; Sp. 980)
 = *F. sporotrichioides* **ATCC 26533** (= T-492; MRC 2319).

This strain (as *F. solani* M-1-1) was originally isolated from moldy soybean hulls collected in 1970 from farms near Obihiro City in Hokkaido, Japan, where outbreaks of "bean hulls poisoning" (see bean hulls poisoning, above) of horses had occurred (Ishii et al., 1971; Ueno et al., 1972a). The fungus was cultured in PSC medium in stationary culture at 25 to 27°C for 12 days (Ishii et al., 1971; Ueno et al., 1972a). Extracts of these cultures inhibited protein synthesis in rabbit reticulocytes. These extracts were also lethal to mice and caused the radiomimetic cellular damage characteristic of trichothecenes (Ishii et al., 1971; Ueno et al., 1972a, 1973a). Extracts of cultures of this strain (as *F. tricinctum* NRRL 6151) on corn grits incubated at 15°C for 3 weeks have also been reported to be lethal to mice by Burmeister et al. (1981).

Mycotoxins. *Fusarium sporotrichioides* (= *F. solani*) M-1-1 has been reported to produce the following seven mycotoxins:

1. *Butenolide.* This strain, as "*F. solani* M-1-1 (ATCC 26533)," was used for the production of the toxic butenolide by Yoshizawa et al. (1979), but this mycotoxin was not detected by TLC in toxic extracts of this strain by Ueno et al. (1972a).

2. *Diacetoxyscirpenol.* *Fusarium solani* M-1-1 was reported to produce trace amounts of diacetoxyscirpenol in PSC medium in stationary culture at 25°C for 12 days by Ishii et al. (1971). The production of diacetoxyscirpenol by this strain has subsequently been confirmed by Ohta et al. (1977), Schmidt et al. (1981a—as *F. solani* Sp. 900), Tatsuno et al. (1973), and Ueno et al., 1972a, b, 1973a, 1975).

3. *HT-2 toxin.* Ueno et al. (1972b, 1973a) reported that *F. solani* M-1-1 produced HT-2 toxin. The yield of crystalline HT-2 toxin obtained from cultures of this strain in PSC medium incubated at 25°C for 20 days was 275 mg/87 ℓ of culture filtrate (Ishii & Ueno, 1981). The production of HT-2 toxin by this strain has also been confirmed by Schmidt et al. (1981a—as *F. solani* Sp. 900) and Yoshizawa et al. (1979, 1980—as *F. sporotrichioides* ATCC 26533).

4. *Neosolaniol* (= *8-Hydroxydiacetoxyscirpenol*). This trichothecene, originally called solaniol, was first isolated and characterized from *F. solani* M-1-1 by Ishii et al. (1971). The name of this compound was changed from solaniol to neosolaniol by Ueno et al. (1972a). The production of neosolaniol by *F. solani* M-1-1 has been confirmed by Ishii & Ueno (1981), Tatsuno et al. (1973), and Ueno et al. (1972a, b, 1973a, b, 1975) and this strain has also been used for the production of neosolaniol used in toxicological studies (Kato et al., 1979; Matsumoto et al., 1978; Ohta et al., 1977; Sato et al., 1978; Yoshizawa et al., 1979, 1980). The yield of neosolaniol obtained by Ishii et al. (1971) and Ueno

et al. (1972a) from stationary cultures of this strain in PSC medium incubated at 25°C for 12 days was 100 mg/20 ℓ, while Ishii & Ueno (1981) isolated 414 mg of crystalline neosolaniol from 87 ℓ of culture material in the same medium incubated at 25°C for 20 days.

5. *NT-1.* This new trichothecene was isolated as a minor metabolite (320 mg/87 ℓ of culture filtrate) from cultures of this strain (as *F. sporotrichioides* M-1-1) in PSC medium incubated at 25°C for 20 days (Ishii & Ueno, 1981).

6. *NT-2.* This new trichothecene was isolated as a minor metabolite (172 mg/87 ℓ of culture filtrate) from cultures of this strain (as *F. sporotrichioides* M-1-1) in PSC medium incubated at 25°C for 20 days (Ishii & Ueno, 1981).

7. *T-2 toxin. Fusarium solani* M-1-1 has been reported to produce T-2 toxin by Burmeister et al. (1981—as *F. tricinctum* NRRL 6251), Ishii & Ueno (1981—as *F. sporotrichioides* M-1-1), Ishii et al. (1971), Tatsuno et al. (1973), and Ueno et al. (1972a, b, 1973a, b, 1975). Reported yields of T-2 toxin from cultures of this strain include the following: 160 mg/20 ℓ of PSC medium in stationary culture at 25 to 27°C for 12 days (Ishii et al., 1971); 2490 mg/87 ℓ in PSC medium at 25°C for 20 days (Ishii & Ueno, 1981); and 1.54 g/11.2 ℓ in PSC medium at 25°C for 7 days (Sato & Amano, 1976—as *F. solani* NHL-F-111). *F. sporotrichioides* (= *F. solani*) M-1-1 was also the highest producer among 12 isolates tested for T-2 toxin production in stationary, shake and, fermentor cultures (Ueno et al., 1975) and has been used for the production of T-2 toxin for use in toxicological investigations (Kato et al., 1979; Masuko et al., 1977; Matsumoto et al., 1978; Ohta et al., 1977; Sato et al., 1975, 1978; Ueno et al., 1978; Yoshizawa & Morooka, 1975a, b; and Yoshizawa et al., 1979, 1980—as *F. sporotrichioides* ATCC 26553).

This strain was referred to as "*F. sporotrichioides* M-1-1" by Ishii & Ueno (1981), who stated that this strain had been classified as *F. sporotrichioides* in a "recent taxonomical study" by Ichinoe (1978a). According to Yoshizawa et al. (1980), this strain "was revised taxonomically by M. Ichinoe, National Institute of Hygienic Sciences, Tokyo, Japan (personal communication)," and these authors referred to this strain as "*F. sporotrichioides* ATCC 26553." Cultures of *F. solani* M-1-1 have been deposited in ATCC as *F. sporotrichioides* ATCC 26533 (ATCC Catalogue of Strains, Fifteenth Ed., 1982) and in NRRL as *F. tricinctum* NRRL 6151 (Burmeister et al., 1981).

Cultures. We have examined the following five subcultures of this strain that have been deposited in the ITFRC: 1) A slant culture that was received in April 1973 from Y. Ueno as "*F. solani* M-1-1"—this culture has been maintained by lyophilization at MRC since 1973 as *F. sporotrichioides* MRC 935, and subcultures of this lyophilized material have also been lyophilized and deposited in BAF as *F. solani* Sp. 900 and the ITFRC as *F. sporotrichioides* T-422; 2) a lyophilized culture that was received on 9 March 1981 from ATCC as "*F. solani* ATCC 26533" (= T-492; MRC 2319); 3) a slant culture that was received on 30 September 1981 from T. Yoshizawa as "*F. sporotrichioides* ATCC 26553" (= T-567; MRC 2478); 4) a slant culture that was received in November 1981 from Y. Ueno as "*F. solani* M-1-1" (= T-568; MRC 2557); and 5) lyophilized cultures (Sp. 934 and Sp. 980) that were received on 8 March 1978 from BAF as "*F. solani* NHL-F-111" (= T-490; MRC 2034)—according to Sato & Amano (1976) this strain was originally obtained from "Dr. Ichinoe, National Institute of Hygienic Sciences, Tokyo," and according to M. Ichinoe (personal communication, September, 1981) this strain is *F. solani* M-1-1.

A detailed examination of these cultures revealed that they all have the following characteristics in common: production of carmine red pigment, polyphialides, two types of microconidia (i.e., spindle-shaped and small pear-shaped), and two types of macroconidia (i.e., straight, fusoid, 1–5 septate macroconidia produced in the aerial mycelium, and curved, apedicellate and pedicellate, 3–5 septate macroconidia produced in sporodochia). These cultures are atypical of *F. sporotrichioides* in several respects, particularly the pear-shaped microconidia, which are smaller and not produced as abundantly as usual, and the prevalence of straight, spindle-shaped, *F. semitectum*-like macroconidia in the aerial mycelium produced on PDA as well as CLA. In fact, W. Gerlach examined subcultures of T-422 (= MRC 935) and identified it as a "degenerate red-pigmented strain of *F. semitectum* var. *majus*." This particular culture, which we have maintained by lyophilization as MRC 935 since 1973, is actually in the best condition of any of these cultures; all of them show some degree of the degeneration that reaches its maximum in T-490 (= MRC 2034), which is completely pionnotal. In our opinion, T-422 (= MRC 935) is *F. sporotrichioides* rather than *F. semitectum,* and the fact that it is somewhat atypical is probably due to cultural degeneration prior to lyophilization in 1973. We consider the other subcultures of this strain that have been received from various sources as cultures of *F. sporotrichioides* that exhibit various stages of degeneration, and conclude that *F. solani* M-1-1 is *F. sporotrichioides*.

Strain 4.13 *F. sporotrichioides* HPB 071178-10 (= T-504; MRC 2182)

This strain was isolated from overwintered wheat in Manitoba, Canada (Scott et al., 1980). Purified extracts of cultures on corn incubated at 14°C for 14 days were toxic to brine shrimp. Analysis of these extracts by TLC revealed the presence of 87 µg of T-2 toxin/g of culture material (Scott et al., 1980). A lyophilized culture of this strain was received on 22 December 1980 from B.J. Blanchfield as "*F. sporotrichioides* HPB 071178-10, identified by CBS."

Strain 4.14 *F. sporotrichioides* HPB 071178-12 (= T-505; MRC 2183)

This strain was isolated from overwintered cribbed corn in Ontario, Canada (Scott et al., 1980). Purified extracts of cultures on corn incubated at 14°C for 14 days were toxic to brine shrimp. Analysis of these extracts by TLC revealed the presence of 380 µg of T-2 toxin/g of culture material (Scott et al., 1980). A lyophilized culture of this strain was received on 22 December 1980 from B.J. Blanchfield as "*F. sporotrichioides?,* HPB 071178-12, identified by CBS."

Strain 4.15 *F. sporotrichioides* HPB 071178-13 (= T-506; MRC 2184)

This strain was isolated from overwintered cribbed corn in Ontario, Canada (Scott et al., 1980). Purified extracts of cultures on corn incubated at 14°C for 14 days were toxic to brine shrimp. Analysis of these extracts by TLC revealed the presence of 1350 µg of T-2 toxin, 34 µg of HT-2 toxin, and 140 µg of neosolaniol/g of culture material. This was the highest T-2 toxin-producing strain found in the study of Scott et al. (1980). Incubation of cultures of this strain on frost-killed corn, barley, or wheat under conditions of fluctuating cold temperatures (one day at 14°C, 7 days at 0°C, 10 days at 14°C, and 5 days at 0°C) did not result in significant increases in the levels of HT-2 toxin and neosolaniol, but the yield of T-2 toxin was much reduced compared to that of cultures of corn incubated at 14°C for 14 days. A lyophilized culture of this strain was received on 22 December 1980 from B.J. Blanchfield as "*F. sporotrichioides?,* HPB 071178-13, identified by CBS."

Strain 4.16 *F. sporotrichioides* HPB 071178-18 (= T-507; MRC 2186)

This strain was isolated from overwintered unharvested corn in Manitoba, Canada (Scott et al., 1980). Purified extracts of cultures on corn incubated at 14°C for 14 days were toxic to brine shrimp. Analysis of these extracts by TLC revealed the presence of 162 µg of T-2 toxin and 20 μg of neosolaniol/g of culture material (Scott et al., 1980). Incubation of cultures of this strain on frost-killed corn under conditions of fluctuating cold temperatures (one day at 14°C, 7 days at 0°C, 10 days at 14°C, and 5 days at 0°C) resulted in reduced levels of T-2 toxin and neosolaniol compared to those produced in corn incubated at 14°C for 14 days, but HT-2 toxin (9 μg/g) was also detected in these cultures. However, incubation of cultures on frost-killed barley or wheat, respectively, under these fluctuating temperature conditions resulted in higher levels of T-2 toxin (340 and 410 μg/g), HT-2 toxin (160 and 160 μg/g), and neosolaniol (120 and 220 μg/g). A lyophilized culture of this strain was received on 22 December 1980 from B.J. Blanchfield as "*F. sporotrichioides* HPB 071178-18, identified by CBS."

Strain 4.17 *F. sporotrichioides* **HPB 071178-20** (= T-508; MRC 2188)
This strain was isolated from overwintered cribbed corn in Ontario, Canada (Scott et al., 1980). Purified extracts of cultures on corn incubated at 14°C for 14 days were toxic to brine shrimp. Analysis of these extracts by TLC revealed the presence of 320 μg of T-2 toxin/g of culture material (Scott et al., 1980). A lyophilized culture of this strain was received on 22 December 1980 from B.J. Blanchfield as "*F. sporotrichioides* HPB 071178-20, identified by CBS."

Strain 4.18 *Fusarium* **sp. HPB 071178-21** (= T-509; MCR 2189)
This strain was isolated from overwintered cribbed corn in Ontario, Canada (Scott et al., 1980). Purified extracts of cultures on corn incubated at 14°C for 14 days were toxic to brine shrimp. Analysis of these extracts revealed the presence of 260 μg of T-2 toxin and 29 μg of neosolaniol/g of culture material (Scott et al., 1980). A slant culture of this strain was received on 22 December 1980 from B.J. Blanchfield as "*Fusarium* sp. HPB 071178-21."

Strain 4.19 *F. sporotrichioides* **HPB 071178-23** (= T-510; MRC 2192)
This strain was isolated from overwintered unharvested corn in Canada (Scott et al., 1980). Purified extracts of cultures on corn incubated at 14°C for 14 days were toxic to brine shrimp. Analysis of these extracts by TLC revealed the presence of 280 μg of T-2 toxin and 23 μg of neosolaniol/g of culture material (Scott et al., 1980). A lyophilized culture of this strain was received on 22 December 1980 from B.J. Blanchfield as "*F. sporotrichioides* HPB 071178-23, identified by CBS."

Strain 4.20 *F. sporotrichioides* **F-38** (= T-249; MRC 2032; Sp. 950; ATCC 34911)
This strain was isolated from millet in Hungary (Szathmary et al., 1976, 1977). This strain was referred to as *F. sporotrichioides* 1973-33, obtained from "Marasas, U.S., corn" by Cullen (1981), but Cullen (personal communication, August 1981) stated that this strain was actually obtained from M. Palyusik in Hungary. The source of the same strain (as *F. sporotrichioides* T-249) was also erroneously given as "Marasas, S. Africa" by Cullen (1981).

According to M. Palyusik (personal communication, 16 December 1977), the unspecified strain of *F. sporotrichioides* used in toxicity studies in geese in Hungary (Palyusik & Koplik-Kovacs, 1975a, b; Palyusik et al., 1968, 1974) was *F. sporotrichioides* F-38. Necrotic oral lesions characteristic of trichothecene toxicoses in poultry were reported for the first time in geese fed culture material of this strain (Palyusik et al., 1968). Inhibition of spermatogenesis in ganders fed cultures of this strain was reported by Palyusik et al. (1974). Culture material of this strain on corn was added to the feed as a 10% ration and fed to one-year-old laying Rhineland geese for 10 days (Palyusik & Koplik-Kovacs, 1975a, b). This diet contained 3 ppm of T-2 toxin as determined by TLC. Egg production stopped completely within 10 days. Clinical signs included feed refusal, loss of weight, and extensive necrotic oral lesions. Pathological changes in the geese that died included emaciation, cachexia, hypoplasia of the ovaries and oviducts, dilation of the heart and myocardial degeneration, hemorrhagic enteritis, and necrotic and diphtheroid lesions in the oral, glottal, and pharyngeal mucosa (Palyusik & Koplic-Kovacs, 1975a, b).

In addition to T-2 toxin, the culture material used in the toxicity study in geese by Palyusik & Koplik-Kovacs (1975a, b) was found to contain three other unidentified trichothecenes by TLC. Palyusik & Koplik-Kovacs concluded that "apart from T-2, other factors share the responsibility for the severe lesions found in this study." The presence of T-2 toxin, together with HT-2 toxin, T-2 tetraol, neosolaniol, and zearalenone, was subsequently confirmed by TLC, GLC, and GC-MS in cultures of this strain (as *F. sporotrichioides* F-38) on corn as well as millet incubated at 23 to 27°C for one week, 5 to 8°C for 2 weeks, and 23 to 27°C for one week (Szathmary et al., 1976, 1977). The maximal yields of T-2 toxin, HT-2 toxin, T-2 tetraol, neosolaniol, and zearalenone detected in various fractions of diethyl ether extracts of cultures on millet and corn were 2300, 1000, 300, 500, and 1500 µg/g, respectively. The production by this strain (as *F. sporotrichioides* 1973-33) of T-2 toxin (171 µg/g), but not of HT-2 toxin and T-2 tetraol, has also been reported in cultures on corn incubated at 24°C for 2 weeks followed by 4 weeks at 8°C (Cullen, 1981). This strain also produced zearalenone (123 µg/g) in cultures on rice incubated at 24°C for 2 weeks followed by 4 weeks at 14°C (Cullen, 1981). However, *F. sporotrichioides* T-249, which is another subculture of the original isolate *F. sporotrichioides* F-38 deposited in the ITFRC, produced only 20 µg/g of T-2 toxin and no zearaleone under identical cultural conditions (Cullen, 1981). Different subcultures of the original isolate *F. sporotrichiodes* F-38 evidently can vary in the degree to which they lose or retain the ability to produce high levels of T-2 toxin and zearalenone.

A slant culture of this strain was received on 16 December 1977 from M. Palyusik as "*F. sporotrichioides* F-38 (= ATCC 34911)." A lyophilized subculture of this strain has also been deposited in BAF as *F. sporotrichioides* Sp. 950.

Strain 4.21 *F. sporotrichioides* **F-14** (= T-245; MRC 2031; Sp. 941; ATCC 34914)
This strain was isolated from millet in Hungary (Szathmary et al., 1976, 1977). Cultures of this strain on millet were incubated at 23 to 27°C for one week, 5 to 8°C for 2 weeks, and again at 23 to 27°C for one week; analyses of diethyl ether extracts by TLC, GLC, and GC-MS revealed the presence of T-2 toxin, HT-2 toxin, T-2 tetraol, and neosolaniol, but no zearalenone (Szathmary et al., 1976, 1977). The maximal yields of T-2 toxin, HT-2 toxin, T-2 tetraol, and neosolaniol detected in various fractions of these extracts were 1,510, 400, 50, and 1,480 µg/g, respectively (Szathmary et al., 1976, 1977). This strain has also been reported to produce T-2 toxin at levels up to 107 mg/kg in cultures on corn incubated at 20°C for 10 days followed by 125 days at 10°C (Müller & Thaler, 1981).

A slant culture of this strain was received on 16 December 1977 from M. Palyusik as "*F. sporotrichioides* F-14 (= ATCC 34914)." A lyophilized subculture of this strain has also been deposited in BAF as *F. sporotrichiodes* Sp. 941.

Strain 4.22 *F. sporotrichioides* **F-26** (= T-248; MRC 2033; Sp. 954; ATCC 36318)
This strain (as *F. sporotrichiodes* F-26) was isolated from millet in Hungary (Szathmary et al., 1977). The origin of this strain was not specified by Szathmary et al. (1976), who referred to strain F-26 as *F. poae*. Cultures of this strain on millet were incubated at 23 to 27°C for one week, 5 to 8°C for 2 weeks, and again at 23 to 27°C for one week; analyses

of diethyl ether extracts by TLC, GLC, and GC-MS revealed the presence of T-2 toxin, HT-2 toxin, and neosolaniol, but no zearalenone (Szathmary et al., 1976, 1977). The maximal yields of T-2 toxin, HT-2 toxin, and neosolaniol detected in various fractions of these extracts were 1,560, 500, and 810 µg/g, respectively (Szathmary et al., 1976, 1977).

A slant culture of this strain was received on 16 December 1977 from M. Palyusik as "*F. sporotrichiodes* F-26 (= ATCC 36318)." A lyophilized subculture of this strain has also been deposited in BAF as *F. sporotrichioides* Sp. 954.

Strain 4.23 *F. sporotrichioides* **72187-12** (= T-544; MRC 2625)
 = *F. tricinctum* **72187-10k**

This strain (as *F. tricinctum*) was isolated from cereal grain in Finland during 1972 when one-third of the Finnish cereal crop was heavily contaminated with *Fusarium* species due to abnormal weather conditions (Nummi, 1977; Nummi et al., 1975a, b). Although the strain was not specified in the above publications, the strain used was "Institute of Plant Pathology, Finland, *F. tricinctum* strain 72187-10k," according to T.M. Enari (personal communication, 17 April 1978). According to an erratum published by Ylimäki et al. (1979), *F. tricinctum* 72187-10k should read *F. sporotrichioides* 72187-12.

Cultures of this strain on a mixture of cereals proved to be toxic to mice and to contain T-2 toxin (Nummi, 1977; Nummi et al., 1975a, b). Subsequently Nummi (1977) reported that "4–10 toxins" could be distinguished in extracts of these cultures by TLC and GLC. In addition to T-2 toxin, HT-2 toxin and neosolaniol have been isolated from cultures of this strain on grain mixture incubated at 6°C for 4 months (Ilus et al., 1977, 1981; Niku-Paavola & Nummi, 1977; Niku-Paavola et al., 1977) and a new trichothecene, 4β, 8α-diacetoxy-12,13-epoxytrichothec-9-ene-3α,15-diol, was isolated and characterized by Ilus et al. (1977). The new compound was called T-1 toxin by Ylimäki et al. (1979), who referred to the producing strain as *F. sporotrichioides* 72187-12 (= *F. tricinctum* 72187-10k).

This strain has also been used for the production of T-2 toxin (Linnainmaa et al., 1979; Norppa et al., 1980; Sorsa et al., 1980; Ylimäki et al., 1979), HT-2 toxin, T-1 toxin, and neosolaniol (Ylimäki et al., 1979) for use in toxicological experiments. According to Ylimäki et al. (1979) cultures on a mixure of oats, wheat, and barley (1:1:1) were first incubated at 20 to 24°C for 2 weeks, then at 5°C for 2 days, and thereafter again at room temperature. T-2 toxin was isolated after 6 weeks' cultivation and the other toxins after 4 months.

A slant culture of this strain was received on 1 October 1981 from A. Ylimäki as "*F. sporotrichioides* 72187-12."

Strain 4.24 *F. moniliforme* **111** (= *F. tricinctum*) **NRRL 3197** (= T-494; MRC 2323)

An unspecified strain of *F. moniliforme* was reported by Prentice et al. (1959) to produce emetic material in Richard's solution. This strain was specified as *F. moniliforme* 111 by Prentice & Dickson (1968), who indicated that this strain was obtained from the Department of Plant Pathology, University of Wisconsin, and had previously been shown by Armolik et al. (1956) to inhibit the germination of barley. The latter authors did not specify *F. moniliforme* 111, but worked with a strain of *F. moniliforme* isolated from deteriorated barley in the U.S.A. According to Prentice & Sloey (1960), *F. moniliforme* 111 increased alpha-amylase activity, but not that of beta-amylase, in barley malt. However, Sloey & Prentice (1962) found that *F. moniliforme* 111 increased the diastatic power and beta-amylase activity, but not alpha-amylase activity as in the previous tests. The latter authors suggested that "the characteristics of the fungi may have changed over the prolonged storage period," thus accounting for the discrepancy.

Shake cultures of *F. moniliforme* 111 in Richard's solution were incubated at room temperature for 12 to 40 days, the culture filtrate extracted with diethyl ether, and the extracts injected into the wing vein of pigeons (Prentice & Dickson, 1968). Two compounds that caused prolonged emesis in pigeons were partially purified from these cultures, but were not chemically characterized. One of these compounds exhibited a toxic effect in addition to its emetic activity and caused the death of the pigeons within 12 hours after injection (Prentice & Dickson, 1968). In contrast to these findings, Ellison & Kotsonis (1973) could not detect any emetic activity in pigeons with ethyl acetate extracts of either corn or Richard's medium cultures of this strain (as *F. moniliforme* NRRL 3197). These authors concluded that this strain had lost its ability to produce emetic material during storage. Ueno et al. (1972b, 1973a) reported that *F. moniliforme* NRRL 3197 was not toxic to mice, did not inhibit protein synthesis in rabbit reticulocytes, and did not produce any known trichothecenes that could be detected by TLC.

Cultures of *F. moniliforme* NRRL 3197 (= Prentice and Dickson No. 111) on corn incubated at 28°C for 13 days were found to induce a feed refusal response in pigs, but no deoxynivalenol, T-2 toxin, HT-2 toxin, or acetyl T-2 toxin could be detected by TLC (Vesonder et al., 1977b). However, subsequent analysis of GLC-MS revealed the presence of deoxynivalenol (1.5 µg/g) and T-2 toxin (33 µg/g) in this culture material (Vesonder et al., 1981a). This is the only available report of deoxynivalenol production by *F. sporotrichioides*. The level of production (1.5 µg/g) is very low, however, and needs to be confirmed.

The taxonomic position of *F. moniliforme* NRRL 3197 was questioned by Vesonder et al. (1981) because it produced deoxynivalenol and T-2 toxin and because "it gave a slow-growing yellowish colony that produced few macroconidia, few chlamydospores, but a broad range in shape of microconidia." A critical examination of the cultures of *F. moniliforme* NRRL 3197 at NRRL, including lyophilized material dating back to receipt of this strain, revealed that NRRL 3197 is "a variant of the species *F. tricinctum* (Cda.) Sacc. and not a strain of *F. moniliforme*" (Vesonder et al., 1981a).

A lyophilized culture of this strain of *F. sporotrichioides* was received on 9 March 1981 from NRRL as "*F. tricinctum* NRRL 3197." In view of the conflicting reports on the toxigenicity of this strain discussed above, and because it seems highly unlikely that a strain of *F. sporotrichioides* would have been identified as *F. moniliforme*, it is possible that the original Prentice & Dickson strain 111 was a mixture of *F. moniliforme* and *F. sporotrichioides*.

Strain 4.25 *F. poae* **2518** (= *F. tricinctum*) **NRRL 3287** (= T-423; MRC 1786)

This strain (as *F. poae* 2518) was obtained from the Canada Department of Agriculture, Winnipeg, according to Prentice & Dickson (1968). The culture was presumably identified by W.L. Gordon, but the exact source is unknown to us.

Treatment of barley with cultures of this strain during malting resulted in increased wort nitrogen content (Prentice & Sloey, 1960), increased alpha-amylase activity, and in "gushing" or excess foaming in beer (Sloey & Prentice, 1962).

Diethyl ether extracts or shake cultures of *F. poae* 2518 in Richard's solution, incubated at room temperature for 12 to 40 days, caused prolonged emesis upon injection in the wing vein of pigeons (Prentice & Dickson, 1968; Prentice et al., 1959). The emetic metabolite was not chemically identified.

The emetic activity of this strain (as *F. poae* NRRL 3287) was re-examined by Ellison &

Kotsonis (1973). Extracts of cultures grown either on corn at 8°C (but not at 25°C) or in Richard's solution at room temperature caused emesis in pigeons at nonlethal concentrations under conditions of oral and intravenous administration. The emetic metabolite in both the corn and liquid cultures was identified as T-2 toxin, and 4 mg/800 g and 49 mg/7 ℓ of pure T-2 toxin was isolated from them, respectively (Ellison & Kotsonis, 1973). The production of both T-2 toxin and HT-2 toxin by this strain (as *F. poae* = *F. tricinctum* NRRL 3287) in shake culture in Richard's solution at room temperature was confirmed by a combination of TLC and fluorodensitometry by Kotsonis & Ellison (1975). These authors also demonstrated that T-2 toxin is a precursor of HT-2 toxin and that T-2 toxin can be converted to HT-2 toxin by cultures of this strain. Subsequently Kotsonis et al. (1975b) reported that this culture material also contained acetyl T-2 toxin in addition to T-2 toxin and HT-2 toxin. Pure acetyl T-2 toxin (7 mg) was isolated from 7 ℓ of culture material in Richard's solution incubated at room temperature for 4 weeks.

Extracts of cultures of this strain in PSC medium incubated at 26°C for 12 days were lethal to mice upon intraperitoneal injection and caused marked hemorrhage in the small intestine and radiomimetic pathological changes in the rapidly dividing cells in the small intestine, spleen, thymus, testes, and bone marrow (Ueno et al., 1972b), 1973a). These extracts also inhibited protein synthesis in rabbit reticulocytes and were shown by TLC to contain T-2 toxin, HT-2 toxin, and neosolaniol (Ueno et al., 1972b, 1973a).

According to Joffe & Palti (1975), *F. poae* NRRL 3287 exhibited "extremely strong toxicity to rabbit skin." However, Joffe (1978b) and Joffe & Yagen (1977) reported that extracts of cultures of this strain on wheat incubated at 12°C for 21 days caused only a slight dermotoxic reaction on rabbit skin and contained a low level (1 mg/10 g of grain) of T-2 toxin.

Cultures of this strain on corn incubated at 28°C for 13 days caused a marked feed refusal response in pigs, but TLC analyses of extracts of the refused corn were negative for T-2 toxin, HT-2 toxin, acetyl T-2 toxin, and fusarenon-X (Vesonder et al., 1977b). However, subsequent analyses of the refused corn by GLC-MS revealed the presence of 30µg of T-2 toxin/g of corn (Vesonder et al., 1981a).

In summary, this strain has been reported to produce the following four trichothecenes: 1) *T-2 toxin* (Ellison & Kotsonis, 1973; Joffe, 1978b; Joffe & Yagen, 1977; Kotsonis & Ellison, 1975; Ueno et al., 1972b, 1973a; Vesonder et al., 1981a), but negative results for T-2 toxin production by this strain have been obtained by TLC analyses of corn cultures by Burmeister et al. (1972—as *F. tricinctum* NRRL 3287) and Vesonder et al. (1977b); 2) *HT-2 toxin* (Kotsonis & Ellison, 1975; Ueno et al., 1972b, 1973a); 3) *Acetyl T-2 toxin* (Kotsonis et al., 1975b); 4) *Neosolaniol* (Ueno et al., 1972b, 1973a).

A lyophilized culture of this strain was received on 23 October 1979 from NRRL as "*F. tricinctum* NRRL 3287." This culture produces well-developed polyphialides and spindle-shaped as well as pear-shaped microconidia. We consider this to be a typical culture of *F. sporotrichioides,* and it is not clear why it was identified as *F. poae* by Joffe & Palti (1975).

Strain 4.26 *Fusarium* **sp. YN-13** (= T-432; MRC 1902)
This strain, as "*F. roseum* isolate No. 2," was isolated from corn from Yankton, South Dakota, USA (Eugenio et al., 1970a). According to C.J. Mirocha (personal communication, 17 March 1980), this is the same isolate that has been referred to as *Fusarium* YN-13 by Christensen et al. (1965) and as *F. sporotrichioides* (= *F. tricinctum*) YN-13 by Joffe & Palti (1975).

According to Christensen et al. (1965), cultures of this strain (as *Fusarium* YN-13) on corn incubated at 20 to 25°C for 3 weeks, followed by 2 weeks at 12°C, caused marked increases in uterine weight of virgin, weanling, female rats.

This strain (as *F. roseum* isolate No. 2) was the second highest producer of zearalenone among 43 isolates of *F. roseum* tested by Eugenio et al. (1970a) and produced 1420 ppm of zearalenone in cultures on polished rice incubated for 2 weeks at 22 to 25°C, followed by eight weeks at 8 to 12°C. Eugenio et al. (1970a) also reported that "30 single-ascospore lines of isolate No. 2" produced from 140 to 4130 ppm of zearalenone on autoclaved corn.

Joffe & Palti (1975) identified isolate YN-13 received from C.J. Mirocha as *F. sporotrichioides;* extracts of cultures of this strain on wheat incubated at 12°C for 21 days were found to be slightly dermotoxic to rabbit skin and to contain 2.0 mg of T-2 toxin/10 g of culture material by Joffe (1978b) and Joffe & Yagen (1977).

A culture preserved on silica gel was received on 17 March 1980 from C.J. Mirocha as *"Fusarium* Yankton, *Fusarium* YN-13." We agree with Joffe & Palti (1975) that this isolate is *F. sporotrichioides,* but we cannot accept that this is the same strain that was found to produce zearalenone by Eugenio et al. (1970a) because the perfect state is unknown in *F. sporotrichioides.* We must therefore conclude that either "*F. roseum* isolate 2" of Eugenio et al. (1970a) was a mixture of an ascospore-producing strain of *F. graminearum* and *F. sporotrichioides* YN-13, or the indication by C.J. Mirocha (personal communication, March 17, 1980) that "*F. roseum* isolate 2" and *F. sporotrichioides* YN-13 are the same strain was an error. In either case, there is no proof that *F. sporotrichioides* YN-13 (= T-432, MRC 1902) produces zearalenone.

Strain 4.27   *F. tricinctum* No. 238 (= T-406; MRC 1657)

This strain was isolated from cranberry in Wisconsin, USA (Bamburg, 1968; Marasas, 1969). Extracts of cultures of this strain in Gregory's medium incubated at 8°C for 14 days were dermotoxic to rat skin, and GLC analysis revealed the presence of 178 mg/ℓ of T-2 toxin (Bamburg, 1968; Marasas, 1969). This strain also produced the toxin butenolide that could be detected by TLC in extracts of cultures on Sabouraud's agar incubated at 15°C for 4 weeks (Marasas, 1969).

A slant culture of this strain was obtained in 1965 from E.B. Smalley as "*F. tricinctum* No. 238." A subculture has been maintained by lyophilization at MRC since 1970 as *F. sporotrichioides* MRC 1657. Subcultures of this lyophilized material have been deposited in the ITFRC as *F. sporotrichioides* T-406.

Strain 4.28   *F. tricinctum* No. 347 (= T-407; MRC 1659)

This strain was isolated from turf grass in Wisconsin, USA (Bamburg, 1968; Marasas, 1969). Extracts of cultures of this strain in Gregory's medium incubated at 8°C for 14 days were dermotoxic to rat skin, and GLC analysis revealed the presence of 61.7 mg/ℓ of T-2 toxin (Bamburg, 1968; Marasas, 1969).

A slant culture of this strain was obtained in 1965 from E.B. Smalley as "*F. tricinctum* No. 347." A subculture has been maintained by lyophilization at MRC since 1970 as *F. sporotrichioides* MRC 1659. Subcultures of this lyophilized material have been deposited in the ITFRC as *F. sporotrichioides* T-407.

Strain 4.29   *F. tricinctum* No. 209 (= T-408; MRC 1660)

This strain was isolated from sorghum in Wisconsin, USA (Bamburg, 1968; Marasas, 1969). Extracts of cultures of this strain in Gregory's medium incubated at 8°C for 14 days were dermotoxic to rat skin, and GLC analysis revealed the presence of 214 mg/ℓ of T-2 toxin (Bamburg, 1968; Marasas, 1969). This strain was the highest producer of T-2 toxin of the 30 "*F. tricinctum*" isolates in the study by Bamburg (1968) and Marasas (1969). This strain also produced the toxin butenolide that could be detected by TLC in extracts of cultures on Sabouraud's agar incubated at 15°C for 4 weeks (Marasas, 1969).

A slant culture of this strain was obtained in 1965 from E.B. Smalley as "*F. tricinctum* No. 209." A subculture has been maintained by lyophilization at MRC since 1970 as *F. sporotrichioides* MRC 1660. Subcultures of this lyophilized material have been deposited in the ITFRC as *F. sporotrichioides* T-408.

Strain 4.30 *F. tricinctum* No. 348 (= T-411; MRC 1663)

This strain was isolated from field corn in Wisconsin, USA (Bamburg, 1968; Marasas, 1969). Extracts of cultures of this strain in Gregory's medium incubated at 8°C for 14 days were dermotoxic to rat skin, and GLC analysis revealed the presence of 212 mg/ℓ of T-2 toxin (Bamburg, 1968; Marasas, 1969).

A slant culture of this strain was obtained in 1965 from E.B. Smalley as "*F. tricinctum* No. 348." A subculture has been maintained by lyophilization at MRC since 1970 as *F. sporotrichioides* MRC 1663. Subcultures of this lyophilized material have been deposited in the ITFRC as *F. sporotrichioides* T-411.

Strain 4.31 *F. tricinctum* ITH-194 (= T-570; MRC 2523)

This strain was isolated from field corn ear rot in Lombardy, Italy (A. Bottalico, personal communication, 19 October 1981). In a study of T-2 toxin production by 47 isolates of *Fusarium* spp. from cereals in Italy, only one isolate of *F. tricinctum* produced T-2 toxin that could be detected by TLC (Bottalico, 1977c). This isolate produced 40 μg/g of T-2 toxin in cultures on corn incubated at 24°C for 15 days followed by 60 days at 10°C. It is not certain whether this T-2 toxin-producing isolate of *F. tricinctum* referred to by Bottalico (1977c) is the same strain as *F. tricinctum* ITH-194 that, "cultured on corn, was found to be a high T-2 toxin producer" (A. Bottalico, personal communication, 19 October 1981).

A slant culture of this strain was received on 19 October 1981 from A. Bottalico as "*F. tricinctum* ITH-194."

Strain 4.32 *F. tricinctum* T-340 (= T-340; MRC 2913)

This strain was obtained from "Booth, England," according to Cullen (1981), who identified it as "*F. tricinctum* sensu Booth." Cultures of this strain on corn incubated at 24°C for 2 weeks followed by 4 weeks at 8°C contained 5950 μg of T-2 toxin/g of dry culture as determined by GLC (Cullen, 1981). This strain was the highest producer of T-2 toxin among the 36 strains of *Fusarium* species in the section Sporotrichiella tested by Cullen (1981).

Substantial increases over the amount of T-2 toxin produced by this strain in cultures on corn (5.9 mg/g) or in surface fermentations in liquid media such as Gregory's (60.0 mg/ℓ) or PSC (27.0 mg/ℓ) were obtained with certain liquid media absorbed into vermiculite (Cullen, 1981). Maximal yields (714 mg/ℓ) were obtained after 24 days incubation at 19°C in a vermiculite-supported, nutritionally complex medium containing soya meal, corn steep liquor, and glucose.

A slant culture of this strain was received from CMI as "*F. tricinctum* CMI 175449." This culture has well developed polyphialides and pear- as well as spindle-shaped microconidia, and we consider it to be a typical representative of *F. sporotrichioides*.

Strain 4.33 *F. tricinctum* T-285 (= T-285; MRC 1709)

This strain was isolated from wheat in the United States according to Cullen (1981), who identified it as "*F. tricinctum* sensu Booth." Cultures of this strain on corn incubated at

24°C for 2 weeks followed by 4 weeks at 8°C contained 1630 µg of T-2 toxin/g of dry culture as determined by GLC (Cullen, 1981). This strain was the third highest producer of T-2 toxin among the 36 strains of *Fusarium* species in the section Sporotrichiella tested by Cullen (1981).

A lyophilized culture of this strain of *F. sporotrichioides* was received on 21 September 1979 from FRC as "*F. sporotrichioides* var. *tricinctum* T-285. Isolated by A.Z. Joffe from wheat, Nebraska."

Strain 4.34   *F. tricinctum* T-288 (= T-288; MRC 1708)
This strain was isolated from rye in the United States according to Cullen (1981), who identified it as "*F. tricinctum* sensu Booth." Cultures of this strain on corn incubated at 24°C for 2 weeks followed by 4 weeks at 8°C contained 510 µg of T-2 toxin/g of dry culture as determined by GLC (Cullen, 1981).

A lyophilized culture of this strain of *F. sporotrichioides* was received on 21 September 1979 from FRC as "*F. sporotrichioides* var. *tricinctum* T-288. Isolated by A.Z. Joffe from rye, North Dakota."

Strain 4.35   *F. tricinctum* M-350 (= T-625; MRC 1606)
This strain (as *F. tricinctum* T-350) was obtained from "Marasas, U.S., corn" according to Cullen (1981), who identified it as "*F. tricinctum* sensu Booth." However, Cullen (personal communication, August 1981) stated that this culture (as *F. tricinctum* M-350) was originally obtained from A.T. Bolton. Ottawa, Canada. Cultures of this strain on corn incubated at 24°C for 2 weeks followed by 4 weeks at 8°C contained 306 µg of T-2 toxin/g of dry culture as determined by GLC (Cullen, 1981). Cultures on rice incubated at 24°C for 2 weeks followed by 4 weeks at 14°C contained 203 µg/g of zearalenone.

A soil culture of this strain was received on 5 May 1978 from E.B. Smalley as "*F. tricinctum* No. 350."

Strain 4.36   *F. tricinctum* F-696 (= T-522; MRC 2489)
This strain was isolated from a grain elevator in the United States according to Cullen (1981), who identified it as "*F. tricinctum* sensu Booth." Cultures of this strain on corn incubated at 24°C for 2 weeks followed by 4 weeks at 8°C contained 180 µg of T-2 toxin/g of dry culture as determined by GLC (Cullen, 1981).

A soil culture of this strain was received on 25 August 1981 from D. Cullen as "*F. tricinctum* F-696."

Strain 4.37   *F. tricinctum* F-14 (= T-521; MRC 2488)
This strain was isolated from a grain elevator in the United States according to Cullen (1981), who identified it as "*F. tricinctum* sensu Booth." Cultures of this strain on corn incubated at 24°C for 2 weeks followed by 4 weeks at 8°C contained 97 µg of T-2 toxin/g of dry culture as determined by GLC (Cullen, 1981). Cultures on rice incubated at 24°C for 2 weeks followed by 4 weeks at 14°C contained 21 µg/g of zearalenone.

A soil culture of this strain was received on 25 August 1981 from D. Cullen as "*F. tricinctum* F-14."

Strain 4.38   *F. tricinctum* T-336 (= T-336; MRC 2912)
This strain was isolated from rye in the United States according to Cullen (1981), who identified it as "*F. tricinctum* sensu Booth." Cultures of this strain on corn incubated at

24°C for 2 weeks followed by 4 weeks at 8°C contained 88 µg of T-2 toxin/g of dry culture as determined by GLC (Cullen, 1981). Cultures on rice incubated at 24°C for 2 weeks followed by 4 weeks at 14°C contained 14 µg/g of zearalenone.

A lyophilized culture of this strain was received from FRC as "*F. sporotrichioides* var. *tricinctum* T-336. Isolated by A.Z. Joffe from rye, USA."

Strain 4.39 *F. sporotrichioides* T-341 (= T-341; MRC 2914)

This strain was obtained from "Booth, England" according to Cullen (1981), who identified it as "*F. sporotrichioides*." Cultures of this strain on corn incubated at 24°C for 2 weeks followed by 4 weeks at 8°C contained 1835 µg of T-2 toxin/g of dry culture as determined by GLC (Cullen, 1981). This strain was the second highest producer of T-2 toxin among the 36 strains of *Fusarium* species in the section Sporotrichiella tested by Cullen (1981). Cultures on rice incubated at 24°C for 2 weeks followed by 4 weeks at 14°C contained 100 µg/g of zearalenone.

A slant culture of this strain was received from CMI as "*F. sporotrichioides* IMI 235359."

Strain 4.40 *F. sporotrichioides* T-290 (T-290; MRC 2910)

This strain was isolated from barley in the United States according to Cullen (1981), who identified it as *F. sporotrichioides*. Cultures of this strain on corn incubated at 24°C for 2 weeks followed by 4 weeks at 8°C contained 289 µg of T-2 toxin/g of dry culture as determined by GLC (Cullen, 1981).

A lyophilized culture of this strain was received from FRC as "*F. sporotrichioides* T-290."

Strain 4.41 *F. sporotrichioides* T-286 (= T-286; MRC 2909)

This strain was isolated from wheat in the United States according to Cullen (1981), who identified it as *F. sporotrichioides*. Cultures of this strain on corn incubated at 24°C for 2 weeks followed by 4 weeks at 8°C contained 177 µg of T-2 toxin/g of dry culture as determined by GLC (Cullen, 1981).

A lyophilized culture of this strain was obtained from FRC as "*F. sporotrichioides* T-286."

Strain 4.42 *F. sporotrichioides* T-306 (= T-306; MRC 2911)

This strain was isolated from wheat in the United States according to Cullen (1981), who identified it as *F. sporotrichioides*. Cultures of this strain on corn incubated at 24°C for 2 weeks followed by 4 weeks at 8°C contained 65 µg of T-2 toxin/g of dry culture as determined by GLC (Cullen, 1981).

A lyophilized culture of this strain was received from FRC as "*F. sporotrichioides* T-306."

Strain 4.43 *F. sporotrichioides* A-1 (= T-520; MRC 2487)

This strain was isolated from an aquatic weed in the United States according to Cullen (1981), who identified it as *F. sporotrichioides*. Cultures of this strain on corn incubated at 24°C for 2 weeks followed by 4 weeks at 8°C contained 20 µg of T-2 toxin/g of dry culture as determined by GLC (Cullen, 1981).

A soil culture of this strain was received on 25 August 1981 from D. Cullen as "*F. sporotrichioides* A-1." This is a badly degenerated, pionnotal culture.

Fusarium chlamydosporum Wollenw. & Reinking

Perfect State: Unknown

Incidence and Distribution

Fusarium chlamydosporum has a higher optimum temperature of growth than the other species in the section Sporotrichiella (Seemüller, 1968) and occurs as a saprophyte in soil and on other substrates in tropical and subtropical areas such as Australia, Central America, India, Iraq, Namibia, South Africa, and the Sonoran Desert in the United States (Booth, 1971; Doidge, 1938; Domsch et al., 1980; Meyer & Frank, 1979; Seemüller, 1968; Wollenwever & Reinking, 1935). It is also commonly associated with seeds of crops such as beans, millet, peanuts, and sorghum in the warmer areas of the world such as Australia, India, and southern Africa (Connole et al., 1981; Domsch et al., 1980; Rabie et al., 1978).

Association with Human and/or Animal Diseases

Onyalai. One of the most toxic species of fungi that was readily isolated from millet (*Pennisetum typhoides* [Burm.] Staph. ex Hubb.) obtained from the households of patients in Namibia (South West Africa) suffering from the hemorrhagic disease onyalai was *F. fusarioides* (Frag. & Cif.) Booth (Rabie et al., 1978). We have identified one of these toxic isolates (MRC 35) as *F. chlamydosporum* (Strain 4.44). Cultures of this strain were toxic to ducklings and rats and the toxic compound was identified as moniliformin (Rabie et al., 1978).

The characteristic pathological changes of onyalai, including hemorrhage and thrombocytopenia (Rabie et al., 1975), were not reported in rats fed toxic culture material of *F. chlamydosporum* (= *F. fusarioides*) MRC 35 (Strain 4.44) for a period of 15 months (Rabie et al., 1978). Moniliformin has also not been reported to cause these changes in experimental animals (Kriek et al., 1977).

It is concluded that there is no experimental evidence available that *F. chlamydosporum* is involved in the etiology of onyalai except for the fact that it has been isolated from millet associated with the disease.

Toxicity to Experimental Animals

Chickens. Extracts of cultures in liquid medium of *F. tricinctum* NRRL A-23377 were assayed for toxicity by crop incubation in day-old De Kalb cockerels (Lansden et al., 1978). We have identified this strain, which was isolated from field-loss peanuts in the United States, as *F. chlamydosporum* (Strain 4.46). A new trichothecene, neosolaniol monoacetate, was isolated from cultures of this strain by means of the cockerel bioassay (Lansden et al., 1978).

Ducklings. Cultures on corn of 13 isolates of *F. fusarioides* from various sources have been reported to be toxic to day-old Pekin ducklings (Rabie et al., 1978). We have identified one of these toxic strains (MRC 35), isolated from millet obtained from households of patients in Namibia (South West Africa) suffering from onyalai disease (see onyalai, above), as *F. chlamydosporum* (Strain 4.44). The sole toxin isolated from this strain, as well as four other strains toxic to ducklings, was moniliformin (Rabie et al., 1978).

Rabbits. According to Joffe (1973a), extracts of cultures of one of six unspecified isolates of "*F. sporotrichioides* Sherb. var. *chlamydosporum* (Wollenw. & Reinking) Joffe" was highly dermotoxic to rabbit skin and caused severe leukocytorrhea and necrosis. Two other isolates caused only a slight reddening of the skin and/or incipient edema, while the remaining isolates caused no skin reaction. Joffe (1974b) reported that *F. sporotrichioides* var. *chlamydosporum* No. 4337 from wheat in Canada was dermotoxic to rabbit skin. We have not seen this strain, which was presumably identified by W.L. Gordon, but consider it unlikely that *F. chlamydosporum* occurs commonly in Canada. In fact, Seemüller (1968) has identified two isolates of "*F. chlamydosporum*" obtained from W.L. Gordon in Canada as *F. equiseti*. The mycotoxin(s) produced by the above dermotoxic strains have not been identified.

Rats. Corn cultures of *F. chlamydosporum* (= *F. fusarioides*) MRC 35 (Strain 4.44), which was isolated from millet obtained from households of patients in Namibia (South West Africa) suffering from onyalai disease (see onyalai, above), have been reported to be toxic to weanling male BD IX rats (Rabie et al., 1978). Diets containing 50% moldy material caused the death of four of eight rats within 10 days, and the weight gains of the survivors were markedly reduced. Although none of the rats that received lower levels of culture material died, diets containing 25% and 10% moldy meal resulted in marked reductions in weight gain (Rabie et al., 1978).

Mycotoxins Produced

Moniliformin. The production of moniliformin by five isolates of *F. fusarioides* in cultures on corn incubated at 25°C for 21 days was reported by Rabie et al. (1978). We have identified one of these strains (MRC 35), isolated from millet obtained from households of patients in Namibia (South West Africa) suffering from onyalai disease (see onyalai, above), as *F. chlamydosporum* (Strain 4.44). Culture material of this strain on corn incubated at 25°C for 21 days was toxic to ducklings and rats and was shown to contain 0.20 g/kg of moniliformin by TLC. The other four strains produced from 0.42 to 0.84 g/kg of moniliformin.

Maximal yields of moniliformin by *F. chlamydosporum* (= *F. fusariodes*) MRC 35 (Strain 4.44) were obtained by incubating corn cultures at 31°C for 7 days (Rabie et al., 1978). The moniliformin content of this culture material was determined as 0.80 g/kg by TLC and as 0.68 g/kg by a UV spectrophotometric method. In another experiment on the effect of incubation time at 31°C on moniliformin production by this strain, it was found that production started on day three and reached a maximum (950 mg/kg) after 12 to 18 days (Rabie et al., 1978).

F. chlamydosporum is the only member of the section Sporotrichiella that has been reported to produce moniliformin.

Neosolaniol Monoacetate. This new trichothecene was isolated from cultures of *F. tricinctum* NRRL A-23377 in liquid medium incubated at 27°C for 32 days (Lansden et al., 1978). We have identified this strain (received as *F. tricinctum* NRRL 6358), which was isolated from field-loss peanuts in the United States, as *F. chlamydosporum* (Strain 4.46). The yield of pure neosolaniol monoacetate obtained from this stain by Lansden et al. (1978) was 250 mg/13 ℓ of culture material.

This is the only strain of *F. chlamydosporum* that has been reported to produce a trichothecene.

T-2 Toxin. None of five strains of *F. fusarioides* tested by Cullen (1981) produced T-2 toxin in cultures on corn incubated at 24°C for 2 weeks followed by 4 weeks at 8°C. We have identified two of these strains that do not produce T-2 toxin (i.e., *F. fusarioides* T-227 and T-428) as *F. chlamydosporum* (Strains 4.44 and 4.45).

Zearalenone. None of five strains of *F. fusarioides* tested by Cullen (1981) produced zearalenone in cultures on rice incubated at 24°C for 2 weeks followed by 4 weeks at 14°C. We have identified two of these strains that do not produce zearalenone (i.e., *F. fusarioides* T-227 and T-428) as *F. chlamydosporum* (Strains 4.44 and 4.45).

Toxigenic Strains in the ITFRC

The following *Fusarium* strains listed under *F. chlamydosporum* in Table 4.1 have been identified as *F. chlamydosporum* according to Nelson et al. (1983).

Strain 4.44.   *F. fusarioides* MRC 35 (= T-428; MRC 35).
This strain was isolated from millet (*Pennisetum typhoides* (Burm.) Staph. ex Hubb.) obtained from households of patients in Namibia (South West Africa) suffering from the hemorrhagic disease onyalai (Rabie et al., 1978).

Culture material of this strain on corn incubated at 25°C for 21 days was toxic to ducklings and caused the death of 4 out of 4 ducklings in an average time of 4 days (Rabie et al., 1978). Cultures incubated at three different temperatures (20, 25, and 31°C) for three different incubation periods (7, 12, and 21 days) were all lethal to ducklings, but less of the material produced at higher temperatures and with longer incubation periods was consumed.

The toxicity of corn cultures incubated at 25°C for 21 days was also tested in rats by incorporating various levels of moldy meal into the diet (Rabie et al., 1978). Diets containing 50% moldy material caused the death of four out of eight rats within 10 days, and the weight gains of the survivors were severely reduced. Although none of the rats that received lower levels of culture material died, diets containing 25 and 10% moldy meal resulted in marked reductions in weight gain.

The sole mycotoxin identified in the culture material toxic to ducklings and rats was moniliformin (Rabie et al., 1978). The level of moniliformin detected by TLC in the culture material incubated at 25°C for 21 days was 0.20 g/kg. Maximal yields of moniliformin by this strain were obtained by incubating corn cultures at 31°C for 7 days. The moniliformin content of this culture material was determined as 0.80 g/kg by TLC and 0.68 g/kg by a UV spectrophotometric method. In another experiment it was found that moniliformin production by this strain at 31°C started on day three and reached a maximum (950 mg/kg) after 12 to 18 days (Rabie et al., 1978).

According to Cullen (1981), this strain (as *F. fusarioides* T-248, but this should read T-428) did not produce T-2 toxin in cultures on corn incubated at 24°C for 2 weeks followed by 4 weeks at 8°C. Cultures on rice incubated at 24°C for 2 weeks followed by 4 weeks at 14°C were also negative for zearalenone.

A lyophilized culture of this strain was received on 22 April 1980 from C.J. Rabie as "*F. fusarioides* MRC 35."

Strain 4.45   *F. fusarioides* T-227 (= T-227; MRC 117)
This strain was obtained from "Marasas, S. Africa, bean hay" according to Cullen (1981), who identified it as *F. fusarioides*. Cultures on corn incubated at 24°C for 2 weeks followed by 4 weeks at 8°C did not contain T-2 toxin that could be detected by GLC

(Cullen, 1981). Cultures on rice incubated at 24°C for 2 weeks followed by 4 weeks at 14°C were also negative for zearalenone.

A lyophilized culture of this strain was received on 15 December 1977 from W.F.O. Marasas as "*F. chlamydosporum* MRC 117."

Strain 4.46   *F. tricinctum* NRRL 6358 (= NRRL A-23377) (= T-502; MRC 2486)

This strain (as *F. tricinctum* NRRL A-23377) was isolated from field-loss peanuts in the United States (Lansden et al., 1978). Cultures in liquid medium were incubated at 27°C for 32 days and purified chloroform extracts assayed for toxicity in day-old cockerels (Lansden et al., 1978). A new trichothecene named neosolaniol monoacetate was isolated and characterized from this culture material at a yield of 250 mg/13 ℓ (Lansden et al., 1978). This is the only strain of *F. chlamydosporum* that has been reported to produce a trichothecene.

A lyophilized culture of this strain was received on 29 March 1978 from NRRL as "*F. tricinctum* NRRL 6358 = NRRL A-23377." This culture was not viable, however, and a slant culture was received on 6 July 1981 from NRRL as "*F. tricinctum* NRRL 6358." This culture is badly degenerated and rapidly becomes pionnotal. It produces carmine red pigment, long chains of chlamydospores that become verrucose, some spindle-shaped microconidia, and a few abnormal 3-septate macroconidia. We were unable to see tyical polyphialides and it was impossible to identify this culture with certainty. We have tentatively identified it as *F. chlamydosporum*. According to the ATCC Catalogue of Strains, Fifteenth Ed., 1982, this culture has also been deposited in ATCC as "*F. tricinctum* ATCC 36939 = NRRL 6358. J. Dorner F6A (NRRL A-23377)." We have not seen this culture and it is not known whether it is in better condition for a positive identification.

Fusarium poae (Peck) Wollenw.

Perfect State: Unknown

Incidence and Distribution

Fusarium poae is predominantly a fungus of temperate areas and is widespread in soils and as a saprophyte or weak parasite on numerous hosts in Europe, Canada and the northern United States, and the USSR (Booth, 1971; CMI Descriptions of Pathogenic Fungi and Bacteria No. 308, 1971; Domsch et al., 1980; Meyer & Frank, 1979; Seemüller, 1968; Wollenweber & Reinking, 1935). As a weak parasite *F. poae* has been implicated in turf disease and head blight of corn and in association with the mite *Siteroptes graminum* in diseases such as silver top of grasses and bud rot of carnations in Europe and North America (Booth, 1971; Seemüller, 1968; Wollenweber & Reinking, 1935). It has also repeatedly been isolated from cereal grains in temperate areas (Gordon, 1952; Flannigan, 1970; Hewett, 1967; Marasas et al., 1979b; Seemüller, 1968; Ylimäki 1981; Ylimäki et al., 1979).

In contrast to *F. sporotrichioides,* which appears to be restricted to temperate and cold areas, and *F. chlamydosporum,* which occurs almost exclusively in tropical and subtropical areas, *F. poae* has a much wider distribution. It has been isolated from cereals that overwintered under snow in Canada (Mills & Frydman, 1980; Scott et al., 1980) and the USSR (Bilai, 1970; Joffe, 1960a, b, 1962, 1965, 1969, 1971, 1973a, 1974b, c, 1978b;

Joffe & Palti, 1974, 1975; Joffe & Yagen, 1977), but also occurs in soils and as a saprophyte on other substrates in warmer areas such as Australia, India, Iraq, and South Africa (Booth, 1971; Domsch et al., 1980).

Association with Human and/or Animal Diseases

Alimentary Toxin Aleukia (ATA). Overwintered cereal grains colonized by *F. poae* and *F. sporotrichioides* have been implicated in outbreaks of ATA in humans in the USSR, and this syndrome is discussed in detail under *F. sporotrichioides* (see *F. sporotrichioides*, ATA).

Cultures of *F. poae* (see also *F. sporotrichioides*, toxicity to experimental animals) isolated from overwintered cereals associated with outbreaks of ATA proved to be highly toxic to a variety of experimental animals. These isolates of *F. poae* have been reported to cause at least some of the characteristic clinical signs and pathological lesions of ATA in a variety of experimental animals including cats, cattle, dogs, guinea pigs, horses, mice, poultry, rabbits, and rats (Bilai, 1970; Joffe, 1960a, b, 1962, 1969, 1971, 1974b, c, 1978a, b; Joffe & Yagen, 1977, 1978; Kurmanov, 1978a, b; Rubinstein, 1960b; Rubinstein et al., 1967; Sarkisov, 1960; Schoental & Joffe, 1974; Schoental et al., 1976, 1979; Yagen & Joffe, 1976; Yagen et al., 1977). It appears from the papers just cited that isolates of *F. poae* (= *F. sporotrichiella* var. *poae*) from overwintered cereals in the USSR are approximately as toxic to experimental animals as isolates of *F. sporotrichioides* (= *F. sporotrichiella* var. *sporotrichioides*) are. However, other Russian investigations have stated that most isolates of *F. sporotrichoides* from the USSR are toxic to highly toxic, while *F. poae* exists as weakly toxic to non-toxic strains (Kvashnina, 1976, 1978, 1979; Rubinstein, 1953, 1960b). In our experience, *F. poae* has a very low degree of toxicity compared to *F. sporotrichioides*, and many isolates are weakly toxic to non-toxic (see Strains 4.49, 4.50, 4.52, 4.53). We have seen only one of the toxic isolates of *F. poae* from the USSR referred to in the above publications, and that isolate, *F. tricinctum* NRRL 3511, we have identified as *F. sporotrichioides* (Strain 4.2). In the absence of cultures it is not possible to determine whether these toxic isolates were in fact *F. poae* or incorrectly identified strains of *F. sporotrichioides*.

The reasons for the current concept of ATA as caused by trichothecenes produced by *F. sporotrichioides* and *F. poae*, rather than by the steroidal compounds (poaefusarin and poaefusariogenin in the case of *F. poae*) that were implicated by Russian scientists, are discussed under *F. sporotrichioides* (see *F. sporotrichioides*, ATA).

Hemorrhagic Syndrome (= Moldy Corn Toxicosis). In our opinion, *F. sporotrichioides* is primarily responsible for outbreaks of this mycotoxicosis in animals such as cattle, pigs, and poultry (see *F. sporotrichioides*, hemorrhagic syndrome). However, *F. poae* has also been isolated from feedstuffs implicated in some of these outbreaks. In Wisconsin, USA, a strain of "*F. tricinctum*" was isolated from toxic corn implicated in field outbreaks of moldy corn toxicosis (Gilgan et al., 1966). We have identified this strain, which was referred to as *F. tricinctum* B24 by Bamburg et al. (1968a) and Smalley et al. (1970), as *F. poae* (Strain 4.47). A rat skin test for dermotoxicity (Gilgan et al., 1966) was used as bioassay to isolate the mycotoxin produced by this strain, and the crystalline toxic compound was identified as diacetoxyscirpenol by Bamburg et al. (1968a). Diacetoxyscirpenol is also known to be produced by other strains of *F. poae* (see mycotoxins produced, below, and Strain 4.51) and is a trichothecene that can cause many of the characteristic lesions of moldy corn toxicosis in experimental animals (see *F. sporotrichioides*, hemorrhagic syndrome). However, two other isolates of *F. tricinctum* from moldy corn in Wisconsin (Nos. 334 and 335) that we have identified as *F. poae* (Strains 4.49 and 4.50) were non-toxic (Bamburg, 1968; Marasas, 1969).

At present it is difficult to assess the role, if any, of *F. poae* in the etiology of field outbreaks of the hemorrhagic syndrome. Although many isolates from moldy feeds may be non-toxic, the fact remains that at least two toxic, diacetoxyscirpenol-producing isolates have been identified as *F. poae* (Strains 4.47 and 4.51). Diacetoxyscirpenol has been reported to occur naturally in feedstuffs (Lafont & Lafont, 1980; Mirocha, 1979; Mirocha et al., 1976; Pathre & Mirocha, 1977; Siegfried, 1977a, b), and may be involved, together with other trichothecenes such as T-2 toxin, in the etiology of the hemorrhagic syndrome.

Urov (Kashin-Beck) Disease. Urov or Kashin-Beck disease is a chronic, disabling, deforming, dystrophic osteoarthrosis involving the peripheral joints and spine that occurs endemically among the Cossacks in the valley of the Urov River in eastern Siberia and has also been reported in North Korea and northern China (Nesterov, 1964; Rubinstein, 1960b). The disease begins slowly and often asymptomatically in children of preschool or school age. In the early stages patients experience pains in some of their joints that become thickened. The disease then develops slowly and chronically and is manifested as a shortening of the long bones, thickening and subsequent deformity of the joints, flexor contractures, and muscular atrophy. Pathological changes occur only in those parts of the growing skeleton that are associated with hyaline cartilage, only endochondral growing portions of the skeleton are involved, and abnormalities are most marked in epiphyses and metaphyses. Progressive dystrophic changes develop in the upper layers of the metaphyses and in articular cartilages, and sclerosis occurs in developing epiphyseal cartilage and metaphyses. No hematological abnormalities or other pathological changes in the internal organs are associated with Urov disease (Nesterov, 1964; Rubinstein, 1960b).

Epidemiological investigations in the USSR revealed certain climatic peculiarities of endemic areas of the disease, i.e., marked temperature changes during the day and the fact that the major portion of the rainfall occurred during late summer or early fall when cereals were maturing and grain harvesting was in progress (Perkel, 1960; Rubinstein, 1960b). These climatic factors were conducive to a high level of *Fusarium* infection in harvested grains. Investigations on the mycoflora of cereal grains in endemic areas in the USSR culminated in the reproduction of the characteristic lesions of Urov disease in puppies (see dogs, below) and rats (see rats, below) with cultures of *F. poae* (Kanshina, 1957; Nesterov, 1964; Perkel, 1957, 1960; Rubinstein, 1953, 1960b). These lesions in the bones and joints could be induced experimentally in puppies and rats only with certain isolates of *F. poae* obtained exclusively from endemic areas of Urov disease and these strains were designated as *F. sporotrichiella* Bilai var. *poae* (Peck) Bilai f. *osteodystrophica* Rubinstein (Rubinstein, 1960b).

Investigations on the chemical nature of the *F. poae* metabolite(s) capable of causing osteodystrophy in experimental animals were reported by Novakovskaya (1960). Marked reductions in the content of certain amino acids were found in grain colonized by two strains of *F. poae* (Nos. 877 and 1179) that "repeatedly caused the signs of Kashin-Beck disease in animals," but these reductions were not found in cultures of *F. poae* No. 1238, which had lost its toxigenic ability. As a result of the transformation of certain amino acids in grain, particularly tyrosine, phenylalanine, tryptophane, arginine, and lysine, by the metabolic activity of certain strains of *F. poae*, the respective amines (i.e., tyramine, phenylethylamine, tryptamine, agmatine, and cadaverine) may be formed, and the simultaneous production of mono- and diamines is also possible. The presence of amines in cultures of the toxic strains of *F. poae* was detected and histamine was chemically identified. According to Novakovskaya (1960), amines such as histamine are neurotropic and exert potent pharmacological effects on the blood vessels, even at very low concentrations. Changes in the vascular system in the metaphyseal-epiphyseal layer of the growth zone of long bones may lead to nutritional disturbances and the formation of necrotic foci in the joint cartilage, with subsequent changes leading to the deformation of the joint (Novakovskaya, 1960).

On the basis of these findings, Russian scientists have concluded that Urov disease is an osteodystrophy of mycotoxic origin caused by certain geographically restricted strains of *F. poae* and that the active metabolites are amines such as histamine (Nesterov, 1964; Novakovskaya, 1960; Perkel, 1957, 1960; Rubinstein, 1953, 1960b). It is not known to us whether the ability of amines such as histamine to cause the characteristic lesions of Urov disease has been proven experimentally. It is also not known whether more recent attempts have been made in the USSR to identify the active metabolite(s) in toxic cultures of *F. poae* from endemic areas of Urov disease.

It is concluded that considerable experimental evidence has been produced in the USSR that Urov disease is caused by certain strains referred to as *F. poae*, but that the mycotoxin(s) involved have not been positively identified and that the etiology of the disease has not been finally resolved.

Human Esophageal Cancer. According to the reviews by Lin & Tang (1980) and Yang (1980), corn meal inoculated with a strain of *F. poae* isolated from cereals in Linxian County in Henan Province, an area with a high esophageal cancer rate in northern China (see *F. moniliforme*, human espophageal cancer), caused papillomas and carcinomas of the forestomach in rats. Carcinoma of the esophagus was not observed in these experiments and there is no proof that *F. poae* is involved in the etiology of this disease.

Toxicity to Experimental Animals

The following discussion is limited to the toxic effects of cultures and extracts of cultures of *F. poae* in experimental animals. Information on the toxic effects of pure trichothecenes such as diacetoxyscirpenol and T-2 toxin, known to be produced by *F. poae*, is given under *F. sporotrichioides* (see *F. sporotrichioides*, toxicity to experimental animals) and is not repeated here.

Brine Shrimp (*Artemia salina*). Extracts of cultures of two isolates of *F. poae* (HPB 071178-7 and -11) from overwintered cereals in Canada were reported to be toxic to brine shrimp by Scott et al. (1980). We have confirmed the identity of isolate HPB 071178-7 as *F. poae* (Strain 4.51). Both of these strains were found to produce diacetoxyscirpenol, but not T-2 toxin or any of the other trichothecenes analyzed for by Scott et al. (1980).

Cats. Cultures on agar or millet of unnumbered isolates of *F. poae* from toxic overwintered cereals associated with outbreaks of ATA in the USSR (see ATA, above, and *F. sporotrichioides*, ATA), have been reported to be lethal to cats (Joffe, 1960a, 1971, 1974c, 1978b). The death of the cats followed a breakdown in blood production manifested by a decline in the number of erythrocytes, leukocytes, and neutrophils and an increase in the number of lymphocytes. In most cases the cats died within 6 to 12 days.

Autopsies revealed marked hyperemia and hemorrhages of internal organs and histopathological changes in the hematopoietic system.

Cattle. According to Kurmanov (1978a), clinical signs induced in cattle by cultures of *F. poae* (as *F. sporotrichiella* var. *poae*) included loss of appetite, atony of the rumen, watery stools, and relative lymphocytosis.

Chickens. According to Kurmanov (1978b), cultures of *F. poae* isolates from the USSR caused a lethal toxicosis in chickens. Clinical signs and pathological changes included loss of appetite, reduced weight gain, cessation of egg production, bloody diarrhea, visceral hemorrhages, leukopenia, and anemia.

Fusarium poae No. 958, originally isolated from overwintered millet associated with outbreaks of ATA (see ATA above, and *F. sporotrichioides*, ATA) in the USSR, was used in a detailed toxicological study in New Hampshire chicks by Joffe & Yagen (1978). This strain is known to produce large amounts of T-2 toxin (2.8 g/kg) in culture on wheat (Joffe, 1978b; Joffe & Yagen, 1977). Culture material for use in the toxicity experiments of Joffe & Yagen (1978) was obtained by inoculating wheat with *F. poae* No. 958 and incubating at 12°C for 21 days. Dry molded grain was added to diets to obtain T-2 toxin levels from 30 to 600 mg/kg feed. Crude ethanol extracts of this culture material were also dosed at levels from 0.46 to 1.2 mg T-2 toxin/kg body weight. Clinical signs in chicks that received diets containing culture material or crude extracts included depression, inappetence, feed refusal, diarrhea, hyperkeratosis of the corner of the beak, and hemorrhages of the mucous membrane of the oral cavity. Hematological changes in these chicks were characterized by severe leuko- and thrombocytopenia and relative decreases of the erythrocyte counts and hemoglobin values. Pathological changes included multiple hemorrhages and necrosis in the intestinal tract and other organs and damage to the hematopoietic system.

The predominant *Fusarium* species isolated by Marasas et al. (1979b) from barley grain intended for malting in Germany was *F. poae*. Cultures on corn of five of these isolates of *F. poae* were incubated at 25°C for 2 weeks followed by 4 weeks at 10°C and tested for toxicity to day-old White Leghorn chicks. Only one of these isolates (*F. poae* Sp. 884) caused any mortality (one out of four chicks in 14 days) or marked reductions in weight gain of the survivors. We have confirmed the identity of this isolate as *F. poae* (Strain 4.52). Analyses of the culture material of this and the other four strains of *F. poae* by TLC were negative for diacetoxyscirpenol, T-2 toxin, and several other Fusarium toxins (Marasas et al., 1979b).

Dogs. A disease characterized by inhibition of bone development and deformation of joints has been induced in puppies with cultures of *F. poae* (as *F. sporotrichiella* var. *poae* f. *osteodystrophica*) isolated from cereals in endemic areas of Urov disease (see Urov disease, above) in the USSR (Kanshina, 1957; Perkel, 1957, 1960; Rubinstein, 1953, 1960b).

Culture material on wheat of *F. poae* No. 1238 was administered to three puppies, either by adding 3 to 4 g of material to the feed per day or by dosing 8 to 10 ml of a suspension (Rubinstein, 1953, 1960b). Two of the three puppies developed clinical signs characterized by reactions of pain to pressure on joints, marked deformity of joints, and cracking noises in the joints. The disease became progressively worse and pronounced contractures of the front extremity joints appeared. The animals evidenced difficulty in movement and, when forced to walk, exhibited a "duck waddle" gait. Both affected puppies were killed 7 weeks after commencement of the experiment. Autopsy revealed marked deformity of the joints, a definite shortening of the bones of the front extremities, curvature of the humerus and femur, and thickening of the epiphyses. The mineral content, including calcium and phosphorus content, of the bones of the affected puppies was lower than in the controls. Histopathologically there was a delay in endochondral growth, most clearly expressed in the long bones. The disturbance in the processes of endochondral ossification led to abnormalities in the formation of long bones in their epiphyseal and metaphyseal areas and to inhibition of growth along the long axis. Changes were noted in the blood vessels that grew into the cartilagineous cells. No hematological disorders or pathological changes in other organs were noted (Rubinstein, 1953, 1960b).

The experiments of Rubinstein (1953, 1960b) were repeated by Perkel (1957, 1960), who found that 4 of 22 isolates of *F. poae* (Nos. 20, 1671, 1677, and 1172) from endemic areas of Urov disease were capable of causing osteodystrophy in puppies. Autopsies were performed on two puppies that died after 24 to 32 days, respectively, of receiving small amounts (5 to 10 g/day) of culture material of one of these isolates of *F. poae*. Histopathological lesions were found to include atrophy of the striated muscle along with deep-seated dystrophic changes, atrophy of the bone marrow in the long bones with osteoporosis of the bony trabeculae, reduction in the number of osteoblasts, and lacunar restoration of bone (Perkel, 1957, 1960). The histopathological changes in these puppies (see also rats, below), characterized by disturbances in endochondral ossification, narrowing and irregularity of the epiphyseal cartilagineous cells, and osteoporosis of the metaphyses, were considered by Kanshina (1957) to be similar to changes in bones of children in endemic areas of Urov disease.

On the basis of these experiments in puppies, as well as in rats, Russian scientists have formulated the hypothesis that Urov disease is an osteodystrophy of mycotoxic origin caused by certain geographically restricted strains of *F. poae* (Kanshina, 1957; Nesterov, 1964; Perkel, 1957, 1960; Rubinstein, 1953, 1960b).

Ducklings. According to Joffe (1978a), cultures of *F. poae* No. 60/9, isolated from millet associated with outbreaks of ATA (see ATA, above) in the USSR, induced vomiting in ducklings. Ducklings fed mixtures containing 2 to 20% wheat cultures of this strain developed pronounced swelling and inflammation of the mouth. On autopsy, hemorrhages were found in many organs and tissues (Joffe, 1978a). *Fusarium poae* No. 60/9 was subsequently shown to produce large amounts of T-2 toxin (20.0 mg/10 g of grain) in cultures on wheat incubated at 12°C for 21 days (Joffe, 1978b; Joffe & Yagen, 1977).

Geese. Cultures on corn of a Hungarian isolate of *F. poae* caused marked inhibition of spermatogenesis involving decreases in sperm volume and azoospermy in ganders (Palyusik et al., 1974).

Guinea Pigs. According to Joffe (1960a, 1971, 1974c, 1978b), culture filtrates and powdered dry mycelium of unspecified isolates of *F. poae* from overwintered cereals associated with ATA (see ATA, above) in the USSR were toxic to guinea pigs. The animals died within 5 to 21 days and autopsies revealed hemorrhages in many organs and tissues.

Horses. According to Joffe (1960a, 1971, 1974c, 1978b), cultures on agar of an unspecified isolate of *F. poae* from overwintered cereals associated with ATA (see ATA, above) in the USSR were toxic to horses. Cultures incubated at low temperatures and subjected to freezing and thawing were more toxic than cultures incubated at room temperature. A small amount (40 g) of the cold-treated culture material caused an acute toxicosis and the horse died after 36 hours. The pathological changes were characteristic of hemorrhagic diathesis.

Mice. Cultures and extracts of cultures of unspecified isolates of *F. poae* from overwintered cereals associated with ATA (see ATA, above) were lethal to white mice (Bilai, 1970; Joffe, 1960a, 1962, 1071, 1974c, 1978b).

The mycotoxicosis caused in mice by an "authentic isolate from overwintered grain" of *F. poae* has been described by Schoental & Joffe (1974). Ethanol extracts of cultures of this strain on wheat incubated for 40 days at temperatures that fluctuated from 7 to 24°C were administered to mice by local application, subcutaneous and intraperitoneal injections, and intragastric intubation. Small doses applied to the skin, esophagus, and stomach had local cytotoxic effects that were followed by regeneration and basal cell hyperplasia of the squamous epithelium. High doses caused death within one or more days regardless of the route of administration. Doses of the order LD_{50-70} caused depletion of the lymphoid tissues and subsequent widespread infections suggestive of an immunosuppressive action (Schoental & Joffe, 1974). Extracts of *F. poae* No. 958, isolated from overwintered cereals in the USSR and known to produce T-2 toxin, HT-2 toxin, diacetoxyscirpenol, and neosolaniol, have also been reported to cause pathological changes in the proliferating tissues of the bone marrow, thymus, spleen, and lymph nodes, resulting in leukopenia and immunosuppression in mice (Frayssinet & Lafont, 1976; Lafarge-Frayssinet et al., 1979; Lafont et al., 1979; Rosenstein et al., 1978, 1979, 1981).

Eight isolates of *F. poae* from moldy animal feeds suspected of causing mycotoxicoses of domestic animals in Finland were tested for toxicity to mice by Korpinen & Ylimäki (1972). Cultures on a mixture of wheat, barley, and oats were incubated at 22°C for 2 weeks followed by another 2 weeks at either 22, 8, or 0°C. Incubation temperature did not have much effect on the toxicity of these cultures to mice. One of the strains of *F. poae* was not toxic at any of the incubation temperatures used, but all the others caused at least some mortality (Korpinen & Ylimäki, 1972).

Fusarium poae NRRL 3287, reported to be toxic to mice by Ueno et al. (1972b, 1973a), is *F. sporotrichioides* (Strain, 4.25).

Pigeons. The strain reported, both as *F. poae* No. 2518 (Prentice & Dickson, 1968; Prentice et al., 1959) and as *F. poae* (= *F. tricinctum*) NRRL 3287 (Ellison & Kotsonis, 1973), to cause emesis in pigeons, is *F. sporotrichioides* (Strain 4.25).

Pigs. *Fusarium poae* NRRL 3287, reported to cause feed refusal in pigs by Vesonder et al. (1977b, 1981a), is *F. sporotrichioides* (Strain 4.25).

Rabbits. According to Joffe (1960a, 1971, 1974c, 1978b), culture filtrates and powdered dry mycelium of unspecified isolates of *F. poae* from overwintered cereals associated with outbreaks of ATA (see ATA, above) in the USSR were acutely toxic to rabbits. The rabbits died within 8 to 24 days and autopsies revealed hemorrhages in organs and tissues.

A skin test involving the application of ether or alcohol extracts of fungal cultures on the shaved skin of rabbits and assessment of the dermotoxic reaction has been used extensively for toxicity screening of *Fusarium,* including *F. poae* (Bilai, 1970; Joffe, 1960a, b, 1969, 1973a, 1974b; Joffe & Palti, 1974; Joffe & Yagen, 1977; Kvashnina, 1976, 1978, 1979; Yagen & Joffe, 1976). In a study of the toxicity of *F. poae* isolated from soil and overwintered cereals associated with outbreaks of ATA in the USSR, 66 strains were grown on various media at temperatures ranging from −5 to 8°C for 25 to 70 days and subjected to freezing and thawing (Joffe, 1960a, b). Extracts of 46 of these strains (70%) were found to be highly toxic and to cause an edematous-hemorrhagic skin reaction; topical application of some of the extracts even caused the death of the treated rabbits. The highest levels of dermotoxicity to rabbit skin in all isolates of *F. poae* tested were found in extracts of cultures incubated at low temperatures (6 to 12°C), although the optimal temperature for vegetative growth was 18 to 24°C (Joffe, 1974b). At all temperatures, cultures incubated in the dark were more toxic than those incubated in the light (Joffe, 1974b). A correlation has been found between the degree of dermotoxicity to rabbit skin of *F. poae* isolates and their phytotoxicity (Joffe, 1973a; Joffe & Palti, 1974). The latter finding is important and implies that other strains of *F. poae* that have been reported to be phytotoxic, e.g., the Canadian isolates of Bolton & Nuttall (1968) and the isolates from Irish oats (Sheridan, 1980), may also be toxic to animals.

A total of 25 isolates of *F. poae* from overwintered cereals associated with ATA in the USSR were tested for dermotoxicity to rabbit skin by Yagen & Joffe (1976). Ethyl alcohol extracts of cultures of 20 strains (80%) on wheat incubated at 12°C for 21 days produced a strong dermotoxic reaction. All the dermotoxic strains produced T-2 toxin as detected by TLC and GLC (Yagen & Joffe, 1976)). Four of the above isolates of *F. poae* (No. 60/9, 396, 792, and 958), which had previously been found to be dermotoxic (Joffe, 1960a), were re-tested for dermotoxicity to rabbit skin by Joffe (1978b) and Joffe & Yagen (1977). All of these isolates were found to be strongly dermotoxic and to produce from 6.2 to 21.0 mg of T-2 toxin/10 g of wheat grain on which they had been cultured at 12°C for 21 days. Pathological examination confirmed that the crude extracts caused the same skin lesions as pure T-2 toxin, i.e., edema, hemorrhage, and necrosis of the epidermis, dermis, and hair follicles (Yagen & Joffe, 1976). There was a positive relationship between the rabbit skin irritation factor and the amount of T-2 toxin detected by TLC and GLC in these extracts, and it could be concluded that T-2 toxin was the dermotoxic compound produced by these strains of *F. poae* (Joffe, 1978b; Joffe & Yagen, 1977; Yagen & Joffe, 1976).

The following dermotoxic strains that have been referred to as *F. poae* by Joffe (1978b), Joffe & Palti (1975) and Joffe & Yagen (1977) are *F. sporotrichioides:* NRRL 3299 (Strain 4.3), Strain T-2 (Strain 4.3), and NRRL 3287 (Strain 4.25).

Rats. A disease characterized by inhibition of bone development, deformation of joints, and muscular atrophy has been induced in rats with cultures of *F. poae* (as *F. sporotrichiella* var. *poae* f. *osteodystrophica*) isolates from cereals in endemic areas of Urov disease (see Urov disease, above) in the USSR (Kanshina, 1957; Perkel, 1957, 1960; Rubinstein, 1953, 1960b). A suspension of culture material on wheat of *F. poae* Nos. 1238 and 1536 was administered in doses of 1 to 2 cc per day to weanling rats (Rubinstein, 1953, 1960b). Experiments usually lasted 2 to 3 months, sometimes 5 to 6 months, and rats were killed during the experiment depending on clinical and radiological signs. In treated rats, pathological changes of a distinctive osteodystrophy were noted. The earlier appearance of synostoses, osteoporosis, changes in the epiphyseal zone, and fragility of the long bones were detected radiologically. At autopsy, contractures of the front extremities, shortening of the femur and tibia, osteoporosis, and muscular atrophy were observed. The mineral content, particularly of calcium and phosphorus, of the bones of treated rats was lower than in controls. Histopathologically there was a marked delay or complete cessation of endochondral development of long bones. Marked increases occurred in resorption of the bony substance in the long bones, the basic substance of the hyaline cartilage of the epiphyseal cartilagineous plate, and the cartilages of the joint surfaces. These resorption processes resulted in the destruction of the bone structure of the spongiosa of the epiphyses and metaphyses. No hematological disorders or pathological changes in other organs were noted (Rubinstein, 1953, 1960b).

The experiments of Rubinstein (1953, 1960b) were repeated by Perkel (1957, 1960), who dosed rats with a suspension of culture material (1 to 2 cc/day) of 20 strains of *F. poae*. Only four isolates (No. 877, 1179, 1862, and 1953) from endemic areas of Urov disease were capable of causing osteodystrophy in rats. Cultures of these strains caused a lag in the growth of the trunk and extremities, shortening of the femora and humeri, contractures of the front extremities, and early synostoses in the long bones. Cultures of *F. poae* No. 877 subjected to marked fluctuations in temperature (10 days at 18 to 20°C followed by periods of 30 to 40 minutes at temperatures of −4 or −5°C for 11 days) induced these characteristic pathological changes in rats within one month.

The histopathological changes in these rats (as well as puppies; see dogs, above), characterized by disturbances in endochondral ossification, narrowing and irregularity of the epiphyseal cartilagineous cells, and osteoporosis of the metaphyses, were considered to be similar to changes in the bones of children in endemic areas of Urov disease by Kanshina (1957). On the basis of these experiments in rats and puppies, Russian scientists have formulated the hypothesis that Urov disease is an osteodystrophy of mycotoxic origin caused by certain geographically restricted strains of *F. poae* (Kanshina, 1957; Nesterov, 1964; Perkel 1957, 1960; Rubinstein, 1953, 1960b).

Cultures of *F. poae* (as *F. sporotrichiella* var. *poae*) No. 1140/26, isolated in the USSR, were incubated on wheat under conditions of several drastic changes in temperature, i.e., 6 days at 24 to 26°C, 9 days at −6 to −8°C, and again for 6 days at 24 to 26°C (Rubinstein et al., 1967). Extracts of these cultures were dosed to rats for 9 months. Pathological changes included hyperkeratosis and the formation of papillomas in the mucosa of the proventriculus.

An unspecified "authentic isolate" of *F. poae* from overwintered cereals associated with outbreaks of ATA (see ATA, above) in the USSR has been reported to be toxic to rats by Schoental & Joffe (1974). Ethanol extracts of cultures of this strain on wheat incubated for 40 days at temperatures that fluctuated from 7 to 24°C were dosed to weanling rats by stomach tube. The lining of the esophagus and of the squamous part of the stomach became ulcerated, keratinized, and desquamated. Animals dosed repeatedly with smaller doses did not appear to have gross lesions, but microscopically there were pockets of distension and invagination in the esophagus and thickened areas in the forestomach lined with multicellular squamous epithelium with basal cell hyperplasia. In rats that died after several weeks, widespread infections suggestive of an immunosuppresive action were noted. According to Schoental et al. (1976), unspecified "chronic lesions" were also found in various organs, including the digestive tract, brain, and sex organs of rats dosed with these extracts of *F. poae* on a long-term basis. Schoental et al. (1979) referred to unpublished results of Schoental & Joffe, who found brain tumors, heart lesions, and pancreatic neoplasms in rats treated chronically with extracts of *F. poae*.

Two other "authentic toxic isolates" of *F. poae* (No. 5627 and 5253) from the USSR were tested for dermotoxicity to rat skin by Szathmary et al. (1976, 1977). Extracts of cultures on millet incubated at 23 to 27°C for one week, 5 to 8°C for 2 weeks, and again at 23 to 27°C for one week were applied to the shaved skin of rats. Only *F. poae* No.

5253 was dermotoxic and chemical analyses of the extracts revealed the presence of T-2 toxin, HT-2 toxin, T-2 tetraol, and neosolaniol (Szathmary et al., 1976, 1977).

A rat skin test was used as bioassay in the isolation of a dermotoxic compound from an unnumbered strain of "*F. tricinctum*" from moldy corn in Wisconsin, USA (Gilgan et al., 1966). We have identified this strain, which was referred to as *F. tricinctum* B24 by Bamburg et al. (1968a) and Smalley et al. (1970), as *F. poae* (Strain 4.47). The dermotoxic compound isolated from this strain was identified as diacetoxyscirpenol by Bamburg et al. (1968a).

Cultures of 10 unnumbered isolates of *F. poae* from cereals in Finland were fed to rats (two rats/strain) for 14 days (Korpinen & Uoti, 1974). We have identified one of these Finnish isolates (*F. poae* 72188), which caused only slight reductions in weight gain in rats, as *F. poae* (Strain 4.53). These isolates of *F. poae* caused slight if any clinical signs and most of the rats survived. Thus *F. poae* was the least toxic of five species of *Fusarium* tested in rats by Korpinen & Uoti (1974).

According to the reviews of Lin & Tang (1980) and Yang (1980), a strain of *F. poae* was isolated from cereals in Linxian county, an area with a high human esophageal cancer rate (see human esophageal center, above) in Henan Province, People's Republic of China. Cultures of this strain on corn meal were incubated at 26°C for 6 to 8 days and fed to 27 Wistar rats for 282 days. Tumors that developed during this period included 27 papillomas and seven carcinomas of the forestomach, and one bladder papilloma.

Mycotoxins Produced

Acetyl T-2 Toxin. The strain reported as *F. poae* NRRL 3287 to produce acetyl T-2 toxin (Kotsonis et al., 1975b), is *F. sporotrichioides* (Strain 4.25).

Butenolide. *Fusarium poae* No. 958, isolated from overwintered cereals associated with the outbreaks of ATA (see ATA, above) in the USSR, has been reported to produce butenolide in culture in modified Czapek medium incubated at 15°C for 9 days (Lafarge-Frayssinet et al., 1979; Lafont et al., 1977; Payen et al., 1978). According to Suzuki et al. (1981a), none of five isolates of *F. poae* from wheat and barley in Japan produced butenolide in cultures on rice.

Diacetoxyscirpenol. *Fusarium poae* No. 958, isolated from overwintered cereals associated with outbreaks of ATA (see ATA, above) in the USSR, has been reported to produce diacetoxyscirpenol in culture in modified Czapek medium incubated at 15°C for 9 days (Lafarge-Frayssinet et al., 1979; Lafont et al., 1977; Payen et al., 1978).

The dermotoxic compound produced by an unnumbered isolate of *F. tricinctum* from moldy corn in Wisconsin, USA, was isolated by means of a rat skin test bioassay (Gilgan et al., 1966). We have identified this strain, which was referred to as *F. tricinctum* B24 by Bamburg et al. (1968a) and Smalley et al. (1970), as *F. poae* (Strain 4.47). The crystalline dermotoxic compound was isolated from cultures of this strain in liquid medium incubated at 7°C for 30 days by Gilgan et al. (1966) and was identified as diacetoxyscirpenol by Bamburg et al. (1968a).

Two isolates of *F. poae* (HPB 071178-7 and -11) from overwintered cereals in Canada have been reported to produce diacetoxyscirpenol in cultures on corn incubated at 14°C (Scott et al., 1980). We have confirmed the identity of isolate HPB 071178-7, which produced 17 μg of diacetoxyscirpenol/g of culture material as detected by TLC, as *F. poae* (Strain 4.51).

None of seven isolates of *F. poae* tested by Cullen (1981) produced diacetoxyscirpenol that could be detected by GLC in cultures on corn incubated at 24°C for 2 weeks followed by 4 weeks at 8°C. We have confirmed the identity of three of these isolates that did not produce diacetoxyscirpenol, i.e., M-355, M-344 and T-280, as *F. poae* (Strains 4.48, 4.49, and 4.54). Suzuki et al. (1980) also reported that none of five Japanese isolates of *F. poae* produced diacetoxyscirpenol.

In conclusion, some but not all isolates in *F. poae* are able to produce diacetoxyscirpenol and the ability may be lost rapidly in culture, as happened with *F. poae* (= *F. tricinctum*) B-24 (Strain 4.47) according to Bamburg (1968), Marasas (1969), and Smalley et al. (1970). In addition to this strain, the only other diacetoxyscirpenol-producing strain of *F. poae* represented in the ITFRC is *F. poae* HPB 071178-7 (Strain 4.51).

HT-2 toxin. *Fusarium poae* No. 958, isolated from overwintered cereals associated with outbreaks of ATA (see ATA, above) in the USSR, has been reported to produce small amounts of HT-2 toxin in culture in modified Czapek medium incubated at 15°C for 9 days (Lafont et al., 1977). Another "authentic toxic isolate" from millet in the USSR, *F. poae* No. 5253, has been reported to produce up to 1600 µg of HT-2 toxin/g of culture on millet incubated at 23 to 27°C for one week, 5 to 8°C for 2 weeks, and again at 23 to 27°C for one week (Szathmary et al., 1976, 1977).

The strain reported as *F. poae* NRRL 3287 to produce HT-2 toxin (Kotsonis & Ellison, 1975; Ueno et al., 1972b, 1973a) is *F. sporotrichioides* (Strain 4.25).

Neosolaniol. *Fusarium poae* No. 958, isolated from overwintered cereals associated with outbreaks of ATA (see ATA, above) in the USSR, has been reported to produce small amounts of neosolaniol in culture in modified Czapek medium incubated at 15°C for 9 days (Lafont et al., 1977; Payen et al., 1978). Another "authentic toxic isolate" from millet in the USSR, *F. poae* No. 5253, has been reported to produce up to 600 µ/g of neosolaniol/g of culture on millet incubated at 23 to 27°C for one week, 5 to 8°C for 2 weeks, and again at 23 to 27°C for one week (Szathmary et al., 1976, 1977).

One strain of *F. poae* out of five isolates from cereals in Japan tested by Suzuki et al. (1980) produced 1.16 µg/g of neosolaniol in culture.

The strain reported as *F. poae* NRRL 3287 to produce neosolaniol (Ueno et al., 1972b, 1973a) is *F. sporotrichioides* (Strain 4.25).

Poin. A crystalline, water soluble substance named poin was isolated from a strain of *F. poae* (as *F. sporotrichiella* var. *poae*) in the USSR by Elpidina (1959, 1960). Doses of 100 to 150 mg/kg of poin were lethal to mice, while lower doses caused leukopenia and also had antineoplastic activity in mice. The structure was not determined, but in view of its biological activity it is probable that poin, like poaefusarin (see ATA, above) contained trichothecenes as contaminants.

T-2 Toxin. In a study of the toxicity of 25 isolates of *F. poae* from overwintered cereals associated with outbreaks of ATA (see ATA, above) in the USSR, Yagen & Joffe (1976) found that all the isolates produced T-2 toxin in cultures on wheat incubated at 12°C for 21 days. The levels of T-2 toxin detected by GLC in purified ethanol extracts of these cultures varied from 0.1 to 21 mg/ml of crude extract. According to Yagen et al. (1977), *F. poae* No. 958 was the most toxic among the 25 isolates of *F. poae* tested and produced 2.8 g of T-2 toxin/kg of culture material on wheat. However, according to the results of Joffe (1978b) and Joffe & Yagen (1977), *F. poae* No. 958 was only the third highest T-2 toxin producer (8.0 mg/10 g) among four isolates from overwintered cereals in the USSR, and *F. poae* No. 60/9 and 396 produced 20.0 and 21.0 mg/10 g of wheat grain, respectively. *Fusarium poae* No. 958 has subsequently been used for the production of T-2 toxin used in toxicological studies (Bertin et al., 1978; Joffe & Yagen, 1978; Lafarge-Frayssinet et al., 1979; Lafont et al., 1977; Payen et al., 1978; Rosenstein et al., 1979; Siriwardana & Lafont, 1978). Another "authentic toxic isolate" from millet in the USSR, *F. poae* No. 5253, has been reported to produce up to 520 µg of T-2 toxin/g of culture on millet incubated at 23 to 27°C for one week, 5 to 8°C for 2 weeks, and again at 23 to 27°C for one week (Szathmary et al., 1976, 1977).

Two North American isolates of *F. poae*, i.e., M-355 and M-334, have been reported to produce small amounts (2.8 and 1.8 µg/g, respectively) of T-2 toxin in cultures on corn incubated at 24°C for 2 weeks followed by 4 weeks at 8°C (Cullen, 1981). We have confirmed the identity of these two T-2 toxin-producing isolates as *F. poae* (Strains 4.48 and 4.49).

The following strains that have been reported in the literature as *F. poae* to produce T-2 toxin, are *F. sporotrichioides*: *F. poae* strain T-2 (Joffe, 1978b; see Strain 4.3), *F. poae* NRRL 3299 (Joffe, 1978b; Joffe & Yagen, 1977; see Strain 4.3), and *F. poae* NRRL 3287 (Ellison & Kotsonis, 1973; Joffe, 1978b; Joffe & Yagen, 1977; Kotsonis & Ellison, 1975; Ueno et al., 1972b, 1973a; Vesonder et al., 1981a; see Strain 4.25). Three isolates of *F. poae* from moldy corn in Wisconsin, USA, i.e., *F. tricinctum* No. 215, M-334, and M-335 (Strains 4.47, 4.49, and 4.50), were found to be negative for T-2 toxin by Bamburg (1968) and Marasas (1969), although *F. tricinctum* No. 215 (= B-24) originally produced diacetoxyscirpenol (Bamburg et al., 1968a). Similarly, two diacetoxyscirpenol-producing strains of *F. poae* from overwintered cereals in Canada, including *F. poae* HPB 071178-7 (Strain 4.51), did not produce T-2 toxin (Scott et al., 1980). Other negative results for T-2 toxin production by *F. poae* included none of 5 isolates from barley in Germany (Marasas et al., 1979b), none of 5 isolates from cereals in Japan (Suzuki et al., 1980), and the finding by Cullen (1981) that two of five strains of *F. poae* produced only very small amounts of T-2 toxin.

In conclusion, there are few reports of the production of T-2 toxin by authenticated strains of *F. poae*, except for the 25 strains from overwintered cereals in the USSR referred to as *F. poae* by Joffe (1978b), Joffe & Yagen (1977), Yagen & Joffe (1976) and Yagen et al. (1977). We have not seen these cultures from the USSR and are unable to confirm their identity. The only two T-2 toxin-producing strains of *F. poae* in the ITFRC (M-355 and M-334, Strains 4.48 and 4.49) have both been reported by Cullen (1981) to produce very small amounts (2.8 and 1.8 µg/g, respectively) of T-2 toxin.

T-2 Tetraol. An "authentic toxic isolate" from millet in the USSR, *F. poae* No. 5253, has been reported to produce 50 µg of T-2 tetraol/g of culture on millet incubated at 23 to 27°C for one week, 5 to 8°C for 2 weeks, and again at 23 to 27°C for one week (Szathmary et al., 1976, 1977).

Zearalenone. *Fusarium poae* has not been reported to produce zearalenone and several authors have reported negative results of zearalenone production by isolates of *F. poae*, e.g., none of 7 by Cullen (1981), none of 41 by Hacking et al. (1976, 1977), one nonproducer by Ichinoe et al. (1977), none of 5 by Marasas et al. (1979b), and niether of 2 by Scott et al. (1980).

Toxigenic Strains in the ITFRC

The *Fusarium* strains listed under *F. poae* in Table 4.1 all have been identified as *F. poae* according to Nelson et al. (1983).

Strain 4.47 F. tricinctum B-24 (= No. 215) (= T-412; MRC 1658)
This strain (as *F. tricinctum* unnumbered) was originally isolated from moldy sweet corn in connection with investigations on moldy corn toxicosis (see hemorrhagic syndrome, above) in Wisconsin, USA (Gilgan et al., 1966). Subsequently this strain has been referred to as *F. tricinctum* No. 215 (Bamburg, 1968; Marasas, 1969) and *F. tricinctum* B-24 (Bamburg et al., 1968a; Smalley et al., 1970). Ethyl acetate extracts of cultures of this strain on Gregory's medium incubated at 7°C for 30 days were assayed for toxicity by a rat skin test (Gilgan et al., 1966). These extracts were highly dermotoxic and caused severe edema, intradermal hemorrhage, and scab formation. The dermotoxic compound was isolated by means of this rat skin bioassay (Gilgan et al., 1966) and identified as diacetoxyscirpenol by Bamburg et al. (1968a). According to Smalley et al. (1970), this strain lost its toxicity and it is not known whether it still produces diacetoxyscirpenol. This strain did not produce T-2 toxin in culture in Gregory's medium incubated at 8°C for 14 days (Bamburg, 1968; Marasas, 1969). It also did not produce butenolide in cultures on Sabouraud's agar incubated at 15°C for 4 weeks (Marasas, 1969).

A slant culture of this strain was received in 1965 from E.B. Smalley as "*F. tricinctum* B-24 (= No. 215)."

Strain 4.48 F. poae M-355 (= T-626; MRC 1607)
This strain was obtained from "Marasas, U.S., corn" according to Cullen (1981), who identified it as *F. poae*. However, Cullen (personal communication, August 1981) stated that this strain was actually isolated from barley grain in Wisconsin during 1968. Cultures of this strain on corn incubated at 24°C for 2 weeks followed by 4 weeks at 8°C contained 2.8 µg/g of T-2 toxin as determined by GLC (Cullen, 1981).

A soil culture of this strain was received on 3 May 1978 from E.B. Smalley as "*F. tricinctum* No. 355."

Strain 4.49 F. poae (= F. tricinctum) M-334 (= T-403; MRC 1605)
This strain (as *F. tricinctum* No. 334) was isolated from sweet corn in Wisconsin, USA (Bamburg, 1968; Marasas, 1969). Extracts of cultures of this strain in Gregory's medium incubated at 8°C for 14 days were not dermotoxic to rat skin and did not contain T-2 toxin that could be detected by GC (Bamburg, 1968; Marasas, 1969). According to Cullen (1981), analysis by GLC of cultures of this strain (as *F. poae* M-334) on corn incubated at 24°C for 2 weeks followed by 4 weeks at 8°C revealed the presence of 1.8 µg/g of T-2 toxin. This strain did not produce butenolide in cultures on Sabouraud's agar incubated at 15°C for 4 weeks (Marasas, 1969) or zearalenone in cultures on rice incubated at 24°C for 2 weeks followed by 4 weeks at 14°C (Cullen, 1981).

A soil culture of this strain was received on 3 May 1978 from E.B. Smalley as "*F. tricinctum* No. 334."

Strain 4.50 F. tricinctum No. 335 (= T-410; MRC 1662)
This strain was isolated from sweet corn in Wisconsin, USA (Bamburg, 1968; Marasas, 1969). Extracts of cultures of this strain in Gregory's medium incubated at 8°C for 14 days were not dermotoxic to rat skin and did not contain T-2 toxin that could be detected by GC (Bamburg, 1968; Marasas, 1969). No butenolide was produced in cultures on Sabouraud's agar incubated at 15°C for 4 weeks (Marasas, 1969).

A slant culture of this strain was received in 1965 from E.B. Smalley as "*F. tricinctum* No. 335."

Strain 4.51 F. poae HPB 071178-7 (= T-503; MRC 2181)

This strain was isolated from overwintered wheat in Manitoba, Canada (Scott et al., 1980). Extracts of cultures of this strain on corn incubated at 14°C were toxic to brine shrimp and contained 17 µg/g of diacetoxyscirpenol as determined by TLC, but no T-2 toxin, HT-2 toxin, neosolaniol, fusarenon-X, nivalenol, deoxynivalenol, acetyldeoxynivalenol, or zearalenone (Scott et al., 1980).

A lyophilized culture of this strain was received on 22 December 1980 from B.J. Blanchfield as "F. poae HPB 071178-7, identified by CBS."

Strain 4.52 F. poae Sp. 884 (= T-247; MRC 1893)

This strain was isolated from barley grain in Germany (Marasas et al., 1979b). Cultures of this strain on corn incubated at 25°C for 2 weeks followed by 4 weeks at 19°C were slightly toxic to day-old chicks, causing the death of one of 4 chicks and marked reductions in weight gain and feed consumption of the survivors (Marasas et al., 1979b). None of the following mycotoxins could be detected in this culture material by TLC: diacetoxyscirpenol, T-2 toxin, HT-2 toxin, neosolaniol, neosolaniol acetate, monoacetoxyscirpenol, deoxynivalenol, acetyldeoxynivalenol, fusarenon-X, moniliformin, and zearalenone (Marasas et al., 1979b).

A lyophilized culture of this strain was received on 2 February 1978 from W.F.O. Marasas as "F. poae Sp. 884."

Strain 4.53 F. poae 72188 (= T-339; MRC 1398)

This strain was one of 10 isolates of F. poae from cereal seed in Finland, according to A. Ylimäki (personal communication, 10 April 1979), that were tested for toxicity to rats by Korpinen & Uoti (1974). Cultures of this strain on a grain mixture were incubated at 20°C for 2 weeks, at 4°C for 5 days, and again at 20°C for 2 weeks, and fed to two rats for 14 days (Korpinen & Uoti, 1974). The 10 isolates of F. poae used in the rat feeding experiment of Korpinen & Uoti (1974) caused no abnormal clinical signs or pathological lesions and only slight reductions in weight gain of rats. Thus this strain can be considered as non-toxic.

A slant culture of this strain was received on 12 August 1978 from A. Ylimäki as "F. poae 72188, barley, Finland."

Strain 4.54 F. poae T-280 (= T-280; MRC 2908)

This strain was isolated from wheat in the United States according to Cullen (1981), who identified it as F. poae. Cultures on corn incubated at 24°C for 2 weeks followed by 8°C for 4 weeks did not contain diacetoxyscirpenol, T-2 toxin, HT-2 toxin, or T-2 tetraol that could be detected by GLC (Cullen, 1981). Cultures on rice were also negative for zearalenone.

A lyophilized culture of this strain was received from FRC as "F. poae T-280."

Fusarium tricinctum (Corda) Sacc.

Perfect State: ? *Gibberella tricincta* El-Gholl, McRitchie, Schoulties & Ridings.

Incidence and Distribution

Published records of the occurrence of F. tricinctum are for the most part unreliable because of confusion with other species in the section Sporotrichiella or the use of the name *F. tricinctum* (Corda) Sacc. emend. Snyder & Hansen. In the following discussion reference will be made only to records that we consider to be reliable for *F. tricinctum* sensu stricto.

Fusarium tricinctum occurs as a saprophyte or weak parasite in soil and on a variety of hosts in Europe, the USSR, and North America (Booth, 1971; Domsch et al., 1980; Marasas et al., 1979b; Seemüller, 1968; Wollenweber & Reinking, 1935). *Fusarium tricinctum* has frequently been isolated from cereal grains in Finland (Enari et al., 1981; Korpinen & Uoti, 1974; Korpinen & Ylimäki, 1972; Ylimäki, 1981; Ylimäki et al., 1979). It has also been isolated from overwintered cereals in the USSR (Joffe, 1960a, b, 1971).

Although *F. tricinctum* appears to be more common in the temperate to cold areas of the world, we have identified the anamorph of *Gibberella tricincta* isolated from leafspots on English ivy (*Hedera helix* L.) in Florida, USA, by El-Gholl et al. (1978) as *F. tricinctum* (Strains 4.60, 4.61, and 4.62). Thus *F. tricinctum* may be more widespread in the warmer areas of the world than is presently recognized.

In the following discussion of the mycotoxicology of *F. tricinctum* sensu stricto, consideration will be given only to papers in which a clear distinction is made between *F. tricinctum* and other species in the section Sporotrichiella and to papers dealing with isolates of *F. tricinctum* represented in the ITFRC. The numerous instances in which the name *F. tricinctum* has been used incorrectly for isolates of *F. sporotrichioides*, *F. poae* and *F. chlamydosporum* (see *F. sporotrichioides*, *F. chlamydosporum*, and *F. poae*) are discussed under these species and will not be repeated here.

Association with Human and/or Animal Diseases

Fusarium tricinctum has not been associated with any human or animal toxicoses.

Toxicity to Experimental Animals

Chickens. Cultures on corn of five isolates of *F. tricinctum* from barley grain obtained from Germany and France were fed to day-old White Leghorn chickens for 14 days (Marasas et al., 1979b). None of these isolates caused any mortality, but all caused some reduction in weight gain and feed consumption. We have confirmed the identity of the two isolates that caused the most pronounced reductions in weight gain, i.e., *F. tricinctum* Sp. 879 and Sp. 880 (Strains 4.55 and 4.56). Culture material of both of these isolates were negative for moniliformin, zearalenone, T-2 toxin, and several other trichothecenes analyzed for by Marasas et al. (1979b).

Guinea Pigs. According to Joffe (1960a, 1971, 1974c, 1978b), culture filtrates and powdered dry mycelium of unspecified isolates of *F. tricinctum* from overwintered cereals associated with ATA (see *F. sporotrichioides*, ATA) in the USSR were toxic to guinea pigs. The animals died within 5 to 21 days and autopsies revealed hemorrhages in many organs and tissues.

Rabbits. According to Joffe (1960a, 1971, 1974c, 1978b), culture filtrates and powered dry mycelium of unspecified isolates of *F. tricinctum* from overwintered cereals associated with ATA (see *F. sporotrichioides*, ATA) in the USSR were toxic to rabbits. The rabbits died within 8 to 24 days and autopsies revealed hemorrhages in organs and tissues.

A total of 22 isolates of *F. tricinctum* from overwintered cereals in the USSR were tested for dermotoxicity to rabbit skin by Joffe (1960a, b). Extracts of cultures of two strains were toxic, one strain was mildly toxic, and the others were non-toxic to rabbit skin. The highest levels of dermotoxicity to rabbit skin of three isolates of *F. tricinctum* (as *F. sporotrichioides* var. *tricinctum*) tested were found in extracts of cultures incubated at 15 and 20°C (Joffe, 1974b). Although the toxicity decreased at higher temperatures, cultures of these isolates of *F. tricinctum* incubated at 30 and 35°C were much more toxic than those of other species of the section Sporotrichiella incubated at these temperatures (Joffe, 1974b).

Fusarium tricinctum NRRL 3509, which was identified as *F. sporotrichioides* Sherb. var. *tricinctum* (Corda) Raillo by Joffe & Palti (1975) and reported to be slightly dermotoxic to rabbit skin by Joffe (1978b) and Joffe & Yagen (1977), is *F. sporotrichioides* (Strain 4.9).

Rats. Cultures of 13 unnumbered isolates of *F. tricinctum* from cereal grains in Finland have been tested for toxicity to rats (Korpinen & Uoti, 1974). Most of the isolates caused significant reductions in weight gain. We have confirmed the identity of two of the isolates used in this experiment, i.e., *F. tricinctum* 72188 and 72183 (Strains 4.57 and 4.58). Rats usually became ill on the fourth or fifth day of feeding and many died before the 10th day. Pathological changes included hemorrhages in the stomach and the small intestine (Korpinen & Uoti, 1974). The mycotoxin(s) produced by these isolates of *F. tricinctum* toxic to rats have not been identified.

Mycotoxins Produced

T-2 Toxin. Cultures on corn of five isolates of *F. tricinctum* have been reported to be negative for T-2 toxin and several other trichothecenes by Marasas et al. (1979b). We have confirmed the identity of two of these isolates that did not produce T-2 toxin, i.e., *F. tricinctum* Sp. 879 and Sp. 880 (Strains 4.55 and 4.56).

According to Siegfried & Langerfeld (1978), *F. tricinctum* IMB 62448 did not produce T-2 toxin in inoculated potato tubers. We have identified this strain as *F. tricinctum* (Strain 4.59).

Cullen (1981) detected T-2 toxin by GC in cultures on corn of nine out of 16 isolates of "*F. tricinctum* sensu Booth." We have identified seven of the nine T-2 toxin-producing strains referred to as *F. tricinctum* available to us (T-340, T-285, T-288, M-350, F-691, F-14 and T-336) as *F. sporotrichioides* (Strains 4.32, 4.33, 4.34, 4.35, 4.36, 4.37, and 4.38). We have not seen the remaining two T-2 toxin-producing strains (1973-76 and M-320) referred to as *F. tricinctum*. One of the non-producing strains, *F. tricinctum* T-346, is also *F. sporotrichioides* (Strain 4.9). We have confirmed the identity of the remaining six isolates that did not produce T-2 toxin (MRC 1399, MRC 1400, MRC 1799, R-P-2, W-P-1, and B-P-1) as *F. tricinctum* (Strains 4.57, 4.58, 4.59, 4.60, 4.61, and 4.62).

The strain identified as *F. sporotrichioides* var. *tricinctum* NRRL 3509 by Joffe & Palti (1975) and reported to produce T-2 toxin by Joffe (1978b) and Joffe & Yagen (1977) is *F. sporotrichioides* (Strain 4.9).

In conclusion, authenticated strains of *F. tricinctum* sensu stricto have not been reported to produce T-2 toxin, and there are no T-2 toxin-producing strains of this species represented in the ITFRC.

Zearalenone. Cultures on corn of five isolates of *F. tricinctum* have been reported to be negative for zearalenone by Marasas et al. (1979b). We have confirmed the identity of two of these isolates that did not produce zearalenone, i.e., *F. tricinctum* Sp. 879 and Sp. 880 (Strains 4.55 and 4.56).

Cullen (1981) detected zearalenone by GC in cultures on rice of four out of 16 isolates of "*F. tricinctum* sensu Booth." Of these four zearalenone-producing strains referred to as *F. tricinctum,* we have identified the three available to use (M-350, F-14 and T-336) as *F. sporotrichioides* (Strains 4.35, 4.37 and 4.38). We have not seen the remaining zearalenone-producing strain, *F. tricinctum* 1973-76. Five of the non-producing strains (*F. tricinctum* T-340, T-285, T-288, F-696, and T-346) are *F. sporotrichioides* (Strains 4.32, 4.33, 4.34, 4.36, and 4.9). We have confirmed the identity of the remaining six isolates that did not produce zearalenone (MRC 1399, MRC 1400, MRC 1799, R-P-1, W-P-1, and B-P-1), as *F. tricinctum* (Strains 4.57, 4.58, 4.59, 4.60, 4.61, and 4.62).

In conclusion, authenticated strains of *F. tricinctum* sensu stricto have not been reported to produce zearalenone, and there are no zearalenone-producing strains of this species represented in the ITFRC.

Toxigenic Strains in the ITFRC

The *Fusarium* strains listed under *F. tricinctum* in Table 4.1 all have been identified as *F. tricinctum* sensu stricto according to Nelson et al. (1983). Toxigenic strains referred to as species of the section Sporotrichiella in the literature but identified as other species by us are listed in Table 4.2.

Strain 4.55   *F. tricinctum* Sp. 879 (= T-429; MRC 1894)

This strain was isolated from barley grain in Germany (Marasas et al., 1979b). Cultures of this strain on corn incubated at 25°C for 2 weeks followed by 4 weeks at 10°C were fed to day-old White Leghorn chickens for 14 days (Marasas et al., 1979b). The culture material caused no mortalities but did cause marked reductions in weight gain and feed consumption (7.7 and 34.0% of the controls, respectively). The culture material was negative for moniliformin, zearalenone, T-2 toxin, and several other trichothecenes analyzed for by Marasas et al. (1979b).

A lyophilized culture of this strain was received on 22 April 1980 from W.F.O. Marasas as "*F. tricinctum* Sp. 879."

Strain 4.56   *F. tricinctum* Sp. 880 (= T-246; MRC 1895)

This strain was isolated from barley grain in Germany (Marasas et al., 1979b). Cultures of this strain on corn incubated at 25°C for 2 weeks followed by 4 weeks at 10°C were fed to day-old White Leghorn chickens for 14 days (Marasas et al., 1979b). The culture material caused no mortalities but did cause marked reductions in weight gain and feed consumption (11.5 and 33.0% of the controls, respectively). The culture material was negative for moniliformin, zearalenone, T-2 toxin, and several other trichothecenes analyzed for by Marasas et al. (1979b).

A lyophilized culture of this strain was received on 16 February 1978 from W.F.O. Marasas as "*F. tricinctum* Sp. 880."

Strain 4.57   *F. tricinctum* 72188 (= T-338; MRC 1399)

This strain was isolated from cereal grains in Finland in 1972 (Korpinen & Uoti, 1974). Cultures on a mixture of wheat, barley, and oats incubated at 20°C for 2 weeks, followed by 4°C for 5 days and 2 weeks at 20°C, were fed to rats for 14 days (Korpinen & Uoti, 1974). The culture material caused significant reductions in weight gain, and some of the rats that died exhibited hemorrhages in the stomach and small intestine (Korpinen & Uoti, 1974). The mycotoxin(s) produced by this strain have not been identified. Cultures of this strain (as *F. tricinctum* MRC 1399) on corn incubated at 24°C for 2 weeks followed by 4 weeks at 8°C did not contain T-2 toxin, HT-2 toxin, T-2 tetraol, or diacetoxyscirpenol that could be detected by GC (Cullen, 1981). Cultures on rice were also negative for zearalenone (Cullen, 1981).

A slant culture of this strain was received on 2 August 1978 from A. Ylimäki as "*F. tricinctum* 72188, isolated from barley, Finland." According to A. Ylimäki (personal communication, 10 April 1979), this is one of the isolates of *F. tricinctum* that was shown to cause reductions in weight gain in rats by Korpinen & Uoti (1974).

Strain 4.58 *F. tricinctum* 72183 (= T-624; MRC 1400)

This strain was isolated from cereal grains in Finland during 1972 (Korpinen & Uoti, 1974). Cultures on a mixture of wheat, barley and oats incubated at 20°C for 2 weeks, followed by 4°C for 5 days and 2 weeks at 20°C, were fed to rats for 14 days (Korpinen & Uoti, 1974). The culture material caused significant reductions in weight gain, and some of the rats that died exhibited hemorrhages in the stomach and small intestine (Korpinen & Uoti, 1974). The mycotoxin(s) produced by this strain have not been identified. Cultures of this strain (as *F. tricinctum* MRC 1400) on corn incubated at 24°C for 2 weeks followed by 4 weeks at 8°C did not contain T-2 toxin, HT-2 toxin, T-2 tetraol, or diacetoxyscirpenol that could be detected by GC (Cullen, 1981). Cultures on rice were also negative for zearalenone (Cullen, 1981). A slant culture of this strain was received on 2 August 1978 from A. Ylimäki as "*F. tricinctum* 72183, isolated from barley, Finland." According to A. Ylimäki (personal communication, 10 April 1979), this is one of the isolates of *F. tricinctum* that was shown to cause reductions in weight gain in rats by Korpinen & Uoti (1974).

Strain 4.59 *F. tricinctum* IMB 62448 (= T-431; MRC 1799)

This strain was obtained from IMB as *F. tricinctum* No. 62448 according to Siegfried & Langerfeld (1978), but the exact source is unknown to us. This strain did not produce T-2 toxin in inoculated potato tubers incubated at 20°C for 3 to 10 weeks (Siegfried & Langerfeld, 1978). These authors apparently wanted to use a strain of *F. tricinctum* as a positive control for T-2 toxin production ("ein T-2 toxin produzent") in potatoes, but unfortunately used this strain of *F. tricinctum* sensu stricto which is not a T-2 toxin producer. Cultures of this strain (as *F. tricinctum* MRC 1799) on corn incubated at 24°C for 2 weeks followed by 4 weeks at 8°C did not contain T-2 toxin that could be detected by GC (Cullen, 1981). Cultures on rice were also negative for zearalenone (Cullen, 1981).

A dried culture of this strain on barley ears was received on 18 April 1979 from W. Gerlach as "*F. tricinctum* No. 62448."

Strain 4.60 *F. tricinctum* R-P-1 (= T-399; MRC 1574)

This strain, as a red-pigmented single-spore isolate, was obtained from a mass culture of *F. tricinctum* isolated from leaf spots on English ivy (*Hedera helix* L.) in Florida, USA (El-Gholl et al., 1978). Perithecia of *Gibberella tricinta* are reportedly produced in pairings of this isolate with either the white (Strain 4.61) or brown (Strain 4.62) single conidial isolates of *F. tricinctum* in propylene-oxide-sterilized Bermuda grass (*Cynodon dactylon* Pers.) stem pieces on water agar (Cullen, 1981; El-Gholl et al., 1978).

Cultures on corn incubated at 24°C for 2 weeks followed by 4 weeks at 8°C did not contain T-2 toxin, HT-2 toxin, T-2 tetraol, or diacetoxyscirpenol that could be detected by GC (Cullen, 1981). Cultures on rice were also negative for zearalenone (Cullen, 1981).

A slant culture of this strain was received on 1 August 1979 from N.E. El-Gholl as "*F. tricinctum*, red-pigmented single conidial isolate."

Strain 4.61 *F. tricinctum* **W-P-1** (= T-701; MRC 1575)

This strain, as a white single-spore isolate, was isolated from the same source as Strain 4.60 by El-Gholl et al. (1978). Cultures on corn and rice incubated as described for Strain 4.60 were negative for T-2 toxin, HT-2 toxin, T-2 tetraol, diacetoxyscirpenol, and zearalenone (Cullen, 1981).

A slant culture of this strain was received on 1 August 1979 from N.E. El-Gholl as "*F. tricinctum,* white single conidial isolate."

Strain 4.62 *F. tricinctum* **B-P-1** (= T-402; MRC 1576)

This strain, as a brown-pigmented single-spore isolate, was isolated from the same source as Strain 4.60 by El-Gholl et al. (1978). Cultures on corn and rice incubated as described for Strain 4.60 were negative for T-2 toxin, HT-2 toxin, T-2 tetraol, diacetoxyscirpenol, and zearalenone (Cullen, 1981).

A slant culture of this strain was received on 1 August 1979 from N.E. El-Gholl as "*F. tricinctum,* brown-pigmented single conidial isolate."

5. Section Roseum

All the species in four sections—Roseum, Arthrosporiella, Gibbosum, and Discolor—of Wollenweber & Reinking (1935) were reduced to the single species, *F. roseum* (Lk.) Snyd. & Hans., by Snyder & Hansen (1945). The use of this name has caused a great deal of confusion in the mycotoxicological literature on *Fusarium* species. In cases where the cultivar names for *F. roseum* proposed by Snyder et al. (1957)—e.g., *F. roseum* (Lk.) Snyd. & Hans. "Avenaceum"—or the varieties proposed by Messiaen & Cassini (1968)—e.g., *F. roseum* (Lk.) Snyd. & Hans. var. *avenaceum* (Sacc.) Snyd. & Hans.—are used for toxigenic species, it is sometimes possible to assign them to species according to Wollenweber & Reinking—e.g., *F. avenaceum* (Fr.) Sacc. However, in the numerous cases where toxigenic *Fusarium* species are referred to simply as "*F. roseum,*" this is obviously impossible. Consequently reports in the literature regarding the toxicity of strains referred to as "*F. roseum*" for which we do not have cultures available will be omitted from the following discussion of the sections Roseum, Arthrosporiella, Gibbosum, and Discolor.

The only species in the section Roseum that has been reported to be toxigenic is *F. avenaceum* (including *F. arthrosporioides*).

Fusarium avenaceum (Fr.) Sacc.

Perfect State: *Gibberella avenacea* Cook

Incidence and Distribution

Fusarium avenaceum has a worldwide distribution as a pathogen that causes root, foot, and ear rots of cereals and a large number of other hosts (Booth, 1971; CMI Descriptions of Pathogenic Fungi and Bacteria No. 25, 1964; Doidge, 1938; Domsch et al., 1980; Meyer & Frank, 1979; Wollenweber & Reinking, 1935). This fungus has been isolated from overwintered cereals in the USSR (Joffe, 1960a, b, 1971) and is particularly prevalent as a seed-borne organism of cereals in temperate areas (Enari et al., 1981; Gordon, 1952; Hacking et al., 1976, 1977; Marasas et al., 1979b; Ylimäki, 1981; Ylimäki et al., 1979).

Association with Human and/or Animal Diseases

Fusarium avenaceum has not been associated with any human and/or animal toxicoses.

Toxicity to Experimental Animals

Brine Shrimp (*Artemia salina*). Purified ethyl acetate extracts of potato tubers inoculated with an isolate of *F. avenaceum* from rotten potatoes in Germany have been reported to be slightly toxic to brine shrimp (Siegfried & Langerfeld, 1978). We have confirmed the identity of this strain as *F. avenaceum* (Strain 5.1). The mycotoxin(s) produced by this strain have not been identified.

Chickens. Purified ethyl acetate extracts of potato tubers inoculated with an isolate of *F. avenaceum* have been reported to be toxic to chicken embryos (Polzhofer & Niehuss, 1980). We have confirmed the identity of this isolate as *F. avenaceum* (Strain 5.1).

Cultures on corn of nine isolates of *F. avenaceum* from barley and corn in Europe were fed to day-old White Leghorn chickens for 14 days by Marasas et al. (1979b). Five of these isolates were toxic and caused the death of two or more out of four chickens. We have confirmed the identity of two of these toxic isolates (Sp. 889 and Sp. 803) as *F. avenaceum* (Strains 5.2 and 5.3). The remaining four isolates also caused marked reductions in weight gain and feed consumption. The sole mycotoxin produced by all nine isolates of *F. avenaceum* was moniliformin, and there was a relationship between toxicity to chicks and moniliformin yield as determined by TLC. Thus *F. avenaceum* Sp. 889 (Strain 5.2) was the most toxic to chicks (4 of 4 dead in a mean time of 5.5 days) and was also the highest producer of moniliformin (760 ppm). The four strains with the lowest degree of toxicity to chickens all produced less than 10 ppm of moniliformin (Marasas et al., 1979b).

Mice. Three isolates of *F. roseum* "Avenaceum" from moldy soybeans associated with "bean hulls poisoning" of horses (see *F. sporotrichioides*, bean hulls poisoning) in Hokkaido, Japan, have been tested for toxicity to mice by Ueno et al. (1972a, 1973a). Extracts of cultures of two of these strains (*F. roseum* "Avenaceum" M-7-1 and M-11-1) in PSC medium were toxic to mice and were also positive in the rabbit reticulocyte assay for inhibition of protein synthesis (Ueno et al., 1972a, 1973a). We have identified *F. roseum* "Avenaceum" M-7-1 as *F. equiseti* (Strain 7.5) and this strain is consequently excluded from *F. avenaceum*. According to Ueno et al. (1973a, 1975), the other strain toxic to mice, *F. roseum* "Avenaceum" M-11-1, produced diacetoxyscirpenol, neosolaniol, and T-2 toxin.

Another strain isolated from bean hulls in Hokkaido and reported (as *Fusarium* sp. M-10-2 by Ueno et al., 1972a, and as *F. moniliforme* M-10-2 by Ueno et al., 1973a) to be toxic to mice, has been identified as *F. avenaceum* (Strain 5.4) by us. Extracts of cultures of this strain in PSC medium were toxic to mice at high concentrations and also caused inhibition of protein synthesis in rabbit reticulocytes. However, none of the trichothecenes analyzed for by Ueno et al. (1973a) could be detected in these extracts by TLC. Thus the mycotoxin(s) produced by this strain of *F. avenaceum* have not been identified.

Rabbits. Extracts of cultures of 3 out of 32 isolates of *F. avenaceum* from overwintered cereals associated with outbreaks of ATA (see *F. sporotrichioides*, ATA) in the USSR were reported to be dermotoxic to rabbit skin by Joffe (1960a, b, 1971). One of 8 isolates of *F. arthrosporioides* Sherb. from the same source was also dermotoxic (Joffe, 1969a, b, 1971). This species was included in *F. avenaceum* by Nelson et al. (1983). The mycotoxins produced by these four dermotoxic strains of *F. avenaceum* have not been identified.

Mycotoxins Produced

Butenolide. According to Suzuki et al. (1981a, b), 3 out of 9 isolates of *F. avenaceum* from cereals in Japan produced butenolide at levels ranging from 0.09 to 1.22 µg/g in cultures on rice as detected by GC with an electron-capture detector. We have not seen these isolates and there are no butenolide-producing isolates of *F. avenaceum* represented in the ITFRC.

Diacetoxyscirpenol. According to Ueno et al. (1972a, 1973a), *F. roseum* "Avenaceum" M-11-1 isolated from moldy soybeans associated with outbreaks of "bean hulls poisoning" of horses (see *F. sporotrichioides*, bean hulls poisoning) in Hokkaido, Japan, produced diacetoxyscirpenol in culture on PSC medium incubated at 27°C for 12 days. The production of low levels of diacetoxyscirpenol by unspecified isolate(s) of *F. avenaceum* has also been reported by Sigg et al. (1965). It has not been possible to verify the identity of the above isolates and there are no diacetoxyscirpenol-producing isolates of *F. avenaceum* represented in the ITFRC.

Moniliformin. Cultures on corn incubated at 25°C for 2 weeks followed by 4 weeks at 10°C of each of 9 isolates of *F. avenaceum* from barley and corn in Europe contained moniliformin at levels ranging from 2 to 760 ppm as determined by TLC (Marasas et al., 1979b). We have confirmed the identity of two of these moniliformin-producing strains (Sp. 889 and Sp. 803) as *F. avenaceum* (Strains 5.2 and 5.3). The highest moniliformin-producing (760 ppm) strain, *F. avenaceum* Sp. 889 (Strain 5.2), was also the most toxic to day-old chickens of the nine isolates of *F. avenaceum* tested by Marasas et al. (1979b).

Neosolaniol. According to Ueno et al. (1972a, 1973a), *F. roseum* "avenaceum" M-11-1 isolated from moldy soybeans associated with outbreaks of "bean hulls poisoning" of horses (see *F. sporotrichioides*, bean hulls poisoning, and *F. moniliforme*, equine leukoencephalomalacia) in Hokkaido, Japan, produced neosolaniol in culture in PSC medium incubated at 27°C for 12 days. This strain has also been reported to produce neosolaniol in shake culture (Ueno et al., 1975). It has not been possible to verify the identity of this isolate and there are no neosolaniol-producing strains of *F. avenaceum* represented in the ITFRC.

T-2 Toxin. According to Ueno et al. (1972a, 1973a), *F. roseum* "Avenaceum" M-11-1 isolated from moldy soybeans associated with outbreaks of "bean hulls poisoning" of horses (see *F. sporotrichioides,* bean hulls poisoning, and *F. moniliforme,* equine leukoencephalomalacia) in Hokkaido, Japan, produced T-2 toxin in culture in PSC medium incubated at 27°C for 12 days. This strain has also been reported to produce T-2 toxin in shake culture (Ueno et al., 1975).

None of 14 isolates of *F. avenaceum* from cereals in Italy produced T-2 toxin that could be detected by TLC in cultures on corn (Bottalico, 1977c). Negative results for the production of T-2 toxin and several other trichothecenes have also been reported for 21 isolates of *F. avenaceum* from wheat and barley in Japan (Suzuki et al., 1980, 1981b).

It has not been possible to verify the identity of the Japanese isolate *F. roseum* "Avenaceum" M-11-1, and there are no T-2 toxin-producing strains of *F. avenaceum* represented in the ITFRC.

Zearalenone. According to Ishii et al. (1974), one Japanese isolate of "*F. roseum* (Avenaceum)" out of two tested produced unspecified levels of zearalenone in cultures on rice incubated at 25°C for 14 days followed by 14 days at 12 to 15°C. It is not known whether this zearalenone-producing isolate was one of the three strains referred to as *F. roseum* "Avenaceum" (M-7-1, M-7-2, and M-11-1) by Ueno et al. (1973a). We have identified one of these strains (M-7-1) as *F. equiseti* (Strain 7.5). We have not seen *F. roseum* "Avenaceum" M-7-2 or M-11-1, which have been reported to produce diacetoxyscirpenol, neosolaniol, and T-2 toxin (Ueno et al., 1973a).

The production of zearalenone by one isolate of *F. avenaceum* from barley grain in the United Kingdom out of 63 tested was reported by Hacking et al. (1976, 1977). This strain produced unspecified levels of zearalenone in cultures on rice incubated at 25°C for 4 weeks followed by 6 weeks at 12°C. A culture of this zearalenone-producing isolate of *F. avenaceum* no longer exists (W.R. Rosser, personal communication, 3 April 1981).

Negative results for zearalenone production have been reported for 19 Italian isolates of *F. avenaceum* by Bottalico (1977b, 1979), for one North American isolate by Caldwell & Tuite (1970) and Caldwell et al. (1970), and for 9 Japanese isolates by Suzuki et al., (1981b).

In conclusion, only two out of approximately 95 isolates of *F. avenaceum* tested have been reported to produce zearalenone. It has not been possible to verify the identity of either of these isolates and there are no zearalenone-producing strains of *F. avenaceum* represented in the ITFRC.

Toxigenic Strains in the ITFRC

The *Fusarium* strains listed in Table 5.1 all have been identified as *F. avenaceum* according to Nelson et al. (1983). Toxigenic strains referred to as *F. avenaceum* in the literature but identified as other species by us are listed in Table 5.2.

Strain 5.1 *F. avenaceum* 80 (= R-6381; MRC 2195)
This strain was isolated from rotten potato tubers in Germany (Siegfried & Langerfeld, 1978). Purified ethyl acetate extracts of potato tubers inoculated with this strain were slightly toxic to brine shrimp (Siegfried & Langerfeld, 1978).

Purified ethyl acetate extracts of potato tubers inoculated with a culture of *F. avena-*

Table 5.1 *Fusarium* Strains of the Section Roseum in the International Toxic *Fusarium* Reference Collection (ITFRC)

Strain No.	ITFRC No.	MRC No.	Published Name(s) and Number(s)	Source	Toxigenicity
Fusarium avenaceum					
5.1	R-6381	2195	*F. avenaceum* 80	E. Langerfeld	a
5.2	R-4606	1413	*F. avenaceum* Sp. 889	W.F.O. Marasas	a,b
5.3	R-4608	1888	*F. avenaceum* Sp. 803	W.F.O. Marasas	a,b
5.4	R-6750	2532	*F. moniliforme* M-10-2	Y. Ueno	a
5.5	R-5147	1374	*F. avenaceum* 72313	A. Ylimäki	a

a = Culture toxic to experimental animals   b = Produces moniliformin

Table 5.2 *Fusarium* Strains Published as *F. avenaceum* but Identified as Other Species in the International Toxic *Fusarium* Reference Collection (ITFRC)

Strain No.	ITFRC No.	MRC No.	Published Name(s) and Number(s)	Source	Present Identification
7.5	R-6768	2561	*F. roseum* "Avenaceum" M-7-1	Y. Ueno	*F. equiseti*

ceum obtained from E. Langerfeld were reported to be toxic to chicken embryos by Polzhofer & Niehuss (1980). Significant mortality rates of chicken embryos were obtained with extracts of visibly infected tuber tissue as well as with tissue adjacent to the visibly rotten parts after incubation of inoculated tubers at 20°C for 8 days. The metabolites toxic to brine shrimp and chicken embryos produced in inoculated potato tubers by this strain have not been identified.

A slant culture of the strain used by Siegfried & Langerfeld (1978) was received on 20 January 1981 from E. Langerfeld as "*F. avenaceum* 80, isolated from potato tuber, northern Germany." It is not certain whether this is the same strain of *F. avenaceum* that was reported to be toxic to chicken embryos by Polzhofer and Niehuss (1980).

Strain 5.2 *F. avenaceum* Sp. 889 (= R-4606; MRC 1413)

This strain was isolated from corn tassels in Germany (Marasas et al., 1979b). Cultures on corn incubated at 25°C for 2 weeks followed by 4 weeks at 10°C caused the death of four out of five day-old White Leghorn chicks in a mean time of 5.5 days (Marasas et al., 1979b). The sole toxin detected by TLC in the culture material toxic to chickens was moniliformin at a level of 760 ppm (Marasas et al., 1979b). The culture material was negative for T-2 toxin, HT-2 toxin, diacetoxyscirpenol, neosolaniol, neosolaniolacetate, monoacetoxyscirpenol, fusarenon-X, deoxynivalenol, acetyldeoxynivalenol, and zearalenone. This strain was the most toxic to chickens and also the highest producer of moniliformin of nine isolates of *F. avenaceum* tested by Marasas et al. (1979b).

A lyophilized culture of this strain was received on 16 February 1978 from W.F.O. Marasas as "*F. avenaceum* Sp. 889." Sub-cultures of this material lyophilized at BAF as *F. avenaceum* Sp. 889 are pionnotal and do not produce carmine red pigment. On the basis of the absence of red pigment, W. Gerlach identified this strain as *F. avenaceum* forma I according to Wollenweber & Reinking (1935). This strain still produces typical macroconidia of *F. avenaceum* and was identified as such by us.

Strain 5.3 *F. avenaceum* Sp. 803 (= R-4608; MRC 1888)

This strain was isolated from barley grain in Germany (Marasas et al., 1979b). Cultures on corn incubated at 25°C for 2 weeks followed by 4 weeks at 10°C caused the death of three out of four day-old White Leghorn chicks in a mean time of 5.6 days (Marasas et al., 1979b). The sole mycotoxin detected by TLC in the culture material toxic to chickens was moniliformin at a level of 250 ppm (Marasas et al., 1979b). The culture material was negative for T-2 toxin, several other trichothecenes, and zearalenone (see Strain 5.2).

A lyophilized culture of this strain was received on 16 February 1978 from W.F.O. Marasas as "*F. avenaceum* Sp. 803." Subcultures of this material lyophilized at BAF as *F. avenaceum* Sp. 803 are semi-pionnotal but still produce carmine red pigment and typical *F. avenaceum* macroconidia.

Strain 5.4 *F. moniliforme* M-10-2 (= R-6750; MRC 2532)

This strain (as *Fusarium* sp. M-10-2) was isolated from moldy soybeans associated with outbreaks of "bean hulls poisoning" of horses in Hokkaido, Japan (Ueno et al., 1972a). Extracts of cultures of this strain in PSC medium incubated at 27°C for 12 days were toxic to mice upon intraperitoneal injection, but were negative in the rabbit reticulocyte assay for inhibition of protein synthesis (Ueno et al., 1972a). This strain was referred to as *F. moniliforme* M-10-2 by Ueno et al. (1973a), who confirmed the toxicity to mice and reported that the toxic extract did not contain any of the trichothecenes analyzed for by TLC.

A slant culture of this strain was received in November 1981 from Y. Ueno as "*F. moniliforme* M-10-2." This culture has degenerated and produces restricted, dark brown colonies without any carmine red pigment. We have identified it as a mutant culture of *F. avenaceum* because of the presence of spindle-shaped macroconidia produced on monophialides in the aerial mycelium and a few sporodochia with typical macroconidia of *F. avenaceum*.

Strain 5.5 *F. avenaceum* 72313 (= R-5147; MRC 1374)

This strain was isolated from cereal grain in Finland during 1972 (Korpinen & Uoti, 1974). Cultures on a mixture of wheat, barley, and oats incubated at 20°C for 2 weeks followed by 5 days at 4°C and 2 weeks at 20°C caused reductions in weight gain of rats (Korpinen & Uoti, 1974). The mycotoxin(s) produced by this strain have not been identified.

A slant culture of this strain was received on 2 August 1978 from A. Ylimäki as "*F. avenaceum* 72313, isolated from wheat, Finland." According to A. Ylimäki (personal communication, 10 April 1979) this was one of the strains of *F. avenaceum* reported to cause reductions in weight gain of rats by Korpinen & Uoti (1974).

6. Section Arthrosporiella

The problems associated with the use of the name *F. roseum* (Lk.) Snyd. & Hans. also apply to section Arthrosporiella (see section Roseum).

The only species in the section Arthrosporiella that has been reported to be toxigenic is *F. semitectum*. However, one strain of each of the following three species included in this section by Wollenweber & Reinking (1935) have also been reported to be toxigenic: *F. anguioides* Sherb., *F. concolor* Reinking, and *F. diversisporum* Sherb. We have identified all three of these strains (Table 6.2) as *F. equiseti* and they are discussed under *F. equiseti*.

Fusarium semitectum Berk. & Rav.

Perfect State: Unknown

Incidence and Distribution

Fusarium semitectum is "extremely common, particularly from tropical and subtropical countries" (Booth, 1971). It occurs as a saprophyte in soil and on decaying plant tissue in the warmer parts of Africa, the Americas, and Asia (Booth, 1971; Doidge, 1938; Domsch et al., 1980; Gordon, 1960a; Meyer & Frank, 1979; Wollenweber & Reinking, 1935). However, F. semitectum is not restricted to the warmer areas of the world and has been isolated from overwintered cereals in the USSR (Joffe, 1960a, b, 1971). It also occurs as a soil saprophyte (Domsch et al., 1980) and in association with cereal grains (Gordon, 1952, 1959; Hacking et al., 1976, 1977; Ylimäki, 1981) in temperate regions of North America and Europe.

Association with Human and/or Animal Diseases

Degnala Disease. This disease of buffaloes and cattle in India and Pakistan, characterized by edematous swelling of the legs and necrosis, gangrene, and sloughing of the extremities, has been attributed to the consumption of rice straw colonized by F. equiseti (Bhat et al., 1978; Irfan, 1971; Kalra et al., 1973, 1977a, b, 1980; Kwatra & Singh, 1973). Degnala disease has been reproduced experimentally with cultures on rice straw of F. equiseti (Kalra et al., 1977a, b, 1980) and this disease is discussed in detail under F. equiseti (see F. equiseti, Degnala disease). However, we have identified two of three isolates of "F. equiseti" from rice straw associated with Degnala disease in India as F. semitectum (Strains 6.1 and 6.2). Consequently it is not certain at present whether this disease is caused by F. equiseti or F. semitectum or both. It is interesting that a similar disease known as "sore foot disease" (see F. equiseti, Degnala disease), which afflicts cattle consuming rice straw in China, has been "proved to be caused by Fusarium equiseti and Fusarium semitectum" (Qin et al., 1981). This disease is characterized by necrosis of the feet, legs, and ears, subcutaneous edema and hemorrhage in the affected areas, congestion of capillaries, thrombosis of arteries, and edema and necrosis of muscles. According to Qin et al. (1981), this disease has been reproduced experimentally in cattle with cultures of F. equiseti and F. semitectum.

Thus there is evidence from India and China that a disease of buffaloes and cattle with certain similarities to ergotism and to "fescue foot" (see F. sporotrichioides, fescue foot) is caused by the consumption of rice straw colonized by F. semitectum and/or F. equiseti. The mycotoxin(s) responsible have not yet been identified, but an Indian isolate of "F. equiseti" has been found to be negative for both butenolide and T-2 toxin (Kalra et al., 1977a, 1980).

Human Esophageal Cancer. According to the reviews by Lin & Tang (1980) and Yang (1980), rats fed corn meal inoculated with a strain of F. semitectum isolated from cereals in Linxian County in Henan Province, an area with a high esophageal cancer rate in northern China (see F. moniliforme, human esophageal cancer), developed lymphosarcoma in the abdominal cavity and papilloma of the bladder. Carcinoma of the esophagus was not observed in these experiments and there is no other evidence that F. semitectum is involved in the etiology of this disease.

Toxicity to Experimental Animals

Brine Shrimp (*Artemia salina*). Culture filtrates of Fusarium sp. HPB 110178-21, isolated from decaying tomatoes in Canada, were found to have some degree of toxicity to brine shrimp, but ethyl acetate extracts had a very low level of toxicity (Harwig et al., 1979). We have identified this strain as F. semitectum (Strain 6.3). The ethyl acetate extracts were mutagenic to Salmonella typhimurium; this was the only Fusarium strain that was found to be mutagenic in the study by Harwig et al. (1979). These extracts did not contain any of the six trichothecenes analyzed for by TLC and the mycotoxin(s) and mutagen(s) produced by this strain have not been identified.

Buffaloes. Cultures on autoclaved rice straw of an unnumbered isolate of *F. equiseti* from toxic rice straw associated with outbreaks of Degnala disease (see Degnala disease, above, and *F. equiseti,* Degnala disease) in India induced clinical signs and pathological lesions similar to field cases of the disease in three buffalo calves (Kalra et al., 1977a, 1980). We have examined three isolates of "*F. equiseti*" from rice straw associated with Degnala disease in India and have identified one of these as *F. equiseti* (Strain 7.1), but the other two proved to be *F. semitectum* (Strains 6.1 and 6.2). It is not known to us which of these three strains, if any, was the isolate of "*F. equiseti*" used in the successful experimental reproduction of Degnala disease in buffaloes by Kalra et al. (1977a, 1980). The strain of "*F. equiseti*" used did not produce butenolide or T-2 toxin, according to Kalra et al. (1977a, 1980), and consequently the chemical nature of the mycotoxin(s) produced in rice straw by this isolate is unknown.

Cattle. According to Qin et al. (1981), a disease of cattle known as "sore foot disease" (see Degnala disease, above, and *F. equiseti,* Degnala disease), which is associated with the consumption of rice straw in China, has been reproduced experimentally with cultures of *F. semitectum* and *F. equiseti.* This disease is characterized by necrosis of the feet and ears and appears to be similar to Degnala disease in India. The chemical nature of the mycotoxin(s) produced in rice straw by these Chinese isolates of *F. semitectum* is unknown but is under study according to Qin et al. (1981).

Chickens. Cultures of an unnumbered isolate of *F. oxysporum* from corn suspected of being toxic to cattle in the United States proved to be highly toxic to day-old chickens (Meronuck et al., 1970). Diets containing 10% culture material of this strain caused marked reductions in weight gain, while diets containing 40% culture material caused 100% mortality. An isolate of *F. oxysporum* has also been reported to cause marked decreases in feed intake, loss of body weight, and a complete cessation of egg production in laying hens (Speers et al., 1972). According to C.J. Mirocha (personal communication, 6 March 1978), the isolates referred to as *F. oxysporum* in the above publications are the same isolate which was referred to as "*F. roseum* (isolate oxyrose)" by Mirocha & Christensen (1974). We have identified this isolate as *F. semitectum* (Strain 6.4). The same strain was referred to as *F. roseum* 'Gibbosum' by Pathre et al. (1976) and Speers et al. (1977), who reported that diets containing either 2.5 or 5.0% culture material resulted in drastically reduced feed consumption, body weight gain, egg production, and egg quality of laying hens. In both treatment groups, necrotic lesions developed on the medial surface of the mandibles, on the dorsal surface of the tongue, and on the skin at the corners of the mouth. Analyses of the culture material revealed the presence of monoacetoxyscirpenol and zearalenone, and the toxicity of this strain to laying hens was attributed solely to monoacetoxyscirpenol (Pathre et al., 1976; Speers et al., 1977).

Guinea Pigs. According to Joffe (1960a, 1971, 1974c, 1978b), culture filtrates and powdered dry mycelium of unspecified isolates of *F. semitectum* from overwintered cereals associated with ATA (see *F. sporotrichioides,* ATA) in the USSR were toxic to guinea pigs. The animals died within five to 21 days and autopsies revealed hemorrhages in many organs and tissues. The mycotoxin(s) produced by these strains of *F. semitectum* toxic to guinea pigs have not been identified.

Ethyl acetate extracts of cultures on rice of *F. incarnatum* (Roberge) Sacc., isolated from moldy sorghum grain in India, have been reported to be dermotoxic to the shaved skin of guinea pigs and to contain T-2 toxin (Bhat & Rukmini, 1980; Rukmini & Bhat, 1978). We have identified this strain as *F. semitectum* (Strain 6.5).

Mice. Extracts of cultures on white corn grits incubated at 15°C for 21 days of two out of eight isolates of *F. semitectum* from toxic fescue hay associated with "fescue foot" (see *F. sporotrichioides*, fescue foot) of cattle in the United States were found to be toxic to mice upon intraperitoneal injection (Burmeister et al., 1981). The toxic extracts of both of these strains contained butenolide as well as T-2 toxin that could be detected by TLC.

Mall et al. (1979) reported that extracts of cultures on wheat of four of 10 isolates of *Fusarium* spp. from wheat in India were mildly toxic to Swiss albino mice upon intraperitoneal injection. We have identified one of these isolates as *F. semitectum* (Strain 6.6). In a continuation of this study, Gupta et al. (1981) found that each of three isolates of *Fusarium* spp. from corn and two of 14 isolates from wheat in India were mildly toxic to mice. We have identified one of the isolates from corn and one from wheat as *F. semitectum* (Strains 6.7 and 6.8). The mycotoxin(s) produced by these Indian isolates toxic to mice have not been identified.

Pigs. Investigations on the use of *F. semitectum* in the production of microbial protein revealed that diets containing culture material of *F. semitectum* IMI 135410 on a defined medium as the sole source of protein caused no ill effects or loss of appetite in pigs, "although these experiments were of far too short duration for assessing possible toxic effects" (Smith et al., 1975).

Rabbits. Extracts of cultures of four out of 27 isolates of *F. semitectum* from overwintered cereals associated with ATA (see *F. sporotrichioides*, ATA) in the USSR were found to be toxic to mildly toxic to rabbit skin (Joffe, 1960a, b, 1971). Culture filtrates and powdered dry mycelium of isolates of *F. semitectum* from overwintered cereals in the USSR have also been reported to be acutely toxic to rabbits (Joffe, 1960a, 1971, 1974c, 1978b). The rabbits died within 8 to 24 days and autopsies revealed hemorrhages in many organs and tissues. The mycotoxin(s) produced by the above isolates of *F. semitectum* toxic to rabbits have not been identified.

Methanol extracts of cultures of an isolate of *F. semitectum* from corn associated with an outbreak of acute aflatoxicosis in humans in India were found to be dermotoxic to rabbit skin (Jain et al., 1980). The Rf value of a spot obtained by TLC was "approximately equal to the Rf of nivalenoldiacetate" (Jain et al., 1980).

Rats. According to the reviews of Lin & Tang (1980) and Yang (1980), a strain of *F. semitectum* was isolated from cereals in Linxian county, an area with a high human esophageal cancer rate (see human esophageal cancer, above) in Henan Province, People's Republic of China. Cultures of this strain on corn meal were incubated at 26°C for 6 to 8 days and fed to 20 Wistar rats for 300 days. Tumors that developed during this period included one lymphosarcoma in the abdominal cavity and one bladder papilloma.

Ethyl acetate extracts of cultures on rice of *F. roseum* 'Gibbosum' isolated from a feed sample suspected of causing abortion and death in dairy cattle and pigs in Minnesota, USA, have been found to be dermotoxic to the shaved skin of rats (Pathre et al., 1976). We have identified this strain as *F. semitectum* (Strain 6.4). Topical application of these extracts to rat skin resulted in hyperkeratosis, petechial hemorrhages and severe gastro-intestinal hemorrhaging. Analyses of this culture material by GC-MS revealed the presence of 1 g/kg of monacetoxyscirpenol and 12 g/kg of zearalenone and the toxicity of the material was attributed to monoacetoxyscirpenol (Pathre et al., 1976). A rat skin test was also used as the bioassay in the isolation and characterization of four trichothecenes produced by this strain in PSC medium (Ishii et al., 1978).

An isolate of F. incarnatum from moldy sorghum grain in India was reported to be toxic to rats by Bhat & Rukmini (1980) and Rukmini & Bhat (1978). We have identified this strain as F. semitectum (Strain 6.5). Ethyl acetate extracts of rice cultures of this strain caused the death of four out of six rats within 7 days upon intraperitoneal injection. Pathological changes were characterized by acute necrotizing inflammatory lesions in the omentum, mesentery, and visceral and parietal peritoneum. Analyses of the toxic extracts revealed the presence of T-2 toxin, which was presumably responsible for the toxicity. Ether extracts of this culture material did not cause uterine hypertrophy in weanling female rats, thus indicating the absence of uterotrophic metabolites such as zearalenone (Bhat & Rukmini, 1980; Rukmini & Bhat, 1978).

Ceruti Scurti et al. (1971) reported that among several species of fungi tested, the only strain that has notable activity in the rat uterotrophic bioassay was F. semitectum 136418 obtained from the Instituto Botanico di Torino. Subsequently Tuttobello et al. (1974) confirmed that a strain of F. semitectum, obtained from the same institute and presumably the same one that was used by Ceruti Scurti et al. (1971), had uterotrophic effects in weanling female rats. The chemical nature of the uterotrophic substance(s) was not determined, but the uterotrophic effects were reportedly equivalent to that of approximately 21 ppm of "zearalanol" in shake cultures in liquid medium and 17.6 ppm in still cultures on corn (Tuttobello et al., 1974).

According to Agarwal & Chauhan (1980), cultures of an unnumbered isolate of F. semitectum fed to weanling albino rats for 60 days caused liver and kidney damage as judged by increases in transaminase enzymes and decreases in total protein and hemoglobin.

Investigations on the use of F. semitectum in the production of microbial protein revealed that diets containing culture material of F. semitectum IMI 135410 on a defined medium fed as the sole source of protein had no ill effects in rats fed over a period of one to 2 years (Smith et al., 1975). These results led Smith et al. (1975) to conclude that "F. semitectum is essentially a safe organism for feeding to simple stomached animals as a protein source."

Turkeys. Cultures of an isolate of F. oxysporum from corn suspected of being toxic to cattle in the United States proved to be highly toxic to day-old turkey poults (Meronuck et al., 1970). Diets containing 10% culture material of this strain caused marked reductions in weight gain, while diets containing 40% culture material caused 100% mortality. According to C.J. Mirocha (personal communication, 6 March 1978), the isolate referred to as F. oxysporum by Meronuck et al. (1970) was referred to as "F. roseum (isolate oxyrose)" by Mirocha & Christensen (1974) and as F. roseum 'Gibbosum' by Pathre et al. (1976). We have identified this strain as F. semitectum (Strain 6.4). Cultures of this strain (as F. roseum 'Gibbosum') on rice were fed to day-old turkey poults at dietary levels of 1.25, 2.5, and 5.0% for 7 days (Pathre et al., 1976). Turkeys in all treatment groups developed bilateral inflammation of the beak and the group that received 5% culture material died within 7 days. Analyses of the culture material revealed the presence of monoacetoxyscirpenol and zearalenone, but the toxicity of this strain to turkeys was attributed solely to monoacetoxyscirpenol (Pathre et al., 1976).

Mycotoxins Produced

4-Acetoxyscirpenediol. This new trichothecene was isolated from *F. roseum* 'Gibbosum' (3-8-66 No. 10) by Ishii et al. (1978). We have identified this strain as *F. semitectum* (Strain 6.4). The yield of pure 4-acetoxyscirpenediol obtained by Ishii et al. (1978) was 15 mg from 8.3 ℓ of culture filtrate of PSC medium incubated at 25°C for 14 days.

Butenolide. Burmeister et al. (1971) detected butenolide by means of TLC in cultures of two isolates of *F. semitectum* (No. 1 and 4) on Sabouraud's agar incubated at 15°C for 21 days. According to Suzuki et al. (1981a, b), four of eight isolates of *F. semitectum* from wheat and barley in Japan produced butenolide at levels ranging from 0.06 to 1.18 µg/g in cultures on rice as detected by GC with an electron-capture detector. We have not seen the above isolates referred to as *F. semitectum* and there are no known butenolide-producing strains of *F. semitectum* represented in the ITFRC.

Diacetoxyscirpenol. An isolate of *F. roseum* 'Gibbosum' from feed suspected of causing abortion and death of cattle and pigs in Minnesota, USA, produced unspecified levels of diacetoxyscirpenol in liquid Czapek's medium, but not in cultures on rice or corn (Pathre et al., 1976). We have identified this strain as *F. semitectum* (Strain 6.4). Diacetoxyscirpenol was also detected by GC-MS in extracts of cultures of this strain in PSC medium incubated at 25°C for 14 days (Ishii et al., 1978).

According to Suzuki et al. (1980, 1981b), eight of 23 isolates of *F. semitectum* from wheat and barley in Japan produced diacetoxyscirpenol at levels ranging from 0.44 to 23.49 µg/g as detected by GC. The highest yield was obtained with *F. semitectum* 5D-14, which also produced nivalenol and neosolaniol (Suzuki et al., 1980).

Fusarenon-X. According to Suzuki et al. (1980, 1981b), eight of 23 isolates of *F. semitectum* from wheat and barley in Japan produced fusarenon-X at levels ranging from 0.09 to 6.00 µg/g as detected by GC. We have not seen these isolates and there are no known fusarenon-X–producing strains of *F. semitectum* represented in the ITFRC.

Monoacetoxysicrpenol. This trichothecene was first isolated from an isolate of *F. roseum* 'Gibbosum' from feed suspected of causing abortion and death of cattle and pigs in Minnesota, USA (Pathre et al., 1976). We have identified this strain as *F. semitectum* (Strain 6.4). Cultures on rice incubated at 28°C for 7 days followed by 21 days at 12°C were found by GC-MS to contain 1 g/kg monoacetoxyscirpenol. This compound was also produced on a solid corn grits medium (Pathre et al., 1976) and in the PSC medium incubated at 25°C for 14 days (Ishii et al., 1978).

The toxicity of this strain of *F. semitectum* (= *F. oxysporum*, = *F. roseum* isolate oxyrose = *F. roseum* 'Gibbosum,' Strain 6.4) to chickens, rats, and turkeys has been attributed solely to monoacetoxyscirpenol (Ishii et al., 1978; Mirocha & Christensen, 1974; Pathre et al., 1976; Speers et al., 1977).

Neosolaniol. According to Suzuki et al. (1980, 1981b), eight of 23 isolates of *F. semitectum* from wheat and barley in Japan produced neosolaniol at levels ranging from 0.59 to 12.87 µg/g as detected by GC. One of the highest-yielding isolates, *F. semitectum* 5D-14, produced 12.83 µg/g of neosolaniol as well as diacetoxyscirpenol and nivalenol (Suzuki et al., 1980). We have not seen these isolates and there are no known neosolaniol-producing strains of *F. semitectum* represented in the ITFRC.

Nivalenol. According to Suzuki et al. (1980, 1981b), eight of 23 isolates of *F. semitectum* from wheat and barley in Japan produced nivalenol at levels ranging from 0.16 to 4.92

μg/g as detected by GC. The highest yield was obtained with *F. semitectum* 5D-14, which also produced diacetoxyscirpenol and neosolaniol (Suzuki et al., 1980). We have not seen these isolates and there are no known nivalenol-producing strains of *F. semitectum* represented in the ITFRC.

Nivalenol diacetate. Analysis by TLC of dermotoxic extracts of an Indian isolate of *F. semitectum* revealed a spot with an Rf value "approximately equal to the Rf value of nivalenoldiacetate" (Jain et al., 1980). The production of nivalenol diacetate by this isolate has not been confirmed by appropriate analytical methods and there are no known nivalenol diacetate-producing strains of *F. semitectum* represented in the ITFRC.

Scirpentriol. *Fusarium semitectum* (= *F. roseum* 'Gibbosum,' Strain 6.4) produced unspecified levels of scirpentriol in cultures on a solid corn grits medium, but not on rice (Pathre et al., 1976). Scirpentriol was also detected by GC-MS in extracts of cultures of this strain in PSC medium incubated at 25°C for 14 days (Ishii et al., 1978).

T-2 Toxin. Burmeister et al. (1971) detected T-2 toxin by means of TLC in cultures of two isolates of *F. semitectum* (Nos. 1 and 4) on white corn grits incubated at 15°C for 21 days.

An isolate of *F. incarnatum* from moldy sorghum grain in India has been reported to produce unspecified levels of T-2 toxin in cultures on rice incubated at 30°C for 7 days (Bhat & Rukmini, 1980; Rukmini & Bhat 1978; Rukmini et al., 1980). We have identified this strain as *F. semitectum* (Strain 6.5). The moldy sorghum grain from which this strain was isolated was reportedly naturally contaminated with T-2 toxin at a level of 24.5 mg/kg (Rukmini et al., 1980). The T-2 toxin both in the naturally contaminated sorghum grain and in the rice cultures of this strain of *F. semitectum* (= *F. incarnatum*) was detected by TLC; the identity of the crystalline toxic compound isolated from these substrates was confirmed as T-2 toxin by melting point, IR spectrum, NMR, and GLC (Bhat & Rukmini, 1980; Rukmini & Bhat, 1978). This strain has also been used for the production of semi-purified T-2 toxin for use in toxicological studies in monkeys (Rukmini et al., 1980).

Negative results for T-2 toxin production have been reported for 23 isolates of *F. semitectum* from wheat and barley in Japan (Suzuki et al., 1980, 1981b).

Zearalenol. *Fusarium roseum* No. S-74-1c, isolated from head-blighted sorghum infected by *F. roseum* 'Gibbosum' and *F. roseum* 'Semitectum' in the United States, produced zearalenol in cultures on either cracked corn or grain sorghum incubated at 25°C for 20 days (Stipanovic & Schroeder, 1975). These authors did not state whether *F. roseum* No. S-74-1c was a strain of *F. roseum* 'Gibbosum' or *F. roseum* 'Semitectum,' and a culture is no longer available to determine its identity (H.W. Schroeder, personal communication, 25 June 1981).

Analysis by TLC, GLC, HPLC, and GC-MS of cultures of *F. semitectum* (= *F. roseum* 'Gibbosum,' strain 6.4) on rice incubated at 25°C for 7 days followed by 30 days at 10 to 12°C revealed the presence of 563 μg/g of alpha zearalenol (Hagler et al., 1979).

The production of zearalenol in culture by the above strains is important because this compound is about three times more estrogenically active than zearalenone in the rat uterotrophic bioassay (Hagler et al., 1979). The fact that at least two *Fusarium* isolates have been shown to produce zearalenol in culture suggested the possibility of its occurrence in animal feeds (Hagler et al., 1979). Subsequently Christensen (1979) reported that "zearalenol has been found in four field samples that were associated with estrogenism and/or other symptoms of toxicosis in farm animals."

Zearalenone. Samples of head-blighted sorghum grain infected by *F. roseum* 'Gibbosum' and *F. roseum* 'Semitectum' in the United States have been reported to be naturally contaminated with zearalenone (Schroeder & Hein, 1975). Eight isolates of *Fusarium* spp. from this sorghum grain produced zearalenone in culture on sorghum grain incubated for 19 days. Zearalenone was detected by TLC and the identity of crystalline material isolated from these cultures was verified by GC-MS. The highest-yielding isolate, *F. roseum* S-74-1c, was grown on cracked corn or grain sorghum at a constant temperature of 25°C for 19 days or at 25°C for 6 days followed by 13 days at 10°C. The yields of zearalenone were obtained by weighing the purified crystals. In contrast to the results of previous studies, this isolate produced considerably more zearalenone (1593 mg/kg) in cultures on corn incubated at a constant temperature of 25°C than at 25°C followed by 10°C (499 mg/kg). This isolate also produced more zearalenone in culture on sorghum (3030 mg/kg) than on corn at 25°C (Schroeder & Hein, 1975). In addition to zearalenone, this isolate also produced zearalenol and 8'-hydroxyzearalenone (Stipanovic & Schroeder, 1975). The above authors did not state whether *F. roseum* S-74-1c was a strain of *F. roseum* 'Gibbosum' or *F. roseum* 'Semitectum' and a culture is no longer available to determine its identity (H.W. Schroeder, personal communication, 25 June 1981).

Cultures of an unnumbered isolate of *F. roseum* 'Gibbosum' on corn incubated at 25°C for 2 weeks followed by 4 weeks at 12°C were incorporated into diets fed to laying hens by Speers et al. (1977). We have identified this strain as *F. semitectum* (Strain 6.4). According to Speers et al. (1977), "analyses of the *F. roseum* 'Gibbosum' emended diet (5% of the total preparation) indicated that the final feed preparation contained 2,500 ppm zearalenone." If this is correct, the zearalenone content of the corn cultures can be calculated as 50 g/kg. Other reported zearalenone yields by this strain include the following: "copious amounts" by Mirocha & Christensen (1974, as *F. roseum* isolate oxyrose); 12 g/kg in cultures on rice incubated at 25°C for 7 days followed by 21 days at 12°C, but only "a small amount" on a solid corn grits medium (Pathre et al., 1976); and 8000 μg/g in cultures on rice incubated at 25°C for 7 days followed by 30 days at 10 to 12°C (Hagler et al., 1979). This strain has also been used in the biosynthesis of (^{14}C) zearalenone from (1–^{14}C) acetate in cultures on rice (Hagler & Mirocha, 1980). In addition to zearalenone, *F. semitectum* (= *F. roseum* 'Gibbosum', Strain 6.4) also produces zearalenol and 8'-hydroxyzearalenone (Hagler et al., 1979).

According to Ichinoe et al. (1977), one out of four isolates of *F. semitectum* from cereals and legumes in Japan produced zearalenone (17 μg/g). Suzuki et al. (1981b) reported that one out of eight Japanese isolates of *F. semitectum* produced zearalenone.

Biological testing of ether extracts of rice cultures of the Indian isolate *F. semitectum* (= *F. incarnatum*, Strain 6.5), in the rat uterotrophic bioassay, indicated that this strain did not produce zearalenone (Bhat & Rukmini, 1980; Rukmini & Bhat, 1978). Negative results for zearalenone production have also been reported for five isolates of *F. semitectum* from barley grain in the United Kingdom (Hacking et al., 1976, 1977) and for five isolates from peanuts imported into Japan (Suzuki et al., 1978).

Toxigenic Strains in the ITFRC

The *Fusarium* strains listed in Table 6.1 all have been identified as *F. semitectum* according to Nelson et al. (1983). Toxigenic strains referred to as members of the section Arthrosporiella in the literature but identified as other species by us are listed in Table 6.2.

Table 6.1 *Fusarium* Strains of the Section Arthrosporiella in the International Toxic *Fusarium* Reference Collection (ITFRC)

Strain No.	ITFRC No.	MRC No.	Published Name(s) and Number(s)	Source	Toxigenicity
Fusarium semitectum					
6.1	R-6606	2610	*F. equiseti* No. 20	R.K. Paul Gupta	a,b?
6.2	R-6607	2611	*F. equiseti* No. 5	R.K. Paul Gupta	a,b?
6.3	R-6355	2191	*Fusarium* sp. HPB 110178-21	B.J. Blanchfield	b
6.4	R-4485	1642	*F. oxysporum* = *F. roseum* isolate oxyrose = *F. roseum* 'Gibbosum'	C.J. Mirocha	a,b,c,d,e,f,g,h
6.5	R-5169	1445	*F. incarnatum*	R.V. Bhat	b,i
6.6	R-6520	2636	*Fusarium* sp. PL W-517	V.C. Vora	b
6.7	R-6983	2804	*Fusarium* sp. PL Z-81	V.C. Vora	b
6.8	R-6984	2806	*Fusarium* sp. PL W-865	V.C. Vora	b

a = Associated with human and/or animal diseases
b = Culture toxic to experimental animals
c = Produces 4-acetoxyscirpenediol
d = Produces diacetoxyscirpenol
e = Produces monoacetoxyscirpenol
f = Produces scirpentriol
g = Produces zearalenol
h = Produces zearalenone
i = Produces T-2 toxin

Table 6.2 *Fusarium* Strains Published as Species in Section Arthrosporiella but Identified as Other Species in the International Toxic *Fusarium* Reference Collection (ITFRC)

Strain No.	ITFRC No.	MRC No.	Published Name(s) and Number(s)	Source	Present Identification
7.12	R-6052	2231	*F. anguioides* NRRL 3020	NRRL	*F. equiseti*
7.13	R-6054	2232	*F. concolor* NRRL 3214	NRRL	*F. equiseti*
7.14	R-6324	2433	*F. diversisporum* NRRL 3211	NRRL	*F. equiseti*

Strain 6.1 *F. equiseti* No. 20 (= R-6066; MRC 2610)

This strain was isolated from toxic rice straw associated with outbreaks of Degnala disease (see Degnala disease, above) of buffaloes and cattle in Haryana, India (Kalra et al., 1973, 1977a, b, 1980; Kwatra & Singh, 1973). According to Kalra et al. (1977a, b, 1980), typical clinical signs and pathological lesions of Degnala disease, including edema of the legs and gangrene of the tail tip, were reproduced experimentally in three buffalo calves with cultures on rice straw of a strain of *F. equiseti* isolated from rice straw in India and identified at CMI.

A slant culture of this strain was received on 26 September 1981 from R.K. Paul Gupta as "*F. equiseti* No. 20, isolated from rice straw associated with Degnala disease." This is a typical culture of *F. semitectum* and so is another isolate from the same source, *F. semitectum* (= *F. equiseti*) No. 5 (Strain 6.2). However, a third strain proved to be *F. equiseti* (Strain 7.1). It is not known to us which of these three strains, if any, was used in the successful experimental reproduction of Degnala disease in buffaloes by Kalra et al. (1977a, b, 1980).

Strain 6.2 F. equiseti No. 5 (= R-6607; MRC 2611)

This strain was isolated from toxic rice straw associated with outbreaks of Degnala disease (see Degnala disease, above) of buffaloes and cattle in India (Kalra et al., 1973, 1977a, b, 1980; Kwatra & Singh, 1973). According to Kalra et al. (1977a, b, 1980), typical clinical signs and pathological lesions of Degnala disease were reproduced experimentally in three buffalo calves with cultures on rice straw of a strain of F. equiseti isolated from rice straw in India and identified at CMI.

A slant culture of this strain was received on 26 September 1981 from R.K. Paul Gupta as "F. equiseti No. 5, isolated from rice straw associated with Degnala disease." This is a typical culture of F. semitectum, and so is another isolate from the same source, F. semitectum (= F. equiseti) No. 20 (Strain 6.1). However, a third strain proved to be F. equiseti (Strain 7.1). It is not known to us which of these three strains, if any, was used in the successful experimental reproduction of Degnala disease in buffaloes by Kalra et al. (1977a, b, 1980).

Strain 6.3 Fusarium sp. HPB 110178-21 (R-6355; MRC 2191)

This strain was isolated from a decaying tomato in Canada (Harwig et al., 1979). Filtrates of cultures of this strain in YES medium incubated at 25°C for 8 to 13 days were slightly toxic to brine shrimp, but ethyl acetate extracts of these cultures had a very low level of toxicity (Harwig et al., 1979). The ethyl acetate extracts were, however, mutagenic to Salmonella typhimurium, and this was the only Fusarium strain that was found to be mutagenic in the study by Harwig et al. (1979). These extracts did not contain any of the six trichothecenes analyzed for by TLC, and the mycotoxin(s) and mutagen(s) produced by this strain have not been identified.

A slant culture of this strain was received on 22 December 1980 from B.J. Blanchfield as "Fusarium sp. HPB 110178-21." This culture is almost completely mycelial, but still produced a few spindle-shaped macroconidia on polyphialides which allowed it to be identified as F. semitectum.

Strain 6.4 F. oxysporum = F. roseum isolate oxyrose = F. roseum 'Gibbosum' (= R-4485; MRC 1642)

This strain (as F. oxysporum) was isolated from a "sample of corn that was suspected of being toxic to cattle" in the United States (Meronuck et al., 1970). The source of this strain (as F. roseum 'Gibbosum') was given as a "feed sample" associated with "abortion and death in dairy cattle and swine" in Minnesota, USA (Pathre et al., 1976).

Diets containing 10% culture material on corn of this strain (as F. oxysporum) caused marked reductions in weight gain of day-old chicks and turkey poults, while diets containing 40% culture material caused 100% mortality (Meronuck et al., 1970). An isolate of F. oxysporum was also reported to cause marked decreases in feed intake, loss of body weight, and a complete cessation of egg production in laying hens (Speers et al., 1972). According to C.J. Mirocha (personal communication, 6 March 1978), the isolates referred to as F. oxysporum by Meronuck et al. (1970) and Speers et al. (1972) are the same isolate which was referred to as "F. roseum (isolate oxyrose)" by Mirocha & Chrstensen (1974) and as F. roseum 'Gibbosum' by Pathre et al. (1976) and Speers et al. (1977). The latter authors reported that diets containing either 2.5 or 5.0% culture material resulted in drastically reduced feed consumption, body weight gain, egg production, and egg quality in laying hens. In both treatment groups, necrotic lesions developed on the medial surface of the mandibles, on the dorsal surface of the tongue, and on the skin at the corners of the mouth (Speers et al., 1977) Cultures on rice were also fed to day-old turkey poults at dietary levels of 1.25, 2.5, and 5.0% for 7 days (Pathre et al., 1976). Turkeys in all treatment groups developed bilateral inflammation of the beak, and the group that received 5% culture material died within 7 days. Analyses of the culture material revealed the presence of monoacetoxyscirpenol and zearalenone, but the toxicity of this strain to chickens and turkeys was attributed to monoacetoxyscirpenol (Pathre et al., 1976; Speers et al., 1977).

Ethyl acetate extracts of cultures of this strain (as *F. roseum* 'Gibbosum') on rice have been found to be dermotoxic to the shaved skin of rats (Pathre et al., 1976). Topical application of these extracts to rat skin resulted in hyperkeratosis, petechial hemorrhages, and severe gastro-intestinal hemorrhaging. Analyses of this culture material by GC-MS revealed the presence of 1 g/kg of monoacetoxyscirpenol and 12 g/kg of zearalenone, but the toxicity of the material was attributed to monoacetoxyscirpenol (Pathre et al., 1976). A rat skin test was also used as the bioassay in the isolation and characterization of four trichothecenes produced by this strain in PSC medium (Ishii et al., 1978).

Mycotoxins. This strain of *F. semitectum* (= *F. oxysporum*, = *F. roseum* isolate oxyrose, = *F. roseum* 'Gibbosum') is known to produce the following six mycotoxins:

4-Acetoxyscirpenediol. This new trichothecene was isolated from this strain, referred to as *F. roseum* 'Gibbosum' (3-8-66 No. 10), by Ishii et al. (1978). The yield of pure 4-acetoxyscirpenediol obtained by Ishii et al. (1978) was 15 mg from 8.3 ℓ of culture filtrate of PSC medium incubated at 25°C for 14 days.

Diacetoxyscirpenol. This strain produced unspecified levels of diacetoxyscirpenol in liquid Czapek's medium, but not in cultures on rice or corn (Pathre et al., 1976). Diacetoxyscirpenol was also detected by GC-MS in extracts of cultures of this strain in PSC medium incubated at 25°C for 14 days (Ishii et al., 1978).

Monoacetoxyscirpenol. This trichothecene was first isolated from this strain by Pathre et al. (1976). Cultures on rice incubated at 28°C for 7 days followed by 21 days at 12°C were found by GC-MS to contain 1 g/kg of monoacetoxyscirpenol. This compound was also produced on a solid corn grits medium (Pathre et al., 1976) and in PSC medium incubated at 25°C for 14 days (Ishii et al., 1978). The toxicity of this strain to chickens, rats and turkeys has been attributed solely to monoacetoxyscirpenol (Ishii et al., 1978; Mirocha & Christensen, 1974; Pathre et al., 1976; Speers et al., 1977).

Scirpentriol. This strain produced unspecified levels of scirpentriol in cultures on a solid corn grits medium, but not on rice (Pathre et al., 1976). Scirpentriol was also detected by GC-MS in extracts of cultures of this strain in PSC medium incubated at 25°C for 14 days (Ishii et al., 1978).

Zearalenol. Analyses by TLC, GLC, HPLC, and GC-MS of cultures of this strain on rice incubated at 25°C for 7 days followed by 30 days at 10 to 12°C revealed the presence of 563 µg/g of alpha zearalenol (Hagler et al., 1979).

Zearalenone. This strain (as *F. roseum* isolate oxyrose) was reported to produce "copious amounts" of zearalenone by Mirocha & Christensen (1974). According to Speers et al. (1977), a diet containing 5% culture material of this strain (as *F. roseum* 'Gibbosum') was found to contain 2,500 ppm of zearalenone in the final feed preparation. If this is correct, the zearalenone content of the corn cultures can be calculated as 50 g/kg. Other reported zearalenone yields by this strain include the following: 12 g/kg in cultures on rice incubated at 25°C for 7 days followed by 21 days at 12°C, but only "a small amount" on a solid corn grits medium (Pathre et al., 1976); and 8000 µg/g in cultures on rice incubated at 25°C for 7 days followed by 30 days at 10 to 12°C (Hagler et al., 1979). This strain has also been used in the biosynthesis of (^{14}C) zearalenone from (1–^{14}C) acetate in cultures on rice (Hagler & Mirocha, 1980).

In addition to zearalenone, this strain (as *F. roseum* 'Gibbosum') also produces zearalenol and 8'-hydroxyzearalenone (Hagler et al., 1979).

A soil culture of this strain was received on 9 December 1977 from C.J. Mirocha as "*F. roseum* 'Gibbosum' isolate Oxy-Rose." We consider this culture to be a typical representative of *F. semitectum*. According to C.J. Mirocha (personal communication, 6 March 1978), this is a culture of the strain that has been referred to either as *F. oxysporum, F. roseum* (isolate oxyrose), or *F. roseum* 'Gibbosum' in the above publications.

On the basis of the high degree of toxicity of this strain of *F. roseum* 'Gibbosum,' which we have identified as *F. semitectum,* Pathre et al. (1976) concluded that "*F. roseum* and not *F. tricinctum* is most commonly associated with corn implicated in field cases of toxicosis," that *F. roseum* is toxic "in all probability by virtue of the production of the scirpene type of trichothecenes such as mono- and diacetoxyscirpenol," and that "*F. roseum* is more important in terms of frequency of occurrence in the production of moldy corn diseases than the T-2 toxin as described by Smalley et al. (1970)." The "*F. roseum*" that Pathre et al. (1976) had in mind when reaching these conclusions was probably *F. graminearum* rather than this strain of *F. semitectum,* which is probably not commonly associated with corn in the United States at all. *Fusarium graminearum* certainly has a high frequency of occurrence in corn, but is not known to produce monoacetoxyscirpenol and rarely produces diacetoxyscirpenol (see *F. graminearum,* mycotoxins produced). This is another example of the serious confusion introduced by the use of the name *F. roseum* and of taxonomic inaccuracies leading to inaccurate conclusions.

Strain 6.5 *F. incarnatum* (= R-5169; MRC 1445)

This strain was isolated from moldy sorghum grain contaminated with T-2 toxin in Hyderabad, India (Bhat & Rukmini, 1980; Rukmini & Bhat, 1978). Ethyl acetate extracts of cultures of this strain on rice incubated at 30°C for 7 days were dermotoxic to the shaved skin of guinea pigs (Bhat & Rukmini, 1980; Rukmini & Bhat, 1978). These extracts also caused the death of four out of six rats within 7 days upon intraperitoneal injection. Pathological changes were characterized by acute necrotizing inflammatory lesions in the omentum, mesentery, and visceral and parietal peritoneum (Bhat & Rukmini, 1980; Rukmini & Bhat, 1978).

The toxic culture material on rice was found to contain unspecified levels of T-2 toxin and the identity of the crystalline toxin isolated from this material was confirmed by melting point, IR spectrum, NMR, and GLC (Bhat & Rukmini, 1980; Rukmini & Bhat, 1978). The T-2 toxin produced by this strain was presumably responsible for its toxicity to guinea pigs and rats. This strain has also been used for the production of semi-purified T-2 toxin used in toxicological studies in monkeys (Rukmini et al., 1980).

Ether extracts of culture material of this strain did not cause uterine hypertrophy in weanling female rats, thus indicating that absence of uterotrophic metabolites such as zearalenone (Bhat & Rukmini, 1980; Rukmini & Bhat, 1978).

A slant culture of this strain was received on 8 February 1979 from R.V. Bhat as "*F. incarnatum.*" We consider this culture to be a typical representative of *F. semitectum*. According to R.V. Bhat (personal communication, 20 July 1978), this strain was identified "as *F. incarnatum* following the classification of Prof. C.V. Subramanian. However, if one follows Dr. Colin Booth, this particular isolate can be placed in *F. semitectum.*" Wollenweber & Reinking (1935) list *F. incarnatum* (Roberge) Sacc. as a synonym of *F. semitec-*

tum Berk & Rav. var. *majus* Wollenw. This variety was included in *F. semitectum* by Nelson et al. (1983).

Strain 6.6 *Fusarium* sp. PL W-517 (= R-6520; MRC 2636)

This strain was isolated from wheat grain in India (Mall et al., 1979). Chloroform extracts of cultures of this strain on wheat incubated at 27 to 28°C for 15 days were toxic to Swiss albino mice upon intraperitoneal injection (Mall et al., 1979). The mycotoxin(s) produced by this strain have not been identified.

A slant culture of this strain of *F. semitectum* was received on 24 August 1981 from V.C. Vora as "*Fusarium* sp. PL W-517, from wheat, toxic to mice."

Strain 6.7 *Fusarium* sp. PL Z-81 (= R-6983; MRC 2804)

This strain was isolated from corn in India (Gupta et al., 1981). Chloroform extracts of cultures of this strain on corn incubated at 28°C for 15 days were mildly toxic to Swiss albino mice upon intraperitoneal injection (Gupta et al., 1981). The mycotoxin(s) produced by this strain have not been identified.

A slant culture of this strain of *F. semitectum* was received on 4 June 1982 from V.C. Vora as "*Fusarium* sp. PL Z-81, from maize, toxic to mice."

Strain 6.8 *Fusarium* sp. PL W-865 (= R-6984; MRC 2806)

This strain was isolated from wheat in India (Gupta et al., (1981). Chloroform extracts of cultures of this strain on wheat incubated at 28°C for 15 days were mildly toxic to Swiss albino mice upon intraperitoneal injection (Gupta et al., 1981). The mycotoxin(s) produced by this strain have not been identified.

A slant culture of this strain of *F. semitectum* was received on 4 June 1982 from V.C. Vora as "*Fusarium* sp. PL W-865, from wheat, toxic to mice."

7. Section Gibbosum

The problems associated with the use of the name *F. roseum* (Lk.) Synd. & Hans. also apply to section Gibbosum (see section Roseum).

Fusarium equiseti (Corda) Sacc. sensu Gordon

Perfect State: *Gibberella intricans,* Wollenw.

Incidence and Distribution

Fusarium equiseti is a cosmopolitan soil saprophyte that is particularly common in subtropical and tropical areas of the world. It is associated with dead and dying plant tissue, seeds, and root, foot, stem, and fruit rots of a great variety of plants (Booth, 1971; Doidge, 1938; Domsch et al., 1980; Gerlach & Ershad, 1970; Gordon, 1956, 1960a; Joffe, 1973b; Joffe & Palti, 1967; Meyer & Frank, 1979; Palti, 1978; Wollenweber & Reinking, 1935). The pathogenic potential of *F. equiseti* to plants adapted to warm climates, such as avocados and cucurbits, may have been underestimated (Joffe & Palti, 1967). However, *F. esquiseti* is not restricted to the warmer areas of the world and has been isolated from overwintered cereals in the USSR (Joffe, 1960a, b, 1971). It also occurs as a soil saprophyte (Domsch et al., 1980) and in association with cereal grains (Gordon, 1952, 1959; Marasas et al., 1979b; Ylimäki, 1981) in temperate regions of North America, Europe, and the USSR.

Association with Human and/or Animal Diseases

Degnala Disease. This disease occurs during winter in buffaloes and cattle fed almost exclusively on rice straw in low-lying, waterlogged rice-growing areas of Pakistan and India. In the Indian states of Haryana and Punjab, the morbidity and mortality among buffaloes, and to a lesser extent among cattle, have been increasing since 1969–70 and the disease appears to be spreading to other areas that have been brought under extensive rice cultivation. The disease is characterized by edematous swelling of the legs and necrosis, gangrene, and sloughing of the extremities. Pathological changes in gangrenous areas are characterized by thickening of the arterial walls and consequent reduction or occlusion of the lumen and thrombus formation. No significant lesions are found in the visceral organs, but focal areas of myocardial necrosis and endocardial hemorrhages occur in association with gelatinous degeneration of pericardial fat (Bhat et al., 1978; Irfan, 1971; Kalra et al., 1973, 1977a, b, 1980; Kwatra & Singh, 1973).

By the feeding of rice straw from farms where field outbreaks had occurred, the characteristic clinical signs and pathological lesions of Degnala disease have been reproduced in 17 buffalo calves by Kalra et al. (1973, 1977a, 1980) and in 16 calves fed during winter by Kwatra & Singh (1973). Subsequently the disease was reproduced experimentally in three buffalo calves (see buffaloes, below) fed cultures on rice straw of an isolate of *F. equiseti* from the toxic rice straw (Kalra et al., 1977a, 1980). The isolate of *F. equiseti* used in the successful experimental reproduction of Degnala disease in buffaloes was also shown to be dermotoxic to rabbit skin (Kalra et al., 1973, 1977a, 1980). This isolate "was sent to Dr Yates of Northern Regional Research Laboratory, Illinois, USA, but no butenolide or T-2 toxin was found to be produced by this culture" (Kalra et al., 1977a, 1980). Thus the mycotoxin(s) produced by this strain have not been identified.

We have identified one isolate from rice straw associated with Degnala disease in India as *F. equiseti* (Strain 7.1). However, two other isolates of "*F. equiseti*" from the same source proved to be *F. semitectum* (Strains 6.1 and 6.2). It is not known to us which of these three strains, if any, was used in the successful experimental reproduction of Degnala disease in three buffalo calves with a pure culture of an isolate of *F. equiseti* by Kalra et al., (1977a, 1980). Consequently it is not certain whether Degnala disease is caused by *F. equiseti* or *F. semitectum* or both.

It is interesting that a similar disease known as "sore foot disease" (see *F. semitectum,* Degnala disease) of cattle consuming rice straw in China, has been "proved to be caused by *Fusarium equiseti* and *Fusarium semitectum*" (Qin et al., 1981). This disease is characterized by necrosis of the feet, legs, and ears; subcutaneous edema and hemorrhage in the affected areas; congestion of capillaries; thrombosis of arteries; and edema and necrosis of muscles. According to Qin et al. (1981), this disease has been produced experimentally in cattle with cultures of *F. equiseti* and *F. semitectum*.

Another similar disease is "Fescue foot" of cattle in the United States (see *F. sporotrichioides,* fescue foot). One of the characteristic pathological lesions of this disease, i.e., necrosis of the tail tip, has been reproduced experimentally in cattle by prolonged daily administration of high doses of butenolide (Grove *et al.,* 1970; Kosuri et al., 1970; Tookey et al., 1972). Butenolide-producing strains of *F. semitectum* (see *F. semitectum,* mycotoxins produced) as well as *F. equiseti* (see mycotoxins produced, below) have been isolated from fescue hay in the United States (Burmeister et al., 1971). We have confirmed the identity of one of these isolates as *F. equiseti* (Strain 7.2)

In conclusion, there is experimental evidence from China, India, and the United States that gangrenous diseases that affect buffaloes and cattle and have certain similarities to ergotism may be caused by fescue hay or rice straw colonized by *F. equiseti* and/or *F. semitectum*. Although there is some indication that the butenolide known to be produced by both species of fungi may be involved, the mycotoxin(s) responsible have not been identified in the naturally contaminated toxic substrates and the etiology of these diseases has not been finally resolved.

Bean Hulls Poisoning. This disease of horses in Hokkaido, Japan (see *F. sporotrichioides*, bean hulls poisoning, and *F. moniliforme*, equine leukoencephalomalacia) has been reproduced experimentally in horses with naturally infected bean hulls (Konishi & Ichijo, 1970b). In a mycotoxicological study of toxic bean hulls from Hokkaido, Ueno et al. (1972a) found that they were extensively colonized by *Fusarium* spp. and that cultures of many of these *Fusarium* isolates were acutely toxic to mice. We have identified the following three isolates from this source as *F. equiseti*: *F. roseum* M-2-5, *F. roseum* 'Culmorum' M-14-2, and *F. roseum* 'Avenaceum' M-7-1 (Strains 7.3, 7.4, and 7.5). All three of these strains caused radiomimetic pathological changes in mice, and two of them (Strains 7.3 and 7.4) produced diacetoxyscirpenol (Ueno et al., 1973a). There is, however, no experimental evidence that "bean hulls poisoning" of horses is a form of trichothecene toxicosis, other than the fact that trichothecene-producing *Fusarium* species, including these diacetoxyscirpenol-producing strains of *F. equiseti*, have been isolated from bean hulls in Hokkaido.

Tibial Dyschondroplasia. A number of etiological factors have been implicated in this bone disease of poultry, and recent work has indicated that mycotoxins produced by "*F. roseum*" can induce this abnormality (Lee & Mirocha, 1981; Walser et al., 1980). We have identified the isolate of "*F. roseum*" from overwintered barley in Alaska that was used in this work as *F. equiseti* (Strain 7.6).

Culture material of this strain was fed to day-old broiler chicks (see chickens, below) in balanced diets containing 2%, 5%, or 10% culture material (Walser et al., 1980). The two higher levels were lethal within one to 3 weeks, but the chicks fed diets containing 2% culture material survived the experimental period of 6 weeks. Tibial dyschondroplasia, characterized by the presence of a cone of cartilage that extended distally from the growth plate of the proximal tibia, was first noted in chicks that died after 8 to 11 days, and was well developed in 90% of the chicks fed 2% culture material (Walser et al., 1980). Cultures of this strain on rice also caused tibial dyschondroplasia in chickens. Purified extracts have been shown to contain 12 components (Lee & Mirocha, 1981), but the mycotoxin(s) produced by this strain capable of causing tibial dyschondroplasia in chickens have not been identified.

The demonstration that a pure culture of *F. equiseti* (= *F. roseum*, Strain 7.6) is capable of causing the characteristic bone lesions seen in field outbreaks of tibial dyschondroplasia in poultry is an important development. The involvement of *F. equiseti* has not been reported in field outbreaks of this disease, however.

Leukemia. A strain of *F. equiseti* was isolated from dust in a house in the United States where a husband and wife had both developed acute myelomonocytic leukemia (Wray et al., 1979). Sterile extracts of cultures of this strain in Czapek-Dox medium depressed the response of guinea pigs to phytohemagglutinin. Previously Wray & O'Steen (1975) had isolated an unidentified *Fusarium* sp. from a house in Georgia, USA, in which four cases of leukemia had occurred, and demonstrated that cultures of this isolate were toxic to ducklings, hamsters, and mice. Wray et al. (1979) concluded that mycotoxin-producing fungi, including *F. equiseti*, may be related to the pathogenesis of leukemia by their immunosuppressive effects. *Fusarium equiseti* is a known producer of diacetoxyscirpenol, which is known to have immunosuppressive effects in animals and man (see mycotoxins produced, below, and *F. sporotrichioides*, hemorrhagic syndrome).

Toxicity to Experimental Animals

Brine Shrimp (*Artemia salina*). Extracts of cultures on nutrient-amended shredded wheat incubated at 25°C for 14 to 21 days of an isolate of *F. equiseti* from cotton in the United States have been reported to be toxic to brine shrimp (Diener et al., 1976). The mycotoxin(s) produced by this strain have not been identified.

Buffaloes. Cultures an autoclaved rice straw incubated at 15°C to 21°C for 4 weeks of an isolate of *F. equiseti* from toxic rice straw associated with outbreaks of Degnala disease (see Degnala disease, above) in India induced clinical signs and pathological lesions similar to field cases of the disease in three buffalo calves (Kalra et al., 1977a, 1980). One calf showed signs of edema in the fetlock area of the forelegs and died after 19 days. The second calf developed edema and evinced pain when pressed in the region of the coronet and above; it died after 62 days. The third calf was sacrificed after 71 days and signs of dry gangrene were noted at the tip of the tail. Histopathological examination of the tail, feet, and ears of these three calves revealed changes simulating those observed in field cases (Kalra et al., 1977a, 1980).

We have examined three isolates of "*F. equiseti*" from rice straw associated with Degnala disease in India and have identified one of these as *F. equiseti* (Strain 7.1), but the other two proved to be *F. semitectum* (Strains 6.1 and 6.2). It is not known to us which of these three strains, if any, was the isolate of "*F. equiseti*" used in the successful experimental reproduction of Degnala disease in buffaloes by Kalra et al. (1977a, 1980). The strain of "*F. equiseti*" used did not produce butenolide or T-2 toxin according to Kalra et al. (1977a, 1980), and consequently the chemical nature of the mycotoxin(s) produced in rice straw by this isolate is unknown.

Cattle. According to Qin et al. (1981), a disease of cattle known as "sore foot disease" (see Degnala disease, above) and associated with the consumption of rice straw in China, has been reproduced experimentally in cattle with cultures of *F. equiseti* and *F. semitectum*. This disease, characterized by necrosis of the feet and ears, appears to be similar to Degnala disease in India. The chemical nature of the mycotoxin(s) produced in rice straw by these Chinese isolates of *F. equiseti* is unknown but is under study, according to Qin et al. (1981).

Chickens. Extracts of cultures on nutrient-amended shredded wheat of an isolate of *F. equiseti* from cotton in the United States have been reported to be toxic to chick embryos (Diener et al., 1976).

Two isolates of *F. equiseti* (Sp. 890 and Sp. 1090) from barley grain in Germany have been reported to be acutely toxic to day-old White Leghorn chicks (Marasas et al., 1979b). We have confirmed the identity of both of these isolates as *F. equiseti* (Strains 7.7 and 7.8). Cultures on corn of these two strains caused the death of 4 of 4 chicks within average times of 6.75 and 5.25 days, respectively. The toxic culture material did not contain moniliformin, zearalenone, or any of nine trichothecenes that could be detected by TLC (Marasas et al., 1979b). Thus the chemical nature of the mycotoxin(s) produced by these two highly toxic strains of *F. equiseti* is unknown.

An isolate of "*F. roseum*" from overwintered barley in Alaska was recently found to cause tibial dyschondroplasia (see tibial dyschondroplasia, above) in chickens (Lee & Mirocha, 1981; Walser et al., 1980). We have identified this isolate of "*F. roseum*" as *F. equiseti* (Strain 7.6). Day-old broiler chicks were fed a nutritionally balanced ration amended with 2%, 5%, or 10% culture material of this strain (Walser et al., 1980). The two higher levels caused 100% mortality within one to 3 weeks, but the chicks fed 2% culture material survived the feeding period of 6 weeks. Pathological changes were noted in the oral cavities, kidneys, and bones of experimental animals. Small foci of necrosis and inflammation were present in the oral and lingual mucosae. Diffuse tubular nephrosis, occasionally with urate deposits, was frequently present. Tibial dyschondroplasia, characterized by the presence of a cone of cartilage that extended distally from the growth plate of the proximal tibia, was first noted in chicks that died after 8 to 11 days, and was well developed in 90% of the chicks fed 2% culture material (Walser et al., 1980). Cultures of this strain of *F. equiseti* (= *F. roseum*, Strain 7.6) on rice also caused tibial dyschondroplasia in chickens. Purified extracts were found to contain 12 components (Lee & Mirocha, 1981), but the mycotoxins produced by this strain capable of causing tibial dyschondroplasia in chickens have not been chemically characterized.

The strain of *F. roseum* 'Gibbosum' reported to be toxic to laying hens by Pathre et al. (1976) and Speers et al. (1977) is *F. semitectum* (Strain 6.4).

Ducklings. According to Martin et al. (1971), cultures of 25 of 39 isolates of *F. equiseti* from human foodstuffs in southern Africa were markedly toxic to day-old ducklings. The mycotoxin(s) produced by these isolates of *F. equiseti* have not been identified.

Guinea Pigs. According to Joffe (1960a, 1971, 1974c, 1978b), culture filtrates and powdered dry mycelium of unspecified isolates of *F. equiseti* from overwintered cereals associated with ATA (see *F. sporotrichioides*, ATA) in the USSR, were toxic to guinea pigs. The animals died within 5 to 21 days and autopsies revealed hemorrhages in many organs and tissues. The mycotoxin(s) produced by these strains of *F. equiseti* toxic to guinea pigs have not been identified.

Insects. Reductions in weight gain, but not mortality, of larvae of the yellow meal worm (*Tenebrio molitor*) were caused by cultures on rice incubated at room temperature followed by 3 to 8 weeks at 10°C of *F. equiseti* isolates DE 7255-59-60 and DE 7255-56-57, which were obtained from the culture collection of W.L. Gordon (G.R.F. Davis et al., 1975). The mycotoxin(s) produced by these two strains of *F. equiseti* have not been identified.

Mice. Extracts of cultures on white corn grits of six isolates of *F. equiseti* from toxic fescue hay associated with outbreaks of "Fescue foot" (see Degnala disease, above) in cattle in the United States have been reported to be toxic to mice upon intraperitoneal injection (Burmeister et al., 1971). We have confirmed the identity of one of these isolates, *F. equiseti* NRRL 5537 (Strain 7.2). All six of these toxic isolates from fescue hay produced T-2 toxin as determined by TLC, and five of them produced butenolide as well (Burmeister et al., 1971).

Many isolates of *Fusarium* spp. from moldy soy beans associated with outbreaks of "bean hulls poisoning" (see bean hulls poisoning, above) of horses in Hokkaido, Japan, have been reported to be toxic to mice upon intraperitoneal injection of extracts of cultures in PSC medium (Ueno et al., 1972a, 1973a). We have identified three of these isolates (*F. roseum* M-2-5, *F. roseum* 'Culmorum' M-14-2, and *F. roseum* 'Avenaceum' M-7-1) that caused radiomimetic pathological changes in the actively dividing cells of the bone marrow, thymus, spleen, testes, and small intestine in mice, as *F. equiseti* (Strains 7.3, 7.4, and 7.5). The presence of diacetoxyscirpenol has been detected by TLC in the toxic-extracts of two of these isolates (Strains 7.3 and 7.4) by Ueno et al., (1973a).

Extracts of cultures in PSC medium of *Fusarium* sp. 5008, isolated from river sediments in Japan, have been reported to be toxic to mice and to cause characteristic radiomimetic pathological changes (Ueno et al., 1977b). We have identified this strain as *F. equiseti* (Strain 7.9). Analyses of the toxic extracts by TLC revealed the presence of diacetoxyscirpenol, neosolaniol, and T-2 toxin (Ueno et al., 1977b).

Mall et al. (1979) reported that extracts of cultures on wheat of four of 10 isolates of *Fusarium* spp. from wheat in India were mildly toxic to mice. We have identified one of these isolates as *F. equiseti* (Strain 7.10). The mycotoxin(s) produced by this strain have not been identified.

Pigeons. Ethyl acetate extracts of cultures on corn incubated at room temperature for one week followed by 4 weeks at 12°C of *F. roseum* 'Equiseti' No. 9, isolated from corn refused by pigs in Wisconsin, USA, induced a limited vomiting response upon force-feeding to pigeons (Kotsonis et al., 1975a). The emetic reaction was difficult to verify because the pigeons vigorously refused the extracts by expectoration. The culture material contained 7.2 µg/g of zearalenone, but no T-2 toxin, HT-2 toxin, or deoxynivalenol that could be detected by TLC (Kotsonis et al., 1975a).

Pigs. Cultures on corn incubated at room temperature for one week followed by 4 weeks at 12°C of *F. roseum* 'Equiseti' No. 9, isolated from corn refused by pigs in Wisconsin, USA, caused a marked feed refusal response and resultant weight loss in pigs (Kotsonis et al., 1975a). Analyses of the culture material by TLC revealed the presence of 7.2 µg/g of zearalenone, but no T-2 toxin, HT-2 toxin, or deoxynivalenol. Thus the mycotoxin(s) responsible for the feed refusal activity of this strain have not been identified.

Rabbits. Extracts of cultures of 51 isolates of *F. equiseti* from overwintered cereals associated with ATA (see *F. sporotrichioides,* ATA) in the USSR were tested for dermotoxicity to rabbit skin; seven were found to be toxic and three mildly toxic (Joffe 1960a, b, 1971). Culture filtrates and powdered dry mycelium of unspecified isolates of *F. equiseti* from overwintered cereals in the USSR have also been reported to be acutely toxic to rabbits (Joffe, 1960a, b, 1971, 1974c, 1978b). The rabbits died within 8 to 24 days and autopsies revealed hemorrhages in many organs and tissues. Cultures of seven isolates of *F. equiseti* from peanuts in Israel on PDA incubated at temperatures from 12 to 35°C all had some degree of dermotoxicity to rabbit skin (Joffe, 1973b). The mycotoxin(s) produced by the above isolates of *F. equiseti* toxic to rabbits have not been identified.

Ethanol extracts of cultures on Sabouraud's agar of a strain of *F. equiseti* (see Strain 7.1) isolated from toxic rice straw associated with Degnala disease (see Degnala disease, above) of buffaloes in India were dermotoxic to rabbit skin and caused the characteristic hemorrhagic skin reaction produced by trichothecenes (Kalra et al., 1973, 1977a, 1980). According to Kalra et al. (1977a, 1980), cultures of this strain were found to be negative for butenolide and T-2 toxin production at NRRL. Consequently the chemical nature of the dermotoxic mycotoxin(s) produced by this strain is unknown.

Methanol extracts of cultures of an isolate of *F. equiseti* from corn associated with an outbreak of acute aflatoxicosis in humans in India were found to be dermotoxic to rabbit skin (Jain et al., 1980). The Rf value of a spot obtained by TLC was "approximately equal to the Rf value of nivalenoldiacetate" (Jain et al., 1980).

Rats. Cultures on corn incubated at room temperature for one week followed by 4 weeks at 12°C of *F. roseum* 'Equiseti' No. 9, isolated from corn refused by pigs in Wisconsin, USA, caused a marked feed-refusal response, weight loss, and uterine hypertrophy in weanling female rats (Kotsonis et al., 1975a). Analyses of the culture material by TLC revealed the presence of 7:2 μg/g of zearalenone, but no T-2 toxin, HT-2 toxin, or deoxynivalenol. Thus the mycotoxin(s) responsible for the feed refusal activity of this strain have not been identified.

The strain reported, as *F. roseum* 'Gibbosum,' to be dermotoxic to rat skin (Ishii et al., 1978; Pathre et al., 1976) is *F. semitectum* (Strain 6.4).

Turkeys. The strain reported, as *F. roseum* 'Gibbosum,' to be toxic to turkey poults (Pathre et al., 1976) is *F. semitectum* (Strain 6.4).

Mycotoxins Produced

Acetoxyscirpenediol (= Monodeacetylanguidin). Loeffler et al. (1967) reported that *F. equiseti* NRRL 3213 produced monodeacetylanguidin (referred to as acetoxyscirpenediol = 4- or 15-acetylscirpentriol by Bamburg & Strong, 1971), an antibiotic with antimitotic activity, in fermentor cultures at 27°C. A lyophilized culture of this strain obtained from NRRL on 9 March 1981 was not viable. We have identified four other strains, i.e., *F. scirpi* NRRL 3212, *F. anguioides* NRRL 3020, *F. concolor* NRRL 3214, and *F. diversisporum* NRRL 3211, that were also found to produce this trichothecene by Loeffler et al. (1967), as *F. equiseti* (Strains 7.11, 7.12, 7.13, and 7.14).

The strain reported, as *F. roseum* 'Gibbosum,' to produce 4-acetoxyscirpenediol (Ishii et al., 1978) is *F. semitectum* (Strain 6.4).

Butenolide. The toxic butenolide was first isolated simultaneously and independently from *F. sporotrichioides* (= *F. nivale*) NRRL 3249 (Strain 4.8) by Yates et al. (1967) and from two isolates of *F. equiseti* by White (1967). The toxic butenolide (as (±)-2-acetamido-2,5-dihydro-5-oxofuran) was isolated from cultures in GAN medium incubated at 25°C for 10 days of two isolates of *F. equiseti* (Plant Disease Division Culture Collection Nos. 63312 and 63313) from *Cortaderia* sp. in New Zealand (White, 1967).

Five isolates of *F. equiseti* (Nos. 16, 18, 20, 21 and 22) from toxic fescue hay associated with outbreaks of "Fescue foot" (see *F. sporotrichioides*, fescue foot) of cattle in the United States have been reported to produce butenolide in cultures on Sabouraud's agar incubated at 15°C for 21 days (Burmeister et al., 1971). We have confirmed the identity of *F. equiseti* NRRL 5537, which was possibly one of these isolates found to produce butenolide by Burmeister et al. (1971), as *F. equiseti* (Strain 7.2).

Negative results for butenolide production have been reported for an isolate of *F. equiseti* from toxic rice straw in India used in the experimental reproduction of Degnala disease (see Degnala disease, above, and Strain 7.1) of buffaloes by Kalra et al. (1977a, 1980) and for eight isolates of *F. equiseti* from wheat and barley in Japan by Suzuki et al. (1981a, b).

Diacetoxyscirpenol (= Anguidin). This trichothecene was first isolated from three strains of *F. equiseti* referred to as *F. equiseti* CMI 35100, *F. scirpi* CMI 45490, and *Gibberella intricans* CBS-Wollenweber strain, by Brian et al. (1961). Diacetoxyscirpenol was produced in GAN medium incubated at 25°C by these three strains at levels of 0.2, 125, and 50 mg/ℓ, respectively. Brian et al. (1961) first reported that crystalline diacetoxyscirpenol was phytotoxic as well as highly toxic to animals upon oral, intraperitoneal, or topical administration. The structure of diacetoxyscirpenol produced by these strains of *F. equiseti* was determined by Dawkins (1966), Dawkins et al. (1965), Flury et al. (1965), and Sigg et al. (1965).

In addition to diacetoxyscirpenol, Brian et al. (1961) also isolated two minor metabolites, nivalenol diacetate and triacetoxyscirpendiol (see nivalenol diacetate and triacetoxyscirpendiol, below) from some of these strains of *F. equiseti*. Both of these minor metabolites were also found to be phytotoxic and highly toxic to animals. The production of diacetoxyscirpenol (as anguidin) in fermentor cultures at 27°C by the following *Fusarium* strains was reported by Loeffler et al. (1967): *F. equiseti* NRRL 3213, *F. scirpi* NRRL 3212, *F. anguioides* NRRL 3020, *F. concolor* NRRL 3214, and *F. diversisporum* NRRL 3211. A lyophilized culture of *F. equiseti* NRRL 3213 obtained from NRRL on 9 March 1981 was not viable, but we have identified the other four diacetoxyscirpenol-producing strains as *F. equiseti* (Strains 7.11, 7.12, 7.13, and 7.14). The isolation of diacetoxyscirpenol from a strain of *F. diversisporum,* presumably NRRL 3211 (Strain 7.14), has also been reported by Flury et al. (1965) and Sigg et al. (1965).

Two *Fusarium* isolates from moldy soybeans associated with "bean hulls poisoning" (see bean hulls poisoning, above) of horses in Hokkaido, Japan, i.e., *F. roseum* M-2-5 and *F. roseum* 'Culmorum' M-14-2, were reported to produce diacetoxyscirpenol in PSC medium incubated at 25 to 27°C for 2 weeks (Ueno et al., 1973a). We have identified these two isolates as *F. equiseti* (Strains 7.3 and 7.4). Another Japanese strain reported to produce diacetoxyscirpenol in PSC medium, *Fusarium* sp. 5008 isolated from river sediments (Ueno et al., 1977b), is also *F. equiseti* (Strain 7.9).

Suzuki et al. (1980, 1981b) detected diacetoxyscirpenol by GC in 16 of 25 unnumbered isolates of *F. equiseti* from wheat and barley in Japan at levels ranging from 0.72 to 86.69 µg/g. Suzuki et al. (1980) also listed six numbered isolates of *F. equiseti* that produced diacetoxyscirpenol. All of these produced fusarenon-X and nivalenol as well, and one of them (No. 3D-10) also produced neosolaniol. A Japanese isolate, *F. equiseti* NHL-F-1121, that produces diacetoxyscirpenol, together with fusarenon-X and nivalenol, according to H. Kurata (personal communication, 29 April 1980), is represented in the ITFRC as *F. equiseti* (Strain 7.15). It is not known to us whether this strain was one of the 16 isolates of *F. equiseti* found to produce diacetoxyscirpenol by Suzuki et al. (1980, 1981b).

The strain reported as *F. roseum* 'Gibbosum' to produce diacetoxyscirpenol (Ishii et al., 1978; Pathre et al., 1976), is *F. semitectum* (Strain 6.4).

In conclusion, diacetoxyscirpenol was first isolated from *F. equiseti* and has since been found to be produced by at least 25 strains of *F. equiseti,* eight of which are represented in the ITFRC (Strains 7.3, 7.4, 7.9, 7.11, 7.12, 7.13, 7.14, and 7.15). Although diacetoxyscirpenol is known to be produced by many isolates of *F. equiseti,* it is apparently not responsible for the toxicity of some strains. For example, two isolates of *F. equiseti* (Strains 7.7 and 7.8) from barley in Germany were highly toxic to chickens, but did not produce diacetoxyscirpenol that could be detected by TLC (Marasas et al., 1979b); the chemical nature of the mycotoxin(s) produced by these strains is still unknown.

Equisetin. Burmeister et al. (1974) reported that *F. equiseti* NRRL 5537 (Strain 7.2) in cultures on corn grits produced equisetin, an antibiotic that inhibits certain gram-positive bacteria and has an intraperitoneal LD_{50} of 63.0 mg/kg in mice; this mycotoxin was produced at levels in excess of 5 g/kg. The structure of equisetin was determined to be a derivative of N-methyl-2,4-pyrollidone by Vesonder et al. (1979a).

Fusarenon-X. Suzuki et al. (1980, 1981b) detected fusarenon-X by GC in 16 of 25 unnumbered isolates of *F. equiseti* from wheat and barley in Japan at levels ranging from 0.21 to 5.52 µg/g. Suzuki et al. (1980) also listed six numbered isolates of *F. equiseti* that produced fusarenon-X, of which the highest producer was No. 18-3; all of these isolates produced diacetoxyscirpenol and nivalenol as well. A Japanese isolate, *F. equiseti* NHL-F-1121, produces fusarenon-X as well as diacetoxyscirpenol and nivalenol, according to H. Kurata (personal communication, 29 April 1980); it is represented in the ITFRC as *F. equiseti* (Strain 7.15). It is not known to us whether this strain was one of the 16 isolates of *F. equiseti* found to produce fusarenon-X by Suzuki et al. (1980, 1981b).

Monoacetoxyscirpenol. The strain reported, as *F. roseum* 'Gibbosum,' to produce monoacetoxyscirpenol (Ishii et al., 1978; Pathre et al., 1976; Speers et al., 1977) is *F. semitectum* (Strain 6.4).

Neosolaniol. Ueno et al. (1977b) detected neosolaniol by TLC In cultures in PSC medium of *Fusarium* sp. 5008 isolated from river sediments in Japan. We have identified this strain, which also produces diacetoxyscirpenol and T-2 toxin, as *F. equiseti* (Strain 7.9).

Suzuki et al. (1980, 1981b) detected neosolaniol by GC in 16 of 25 isolates of *F. equiseti* from wheat and barley in Japan at levels ranging from 0.44 to 8.92 µg/g.

Nivalenol. Suzuki et al. (1980, 1981b) detected nivalenol by GC in 16 of 25 unnumbered isolates of *F. equiseti* from wheat and barley in Japan at levels ranging from 0.42 to 12.00 µg/g. Suzuki et al. (1980) also listed six numbered isolates of *F. equiseti* that produced nivalenol, of which the highest producer was No. 18-3; all of these isolates produced diacetoxyscirpenol and fusarenon-X as well. A Japanese isolate, *F. equiseti* NHL-F-1121, produces nivalenol, as well as diacetoxyscirpenol and fusarenon-X, according to H. Kurata (personal communication, 29 April 1980); it is represented in the ITFRC as *F. equiseti* (Strain 7.15). It is not known to us whether this strain was one of the 16 isolates of *F. equiseti* found to produce nivalenol by Suzuki et al. (1980, 1981b).

Nivalenol Diacetate. The minor phytotoxic and toxic metabolite $C_{19}H_{24}O_9$ was isolated from cultures of *F. scirpi* CMI 45490 and *G. intricans* CBS-Wollenweber strain in GAN medium (2 mg/ℓ) by Brian et al. (1961), and was structurally characterized by Dawkins et al. (1965), Grove (1970a), and Tidd (1967); it has been identified by Tatsuno et al. (1970) as nivalenol diacetate.

Analysis by TLC of dermotoxic extracts of an Indian isolate of *F. equiseti* revealed a spot with an Rf value "approximately equal to the Rf value of nivalenoldiacetate" (Jain et al., 1980). The production of nivalenol diacetate by this strain has not been confirmed by appropriate analytical methods.

There are no known nivalenol-diacetate-producing strains of *F. equiseti* represented in the ITFRC.

Scirpentriol. The strain reported, as *F. roseum* 'Gibbosum,' to produce scirpentriol (Ishii et al., 1978; Pathre et al., 1976) is *F. semitectum* (Strain 6.4).

Triacetoxyscirpendiol. This trichothecene was isolated as a minor metabolite $C_{21}H_{28}O_{10}$ from *F. scirpi* CMI 45490 by Brian et al. (1961). The yield of this metabolite in GAN medium incubated at 25°C was 1 mg/ℓ. The crystalline compound was reported to be phytotoxic as well as toxic to animals by Brian et al. (1961) and the structure was determined by Grove (1970b).

T-2 Toxin. Six isolates of *F. equiseti* (Nos. 6, 16, 18, 20, 21 and 22) from toxic fescue hay associated with "Fescue foot" (see *F. sporotrichioides,* fescue foot) of cattle in the United States have been reported to produce T-2 toxin as detected by TLC in cultures on white corn grits incubated at 15°C for 21 days (Burmeister et al., 1971). Subsequently Burmeister et al. (1972) listed five isolates (Nos. 1 to 5) of *F. equiseti* from the same source that produced T-2 toxin in cultures on the same medium. We have confirmed the identity of one of these T-2 toxin-producing isolates from toxic fescue hay, i.e., *F. equiseti* NRRL 5537 (Strain 7.2).

Ueno et al. (1977b) detected T-2 toxin by TLC in cultures of *Fusarium* sp. 5008 in PSC medium incubated at 25 to 27°C for 2 weeks. We have identified this strain as *F. equiseti* (Strain 7.9).

Negative results for T-2 toxin production have been reported for an isolate of *F. equiseti* from toxic rice straw in India used in the experimental reproduction of Degnala disease (see Degnala disease, above, and Strain 7.1) in buffaloes by Kalra et al. (1977a, 1980). Several other authors have also reported negative results for T-2 toxin production by isolates of F. equiseti: none of 3 by Bottalico (1977c) and Bottalico et al. (1981); neither of 2 (Strains 7.7 and 7.8) by Marasas et al. (1979b); and none of 25 by Suzuki et al. (1980).

In conclusion, T-2 toxin has been detected by TLC in 7 out of approximately 30 isolates of *F. equiseti* that have been analyzed. The production of T-2 toxin by authenticated isolates of *F. equiseti* needs to be confirmed by more reliable analytical methods.

Zearalenol. *Fusarium roseum* No. S-74-1c, isolated from head-blighted sorghum infected by *F. roseum* 'Gibbosum' and *F. roseum* 'Semitectum' in the United States, produced zearalenol in cultures on either cracked corn or grain sorghum incubated at 25°C for 20 days (Stipanovic & Schroeder, 1975). These authors did not state whether *F. roseum* No. S-74-1c was a strain of *F. roseum* 'Gibbosum' or *F. roseum* 'Semitectum,' and a culture is no longer available to determine its identity (H.W. Schroeder, personal communication, 25 June 1981).

The strain reported, as *F. roseum* 'Gibbosum,' to produce zearalenol (Hagler et al., 1979) is *F. semitectum* (Strain 6.4).

Zearalenone. Caldwell et al. (1970) detected zearalenone by TLC in cultures on corn incubated at 16°C for 3 weeks of 3 of 6 isolates of *F. roseum* 'Equiseti' at levels of 0.6 to 2.0 µg/g, and in one isolate of *F. roseum* 'Gibbosum' at levels of 115 to 175 µg/g.

Fusarium equiseti No. 577, isolated from wheat in Canada, produced zearalenone within 5 days in cultures on YES medium incubated at room temperature (Scott et al. 1972). Zearalenone was detected by TLC and its identity confirmed by GC. This strain also produced a more polar compound which was tentatively identified by MS as hydroxyzearalenone (Scott et al., 1972).

Cultures on corn incubated at room temperature for one week followed by 4 weeks at 12°C of *F. roseum* 'Equiseti' isolated from corn refused by pigs in Wisconsin, USA, caused uterine hypertrophy in female rats (Kotsonis et al., 1975a). Analysis of this culture material by TLC revealed the presence of 7.2 µg/g of zearalenone.

Samples of head-blighted sorghum grain infected by *F. roseum* 'Gibbosum' and *F. roseum* 'Semitectum' in the United States have been reported to be naturally contaminated with zearalenone (Schroeder & Hein, 1975). Eight isolates of *Fusarium* spp. from this sorghum grain produced zearalonone in culture on sorghum grain incubated for 19 days. Zearalenone was detected by TLC and the identity of crystalline material isolated from these cultures was verified by GC-MS. The highest yielding isolate, *F. roseum* S-74-1c, was grown on cracked corn or grain sorghum at a constant temperature of 25°C for 19 days or at 25°C for 6 days followed by 13 days at 10°C. The yields of zearalenone were obtained by weighing the purified crystals. In contrast to the results of previous studies, this isolate produced considerably more zearalenone (1593 mg/kg) in cultures on corn incubated at a constant temperature of 25°C than at 25°C followed by 10°C (499 mg/kg). This isolate also produced more zearalenone in culture on sorghum (3030 mg/kg) than on corn at 25°C (Schroeder & Hein, 1975). In addition to zearalenone, this isolate also produced zearalenol and 8'-hydroxyzearalenone (Stipanovic & Schroeder, 1975). The above authors did not state whether *F. roseum* S-74-1c was a strain of *F. roseum* 'Gibbosum' or *F. roseum* 'Semitectum' and a culture is no longer available to determine its identity (H.W. Schroeder, personal communication, 25 June 1981).

Ichinoe et al. (1977) reported that 3 of 12 Japanese isolates of *F. equiseti* produced zearalenone at levels ranging from 0.7 to 5.0 µg/g in cultures on rice. Zearalenone has also been detected by GC at a level of 4 µg/g in cultures of one of 4 isolates of *F. equiseti* from peanuts imported into Japan (Suzuki et al., 1978).

According to Connole et al. (1981), zearalenone was produced in culture at levels ranging from 3.3 to 9.2 mg/kg by 2 of 5 isolates of *F. equiseti* from animal feeds in Australia.

The following two isolates, which were reported as members of the section Gibbosum that produced zearalenone, we have identified as other species (Table 7.2): *F. roseum* 'Gibbosum' A-O-2, reported to produce zearalenone by Ishii et al. (1974), is *F. graminearum* (Strain 8.31); *F. roseum* 'Gibbosum,' reported to produce zearalenone by Hagler & Mirocha (1980), Hagler et al. (1979), Pathre et al. (1976) and Speers et al. (1977), is *F. semitectum* (Strain 6.4).

Negative results for zearalenone production have been reported for two Italian isolates of *F. equiseti* by (Bottalico, 1977b), one Austrian isolate by Bottalico et al. (1981), three North American isolates by Caldwell et al. (1970), three Australian isolates by Connole et al. (1981), two German isolates (Strains 7.7 and 7.8) by Marasas et al. (1979b), and 20 Japanese isolates by Ichinoe et al. (1977), Ishii et al. (1974), and Suzuki et al. (1978, 1981b).

In conclusion, negative results were reported for 30 of the 52 isolates of *F. equiseti* tested for zearalenone production, and 22 isolates were reported as zearalenone producers. We have identified 2 of these 22 as other species, and we have not seen any of the remaining 20 isolates. At present there are no known zearalenone-producing strains of *F. equiseti* represented in the ITFRC, and the production of zearalenone by authenticated isolates of *F. equiseti* remains to be confirmed.

Toxigenic Strains in the ITFRC

The *Fusarium* strains listed under *F. equiseti* in Table 7.1 all have been identified as *F. equiseti* according to Nelson et al. (1983). Toxigenic strains referred to as members of the section Gibbosum but identified as other species by us are listed in Table 7.2.

Table 7.1 Fusarium Strains of the Section Gibbosum in the International Toxic Fusarium Reference Collection (ITFRC)

Strain No.	ITFRC No.	MRC No.	Published Name(s) and Number(s)	Source	Toxigencity
Fusarium equiseti					
7.1	R-6605	2609	*F. equiseti* No. 1/81	R.K. Paul Gupta	a,b?
7.2	R-6325	2483	*F. equiseti* (= *F. roseum* 'Gibbosum') NRRL 5537	NRRL	a,b,c,d?,e
7.3	R-6765	2558	*F. roseum* M-2-5	Y. Ueno	a,b,f
7.4	R-6766	2559	*F. roseum* 'Culmorum' M-14-2	Y. Ueno	a,b,f
7.5	R-6768	2561	*F. roseum* 'Avenaceum' M-7-1	Y. Ueno	a,b
7.6	R-6137	2330	*F. roseum* Alaska 2-2	C.J. Mirocha	b
7.7	R-4482	1891	*F. equiseti* Sp. 890	W.F.O. Marasas	b
7.8	R-4783	1892	*F. equiseti* Sp. 1090	W.F.O. Marasas	b
7.9	R-6784	2568	*Fusarium* sp. 5008	Y. Ueno	b,c,f,g
7.10	R-6521	2637	*Fusarium* sp. PL W-741	V.C. Vora	b
7.11	R-6336	2435	*F. scirpi* NRRL 3212	NRRL	f,h
7.12	R-6053	2231	*F. anguioides* NRRL 3020	NRRL	f,h
7.13	R-6054	2232	*F. concolor* NRRL 3214	NRRL	f,h
7.14	R-6324	2433	*F. diversisporum* NRRL 3211	NRRL	f,h
7.15	R-5468	1961	*F. equiseti* NHL-F-1121	H. Kurata	f,i,j
Fusarium acuminatum					
7.16	R-5319	1783	*F. acuminatum* (= *F. roseum* 'Acuminatum') NRRL 6227	NRRL	b,c,k
7.17	R-6379	2190	*F. sulphureum* HPB 110178-19	B.J. Blanchfield	b,c,g,l
7.18	R-5701	2017	*F. heterosporum* F-77-1A	R.J. Cole	c,l,m,n
7.19	R-5466	1959	*F. acuminatum* NHL-F-1025	H. Kurata	c,f,g,l

a = Associated with human and/or animal diseases
b = Culture toxic to experimental animals
c = Produces T-2 toxin
d = Produces butenolide
e = Produces equisetin
f = Produces diacetoxyscirpenol (= anguidin)
g = Produces neosolaniol
h = Produces acetoxyscirpendiol (= monodeacetylanguidin)
i = Produces fusarenon-X
j = Produces nivalenol
k = Produces cyclic peptide "swelling factor"
l = Produces HT-2 toxin
m = Produces T-2 tetraol
n = Produces three new trichothecenes

Strain 7.1 *F. equiseti* No. 1/81 (= R-6605; MRC 2609)

This strain was isolated from toxic rice straw associated with Degnala disease (see Degnala disease, above) of buffaloes and cattle in Haryana, India (Kalra et al., 1973, 1977a, b, 1980). The characteristic clinical signs and pathological lesions of Degnala disease have been reproduced experimentally in three buffalo calves (see buffaloes, above) fed cultures on autoclaved rice straw incubated at 15°C to 21°C for 4 weeks of an isolate of *F. equiseti* from toxic rice straw associated with field outbreaks of the disease in India (Kalra et al., 1977a, 1980). One calf showed signs of edema in the fetlock area of the forelegs and died after 19 days. The second calf developed edema and evinced pain when pressed in the region of the coronet and above; it died after 62 days. The third calf was sacrificed after 71 days and signs of dry gangrene were noted at the tip of the tail. Histopathological examination of the tail, feet, and ears of these three calves revealed changes simulating those observed in field cases (Kalra et al., 1977a, 1980). The isolate of *F. equiseti* that was used in the successful experimental reproduction of Degnala disease in buffaloes was also shown to be dermotoxic to rabbit skin (Kalra et al., 1973,

Table 7.2 *Fusarium* Strains Published as Members of the Section Gibbosum but Identified as Other Species in the International Toxic *Fusarium* Reference Collection (ITFRC)

Strain No.	ITFRC No.	MRC No.	Published Name(s) and Number(s)	Source	Present Identification
6.4	R-4485	1642	*F. oxysporum* = *F. roseum* isolate oxyrose = *F. roseum* 'Gibbosum'	C.J. Mirocha	*F. semitectum*
8.31	R-6802	2606	*F. roseum* 'Gibbosum' A-O-2	Y. Ueno	*F. graminearum*

1977a, 1980). However, this isolate "was sent to Dr Yates of Northern Regional Research Laboratory, Illinois, USA, but no butenolide or T-2 toxin was found to be produced by this culture" (Kalra et al., 1977a, 1980). Thus the mycotoxin(s) produced by this strain have not been identified.

A slant culture of *F. equiseti* Strain 7.1 was received on 26 september 1981 from R.K. Paul Gupta as "*F. equiseti* No. 1/81, isolated from rice straw associated with Degnala disease." We consider this culture to be a typical representative of *F. equiseti*. However, two other isolates of "*F. equiseti*" from the same source proved to be *F. semitectum* (Strains 6.1 and 6.2). It is not known to us which of these three strains, if any, was used in the successful experimental reproduction of Degnala disease in three buffalo calves with a pure culture of an isolate of *F. equiseti* (identified at CMI) by Kalra et al. (1977a, 1980). Consequently it is not certain whether *F. equiseti* Strain 7.1 was actually used in the successful experimental reproduction of Degnala disease or not.

Strain 7.2 *F. equiseti* (= *F. roseum* 'Gibbosum') **NRRL 5537** (= R-6325; MRC 2483) This strain was isolated from toxic fescue hay associated with outbreaks of "Fescue foot" (see *F. sporotrichioides*, fescue foot) in the United States (Burmeister et al., 1974).

Extracts of cultures on white corn grits incubated at 15°C for 21 days of six isolates of *F. equiseti* from toxic fescue hay (Nos. 6, 16, 18, 20, 21 and 22) have been reported to be toxic to mice upon intraperitoneal injection (Burmeister et al., 1971). All six of these toxic isolates from fescue hay produced T-2 toxin as determined by TLC, and five of them produced butenolide as well (Burmeister et al., 1971). Subsequently Burmeister et al. (1972) listed five isolates of *F. equiseti* (Nos. 1 to 5) from the same source that produced T-2 toxin in culture on the same medium. According to J.J. Ellis (personal communication, 25 June 1981), *F. equiseti* NRRL 5537 was "a *F. equiseti* from fescue hay that produced T-2 toxin." It is not known to us which number, if any, was used for this strain by Burmeister (1971, 1972) and whether *F. equiseti* Strain 7.2 also produces butenolide or not.

Burmeister et al. (1974) reported that *F. equiseti* NRRL 5537, in culture on white corn grits incubated at room temperature for 3 to 4 weeks, produced an antibiotic, equisetin, at levels in excess of 5 g/kg. The structure of equisetin produced by this strain was determined to be a derivative of N-methyl-2,4-pyrollidone by Vesonder et al. (1979a).

A slant culture of this strain was received on 25 June 1981 from NRRL as "*F. roseum* 'Gibbosum' NRRL 5537." This culture was degenerated and semi-pionnotal, but still produced numerous chlamydospores and some typical macroconidia in sporodochia on CLA that enabled us to identify it as *F. equiseti*.

Strain 7.3 *F. roseum* M-2-5 (= R-6765; MRC 2558) This strain (as *Fusarium* sp. M-2-5) was isolated from moldy soybeans associated with outbreaks of "bean hulls poisoning" (see bean hulls poisoning, above) of horses in

Hokkaido, Japan (Ueno et al., 1972a). Extracts of cultures of this strain in PSC medium incubated at 25°C for 2 weeks were lethal to mice and caused radiomimetic pathological changes in the actively dividing cells of the bone marrow, thymus, spleen, testes, and small intestine (Ueno et al., 1972a, 1973a). Analyses by TLC of the toxic extract of this strain (as *F. roseum* M-2-5) revealed the presence of diacetoxyscirpenol (Ueno et al., 1973a).

A slant culture of this strain was received in November 1981 from Y. Ueno as "*F. roseum* M-2-5." The macroconidia produced by this strain on CLA have very long, whip-like apical cells resembling *F. scirpi* Lambotte & Fautr. var. *filiferum* (Preuss) Wollenw. This variety was included in *F. equiseti* by Nelson et al. (1983) and this culture was identified as *F. equiseti*.

Strain 7.4 *F. roseum* 'Culmorum' M-14-2 (= R-6766; MRC 2559)

This strain (as *Fusarium* sp. M-14-2) was isolated from moldy soybeans associated with outbreaks of "bean hulls poisoning" (see bean hulls poisoning, above) of horses in Hokkaido, Japan (Ueno et al., 1972a). Extracts of cultures of this strain in PSC medium incubated at 25 to 27°C for 2 weeks were lethal to mice and caused radiomimetic pathological changes in the actively dividing cells of the bone marrow, thymus, spleen, testes, and small intestine (Ueno et al., 1972a, 1973a). Analyses by TLC of the toxic extract of this strain (as *F. roseum* 'Culmorum' M-14-2) revealed the presence of diacetoxyscirpenol (Ueno et al., 1973a).

A slant culture of this strain was received in November 1981 from Y. Ueno as "*F. roseum* 'Culmorum' M-14-2." This strain of *F. equiseti* is identical to Strain 7.3 and the macroconidia also have very long, whip-like apical cells.

Strain 7.5 *F. roseum* 'Avenaceum' M-7-1 (= R-6768; MRC 2561)

This strain (as *Fusarium* sp. M-7-1) was isolated from moldy soybeans associated with outbreaks of "bean hulls poisoning" (see bean hulls poisoning, above) of horses in Hokkaido, Japan (Ueno et al., 1972a). Extracts of cultures of this strain in PSC medium incubated at 27°C for 12 days were lethal to mice and caused radiomimetic pathological changes in the dividing cells of the bone marrow, thymus, spleen, testes, and small intestine and also inhibited protein synthesis in rabbit reticulocytes (Ueno et al., 1972a); however, extracts of this strain (as *F. roseum* 'Avenaceum' M-7-1) were found to be non-toxic to mice by Ueno et al. (1973a). Although these extracts were still positive in the rabbit reticulocyte assay, no trichothecenes could be detected by TLC.

A slant culture of this strain was received in November 1981 from Y. Ueno as "*F. roseum* 'Avenaceum' M-7-1." This culture of *F. equiseti* is identical to Strains 7.3 and 7.4 and the macroconidia also have very long, whip-like apical cells.

Strain 7.6 *F. roseum* Alaska 2-2 (= R-6137; MRC 2330)

This strain was isolated from overwintered barley in Alaska (Lee & Mirocha, 1981). Day-old broiler chicks were fed a nutritionally balanced ration amended with 2%, 5%, or 10% culture material of this strain (Walser et al., 1980). The two higher levels caused 100% mortality within one to 3 weeks, but the chicks fed 2% culture material survived the feeding period of 6 weeks. Pathological changes were noted in the oral cavities, kidneys and bones of experimental animals. Small foci of necrosis and inflammation were present in the oral and lingual mucosae. Diffuse tubular nephrosis, occasionally with urate deposits, was frequently present. Tibial dyschondroplasia, characterized by the presence of a cone of cartilage that extended distally from the growth plate of the proximal tibia, was first noted in chicks that died after 8 to 11 days, and was well developed in 90% of the chicks fed 2% culture material (Walser et al., 1980). Cultures of this strain of *F. equiseti* (= *F. roseum*) on rice also caused tibial dyschondroplasia in chickens and purified extracts were found to contain 12 components (Lee & Mirocha, 1981); however, the mycotoxins produced by this strain capable of causing tibial dyschondroplasia in chickens have not been chemically characterized.

A silica gel culture of this strain was received on 4 October 1981 from C.J. Mirocha as "*F. roseum* Alaska 2-2." This culture was badly degenerated and almost completely mycelial, but still produced numerous chlamydospores and some typical macroconidia in sporodochia on CLA that allowed it to be identified as *F. equiseti*.

Strain 7.7   *F. equiseti* Sp. 890 (= R-4482; MRC 1891)

This strain was isolated from barley in Germany (Marasas et al., 1979b). Cultures of this strain on corn incubated at 25°C for 2 weeks followed by 4 weeks at 10°C caused the death of 4 of 4 day-old chicks within an average time of 6.75 days (Marasas et al., 1979b). Analyses of the toxic culture material by TLC for moniliformin, zearalenone, and nine trichothecenes, including diacetoxyscirpenol and T-2 toxin, were all negative (Marasas et al., 1979b). Thus the chemical nature of the mycotoxin(s) produced by this highly toxic strain is unknown.

A lyophilized culture of this strain was received on 16 February 1978 from W.F.O. Marasas as "*F. equiseti* Sp. 890."

Strain 7.8   *F. equiseti* Sp. 1090 (= R-4783; MRC 1892)

This strain was isolated from barley grain in Germany (Marasas et al., 1979b). Cultures of this strain on corn incubated at 25°C for 2 weeks followed by 4 weeks at 10°C caused the death of 4 of 4 day-old chicks within an average time of 5.25 days (Marasas et al., 1979b). Analyses of the toxic culture material by TLC for moniliformin, zearalenone, and nine trichothecenes, including diacetoxyscirpenol and T-2 toxin, were all negative (Marasas et al., 1979b). Thus the chemical nature of the mycotoxin(s) produced by this highly toxic strain is unknown.

A lyophilized culture of this strain was received on 16 February 1978 from W.F.O Marasas as "*F. equiseti* Sp. 1090."

Strain 7.9   *Fusarium* sp. 5008 (= R-6784; MRC 2568)

This strain was isolated from river sediments in Japan (Ueno et al., 1977b). Extracts of cultures in PSC medium incubated at 25 to 27°C for 2 weeks were lethal to mice and caused radiomimetic pathological changes in the actively dividing cells of the bone marrow, thymus, spleen, testes, and small intestine (Ueno et al., 1977b). These extracts also markedly inhibited protein synthesis in rabbit reticulocytes. Analyses of the toxic extract by TLC revealed the presence of diacetoxyscirpenol, neosolaniol, and T-2 toxin (Ueno et al., 1977b).

A slant culture of this strain was received in November 1981 from Y. Ueno as "*Fusarium* sp. 5008." This strain is atypical of *F. equiseti* in that it produces red pigment as well as numerous microconidia on PDA. However, it also produces numerous chlamydospores and typical *F. equiseti* macroconidia in sporodochia in CLA. This strain resembles *F. scirpi* Lambotte & Fautr. var. *compactum* Wollenw. Nelson et al. (1983) stated that strains of this type "appear to be a distinct group," but provisionally placed them under *F. equiseti*.

Strain 7.10   *Fusarium* sp. PL W-741 (= R-6521; MRC 2637)

This strain was isolated from wheat in India (Mall et al., 1979). Extracts of cultures of this strain on wheat incubated at 27 to 28°C for 15 days were mildly toxic to mice. The mycotoxin(s) produced by this strain have not been identified.

A slant culture of this strain was received on 24 August 1981 from V.C. Vora as "*Fusarium* sp. PL W-741, from wheat, toxic to mice." This culture is pionnotal, but still produces some typical *F. equiseti* macroconidia in sporodochia on CLA.

Strain 7.11   *F. scirpi* NRRL 3212 (= R-6336; MRC 2435)

This strain, from an unknown source, has been reported to produce diacetoxyscirpenol (= anguidin) and acetoxyscirpenediol (= monodeacetylanguidin) in liquid medium in fermentor cultures incubated at 27°C (Loeffler et al., 1967).

A slant culture of this strain was received on 6 July 1981 from NRRL as "*F. scirpi* NRRL 3212." This culture is semipionnotal and produces numerous microconidia and chlamydospores, but also some typical *F. equiseti* macroconidia in sporodochia on CLA.

Strain 7.12   *F. anguioides* NRRL 3020 (= R-6053; MRC 2231)

This strain, from an unknown source, has been reported to produce diacetoxyscirpenol (= anguidin) and acetoxyscirpenediol (= monodeacetylanguidin) in liquid medium in fermentor cultures incubated at 27°C (Loeffler et al., 1967).

A lyophilized culture of this strain of *F. equiseti* was received on 9 March 1981 from NRRL as "*F. anguioides* NRRL 3020."

Strain 7.13   *F. concolor* NRRL 3214 (= R-6054; MRC 2232)

This strain, from an unknown source, has been reported to produce diacetoxyscirpenol (= anguidin) and acetoxyscirpenediol (= monodeacetylanguidin) in liquid medium in fermentor cultures incubated at 27°C (Loeffler et al., 1967).

A lyophilized culture of this strain of *F. equiseti* was received on 9 March 1981 from NRRL as "*F. concolor* NRRL 3214."

Strain 7.14   *F. diversisporum* NRRL 3211 (= R-6324; MRC 2433)

This strain, from an unknown source, has been reported to produce diacetoxyscirpenol (= anguidin) and acetoxyscirpenediol (= monodeacetylanguidin) in liquid medium in fermentor cultures incubated at 27°C (Loeffler et al., 1967). The isolation of diacetoxyscirpenol from an isolate of *F. diversisporum,* presumably this one, has also been reported by Flury et al. (1965) and Sigg et al. (1965).

A lyophilized culture of this strain of *F. equiseti* was received on 9 March 1981 from NRRL as "*F. diversisporum* NRRL 3211."

Strain 7.15   *F. equiseti* NHL-F-1121 (= R-5468; MRC 1961)

A slant culture of this strain was received on 29 April 1980 from H. Kurata as "*F. equiseti* NHL-F-1121, from barley grain, Japan. Produces diacetoxyscirpenol, fusarenon-X and nivalenol." We consider this culture to be a typical representative of *F. equiseti.*

Suzuki et al. (1980, 1981b) detected diacetoxyscirpenol, fusarenon-X, and nivalenol by GC in 16 of 25 unnumbered isolates of *F. equiseti* from wheat and barley in Japan at levels ranging from 0.72 to 86.69 µg/g, 0.21 to 5.52 µg/g, and 0.42 to 12.00 µg/g, respectively. Suzuki et al. (1980) also listed six isolates of *F. equiseti* (Nos. 3D-10, 6W-9, 10D-18, 10W-9, 11W-8, and 18-3) that produced a combination of diacetoxyscirpenol, fusarenon-X, and nivalenol. It is not known to us whether *F. equiseti* NHL-F-1121, which also produces this combination of trichothecenes according to H. Kurata (personal communication, 29 April 1980), was one of these strains analyzed by Suzuki et al. (1980, 1981b) or not.

Fusarium acuminatum Ell. & Ev. sensu Gordon

Perfect State: *Gibberella acuminata* Wollenw.

Incidence and Distribution

Fusarium acuminatum occurs world-wide as a soil saprophyte and as a secondary invader associated with root, foot, and stem rots of a great variety of plants in temperate as well as tropical areas (Booth, 1971; Doidge, 1938; Domsch et al. 1980; Gerlach & Ershad, 1970; Meyer & Frank, 1979; Wollenweber & Reinking, 1935). *Fusarium acuminatum* has also been isolated from overwintered cereals in the USSR (Joffe, 1960a, b, 1971) and from cereal grains in Canada and Europe (Gordon, 1952; Ylimäki, 1981).

Association with Human and/or Animal Diseases

Fusarium acuminatum has not been associated with any human or animal toxicoses.

Toxicity to Experimental Animals

Brine Shrimp (*Artemia salina*). Culture filtrates and ethyl acetate extracts of cultures of *F. sulphureum* HPB 110178-19 in YES medium were highly toxic to brine shrimp (Harwig et al., 1979). We have identified this strain, which was isolated from a decaying tomato in Canada, as *F. acuminatum* (Strain 7.17). Ethyl acetate extracts of these cultures were also mutagenic to *Salmonella typhimurium*. Analyses of these extracts by TLC revealed the presence of unspecified levels of T-2 toxin, and the identity of crystalline material isolated from the cultures in YES medium was confirmed by GC-MS (Harwig et al., 1979). Extracts of cultures of this strain on corn have also been reported to be toxic to brine shrimp and to contain 620 µg/g of T-2 toxin (Scott et al., 1980). The T-2 toxin produced by this strain was presumably responsible for its toxicity to brine shrimp, but the chemical nature of the mutagen(s) is unknown.

Chickens. Chloroform extracts of cultures on shredded wheat supplemented with mycological broth and YES medium of an isolate of *F. heterosporum* from *Claviceps paspali* honeydew on *Paspalum distichum* in Georgia, USA, were highly toxic to day-old chickens (Cole et al., 1981). We have identified this strain as *F. acuminatum* (Strain 7.18). The toxicity to chickens was used as bioassay in the isolation of six mycotoxins (i.e., T-2 toxin, HT-2 toxin, T-2 tetraol, and three new trichothecenes) from cultures of this strain (Cole et al., 1981).

Insects. Cultures on rice of *F. roseum* DE 7246 were reported to cause growth retardation of larvae of the yellow meal worm (*Tenebrio molitor*) by G.R.F. Davis et al. (1975). The optimal temperature for the production of metabolites that caused growth retardation and increased mortality of larvae by this strain in cultures on rye was 20°C (Davis & Smith, 1977). Subsequently Davis & Smith (1981) stated that *F. roseum* DE 7246, which had been isolated from *Myriosclerotinia borealis* (Bub. & Vleug.) Kohn on grasses in Canada, was identified by R.A. Shoemaker as *F. acuminatum*; they also found that the optimal temperature for toxin production by this strain in cultures on rye, as indicated by reduction in weight gain of larvae, was 24°C and that cultures incubated in the dark at all temperatures caused greater inhibition than cultures incubated in the light. The mycotoxin(s) produced by *F. acuminatum* (= *F. roseum*) DE 7246 have not been identified.

Mice. Extracts of cultures on corn grits of *F. roseum* 'Acuminatum' NRRL 6227 have been reported to be lethal to mice upon intraperitoneal injection (Burmeister et al., 1981). We have identified this strain that was isolated from fescue hay associated with outbreaks of "Fescue foot" (see *F. sporotrichioides,* fescue foot) of cattle in Missouri, USA, as *F. acuminatum* (Strain 7.16). Analyses of the toxic extracts revealed the presence of T-2 toxin, which was presumably responsible for the toxicity of this strain to mice (Burmeister et al., 1981). This strain is also known to produce a cyclic peptide antibiotic that causes marked swelling of *Penicillium* conidia (Burmeister et al., 1977, 1981).

Rabbits. Extracts of cultures of one out of six isolates of *F. acuminatum* [as *F. scirpi* Lamb. & Fautr. var. *acuminatum* (Ell. & Ev.) Wollenw.] from overwintered cereals associated with ATA (see *F. sporotrichioides,* ATA) in the USSR, have been found to be toxic to rabbit skin (Joffe 1960a, b, 1971). The mycotoxin(s) produced by this dermotoxic strain have not been identified.

According to Jain et al. (1980), extracts of cultures of an isolate of *F. acuminatum* from corn associated with an outbreak of aflatoxicosis in humans in India were dermotoxic to rabbits. Analysis of these extracts by TLC revealed that the "Rf value of *F. acuminatum* approximately resembled that of diacetoxyscirpenol" (Jain et al., 1980).

Mycotoxins Produced

Butenolide. According to Suzuki et al. (1981a, b), three out of three isolates of *F. acuminatum* from wheat and barley in Japan produced butenolide at levels ranging from 1.63 to 9.11 µg/g in cultures on rice as detected by GC with an electron-capture detector. We have not seen the isolates and there are no known butenolide-producing strains of *F. acuminatum* represented in the ITFRC.

Cyclic Peptide "Swelling Factor." Burmeister et al. (1977) reported that *F. acuminatum* NRRL 6227 (Strain 7.16) produced an antifungal antibiotic, a cyclic peptide, that caused marked swelling of conidia of *Penicillium* spp. while inhibiting germination. The production of the "swelling factor" by this strain was confirmed by Burmeister et al. (1981), who found that 23 of 132 *Fusarium* strains, belonging to several species, also produced this compound.

Deoxynivalenol. According to Morooka et al. (1980), two isolates of *F. acuminatum* from wheat and barley in southern Japan produced deoxynivalenol as well as 3-acetyldeoxynivalenol, while a third isolate produced only traces of deoxynivalenol. On the other hand, Suzuki et al. (1980, 1981b) reported that four isolates of *F. acuminatum* from wheat and barley in Japan did not produce deoxynivalenol that could be detected by GC. We have not seen these isolates and there are no known deoxynivalenol-producing strains of *F. acuminatum* represented in the ITFRC.

Diacetoxyscirpenol. According to H. Kurata (personal communication, 29 April 1980), *F. acuminatum* NHL-F-1025 (Strain 7.19), isolated from barley in Japan, produces diacetoxyscirpenol together with neosolaniol, T-2 toxin, and HT-2 toxin. Suzuki et al. (1980, 1981b) reported than 3 of 4 Japanese isolates of *F. acuminatum* produced neosolaniol and T-2 toxin, but none of them produced diacetoxyscirpenol that could be detected by GC. We are not aware of any published reports of diacetoxyscirpenol production by Japanese isolates of *F. acuminatum.*

According to Jain et al. (1980) extracts of cultures of an Indian isolate of *F. acuminatum* were dermotoxic to rabbit skin and analyses of these extracts by TLC revealed that the "Rf values of *F. acuminatum* approximately resembled that of diacetoxyscirpenol." The production of diacetoxyscirpenol by this isolate has not been confirmed by appropriate analytical techniques.

3α, 4β-Dihydroxy-15-acetoxy-8α-(3-hydroxy-3-methylbutyryloxy)-12,13-epoxytrichothec-9-ene. This new trichothecene was isolated and characterized from cultures of an isolate of *F. heterosporum* on shredded wheat supplemented with mycological broth and YES medium incubated at 27 to 30°C for 3 weeks (Cole et al., 1981). We have identified this strain, which was isolated from *Claviceps paspali* honeydew on *Paspalum distichum* in Georgia, USA, as *F. acuminatum* (Strain 7.18).

HT-2 Toxin. Tomatoes were inoculated with *F. sulphureum* HPB 110178-19, which was isolated from a decaying tomato in Canada, and incubated at either 15 or 25°C until the desired stage of rot developed (Harwig et al., 1979). We have identified this strain as *F. acuminatum* (Strain 7.17). Analyses by TLC of these inoculated tomatoes revealed that those incubated at 25°C contained only traces (<1.5 μg/g) of HT-2 toxin and T-2 toxin. However, inoculated tomatoes incubated at 15°C contained HT-2 toxin at levels ranging from 1.9 to 37.8 μg/g and also contained T-2 toxin and neosolaniol. The identity of these compounds was confirmed by GC-MS of purified material isolated from the inoculated tomatoes (Harwig et al., 1979). Although *F. acuminatum* (= *F. sulphureum*) HPB 110178-19 (Strain 7.17) produced T-2 toxin in YES medium incubated at 25°C (Harwig et al., 1979) and in cultures on corn grits incubated at 14°C (Scott et al., 1980), no HT-2 toxin could be detected by TLC in these cultures.

Cole et al. (1981) isolated and identified HT-2 toxin from cultures of an unnumbered isolate of *F. heterosporum* on shredded wheat supplemented with mycological broth and YES medium and incubated at 27 to 30°C for 3 weeks. We have identified this strain as *F. acuminatum* (Strain 7.18).

According to H. Kurata (personal communication, 29 April 1980), *F. acuminatum* NHL-F-1025 (Strain 7.19), isolated from barley in Japan, produces HT-2 toxin together with T-2 toxin, diacetoxyscirpenol, and neosolaniol. We are not aware of any published reports on the production of HT-2 toxin by Japanese isolates of *F. acuminatum*.

Neosolaniol. Tomatoes inoculated with *F. acuminatum* (= *F. sulphureum*) HPB 110178-19 (Strain 7.17), isolated from a decaying tomato in Canada, and incubated at 15°C for 6 to 20 days, contained neosolaniol at levels ranging from 1.5 to 5.6 μg/g and also contained T-2 toxin and HT-2 toxin (Harwig et al., 1979). The levels of neosolaniol in inoculated tomatoes were determined by TLC and the identity of purified material confirmed by GC-MS.

According to Suzuki et al. (1980, 1981b), three out of four isolates of *F. acuminatum* from wheat and barley in Japan produced neosolaniol at levels ranging from 2.97 to 184.31 μg/g as determined by GC. *Fusarium acuminatum* NHL-F-1025 (Strain 7.19), isolated from barley in Japan, produces neosolaniol together with diacetoxyscirpenol, T-2 toxin, and HT-2 toxin, according to H. Kurata (personal communication, 29 April 1980). It is not known to us whether this strain was one of the three isolates of *F. acuminatum* found to produce neosolaniol by Suzuki et al. (1980, 1981b).

3α,4β,8α-Trihydroxy-15-acetoxy-12,13-epoxytrichothec-9-ene. This new trichothecene was isolated and characterized from cultures of an unnumbered isolate of *F. heterosporum* on shredded wheat supplemented with mycological broth and YES medium incubated at 27 to 30°C for 3 weeks (Cole et al., 1981). We have identified this strain, which was isolated from *Claviceps paspali* honeydew on *Paspalum distichum* in Georgia, USA, as *F. acuminatum* (Strain 7.18).

3α,4β,15-Trihydroxy-8α-(3-hydroxy-3-methylbutyryloxy)-12,13-epoxy-trichothec-9-ene. This new trichothecene was isolated and characterized from cultures of *F. acuminatum* (= *F. heterosporum,* Strain 7.18) on shredded wheat supplemented with mycological broth and YES medium incubated at 27 to 30°C for 3 weeks (Cole et al., 1981).

T-2 Toxin. *Fusarium acuminatum* NRRL 6227 (Strain 7.16) produced unspecified levels of T-2 toxin (TLC) in cultures on corn grits incubated at 15°C for 3 weeks (Burmeister et al., 1981).

Harwig et al. (1979) detected T-2 toxin by TLC in cultures of *F. acuminatum* (= *F. sulphureum*) HPB 110178-19 (Strain 7.17), isolated from a decaying tomato in Canada, in YES medium incubated at 25°C for 8 to 13 days and also in inoculated tomatoes. The inoculated tomatoes incubated at 25°C for 6 to 20 days contained only traces (<1.5 μg/g) of T-2 toxin and HT-2 toxin, but tomatoes incubated at 15°C contained T-2 toxin at levels ranging from 0.38 to 37.5 μg/g and also contained HT-2 toxin and neosolaniol. The identity of purified toxic material isolated from these cultures was confirmed as T-2 toxin by GC-MS (Harwig et al., 1979). This strain has also been reported to produce 620 μg/g of T-2 toxin as determined by TLC in cultures on corn grits incubated at 14°C (Scott et al., 1980).

Cole et al. (1981) isolated and identified T-2 toxin from cultures of *F. acuminatum* (= *F. heterosporum,* Strain 7.18) on shredded wheat supplemented with mycological broth and YES medium incubated at 27 to 30°C for 3 weeks.

According to Suzuki et al. (1980, 1981b), three out of four unnumbered isolates of *F. acuminatum* from wheat and barley in Japan produced T-2 toxin at levels ranging from 8.35 to 373.15 μg/g as determined by GC. *Fusarium acuminatum* NHL-F-1025 (Strain 7.19), isolated from barley in Japan, produces T-2 toxin together with diacetoxyscirpenol, HT-2 toxin, and neosolaniol, according to H. Kurata (personal communication, 29 April 1980). It is not known to us whether this strain was one of the three Japanese isolates of *F. acuminatum* found to produce T-2 toxin by Suzuki et al. (1980, 1981b).

T-2 Tetraol. Cole et al. (1981) isolated and identified T-2 tetraol from cultures of *F. acuminatum* (= *F. heterosporum,* Strain 7.18) on shredded wheat supplemented with mycological broth and YES medium incubated at 27 to 30°C for 3 weeks.

Toxigenic Strains in the ITFRC

The following strains listed under *F. acuminatum* in Table 7.1 all have been identified as *F. acuminatum* according to Nelson et al. (1983).

Strain 7.16 *F. acuminatum* (= *F. roseum* 'Acuminatum') **NRRL 6227** (= R-5319; MRC 1783)

This strain (as *F. roseum* 'Acuminatum' NRRL 6227) was isolated from fescue hay associated with outbreaks of "Fescue foot" (see *F. sporotrichioides,* fescue foot) of cattle in Missouri, USA (Burmeister et al., 1981). Extracts of cultures of this strain on corn grits incubated at 15°C for 3 weeks were lethal to mice upon intraperitoneal injection (Burmeister et al, 1981). Analyses of the toxic extracts by TLC revealed the presence of unspecified levels of T-2 toxin, which was presumably responsible for the toxicity of this strain to mice (Burmeister et al., 1981). This strain is also known to produce a cyclic peptide "swelling factor" that is an antifungal antibiotic which causes marked swelling of conidia of *Penicillium* spp. while inhibiting germination (Burmeister et al., 1977, 1981).

A lyophilized culture of this strain was received on 23 October 1979 from NRRL as "*F. acuminatum* NRRL 6227." We consider this to be a typical culture of *F. acuminatum*.

Strain 7.17   *F. sulphureum* HPB 110178-19 (= R-6379; MRC 2190)

This strain was isolated from a decaying tomato in Canada (Harwig et al., 1979). Culture filtrates and ethyl acetate extracts of cultures of this strain in YES medium incubated at 25°C for 8 to 13 days were highly toxic to brine shrimp (Harwig et al., 1979). The presence of T-2 toxin in these extracts was detected by TLC and the identity of purified material isolated from the cultures in YES medium was confirmed by GC-MS. Extracts of cultures of this strain on corn incubated at 14°C have also been found to be toxic to brine shrimp and to contain 620 µg/g of T-2 toxin (Scott et al., 1980). The T-2 toxin produced by this strain was presumably responsible for its toxicity to brine shrimp. Ethyl acetate extracts of cultures of YES medium were also mutagenic to *Salmonella typhimurium* (Harwig et al., 1979). The chemical nature of the mutagen(s) produced by this strain is unknown.

Tomatoes inoculated with this strain were incubated at 15 or 25°C for 6 to 20 days until the desired stage of rot developed and were then analyzed by TLC for seven trichothecenes and zearalenone (Harwig et al., 1979). Tomatoes incubated at 25°C contained only traces (<1.5 µg/g) of T-2 toxin and HT-2 toxin. Inoculated tomatoes incubated at 15°C contained T-2 toxin (0.38 to 37.5 µg/g), HT-2 toxin (1.9 to 37.8 µg/g), and neosolaniol (1.5 to 5.6 µg/g). The identity of these trichothecenes detected by TLC was confirmed by GC-MS of purified compounds isolated from the inoculated tomatoes (Harwig et al., 1979).

A lyophilized culture of this strain was received on 22 December 1980 from B.J. Blanchfield as "*F. sulphureum* HPB 110178-19." This culture, which was identified as *F. sulphureum* by CBS according to Harwig et al. (1979), proved to be very difficult to identify. Some subcultures from the original were brown and others red on PDA. Both types produced microconidia, chlamydospores, and relatively short curved macroconidia on PDA and much longer and thinner macroconidia in sporodochia on CLA. Although the brown cultures may be confused with *F. equiseti* and the red ones with *F. scirpi* var. *compactum* (see Nelson et al., 1983), we consider this strain to be *F. acuminatum* with microconidia.

Strain 7.18   *F. heterosporum* F-77-1A (= R-5701; MRC 2017)

This strain was isolated from *Claviceps paspali* honeydew on *Paspalum distichum* L. in Georgia, USA (Cole et al., 1981). Chloroform extracts of cultures of this strain on shredded wheat supplemented with mycological broth and YES medium and incubated at 27 to 30°C for 3 weeks were highly toxic to day-old chickens (Cole et al., 1981). The toxicity to chickens was used as bioassay in the isolation and characterization of the following six trichothecenes from these cultures by Cole et al. (1981): T-2 toxin, HT-2 toxin, T-2 tetraol, and three new trichothecenes, $3\alpha,4\beta$,dihydroxy-15-acetoxy-8α-(3-hydroxy-3-methylbutyryloxy)-12,13-epoxytrichothec-9-ene; $3\alpha,4\beta$,15-trihydroxy-8α-(3-hydroxy-3-methylbutyryloxy)-12,13-epoxy-trichothec-9-ene; and $3\alpha,4\beta,8\alpha$-trihydroxy-15-acetoxy-12,13-epoxytrichothec-9-ene.

A plate culture of this strain was received on 7 October 1980 from R.J. Cole as "*F. heterosporum* F-77-1A." This culture is badly degenerated, semi-pionnotal, and difficult to identify with certainty. Subcultures of the original on PDA were brown without any red pigment, and produced microconidia, curved macroconidia, and numerous chlamydospores. Detailed comparisons of this strain with authentic strains of F. heterosporum revealed that it is definitely not F. heterosporum. We consider this strain to be a badly degenerated, semi-pionnotal strain of F. acuminatum that has lost its ability to produce red pigment.

According to Cole et al. (1981), colonization by F. heterosporum of the stromata of Claviceps paspali infecting Paspalum spp. is extremely common (30 to 50% of stromata) in Georgia and throughout most of the southern United States. We do not know whether the above toxigenic isolate that we have identified as F. acuminatum is representative of this species of Fusarium that colonizes ergot sclerotia in the southern United States or not. It is also not known to us how many isolates of this fungus referred to as F. heterosporum by Cole et al. (1981) have been found to be toxigenic. We conclude that more isolates of the Fusarium species colonizing ergot sclerotia in the southern United States should be examined in order to determine its identity and toxigenic potential.

Strain 7.19 F. acuminatum NHL-F-1025 (= R-5466; MRC 1959)
This strain was received on 19 April 1980 from H. Kurata as "F. acuminatum NHL-F-1025, isolate from barley grain, Japan. Produces T-2 toxin, HT-2 toxin, diacetoxyscirpenol, and neosolaniol." We consider this culture to be a typical representative of F. acuminatum.

According to Suzuki et al. (1980, 1981b), three out of four unnumbered isolates of F. acuminatum from wheat and barley in Japan produced T-2 toxin at levels ranging from 8.35 to 373.15 µg/g and neosolaniol at levels ranging from 2.97 to 184.31 µg/g as determined by GC. However, none of these isolates of F. acuminatum produced diacetoxyscirpenol, deoxynivalenol, fusarenon-X, or nivalenol (Suzuki et al., 1980, 1981b).

It is not known to us whether F. acuminatum NHL-F-1025 was one of the isolates of F. acuminatum reported to produce T-2 toxin and neosolaniol by Suzuki et al. (1980, 1981b). We are also not aware of any published reports on the production of diacetoxyscirpenol or HT-2 toxin by Japanese isolates of F. acuminatum.

8. Section Discolor

The problems associated with the use of the name *F. roseum* (Lk.) Snyd. & Hans. also apply to section Discolor (see section Roseum). *Fusarium heterosporum, F. sambucinum* (including *F. sulphureum*), *F. culmorum*, and *F. graminearum* have been reported to be toxigenic.

Fusarium heterosporum Nees

Perfect State: *Gibberella gordonii* Booth

Incidence and Distribution

Fusarium heterosporum is a cosmopolitan fungus that characteristically causes head blights of cereals and grasses and also occurs in association with the *Sphacelia* state and sclerotia of *Claviceps* spp. (Booth, 1971; Doidge, 1938; Gordon, 1959, 1960a; Meyer & Frank, 1979; Wollenweber & Reinking, 1935). The fungus appears to be particularly widespread throughout Africa on millet, and there are also numerous records of its occurrence on seeds (Booth, 1971). In Georgia and the southern United States, *F. heter-*

osporum has been reported to be very prevalent as a parasite of the sclerotia of *Claviceps paspali* infecting *Paspalum* spp. (Cole et al., 1981). However, we have identified one isolate from this source, reported to be toxigenic by Cole et al. (1981), as *F. acuminatum* (Strain 7.18).

Association with Human and/or Animal Diseases

Fusarium heterosporum has not been associated with any human or animal toxicoses.

Toxicity to Experimental Animals

Brine Shrimp. According to Diener et al. (1976), extracts of cultures on nutrient-amended shredded wheat of an isolate of *F. heterosporum* from cotton in the United States were not highly toxic to brine shrimp.

Chickens. According to Diener et al. (1976), extracts of cultures on nutrient-amended shredded wheat of an isolate of *F. heterosporum* from cotton in the United States were not highly toxic to chick embryos.

The isolate of *F. heterosporum* from honeydew of *Claviceps paspali* on *Paspalum distichum* in Georgia, USA, reported to be highly toxic to chickens by Cole et al. (1981) is *F. acuminatum* (Strain 7.18).

Rats. According to Diener et al. (1976), extracts of cultures on nutrient-amended shredded wheat of an isolate of *F. heterosporum* from cotton in the USA were not highly toxic to rats.

Mycotoxins Produced

The isolate of *F. heterosporum* from honeydew of *Claviceps paspali* on *Paspalum distichum* in Georgia, USA, which was reported to produce T-2 toxin, HT-2 toxin, T-2 tetraol, and three new trichothecenes by Cole et al. (1981) is *F. acuminatum* (Strain 7.18). The mycotoxin(s) produced by the isolate of *F. heterosporum* from cotton in the United States were not identified by Diener et al. (1976). Consequently information is not available at present on the chemical nature of the mycotoxin(s), if any, produced by *F. heterosporum*.

Toxigenic Strains in the ITFRC

The isolate of *F. heterosporum* reported to be toxigenic by Cole et al. (1981) has been identified as *F. acuminatum* (Strain 7.18) by us (Table 8.2), and there are no toxigenic strains of *F. heterosporum* represented in the ITFRC.

Fusarium sambucinum Fuckel

Perfect State: *Gibberella pulicaris* (Fr.) Sacc.

Incidence and Distribution

Fusarium sambucinum is a cosmopolitan soil saprophyte and is also known as a pathogen that causes cankers of woody trees, root and seedling rots of cereals and a wide range of other hosts, and storage rots of fruit and potatoes (Booth, 1971; CMI Descriptions of Pathogenic Fungi and Bacteria No. 385, 1973; Doidge, 1938; Domsch et al., 1980; Meyer & Frank, 1979; Wollenweber & Reinking, 1935). As *F. sulphureum,* it is the cause of a serious storage rot of potatoes in Europe, Iran, and North America (Booth, 1971; Gerlach & Ershad, 1970). *Fusarium sambucinum* has been isolated from overwintered cereals in the USSR (Joffe, 1960a, b, 1971), but has rarely been recorded on cereal seed in Canada (Gordon, 1952) and Europe (Hacking et al., 1976, 1977; Ylimäki, 1981; Ylimäki et al., 1979).

Although *F. sambucinum* is common in the temperate and Mediterranean regions of the northern hemisphere, it is less frequently encountered in the southern hemisphere, where it may have been introduced (CMI Descriptions of Pathogenic Fungi and Bacteria, No. 385, 1973).

Association with Human and/or Animal Diseases

Human Esophageal Cancer. The dominant fungus isolated from potatoes collected in a high-incidence area of esophageal cancer in Gonbad, Iran, was *F. sulphureum* (P.S. Steyn et al., 1978). We have identified one of these isolates, i.e., *F. sulphureum* MRC 514, as *F. sambucinum* (Strain 8.1). Cultures of this isolate on corn were acutely toxic to ducklings and caused acute and chronic toxicoses with gross and histopathological lesions similar to those caused by trichothecenes in rats. Four trichothecenes—i.e., diacetoxyscirpenol, monoacetoxyscirpenol, triacetoxyscirpenol, and 4-acetoxyscirpenediol—were identified in the toxic culture material (P.S. Steyn et al., 1978). Three other isolates of *F. sulphureum* (BBA 11124, 11125, and 11126) from potatoes in Iran were also shown to be toxic to ducklings and to produce diacetoxyscirpenol by P.S. Steyn et al. (1978). We have identified *F. sulphureum* BBA 11124 as *F. sambucinum* (Strain 8.2). No epidemiological evidence has been presented that potatoes contaminated with trichothecenes produced by *F. sambucinum* may be involved in the etiology of esophageal cancer in Iran.

Toxicity to Experimental Animals

Brine Shrimp. Extracts of potato tubers inoculated with an isolate of *F. sulphureum* from rotten potato tubers in Germany have been reported to be highly toxic to brine shrimp (>80% mortality in 4 hours) by Siegfried & Langerfeld (1978). We have identified this isolate as *F. sambucinum* (Strain 8.3). The mycotoxin(s) produced in inoculated potato tubers by this strain, which was the most toxic *Fusarium* spp. isolated from potato tubers by Siegfried & Langerfeld (1978), have not been identified.

Extracts of cultures on corn of *Fusarium* sp. HPB 071178-19, isolated from cribbed corn in Canada, were toxic to brine shrimp (Scott et al., 1980). We have identified this strain as *F. sambucinum* (Strain 8.4). Traces of zearalenone (up to 3µg/g) were detected by TLC in cultures of this strain, but the toxic extracts were negative for diacetoxyscirpenol, T-2 toxin, and all the other trichothecenes analyzed for by Scott et al. (1980). Consequently the chemical nature of the mycotoxin(s) produced by this strain toxic to brine shrimp is unknown.

Fusarium sulphureum HPB 110178-19, reported to be toxic to brine shrimp (Harwig et al., 1979; Scott et al., 1980) and mutagenic in *Salmonella typhimurium* (Harwig et al., 1979), is *F. acuminatum* (Strain 7.17).

Ducklings. P.S. Steyn et al. (1978) reported that cultures on corn of five isolates of *F. sulphureum* were acutely toxic to ducklings and caused the mortality of 4 of 4 ducklings within average times of 3 to 5 days. We have identified *F. sulphureum* MRC 514, which was isolated from potatoes collected in a high incidence area of human esophageal cancer (see human esophageal cancer, above) in Iran, and *F. sulphureum* MRC 846 (= BBA 11124), which was also isolated from potatoes in Iran, as *F. sambucinum* (Strains 8.1 and 8.2). Diacetoxyscirpenol was isolated from cultures of all five of these isolates and was presumably responsible for their toxicity to ducklings (P.S. Steyn et al., 1978). In addition to diacetoxyscirpenol, *F. sambucinum* (= *F. sulphureum*) MRC 514 also produced three other trichothecenes (P.S. Steyn et al., 1978).

Guinea Pigs. According to Joffe (1960a, 1971, 1974c, 1978b), culture filtrates and powdered dry mycelium of unspecified isolates of *F. sambucinum* from overwintered cereals associated with ATA (see *F. sporotrichioides*, ATA) in the USSR were toxic to guinea pigs. The animals died within 5 to 21 days and autopsies revealed hemorrhages in many organs and tissues. The mycotoxin(s) produced by these strains of *F. sambucinum* toxic to guinea pigs have not been identified.

Extracts of cultures in Czapek-Dox medium of an isolate of *F. sambucinum* from the "back yard" of a house in Georgia, USA, in which four cases of leukemia (see *F. equiseti*, leukemia) had occurred, depressed the response of guinea pigs to phytohemagglutinin (Wray et al., 1979).

Mice. Burmeister et al. (1981) reported that extracts of cultures on corn grits of *F. roseum* 'Sambucinum' NRRL 6478, isolated from pasture soil in Iceland, were toxic to mice upon intraperitoneal injection. No T-2 toxin could be detected by TLC in these toxic extracts by Burmeister et al. (1981) and the chemical nature of the mycotoxin(s) produced by this strain toxic to mice is unknown. This strain has been reported to produce a cyclic peptide antibiotic (see *F. acuminatum*, cyclic peptide) that caused marked swelling of *Penicillium* conidia while inhibiting germination (Burmeister et al., 1981).

Rabbits. Extracts of cultures of one out of 16 isolates of *F. sambucinum* from overwintered cereals associated with ATA (see *F. sporotrichioides*, ATA) in the USSR were found to be toxic to rabbit skin, and those of another isolate were mildly toxic (Joffe, 1960a, b, 1971). Culture filtrates and powdered dry mycelium of unspecified isolates of *F. sambucinum* from overwintered cereals in the USSR have also been reported to be acutely toxic to rabbits (Joffe, 1960a, b, 1971, 1974c, 1978b). The rabbits died within 8 to 24 days and autopsies revealed hemorrhages in many organs and tissues. The mycotoxin(s) produced by these isolates of *F. sambucinum* toxic to rabbits have not been identified.

Mower et al. (1975) examined the possibility that *F. roseum* 'Sambucinum' R-13c, a hyperparasite isolated from ergot sclerotia in California, USA, may degrade certain ergot alkaloids to hallucinogenic compounds such as d-lysergic acid diethylamide (LSD). The rabbit hyperthermia assay was used to test for the presence of such psychotropic compounds by the intravenous injection of ergotamine digested by this strain. No hyperthermic response could be elicited, indicating the absence of hallucinogenic indole derivatives such as LSD. Feeding of the digested ergotamine to rabbits for 30 days caused no signs of toxicity, thus indicating that *F. roseum* 'Sambucinum' R-13c had degraded the ergotamine to less toxic compounds (Mower et al., 1975).

Rats. According to Tuttobello et al. (1974), three strains of *F. sambucinum* (CBS 22931, CBS 26053, and CBS 18529) had uterotrophic effects in virgin weanling female rats fed cultures in liquid medium or on corn. The chemical nature of the active compound(s) was not determined, but the uterotrophic effects were equivalent to those of 12.8 and 38.0

ppm of "zearalanol" in the liquid cultures and 13.8 to 16.0 ppm in the corn cultures (Tuttobello et. al., 1974).

P.S. Steyn et al. (1978) reported that cultures on corn of *F. sulphureum* MRC 514, isolated from potatoes collected in a high-incidence area of human esophageal cancer (see human esophageal cancer, above) in Iran, caused acute and chronic toxicoses in rats when fed at different dietary levels. We have identified this strain as *F. sambucinum* (Strain 8.1). The gross and histopathological lesions observed in rats were "similar to those caused by the trichothecene mycotoxins" (P.S. Steyn et al., 1978). The toxic culture material was shown to contain large amounts of diacetoxyscirpenol (3.02 g/5 kg) as well as three other trichothecenes which were isolated by means of a rat skin test (P.S. Steyn et al., 1978).

Mycotoxins Produced

Acetoxyscirpenediol (= Monodeacetylanguidin). Loeffler et al. (1967) reported that *F. sambucinum* ATCC 11852 produced monodeacetylanguidin [referred to as acetoxyscirpenediol (= 4 or 15 acetylscirpentriol) by Bamburg & Strong, 1971], an antibiotic with antimitotic activity, in fermentor cultures incubated at 27°C for 145 hours. We have confirmed the identity of this strain as *F. sambucinum* (Strain 8.5).

P.S. Steyn et al. (1978) isolated 4-acetoxyscirpenediol from cultures on corn incubated at 25°C for 21 days of *F. sulphureum* MRC 514 which was isolated from potatoes collected in a high-incidence area of human esophageal cancer (see human esophageal cancer, above) in Iran. We have identified this strain as *F. sambucinum* (Strain 8.1). The yield of 4-acetoxyscirpenediol obtained by P.S. Steyn et al. (1978) from 5 kg of culture material was 600 mg of oil that could not be induced to crystallize.

Acetylneosolaniol (= Neosolaniol monoacetate). This new trichothecene was isolated from cultures of *F. roseum* V-18 in PSC medium incubated at 25°C for 14 days by Ishii et al. (1978). We have identified this strain as *F. sambucinum* (Strain 8.7). The yield of 8-acetylneosolaniol obtained from this strain, which also produced diacetoxyscirpenol, neosolaniol, and T-2 toxin, was not given by Ishii et al. (1978).

Butenolide. According to Suzuki et al. (1981a, b), a single strain of *F. sulphureum* isolated from wheat (or barley) in Japan produced 44.5 µg/g of butenolide in cultures on rice as detected by GC with an electron-capture detector. We have not seen this isolate, and there are no known butenolide-producing strains of *F. sambucinum* represented in the ITFRC.

Diacetoxyscirpenol. Sigg et al. (1965) isolated 2.9 g of crystalline diacetoxyscirpenol from 10ℓ of culture material of *F. sambucinum* ATCC 11852 (Strain 8.5) in liquid medium incubated in a fermentor at 27°C for 30 hours. Loeffler et al. (1967) also isolated diacetoxyscirpenol at unspecified yields from similar fermentor cultures incubated for 145 hours. The structure of diacetoxyscirpenol isolated from this strain was determined by Sigg et al. (1965) and Flury et al. (1965).

Diacetoxyscirpenol has been isolated from cultures on corn incubated at 25°C for 21 days of five isolates of *F. sulphureum* that were acutely toxic to ducklings (P.S. Steyn et al., 1978). We have identified *F. sulphureum* MRC 514, which was isolated from potatoes in a high-incidence area of human esophageal cancer (see human esophageal cancer, above) in Iran, and *F. sulphureum* MRC 846 (= BBA 11124), which was also isolated from potatoes in Iran, as *F. sambucinum* (Strains 8.1 and 8.2). The yield of crystalline diacetoxyscirpenol isolated from *F. sambucinum* (= *F. sulphureum*) MRC 514 (Strain 8.1) was 3.02 g/5 kg of culture material (P.S. Steyn et al., 1978).

Claridge & Schmitz (1979) used *Fusarium* sp. C-37410 for the production of diacetoxyscirpenol by fermentation. We have identified this strain, the source of which is unknown to us, as *F. sambucinum* (Strain 8.6).

The production of diacetoxyscirpenol by *F. roseum* V-18, isolated from corn in Minnesota, USA, was reported by Ishii et al. (1978). We have identified this strain as *F. sambucinum* (Strain 8.7). The yield of diacetoxyscirpenol obtained by Ishii et al. (1978) was 208 mg from 6.6 ℓ of culture material of this strain in PSC medium incubated at 25°C for 14 days.

In conclusion, it seems that *F. sambucinum* (= *F. sulphureum*) is a good producer of diacetoxyscirpenol; there are five known producing strains represented in the ITFRC (Table 8.1), and at least two of these (Strains 8.1 and 8.5) have been reported to produce large amounts of diacetoxyscirpenol in liquid cultures as well as on corn.

Fusarenon-X. According to Suzuki et al. (1980, 1981b) a single isolate of *F. sulphureum* from wheat (or barley) in Japan produced 5.01 µg/g of fusarenon-X and also produced nivalenol as detected by GC. We have not seen this isolate, and there are no known fusarenon-X–producing strains of *F. sambucinum* represented in the ITFRC.

HT-2 Toxin. *Fusarium sulphureum* HPB 110178-19, reported to produce HT-2 toxin by Harwig et al. (1979), is *F. acuminatum* (Strain 7.17).

Monoacetoxyscirpenol. P.S. Steyn et al. (1978) isolated 380 mg of crystalline monoacetoxyscirpenol from 5 kg of culture material of *F. sambucinum* (= *F. sulphureum*) MRC 514 (Strain 8.1) on corn incubated at 25°C for 21 days.

Neosolaniol. The production of neosolaniol by *F. roseum* V-18 was reported by Ishii et al. (1978). We have identified this strain, which was isolated from corn in Minnesota, USA, as *F. sambucinum* (Strain 8.7). The yield of neosolaniol obtained by Ishii et al. (1978) was 16 mg from 6.6 ℓ of culture material of this strain in PSC medium incubated at 25°C for 14 days. *Fusarium sulphureum* HPB 110178-19, reported to produce neosolaniol by Harwig et al. (1979), is *F. acuminatum* (Strain 7.17).

Nivalenol. According to Suzuki et al. (1980, 1981b), a single isolate of *F. sulphureum* from wheat (or barley) in Japan produced 1.38 µg/g of nivalenol and also produced fusarenon-X as detected by GC. We have not seen this isolate and there are no known nivalenol-producing strains of *F. sambucinum* represented in the ITFRC.

Triacetoxyscirpenol. P.S. Steyn et al. (1978) isolated 150 mg of crystalline triacetoxyscirpenol from 5 kg of culture material of *F. sambucinum* (= *F. sulphureum*) MRC 514 (Strain 8.1). on corn incubated at 25°C for 21 days.

T-2 Toxin. Ishii et al. (1978) detected unspecified levels of T-2 toxin by TLC and GC-MS in cultures of *F. roseum* V-18 in PSC medium incubated at 25°C for 14 days. We have identified this strain, which was isolated from corn in Minnesota, USA, as *F. sambucinum* (Strain 8.7). *Fusarium sulphureum* HPB 110178-19, reported to produce T-2 toxin by Harwig et al. (1979) and Scott et al. (1980), is *F. acuminatum* (Strain 7.17).

Zearalenone. According to Tuttobello et al. (1974), liquid cultures or culture material on corn of three strains of *F. sambucinum* (CBS 22931, CBS 26053, and CBS 18529) had uterotrophic effects in female rats. The chemical nature of the active compound(s) was not determined, but the uterotrophic effects were equivalent to those of 12.8 to 38.0 ppm of "zearalanol" in the liquid cultures and 13.8 and 16.0 ppm in the corn cultures (Tuttobello et al., 1974).

One out of three isolates of *F. sambucinum* var. *coeruleum* from barley grain in the UK has been reported to produce unspecified levels of zearalenone as detected by TLC in cultures on rice incubated at 25°C for 4 weeks followed by 6 weeks at 12°C (Hacking et al., 1976, 1977).

Traces of zearalenone (up to 3 µg/g) have been detected by TLC in culture on corn incubated at 14°C of *Fusarium* sp. HPB 071178-19, which was isolated from cribbed corn in Canada (Scott et al., 1980). We have identified this isolate as *F. sambucinum* (Strain 8.4).

Negative results for zearalenone production have been reported for one North American isolate of *F. sambucinum* by Caldwell et al. (1970) and one Japanese isolate by Suzuki et al. (1978).

In conclusion, only two strains of *F. sambucinum* have been reported to produce zearalenone, and one of these, which produces only traces of zearalenone, is represented in the ITFRC as *F. sambucinum* HPB 071178-19 (Strain 8.4).

Toxigenic Strains in the ITFRC

The *Fusarium* Strains listed under *F. sambucinum* in Table 8.1 all have been identified as *F. sambucinum* according to Nelson et al. (1983). Toxigenic strains that have been referred to as members of the section Discolor but identified as other species by us are listed in Table 8.2.

Strain 8.1   *F. sulphureum* MRC 514 (= R-5389; MRC 514)

This strain was isolated from potatoes collected in a high-incidence area of human esophageal cancer (see human esophageal cancer, above) in Gonbad, Iran (P.S. Steyn et al., 1978). The statement that this is a "South African isolate" (Harwig et al., 1979) is incorrect.

Cultures of this strain on corn incubated at 25°C for 21 days were acutely toxic to day-old ducklings and caused the death of 4 of 4 ducklings within an average time of 5 days (P.S. Steyn et al., 1978). This culture material also caused acute and chronic toxicoses in rats when fed at different dietary levels. The gross and histopathological lesions observed in rats "were similar to those caused by the trichothecene mycotoxins" (P.S. Steyn et al., 1978).

Four trichothecenes were isolated and characterized by P.S. Steyn et al. (1978) from the toxic culture material by means of a rat skin test bioassay. The main toxin was identified as diacetoxyscirpenol and 3.02 g of crystalline diacetoxyscirpenol was isolated from 5 kg of culture material. The three other trichothecenes isolated by P.S. Steyn et al. (1978) were monacetoxyscirpenol (380 mg), triacetoxyscirpenol (150 mg), and 4-acetoxyscirpenediol. The last compound was isolated as 600 mg of oil that could not be induced to crystallize.

A lyophilized culture of this strain was received on 22 April 1980 from C.J. Rabie as "*F. sulphureum* MRC 514." According to P.S. Steyn et al. (1978), this culture was identified as *F. sulphureum* by C. Booth. This sulphur yellow culture is in good condition, and we agree that it could be identified as *F. sulphureum* according to Booth (1971). However, *F. sulphureum* was included in *F. sambucinum* by Nelson et al. (1983) because of the characteristic macroconidial shape, and we have identified this culture accordingly.

140 Discolor (8)

Table 8.1. *Fusarium* Strains of the Section Discolor in the International Toxic *Fusarium* Reference Collection (ITFRC)

Strain No.	ITFRC No.	MRC No.	Published Name(s) and Number(s)	Source	Toxigenicity
Fusarium sambucinum					
8.1	R-5389	514	*F. sulphureum* MRC 514	C.J. Rabie	a,b,c,d,e,f
8.2	R-5390	846	*F. sulphureum* MRC 846 (= BBA 11124)	W. Gerlach	b,c
8.3	R-6380	2193	*F. sulphureum* 722	E. Langerfeld	b
8.4	R-6354	2187	*Fusarium* sp. HPB 071178-19	B.J. Blanchfield	b,g
8.5	R-6112	2320	*F. sambucinum* ATCC 11852	ATCC	c,f
8.6	R-6239	2394	*Fusarium* sp. C-37410	C.A. Claridge	c
8.7	R-5455	1903	*Fusarium roseum* V-18	C.J. Mirocha	c,h,i,j
Fusarium culmorum					
8.8	R-5452	939	*F. culmorum* M 6.2/66	M.E. di Menna	a,b
8.9	R-6382	2197	*F. culmorum* BBA 62183	E. Langerfeld	b
8.10	R-4607	1889	*F. culmorum* Sp. 885	W.F.O. Marasas	b,g,k,l
8.11	R.5391	1890	*F. culmorum* Sp. 887	W.F.O. Marasas	b,g,k,l
8.12	R-5145	1369	*F. culmorum* 72196	A. Ylimäki	b,g
8.13	R-5146	1371	*F. culmorum* 72313-20	A. Ylimäki	b,g
8.14	R-5456	1905	*F. culmorum* 7289	C.J. Mirocha	a,k
8.15	R-5321	1787	*F. culmorum* 3737 = NRRL 3288	NRRL	b,g,k
8.16	R-5251	1623	*F. culmorum* Washington	C.J. Mirocha	a,g,k
8.17	R-5797	2064	*F. culmorum* ITM-122	A. Bottalico	k
8.18	R-6353	2434	*F. culmorum* No. 34 = *Calonectria nivalis* CMI 14764 = *Micronectriella nivalis* ATCC 26559	ATCC	b,l,m,n,o
Fusarium graminearum					
8.19	R-5318	1782	*Gibberella zeae* NRRL 2830 = ATCC 20272	NRRL	a?,b,g,k
8.20	R-5250	1621	*Fusarium* No. 5	C.J. Mirocha	a,b,g
8.21	R-6136	2329	*F. roseum* 'Graminearum' Mapleton No. 10	C.J. Mirocha	a,b,g
8.21	R-5453	1806	= *Gibberella zeae* ATCC 24689 (= JRL Map. 10)	ATCC	
8.21	R-5454	1904	= *Gibberella zeae* ATCC 24688 (= Map. 10-Ascospore 55)	C.J. Mirocha	
8.22	R-5796	1946	*F. roseum* No. 117 = ATCC 28114	ATCC	a,b,k,l,p,q
8.23	R-5320	1785	*F. graminearum* NRRL 5883	NRRL	a,b,g,k,r
8.24	R-5316	1693	*F. graminearum* NRRL 5864	NRRL	a?,b?,g,k
8.25	R-6055	2233	*F. graminearum* NRRL 6450	NRRL	a,k
8.26	R-6337	2436	*F. graminearum* NRRL 6451	NRRL	a,k
8.27	R-6056	2234	*F. graminearum* NRRL 6452	NRRL	a,k
8.28	R-6796	2581	*Gibberella zeae* Ishii	Y. Ueno	a,b,g
8.29	R-6795	2580	*F. rigidiusculum* M-1-3	Y. Ueno	a,b,c,h,i
8.30	R-6798	2583	*F. roseum* 'Culmorum' M-3-2	Y. Ueno	a,b,g
8.31	R-6802	2606	*F. roseum* 'Gibbosum' A-O-2	Y. Ueno	g
8.32	R-6801	2605	*F. tricinctum* A-R-5	Y. Ueno	h,i
8.33	R-6777	2563	*F. tricinctum* A-B-2 = Abashiri-2	Y. Ueno	b,c,g,h,i,s
8.34	R-5467	1960	*F. graminearum* NHL-F-1118	H. Kurata	g,k,l
8.35	R-5469	1963	*F. graminearum* NHL-F-1104	H. Kurata	t,u
8.36	R-5470	1976	*F. graminearum* NHL-F-1112	H. Kurata	g,t,u
8.37	R-6800	120	*F. graminearum* MRC 120	W.F.O. Marasas	b,g
8.38	R-4838	1646	*F. graminearum* F-59 = ATCC 34909	M. Palyusik	b,g,k
8.39	R-4613	2940	*F. culmorum* F-79 = ATCC 34910	M. Palyusik	b,g
8.40	R-6767	2560	*F. roseum* 'Culmorum' 70-K-11	Y. Ueno	b,c,h,i,s
8.41	R-5317	1781	*F. tricinctum* NRRL 5908	NRRL	b,h
8.42	R-4057	121	*F. graminearum* MRC 121	W.F.O. Marasas	b,g,k
8.43	R-4053	460	*F. graminearum* MRC 460	W.F.O. Marasas	b,g,k
8.44	R-5176	1391	*F. graminearum* 72322	A. Ylimäki	b
8.45	R-5953	1393	*F. graminearum* 7137	A. Ylimäki	b,k
8.46	R-6574	2638	*Gibberella zeae* B-507	L.P. Hart	k

Table 8.1. (Continued)

Strain No.	ITFRC No.	MRC No.	Published Name(s) and Number(s)	Source	Toxigenicity
8.47	R-6575	2639	*Gibberella zeae* B-601	L.P. Hart	k
8.48	R-6576	2640	*Gibberella zeae* W-8	L.P. Hart	g,k
8.49	R-5798	2065	*F. graminearum* ITM 126	A. Bottalico	k
8.50	R-4610	3081	*F. graminearum* F-107 (= Sp. 898)	J.M. Möller	g
8.51	R-4611	3082	*F. graminearum* F-110 (= Sp. 899)	J.M. Möller	g
8.52	R-6182	2393	*F. graminearum* G-73-2	J.C. Sutton	g

a = Associated with human and/or animal diseases
b = Culture toxic to experimental animals
c = Produces diacetoxyscirpenol
d = Produces monoacetoxyscirpenol
e = Produces triacetoxyscirpenol
f = Produces acetoxyscirpenediol
g = Produces zearalenone
h = Produces T-2 Toxin
i = Produces neosolaniol
j = Produces acetylneosolaniol
k = Produces deoxynivalenol
l = Produces acetyldeoxynivalenol
m = Produces calonectrin and 15-deacetylcalonectrin
n = Produces culmorin
o = Produces 2-acetylquinazolin-4(3H)-one
p = Produces butenolide
q = Produces diacetyldeoxynivalenol
r = Produces 4-acetamido-2-butenoic acid
s = Produces HT-2 toxin
t = Produces fusarenon-X
u = Produces nivalenol

Strain 8.2 *F. sulphureum* BBA 11124 (= R-5390; MRC 846)
This strain was isolated from potatoes in Iran by W. Gerlach in 1968 (P.S. Steyn et al., 1978). Cultures of this strain on corn incubated at 25°C for 21 days were acutely toxic to day-old ducklings and caused the death of 4 of 4 ducklings in an average time of 3 days (P.S. Steyn et al., 1978). Diacetoxyscirpenol was isolated from the toxic culture material at unspecified yields by P.S. Steyn et al. (1978) and was presumably responsible for the toxicity to ducklings.

A slant culture of this strain was received in 1977 from W. Gerlach as "*F. sulphureum* BBA 11124." This culture is sulphur yellow and could be identified as *F. sulphureum* according to Booth (1971). However, *F. sulphureum* was included in *F. sambucinum* by Nelson et al. (1983), and we have identified this culture accordingly.

Strain 8.3 *F. sulphureum* 722 (R-6380; MRC 2193)
This strain was isolated from rotten potatoes in Germany (Siegfried & Langerfeld, 1978). Extracts of potato tubers inoculated with this strain and incubated at either 6, 15, or 20°C were highly toxic to brine shrimp and caused over 80% mortality in 4 hours (Siegfried & Langerfeld, 1978). No diffusion of the toxin(s) into healthy tuber tissue adjacent to the infected parts could be demonstrated. The mycotoxin(s) produced by this strain have not been identified.

A slant culture of this strain was received on 20 January 1981 from E. Langerfeld as "*F. sulphureum* 722." This culture is sulphur yellow and could be identified as *F. sulphureum* according to Booth (1971). However, *F. sulphureum* was included in *F. sambucinum* by Nelson et al. (1983), and we have identified this culture accordingly.

Strain 8.4 *Fusarium* sp. HPB 071178-19 (=R-6354; MRC 2187)
This strain was isolated from cribbed corn in Canada (Scott et al., 1980). Extracts of cultures of this strain on corn incubated at 14°C were toxic to brine shrimp (Scott et al., 1980). Analyses of these toxic extracts by TLC were negative for diacetoxyscirpenol, T-2 toxin, and several other trichothecenes, but revealed the presence of traces (up to 3

Table 8.2. *Fusarium* Strains Published as Members of the Section Discolor but Identified as Other Species in the International Toxic *Fusarium* Reference Collection (ITFRC)

Strain No.	ITFRC No.	MRC No.	Published Name(s) and Number(s)	Source	Present Identification
7.18	R-5701	2017	F. heterosporum F-77-1A	R.J. Cole	F. acuminatum
7.17	R-6379	2190	F. sulphureum HPB 110178-19	B.J. Blanchfield	F. acuminatum
7.4	R-6766	2559	F. roseum "culmorum" M-14-2	Y. Ueno	F. equiseti

µg/g) of zearalenone (Scott et al., 1980). Thus the mycotoxin(s) produced by this strain toxic to brine shrimp have not been identified.

A lyophilized culture of this strain was received on 13 July 1981 from B.J. Blanchfield as "*Fusarium* sp. HPB 071178-19." This is a typical, red-pigmented culture of *F. sambucinum*.

Strain 8.5 *F. sambucinum* ATCC 11852 (= R-6112; MRC 2320)

This strain is not listed under *F. sambucinum* in the ATCC Catalogue of Strains, Fifteenth Ed. (1982), and the source is unknown to us.

Sigg et al. (1965) isolated 2.9 g of crystalline diacetoxyscirpenol (= anguidin) from 10ℓ of culture filtrate of this strain in liquid medium incubated in a fermentor of 27°C for 30 hours. Loeffler et al. (1967) also isolated diacetoxyscirpenol at unspecified yields from similar cultures incubated for 145 hours. The structure of diacetoxyscirpenol isolated from this strain of *F. sambucinum* was determined by Flury et al. (1965) and Sigg et al. (1965). A related trichothecene, monodeacetylanguidin, was also isolated from fermentor cultures of this strain. This compound was referred to as acetoxyscirpenediol = 4 or 5 acetylscirpentriol by Bamburg & Strong (1971).

A lyophilized culture of this strain was received on 9 March 1981 from ATCC as "*F. sambucinum* ATCC 11852." This culture was badly degenerated and pionnotal, but still produced some typical macroconidia on CLA, allowing it to be identified as *F. sambucinum*.

Strain 8.6 *Fusarium* sp. C-37410 (= R-6239; MRC 2394)

The source of this strain, which was used for the production of diacetoxyscirpenol (= anguidin) in fermentor cultures by Claridge & Schmitz (1975), is unknown to us.

A slant culture of this strain was received on 1 June 1981 from C.A. Claridge as "*Fusarium* sp. C-37410-L(A)." This culture was badly degenerated and pionnotal, but still produced some typical macroconidia on CLA that allowed it to be identified as *F. sambucinum*.

Strain 8.7 *Fusarium roseum* V-18 (= R-5455; MRC 1903)

This strain was isolated from harvested corn in Minnesota USA (Ishii et al., 1978). Four toxic compounds were isolated from the culture filtrate (6.6 ℓ) of *F. roseum* V-18 in PSC medium incubated at 25°C for 14 days (Ishii et al., 1978). Three of these were identified by TLC and GC as diacetoxyscirpenol (208 mg), neosolaniol (16 mg), and T-2 toxin. The fourth compound was characterized as the new trichothecene 8-acetylneosolaniol (= neosolaniol monoacetate) by Ishii et al. (1978), but the yield was not stated.

A silica gel culture of this strain was received on 17 March 1980 from C. J. Mirocha as "*F. roseum* V-18." This culture was badly degenerated, pionnotal, and impossible to identify with absolute certainty. The macroconidia produced in CLA are narrower and have thinner septa than those of *F. culmorum*. The apical cell is sharply curved and constricted and microconidia are produced on PDA. On the basis of these characteristics we have identified this strain as *F. sambucinum* according to Nelson et al. (1983).

Fusarium culmorum (W.G. Smith) Sacc.

Perfect State: Unknown.

Incidence and Distribution

Fusarium culmorum has a world-wide distribution as a soil inhabitant and as a serious pathogen that causes root, foot, stem, and ear rots of cereals and a large number of other hosts (Booth, 1971; CMI Descriptions of Pathogenic Fungi and Bacteria, No. 26, 1964; Doidge, 1938; Domsch et al., 1980; Meyer & Frank, 1979; Wollenweber & Reinking, 1935). *Fusarium culmorum* has been isolated from overwintered cereals in the USSR (Joffe, 1960a, b, 1971) and is particularly prevalent as a seed-borne organism of cereals in temperate areas (Enari et al., 1981; Gerlach & Ershad, 1970; Gordon, 1952; Hacking et al., 1976, 1977; Marasas et al., 1979b; Ylimäki, 1981; Ylimäki et al., 1979).

Association with Human and/or Animal Diseases

Moldy Feed Toxicosis. Two field outbreaks in Victoria, Australia, of a toxicosis in dairy cattle, characterized by loss of appetite, decreased milk production, scouring and staggering, were attributed by Fisher et al. (1967) to the consumption of moldy corn. Among several species of *Fusarium* isolated from the incriminated corn, only two isolates of *F. culmorum* were dermotoxic to rabbit skin and caused a severe hemorrhagic skin reaction and death. We have confirmed the identity of one of these isolates as *F. culmorum* (Strain 8.8). The mycotoxin(s) produced by these two dermotoxic strains were not identified by Fisher et al. (1967) and there are not analytical results to support the statement by Wright (1968) that "it is likely that the toxin involved is diacetoxyscirpenol." In addition to *F. culmorum,* the incriminated corn samples were also infected by *F. graminearum* (Fisher et al., 1967), which may also have contributed to their toxicity.

Reductions in milk yield together with diarrhea have also been recorded in dairy cows grazing rye grass pastures in France (Moreau, 1972). The dominant fungus isolated from the incriminated forage was *F. culmorum* (as *F. roseum* var. *culmorum*), but no experimental evidence was presented to support the claim that this fungus was responsible for the clinical signs observed.

Estrogenic Syndrome (= Hyperestrogenism, = Vulvo-Vaginitis). The predominant *Fusarium* spp. isolated from corn or other feedstuffs contaminated with zearalenone and associated with field outbreaks of the estrogenic syndrome in pigs, are members of the section Discolor (Mirocha et al., 1976; Vesonder et al., 1981b). In many reports on these field outbreaks, the *Fusarium* species involved are unfortunately referred to as "*F. roseum*" and it is consequently not possible to determine whether the estrogenic metabolite, zearalenone, was produced by *F. culmorum, F. graminearum,* or both. In most outbreaks of hyperestrogenism in pigs associated with moldy corn in the United States and elsewhere, the main zearalenone-producing species is probably *F. graminearum,* and the estrogenic syndrome is discussed in detail under *F. graminearum* (see *F. graminearum,* estrogenic syndrome). However, *F. culmorum* is also a well-known producer of zearalenone (see zearalenone, below) and may well be the primary producer of zearalenone in certain field outbreaks of the estrogenic syndrome. Such a zearalenone-producing strain of *F. culmorum* has been associated with an outbreak of porcine hyperestrogenism in Washington, USA (Mirocha et al., 1976). In Europe and the UK, *F. culmorum* is the predominant *Fusarium* species associated with barley grain (Enari et al., 1981; Hacking et al., 1976, 1977; Ylimäki, 1981; Ylimäki et al., 1979). Isolates of *F. culmorum* from barley in these areas have been shown to produce zearalenone in culture (Hacking et al., 1976, 1977; Mannio & Enari, 1973; Marasas et al., 1979b; Miller et al., 1973; Roine et al., 1971; Ylimäki et al., 1979) and also in barley inoculated in the field (Enari et al., 1981; Gross & Robb, 1975; Ylimäki et al., 1979). *Fusarium*-infected barley has been implicated in outbreaks of porcine hyperestrogenism in Ireland (McErlean, 1952) and Germany (Barnikol et al., 1981), a clinical condition in sows characterized by the production of litters with stillborn pigs or pigs with a "splayleg" incoordination of the hind limbs in Scotland (Miller et al., 1973), and infertility in pigs in Scotland (Gross & Robb, 1975). Fertility disturbances in dairy cattle in Finland have also been attributed to feed from which *F. culmorum* and *F. graminearum* were isolated (Roine et al., 1971).

Feed Refusal and Emetic Syndromes. The same considerations pointed out under the estrogenic syndrome (see estrogenic syndrome, above) apply to the feed refusal and emetic syndromes in pigs, which are discussed in detail under *F. graminearum* (see *F. graminearum*, feed refusal and emetic syndromes). *F. culmorum* is also known to produce deoxynivalenol (see deoxynivalenol, below) and may well be the primary producer of this trichothecene associated with feed refusal and emetic activity in certain field outbreaks.

Toxicity to Experimental Animals

Brine Shrimp (*Artemia salina*). Extracts of potato tubers inoculated with an isolate of *F. culmorum* from rotten potato tubers in Germany were found to be slightly toxic to brine shrimp (Siegfried & Langerfeld, 1978). We have confirmed the identity of this isolate as *F. culmorum* (Strain 8.9). The mycotoxin(s) produced by this strain have not been identified.

Chickens. Cultures on corn of five isolates of *F. culmorum* from corn stems and barley grain in Germany had a low degree of toxicity to day-old White Leghorn chickens and caused the death of none to two out of four birds within 14 days (Marasas et al., 1979b). We have confirmed the identity of the two most toxic strains (*F. culmorum* Sp. 885 and Sp. 887), which both caused the death of two out of four chickens as well as marked reductions in feed consumption and weight gain of the survivors, as *F. culmorum* (Strains 8.10 and 8.11). Analyses by TLC of the culture material on corn revealed that four of the five strains of *F. culmorum* produced zearalenone, and three of the five produced deoxynivalenol and acetyldeoxynivalenol at levels ranging from 320 to 1400 ppm, 1 to 15 ppm, and 1 to 2 ppm, respectively (Marasas et al., 1979b). The deoxynivalenol produced by *F. culmorum* Sp. 885 and Sp. 887 (Strains 8.10 and 8.11) at levels of 7 and 15 ppm, respectively, was presumably responsible for the reduced feed consumption of the culture material by chickens and the resultant loss in weight gain and death of some birds.

Geese. The Hungarian isolate of *F. culmorum* reported to cause non-significant reductions in egg production and egg fertility in geese (Palyusik & Koplik-Kovacs, 1975a, b) was *F. culmorum* F-79, according to M. Palyusik (personal communication, 16 December 1977). We have identified this strain as *F. graminearum* (Strain 8.39).

Mice. During 1972 in Finland, barley naturally infected by *F. culmorum* in the field was acutely toxic to mice, and mice fed this grain died within 9 to 11 days and showed "clear signs of *Fusarium* toxicosis" (Mannio & Enari, 1973). Barley inoculated with an unnumbered Finnish isolate of *F. culmorum* was also acutely toxic to mice and caused 100% mortality within 7 to 9 days. The strain used in these toxicological studies was *F. culmorum* 72196, according to T.M. Enari (personal communication, 17 April 1978). We have confirmed the identity of this strain as *F. culmorum* (Strain 8.12). The malt produced from the barley inoculated with this strain by Mannio & Enari (1973) was highly toxic to mice, and the animals, which died within 7 to 9 days, showed marked signs of gastrointestinal inflammation. The dried "spent grains" that remained behind after this malt had been used for brewing was also acutely toxic to mice and consequently unacceptable as animal feed, but the beer itself was not toxic (Mannio & Enari, 1973). According to A. Ylimäki (personal communication, 10 April 1979), *F. culmorum* 72196 (Strain 8.12) was one of the Finnish isolates of *F. culmorum* reported to produce zearalenone by Niku-Paavola & Nummi (1977) and Niku-Paavola et al. (1977). The mycotoxin(s) responsible for the toxicity to mice of cultures of this strain on barley, and of the malt prepared from these cultures, have not been identified. In view of the toxicity of *F. culmorum* and other *Fusarium* spp. infecting barley and the problems caused by these fungi in malting and brewing, the maximal level of visible *Fusarium* infection in barley used for malting in Finland is 0.06% (Mannio & Enari, 1973).

Ueno et al. (1973a) reported that extracts of cultures of eight isolates of *F. roseum* 'Culmorum' were toxic to mice. We have identified two of these (M-3-2 and 70-K-11) as *F. graminearum* (Strains 8.30 and 8.40), and one (M-14-2) as *F. equiseti* (Strain 7.4).

Pigeons. Prolonged emesis was induced in pigeons injected in the wing vein with extracts of cultures in Richards medium of *F. culmorum* No. 3737 (Prentice & Dickson, 1968; Prentice et al., 1959). We have confirmed the identity of this strain as *F. culmorum* (Strain 8.15). However, no emetic activity could be demonstrated in pigeons with extracts of cultures of this strain (as *F. culmorum* NRRL 3288) in Richards medium or on corn by Ellison & Kotsonis (1973). In culture on corn, *F. culmorum* NRRL 3288 (Strain 8.15) was subsequently shown to produce approximately 40 mg/kg of deoxynivalenol, but not T-2 toxin, HT-2 toxin, acetyl T-2 toxin, or fusarenon-X (Vesonder et al., 1977b, 1981b). The deoxynivalenol produced by this strain was presumably responsible for the emetic activity in pigeons reported by Prentice & Dickson (1968) and Prentice et al. (1959).

Pigs. Cultures on corn of *F. culmorum* NRRL 3288 (Strain 8.15) caused a marked feed refusal response (82%) in pigs (Vesonder et al., 1977b). The rejected culture material contained approximately 40 mg/kg of deoxynivalenol as determined by TLC, but no T-2 toxin, HT-2 toxin, acetyl T-2 toxin, or fusarenon-X. Analyses by GLC of cultures of this strain on corn subsequently revealed the presence of 39 μg/g of deoxynivalenol and unspecified levels of zearalenone (Vesonder et al., 1981b). The deoxynivalenol produced by this strain was presumably responsible for its feed refusal activity in pigs.

Rabbits. Two out of 16 isolates of *F. culmorum* from overwintered cereals associated with ATA (see *F. sporotrichioides*, ATA) in the USSR were found to be dermotoxic to rabbit skin, and a third isolate was mildly toxic (Joffe, 1960a, b, 1971). Ether extracts of cultures on corn of two unnumbered isolates of *F. culmorum* from moldy corn associated with outbreaks of moldy feed toxicosis (see moldy feed toxicosis, above) of dairy cows in Australia were found to be dermotoxic to rabbit skin (Fisher et al., 1967). We have confirmed the identity of one of these isolates as *F. culmorum* (Strain 8.8). The extracts caused hemorrhagic skin lesions followed by death within 5 to 7 days. The mycotoxin(s) produced by these dermotoxic strains of *F. culmorum* have not been identified.

Rats. Twelve isolates of *Fusarium* spp. from corn in the United States were incubated on corn at 20 and 25°C for 3 weeks followed by 2 weeks at 12°C. These isolates caused increases of five to eight times in the weight of the uterus of weanling female rats (Christensen et al., 1965). According to Christensen et al. (1965), "all or most of these isolates of *Fusarium* tentatively appear to be *F. culmorum* or *F. graminearum.*" The uterotrophic metabolite, zearalenone (= F-2), was isolated and partially characterized from *Fusarium* sp. No. 5 (= *F. graminearum*, Strain 8.20) and was presumably also responsible for the uterotrophic activity of the other *Fusarium* isolates.

A strain of *F. culmorum* isolated from feed suspected of causing infertility in dairy cows (see estrogenic syndrome, above) in Finland caused a significant increase in uterine weight when cultures on a mixture of oats, barley and wheat grain were fed to weanling female rats (Roine et al., 1971). This culture material was found to contain 3.19 mg/kg of zearalenone.

According to Tuttobello et al. (1974), cultures in liquid media as well as on corn of two strains of *F. culmorum* (ISS Roma and Waksman) had uterotrophic effects in weanling female rats. The active compound was not identified, but the uterotrophic activity of the two strains was equivalent to approximately 21.7 to 23.0 ppm of "zearalanol" in the liquid media and 12.0 to 14.4 ppm in cultures on corn.

Szathmary et al. (1976, 1977) reported that cultures on millet of two isolates of *F. culmorum* (F-79 and F-247) from millet in Hungary caused uterine hypertrophy in female weanling rats. We have identified Strain F-79 as *F. graminearum* (Strain 8.39). Analyses by GC-MS of the culture material of *F. culmorum* F-247 revealed the presence of 3000 µg/g of zearalenone (Szathmary et al., 1976, 1977). This strain was not dermotoxic to rat skin and did not produce any trichothecenes that could be detected by GC-MS.

Cultures on corn inoculated with a mixture of Hungarian isolates of *F. culmorum* and *F. graminearum* contained 34 µg/g of zearalenone, which was used in a study of the effects of zearalenone on the reproductive system and fertility of male and female rats (Ruzsas et al., 1979). Diets containing 50% of culture material caused significant reductions in the fertility rates of both males and females, as well as decreases in gonadal weights and disturbances in follicular maturation and spermatogenesis. Consumption of the diet containing zearalenone by female rats during pregnancy and lactation resulted in permanent changes in the reproductive organs, disorders in vaginal cycling, and disturbed fertility in the offspring. The possibility has to be considered that metabolites other than zearalenone were produced in culture by these isolates of *F. culmorum* and *F. graminearum*, and that these may have been responsible for some of the effects observed in rats by Ruzsas et al. (1979).

Cultures of 17 isolates of *F. culmorum* from cereals in Finland were fed to rats (two rats/strain) for 14 days (Korpinen & Uoti, 1974). We have confirmed the identity of one of these isolates, which caused marked reduction in weight gain and some mortality of rats, as *F. culmorum* (Strain 8.13). This strain is known to produce zearalenone (see Strain 8.13), but the mycotoxin(s) responsible for its toxicity to rats have not been identified.

Blight & Grove (1974) reported that ethyl acetate extracts of cultures in Czapek-Dox medium of *F. culmorum* No. 34 caused a positive necrotic skin reaction in rats. We have confirmed the identity of this strain as *F. culmorum* (Strain 8.18). The following three compounds were isolated from the toxic culture material of this strain by Blight & Grove (1974): 2-acetyl-quinazolin-4(3H)-one, culmorin, and acetyldeoxynivalenol. The dermotoxicity of this strain was presumably due to acetyldeoxynivalenol, but possibly also to calonectrin and 15-deacetylcalonectrin, which are also known to be produced (Gardner et al., 1972) by this strain (as *Calonectria nivalis* CMI 14764). Chloroform extracts of cultures in GAN medium incubated at 25°C for 14 days of *F. culmorum* No. 53, isolated from a dairy pasture in Ireland, were also reported to cause a necrotic skin reaction in rats by Blight & Grove (1974). The toxic culture material contained culmorin, which was not responsible for the dermotoxicity, and the trichothecene diacetylnivalenol (Blight & Grove, 1974).

Mycotoxins Produced

Acetyldeoxynivalenol (=Deoxynivalenol monoacetate). Blight & Grove (1974) first isolated and characterized this trichothecene (= 3α-acetoxy-12,13-epoxy-7α,15-dihydroxytrichothec-9-en-8-one) from cultures of *F. culmorum* No. 34 (= *Calonectria nivalis* CMI 14764) in Czapek-Dox medium incubated at 25°C for 10 days. We have confirmed the identity of this strain as *F. culmorum* (Strain 8.18). The yield of acetyldeoxynivalenol obtained by Blight & Grove (1974) was 52 mg/ℓ

Three isolates of *F. culmorum* from barley grain in Germany produced low levels (1 to 2 ppm) of acetyldeoxynivalenol as determined by TLC in cultures on corn incubated at 25°C for 2 weeks followed by 4 weeks at 10°C (Marasas et al., 1979b). We have confirmed the identity of two of these isolates, i.e., *F. culmorum* Sp. 885 and Sp. 887 (Strains 8.10 and 8.11).

2-Acetylquinazolin-4(3H)-one. Blight & Grove (1974) isolated this compound at a yield of 0.6 mg/ℓ from cultures of *F. culmorum* No. 34 (= *Calonectria nivalis* CMI 14764) in Czapek-Dox medium incubated at 25°C for 10 days. We have confirmed the identity of this strain as *F. culmorum* (Strain 8.18).

Calonectrins. The trichothecenes calonectrin and 15-deacetylcalonectrin were first isolated and characterized by Gardner et al. (1972) from cultures of *Calonectria nivalis* CMI 14764 in Czapek-Dox medium incubated in stationary culture for 21 days. We have identified this strain as *F. culmorum* (Strain 8.18). These two trichothecenes have only been isolated from this strain, which was originally isolated (as *F. culmorum* CMI 14764) from culms of wheat in New Zealand, according to the Catalogue of the Culture Collection of the CMI, Seventh Ed. (1975). Although Blight & Grove (1974) already stated that "the organism described as *Calonectria nivalis* (= *F. nivale*) which produces calonectrin and 15-de-O-acetylcalonectrin, is now considered to be *F. culmorum*," the incorrect name introduced by Gardner et al. (1972) has persisted in the literature and this strain is still listed as *Micronectriella nivalis* ATCC 26559 in the ATCC Catalogue of Strains, Fifteenth Ed. (1982). This taxonomic error is probably the main reason for the fact that the calonectrins have been isolated from only one strain, and the examination of the other strains of *F. culmorum* for the production of these trichothecenes is clearly indicated.

Culmorin. Gardner et al. (1972) isolated 50 mg of the sesquiterpene culmorin from 79ℓ of culture material of *Calonectria nivalis* CMI 14754 in Czapek-Dox medium incubated for 21 days. We have identified this strain as *F. culmorum* (Strain 8.18). Culmorin has also been isolated at a yield of 0.9 mg/ℓ from cultures of this strain (as *F. culmorum* No. 34) in Czapek-Dox medium incubated at 25°C for 10 days (Blight & Grove, 1974). The latter authors also isolated culmorin (20 mg/ℓ) from cultures of *F. culmorum* No. 53 in GAN medium incubated at 25°C for 14 days and stated that it was not responsible for the dermotoxicity of the culture material to rat skin.

Deoxynivalenol. The production of deoxynivalenol by *F. culmorum* was first reported by Mirocha et al. (1976). An isolate of *F. culmorum* (Strain 8.16) associated with an outbreak of porcine hyperestrogenism in Washington, USA, was found to produce unspecified levels of deoxynivalenol as detected by GC-MS in cultures on rice incubated at room temperature for 1 week followed by 4 weeks at 14°C.

Vesonder et al. (1977b) reported that cultures on corn of *F. culmorum* NRRL 3288 (Strain 8.15) incubated at 28°C for 13 days were refused by pigs and contained approximately 40 mg/kg of deoxynivalenol as detected by TLC. Subsequent analyses by GLC of cultures on corn of this strain incubated at 28°C for 30 days revealed the presence of 39 μg/g of deoxynivalenol, and 35 μg/g of 90% pure deoxynivalenol was isolated from this material (Vesonder et al., 1981b).

Pathre & Mirocha (1978) reported that *F. culmorum* 7289 (Strain 8.14) produced 125 μg/g of deoxynivalenol in cultures on rice incubated at 28°C for 7 days followed by 21 days at 12°C and 20 mg/ℓ in PSC liquid medium incubated in stationary culture at 21°C for 21 days. Deoxynivalenol analyses were done by GLC and confirmed by GC-MS.

Three isolates of *F. culmorum* from barley grain in Germany were reported to produce deoxynivalenol at levels ranging from 1 to 15 ppm as determined by TLC in cultures on corn incubated at 25°C for 2 weeks followed by 4 weeks at 10°C (Marasas et al., 1979b). We have confirmed the identity of two of these isolates, *F. culmorum* Sp. 885 and Sp. 887, that produced 7 and 15 ppm of deoxynivalenol, respectively, as *F. culmorum* (Strains 8.10 and 8.11).

Palmisano et al. (1981) reported that *F. culmorum* ITM-122 (Strain 8.17) produced deoxynivalenol in cultures on corn incubated at 27°C for 4 weeks. The levels of deoxynivalenol detected in this culture material by GC and by differential-pulse polarography were 1.59 μg/g and 1.39 μg/g, respectively.

In conclusion, six isolates of *F. culmorum* represented in the ITFRC (Strains 8.10, 8.11, 8.14, 8.15, 8.16, and 8.17) have been reported to produce deoxynivalenol at levels ranging from 1 to 125 μg/g. Together with *F. graminearum* (see *F. graminearum*, deoxynivalenol), *F. culmorum* has to be considered as an important producer of deoxynivalenol that may be involved in field outbreaks of the feed refusal and emetic syndromes (see feed refusal and emetic syndromes, above) in pigs.

Diacetoxyscirpenol. *Fusarium roseum* 'Culmorum' isolates M-14-2 and 70-K-11, reported to produce diacetoxyscirpenol by Ueno et al. (1973a), have been identified as *F. equiseti* (Strain 7.4) and *F. graminearum* (Strain 8.40), respectively.

Diacetylnivalenol. Blight & Grove (1974) reported that *F. culmorum* No. 53 produced 2 mg/ℓ of diacetylnivalenol in culture in GAN medium incubated at 25°C for 14 days. We have not seen this strain and there are no known diacetylnivalenol-producing strains of *F. culmorum* represented in the ITFRC.

HT-2 Toxin. *Fusarium roseum* 'Culmorum' 70-K-11, reported to produce HT-2 toxin by Ueno et al. (1973a), is *F. graminearum* (Strain 8.40).

Neosolaniol. Ueno et al. (1973a) reported that five isolates of *F. roseum* 'Culmorum' produced neosolaniol as detected by TLC in cultures in PSC medium incubated at 25 to 27°C for 2 weeks. We have identified one of these isolates, *F. roseum* 'Culmorum' 70-K-11, which was also reported to produce neosolaniol by Ueno et al. (1972b, 1975), as *F. graminearum* (Strain 8.40). We have not seen the other four isolates reported to produce neosolaniol by Ueno et al. (1973a), and there are no known neosolaniol-producing strains of *F. culmorum* represented in the ITFRC.

T-2 Toxin. Ueno et al. (1973a) reported that five isolates of *F. roseum* 'Culmorum' out of 13 tested produced T-2 toxin as detected by TLC in cultures in PSC medium incubated at 25 to 27°C for 2 weeks. We have identified one of these isolates, *F. roseum* 'Culmorum' 70-K-11, which was also reported to produce T-2 toxin by Ueno et al. (1972b, 1975), as *F. graminearum* (Strain 8.40).

Negative results for T-2 toxin production have been reported for the following four strains of *F. culmorum* represented in the ITFRC: Sp. 885 and Sp. 887 (Strains 8.10 and 8.11), by Marasas et al. (1979b); NRRL 3288 (Strain 8.15), by Vesonder et al. (1977b), and the isolate of *F. culmorum* from Washington, USA (Strain 8.16), by Mirocha et al. (1976). In addition, negative results have also been reported for two Hungarian isolates of *F. culmorum*, i.e., F-79 (= *F. graminearum*, Strain 8.39) and F-247, by Szathmary et al. (1976, 1977), and for 11 isolates of *F. culmorum* from cereals in Italy by Bottalico (1977c).

In conclusion, only five out of approximately 30 isolates of *F. culmorum* tested have been reported to produce T-2 toxin. One of these five isolates, *F. roseum* 'Culmorum' 70-K-11 (Ueno et al., 1973a), is *F. graminearum* (Strain 8.40). We have not seen the other four isolates reported to produce T-2 toxin by Ueno et al. (1973a) and there are no known T-2 toxin–producing strains of *F. culmorum* represented in the ITFRC.

Zearalenone (= F-2). Christensen et al. (1965) reported that cultures on corn of 12 isolates of *Fusarium* spp. from corn in the United States caused increases in the weight of the uterus in weanling female rats. These authors stated that "all or most of these isolates of *Fusarium* tentatively appear to be *F. culmorum* or *F. graminearum*." The estrogenic metabolite zearalenone (= F-2) was isolated from cultures of *Fusarium* sp. No. 5 (= *F. graminearum*, Strain 8.20) but it was not stated whether this compound was detected chemically in the other 11 isolates that caused uterine hypertrophy. These 11 isolates were also not identified and consequently it is not possible to decide whether Christensen et al. (1965) demonstrated zearalenone production by *F. culmorum*.

The production of zearalenone by *F. culmorum* was first reported by Caldwell et al. (1970), who found that three out of three isolates of *F. roseum* 'Culmorum' from the United States and Canada "collected by W.C. Snyder and R.J. Cook (also includes ATCC 15620)" produced zearalenone at levels ranging from 1 to 210 µg/g as determined by TLC in cultures on corn incubated at 16°C for 3 weeks. It is not known whether these three strains included *F. roseum* 'Culmorum' FR 22, which was reported by Caldwell & Tuite (1970) to produce 300 ppm of zearalenone in cultures on corn incubated at 24°C for 2 weeks followed by 8 weeks at 12°C. *Fusarium roseum* 'Culmorum' FR 22 also produced 0.5 to 3.5 ppm of zearalenone in corn ears inoculated in the field (Caldwell & Tuite, 1970). This strain (as *F. culmorum* FR 22) was subsequently shown to produce 0.02 mg/g of zearalenone in cultures on corn incubated for 8 weeks, compared to 25 mg/ℓ in stationary cultures in a liquid starch glutamate medium incubated at 24 to 28°C for 2 weeks (Bacon et al., 1977). The latter authors also found that *F. culmorum* No. 182, isolated from poultry feed, produced 0.095 mg/g of zearalenone in cultures on corn compared to 42 mg/ℓ in liquid medium.

A strain of *F. culmorum* isolated from feed suspected of causing infertility in cattle in Finland (see estrogenic syndrome, above) produced 3.19 mg/kg of zearalenone in cultures on a mixture of oats, barley, and wheat grain incubated at room temperature for one month (Roine et al., 1971).

Isolates of *F. culmorum* from cereals in Finland have been reported to produce unspecified levels of zearalenone in culture (Niku-Paavola & Nummi, 1977; Niku-Paavola et al., 1977). We have confirmed the identity of two of these isolates, i.e., *F. culmorum* 72196 and 72313-20 (Strains 8.12 and 8.13). The latter strain has also been used for the production of zearalenone for use in toxicological studies by incubating cultures on a mixture of oats, barley, and wheat grain at 20 to 24°C for 2 weeks, followed by 2 days at 5°C and 4 months at room temperature (Ylimäki et al., 1979). *Fusarium culmorum* 72313-20 (Strain 8.13) has also been used, together with other *Fusarium* isolates, to inoculate barley in the field in Finland in a study on the accumulation of zearalenone in barley during storage after harvest (Enari et al., 1981).

An unnumbered strain of *F. culmorum* isolated from barley in England and identified at CMI has been used for the production of zearalenone for use in toxicological studies by incubating cultures on rice at room temperature for 6 weeks followed by 1 week at 15°C (Miller et al., 1973).

The production of zearalenone by an isolate of *F. culmorum* from barley associated with a case of infertility (see estrogenic syndrome, above) of pigs in Scotland was studied by Gross & Robb (1975). Barley grains were adjusted to moisture contents of 25, 35, or 45%, inoculated with this strain, incubated at 25°C for 2 weeks, and stored at 10, 15, or 25°C; zearalenone determinations by TLC were done after 7, 10, or 19 weeks. Maximal levels of zearalenone (208 μg/g) were detected in barley adjusted to a moisture content of 45% and stored at 15°C for 7 to 10 weeks. Zearalenone was not detected in freshly harvested barley inoculated in the field with this strain, but low levels (2.1 to 4.4 μg/g) appeared after storage of the harvested barley for 12 weeks or longer (Gross & Robb, 1975).

Hacking et al. (1976, 1977) reported that *F. culmorum* was the most prevalent *Fusarium* sp. associated with barley grain in the UK and that 270 out of 431 isolates of *F. culmorum* (63%) produced zearalenone that could be detected by TLC in cultures on rice incubated at 25°C for 4 weeks followed by 6 weeks at 12°C.

An isolate of *F. culmorum* (Strain 8.16) associated with an outbreak of porcine hyperestrogenism (see estrogenic syndrome, above) in Washington, USA, has been reported to produce unspecified levels of zearalenone detected by GC-MS in cultures on rice incubated at 24 to 27°C for 1 week followed by 4 weeks at 14°C (Mirocha et al., 1976).

The production of zearalenone by three out of four Japanese isolates of *F. roseum* 'Culmorum' in cultures on rice incubated at 25°C for 14 days followed by 14 days at 12 to 15°C has been reported by Ishii et al. (1974) and Ueno (1973b). According to Ichinoe et al. (1977), four out of seven Japanese isolates of *F. culmorum* produced zearalenone at levels of 1.5 to 441 μg/g in cultures on rice. The Japanese isolate reported (as *F. roseum* M-3-2) to produce zearalenone by Ishii et al. (1974), Ueno (1973b), and Ueno et al. (1974b, 1977b), and referred to as *F. roseum* 'Culmorum' by Ueno et al. (1973a), is *F. graminearum* (Strain 8.30).

Fusarium culmorum F-247, isolated from millet in Hungary, has been reported to produce zearalenone at levels up to 3000 μg/g as detected by GC-MS in cultures on millet incubated at 23 to 27°C for 1 week followed by 2 weeks at 5 to 8°C and 1 week at

23 to 27°C (Szathmary et al., 1976, 1977). Another Hungarian isolate, *F. culmorum* F-79, which has been reported to produce zearalenone by Kovacs et al. (1975), Palyusik & Koplik-Kovacs (1975a, b), Sandor et al. (1980), and Szathmary et al. (1976, 1977), is *F. graminearum* (Strain 8.39).

Unspecified numbers of unnumbered Hungarian isolates of *F. culmorum* have also been reported to produce unspecified levels of zearalenone by Lasztity & Wöller (1975a, b, c, 1977). Corn inoculated with a mixture of Hungarian isolates of *F. culmorum* and *F. graminearum* contained 34 μg/g of zearalenone and was used in a study on the effects of zearalenone on the fertility of rats by Ruzsas et al. (1979).

According to Tuttobello et al. (1974), cultures in liquid medium as well as on corn of two strains of *F. culmorum* (ISS Roma and Waksman) had uterotrophic effects in weanling female rats. The active compound was not identified, but the uterotrophic activity of the two strains was equivalent to approximately 21.7 and 23.0 ppm of "zearalanol" in the liquid media and 12.0 and 14.4 ppm in the cultures on corn.

Two isolates of *F. culmorum* from foot rot of wheat in Italy and one from Germany produced zearalenone at levels ranging from 24 to 256 μg/g as determined by TLC in cultures on corn incubated at 25°C for 2 weeks followed by 8 weeks at 12°C (Bottalico, 1975; Bottalico & Piglionica, 1977). Subsequently Bottalico (1977b, 1978b, 1979) reported that 15 of 16 isolates of *F. culmorum* from cereals in Italy produced zearalenone at levels ranging from 2.3 to 6336.0 μg/g as determined by TLC in cultures on corn incubated at 24°C for 15 days followed by 60 days at 12°C. The unnumbered strain of *F. culmorum* that produced 6336.0 μg/g of zearalenone was the highest producer among 66 isolates belonging to eight *Fusarium* species in the study by Bottalico (1977b).

Fusarium culmorum F-104, isolated from feedstuffs in Germany, has been reported to produce 325 mg/kg of zearalenone in cultures on corn incubated at 25°C for 21 days followed by 14°C for 14 days (Möller et al., 1978). This strain was found to produce up to 1352 mg/kg of zearalenone in cultures on corn adjusted to 32% moisture content and incubated for 10 days at 20°C, 31 days at 10°C, and finally 111 days at 20°C (Müller & Thaler, 1981). These authors also reported that *F. culmorum* ATCC 34910 produced up to 222 mg/kg of zearalenone in similar cultures on corn. Neither of these two strains produced zearalenone in cultures on corn amended with propionic acid during an incubation period of 4 months (Müller & Thaler, 1981).

Four out of five isolates of *F. culmorum* from barley grain in Germany produced zearalenone at levels ranging from 320 to 1500 ppm as determined by TLC in cultures on corn incubated at 25°C for 2 weeks followed by 4 weeks at 10°C (Marasas et al., 1979b). We have confirmed the identity of two of these isolates that produced 1400 and 320 ppm of zearalenone, respectively, i.e., *F. culmorum* Sp. 885 and 887 (Strains 8.10 and 8.11).

Vesela et al. (1981) reported that 11 of 11 isolates of *F. culmorum* from wheat and pelleted feeds naturally contaminated with zearalenone in Czechoslovakia produced unspecified levels of zearalenone in culture.

Fusarium culmorum NRRL 3288 (Strain 8.15) produced unspecified levels of zearalenone in cultures on corn incubated at 28°C for 30 days (Vesonder et al., 1981b).

In conclusion, at least 325 isolates of *F. culmorum* have been reported to produce zearalenone in culture at levels ranging from 1 to 6336 μg/g. Six of these zearalenone-producing isolates (Strains 8.10, 8.11, 8.12, 8.13, 8.15, and 8.16) are represented in the ITFRC. Together with *F. graminearum* (see *F. graminearum*, zearalenone), *F. culmorum* has to be considered as an important producer of zearalenone that may be involved in field outbreaks of the estrogenic syndrome (see estrogenic syndrome, above) in pigs and other animals.

Toxigenic Strains in the ITFRC

The *Fusarium* strains listed under *F. culmorum* in Table 8.1 all have been identified as *F. culmorum* according to Nelson et al. (1983). Toxigenic strains that have been referred to as members of the section Discolor but identified as other species by us are listed in Table 8.2.

Strain 8.8   *F. culmorum* M6.2/66 (= R-5452; MRC 939)

This strain was isolated from corn implicated in an outbreak of moldy feed toxicosis (see moldy feed toxicosis, above) of dairy cows in Victoria, Australia (Fisher et al., 1967). Ether extracts of cultures of this strain on corn incubated at 21°C for 2 weeks were strongly dermotoxic to rabbit skin and caused a hemorrhagic skin reaction and death within 5 to 7 days (Fisher et al., 1967). The mycotoxin(s) produced by this dermotoxic strain have not been identified and there is no analytical data to support the statement by Wright (1968) that "it is likely that the toxin involved is diacetoxyscirpenol."

A slant culture of this strain was received on 24 June 1970 from M.E. di Menna (who had received it in 1966 from E.E. Fisher) as "*F. culmorum* M6.2/66." This strain, which was originally identified by C. Booth, according to Fisher et al. (1967), is badly degenerated and pionnotal. However, it still produced some typical macroconidia on CLA that allowed it to be identified as *F. culmorum*.

Strain 8.9   *F. culmorum* BBA 62183 (= R-6382; MRC 2197)

This strain (unnumbered) was isolated from a rotten potato tuber in Germany (Siegfried & Langerfeld, 1978). Extracts of potato tubers inoculated with this strain and incubated at 20°C were slightly toxic to brine shrimp (Siegfried & Langerfeld, 1978). The mycotoxin(s) produced by this strain have not been identified.

A slant culture of this strain was received on 20 January 1981 from E. Langerfeld as "*F. culmorum* BBA 62183."

Strain 8.10   *F. culmorum* Sp. 885 (= R-4607; MRC 1889)

This strain was isolated from barley grains in Germany (Marasas et al., 1979b). Cultures of this strain on corn incubated at 25°C for 2 weeks followed by 4 weeks at 10°C caused the death of two out of four day-old ducklings and marked reductions in feed consumption and weight gain of the survivors (Marasas et al., 1979b). Analyses of the toxic culture material by TLC revealed the presence of 1 ppm of acetyldeoxynivalenol, 7 ppm of deoxynivalenol, and 1,400 ppm of zearalenone.

A lyophilized culture of this strain was received on 16 February 1978 from W.F.O. Marasas as "*F. culmorum* Sp. 885."

Strain 8.11   *F. culmorum* Sp. 887 (= R-5391; MRC 1890)

This strain was isolated from barley grain in Germany (Marasas et al, 1979b). Cultures of this strain on corn incubated at 25°C for 2 weeks followed by 4 weeks at 10°C caused the death of two out of four day-old ducklings and marked reductions in feed consumption and weight gain of the survivors (Marasas et al., 1979b). Analyses of the toxic culture material by TLC revealed the presence of 2 ppm of acetyldeoxynivalenol, 15 ppm of deoxynivalenol, and 320 ppm of zearalenone.

A lyophilized culture of this strain was received on 22 April 1980 from W.F.O. Marasas as "*F. culmorum* Sp. 887."

Strain 8.12   *F. culmorum* 72196 (= R-5145; MRC 1369)

This strain was isolated from wheat in Finland, according to A. Ylimäki (personal communication, 10 April 1979). Barley inoculated with a Finnish isolate of *F. culmorum* was acutely toxic to mice and caused 100% mortality within 7 to 9 days (Mannio & Enari, 1973). According to T.M. Enari (personal communication, 17 April 1978), the strain found to be toxic to mice by Mannio & Enari (1973) was *F. culmorum* 72196. The malt produced from the barley inoculated with this strain was also highly toxic to mice, and the mice died within 7 to 9 days, showing marked signs of gastro-intestinal inflammation. The dried "spent grains" that remained behind after this malt had been used for brewing were also highly toxic to mice, but the beer itself was not toxic (Mannio & Enari, 1973).

According to A. Ylimäki (personal communication, 10 April 1979), this strain was found to produce zearalenone by Niku-Paavola & Nummi (1977) and Niku-Paavola et al. (1977). However, the mycotoxin(s) responsible for the toxicity to mice of both cultures of this strain on barley and the malt prepared from these cultures have not been identified.

A slant culture of this strain was received on 2 August 1978 from A. Ylimäki as "*F. culmorum* 72196."

Strain 8.13   *F. culmorum* 72313-20 (= R-5146; MRC 1371)

This strain was isolated from wheat in Finland during 1972 and was one of the 17 unnumbered isolates of *F. culmorum* from cereals in Finland in the toxicity study in rats by Korpinen & Uoti (1974), according to A. Ylimäki (personal communication, 10 April 1979). Cultures of this strain on a mixture of wheat, barley, and oats grain incubated at 20°C for 2 weeks, followed by 5 days at 4°C and for 2 weeks at 20°C, caused marked reductions in weight gain of rats (Korpinen & Uoti, 1974).

According to T.M. Enari (personal communication, 17 April 1978), *F. culmorum* 72313-20 was one of the unnumbered isolates of *F. culmorum* found to produce zearalenone by Niku-Paavola & Nummi (1977) and Niku-Paavola et al. (1977). This strain has been used for the production of zearalenone for use in toxicological studies by incubating cultures on a mixture of wheat, barley, and oats grain at 20 to 24°C for 2 weeks, followed by 2 days at 5°C and 4 months at room temperature (Ylimäki et al., 1979). This strain, together with other *Fusarium* isolates, has also been used to inoculate barley in the field in Finland to study the accumulation of zearalenone in barley during storage after harvest (Enari et al., 1981).

The mycotoxin(s) responsible for the toxicity of this strain to rats have not been identified.

A slant culture of this strain was received on 10 April 1979 from A. Ylimäki as "*F. culmorum* 72313-20."

Strain 8.14   *F. culmorum* 7289 (= R-5456; MRC 1905)

This strain was isolated from "feedstuffs associated with mycotoxicoses in farm animals," presumably in Finland (Pathre & Mirocha, 1978). This strain produced 125 μg/g of deoxynivalenol in cultures on rice incubated at 25°C for 7 days followed by 12°C for 21 days (Pathre & Mirocha, 1978). In liquid PSC medium incubated in stationary culture at 12°C for 21 days, the yield of deoxynivalenol was 25 mg/ℓ. Deoxynivalenol analyses were done by GLC and confirmed by GC-MS.

A silica gel culture of this strain was received on 17 March 1980 from C.J. Mirocha as "*F. roseum* 'Culmorum' 7289, ex Korpinen."

Strain 8.15   *F. culmorum* 3737 = NRRL 3288 (= R-5321; MRC 1787)

This strain (as *F. culmorum* 3737) was obtained from the Canadian Department of Agriculture in Winnipeg, according to Prentice & Dickson (1968). It was presumably identified by W.L. Gordon, but the exact source is unknown to us. According to Ellison & Kotsonis (1973) and Vesonder et al. (1977b, 1981b), this strain has been deposited at NRRL as *F. culmorum* NRRL 3288.

Extracts of cultures of this strain in Richard's medium incubated in shake culture at room temperature for 12 to 40 days caused prolonged emesis in pigeons upon injection in the wing vein (Prentice & Dickson, 1968; Prentice et al., 1959). In contrast to these findings, Ellison & Kotsonis (1973) observed no emetic activity in pigeons when ethyl acetate extracts of this strain (as *F. culmorum* NRRL 3288) in Richard's medium incubated at 25°C or on corn incubated at 8°C or 25°C were either force-fed to or injected into the wing vein of pigeons.

Subsequently cultures of *F. culmorum* NRRL 3288 on corn incubated at 28°C for 13 days were found to induce a marked feed refusal response (82%) in pigs (Vesonder et al., 1977b). Analyses of the refused culture material by TLC revealed the presence of approximately 40 mg/kg of deoxynivalenol, but no T-2 toxin, HT-2 toxin, acetyl T-2 toxin, or fusarenon-X. Cultures of this strain on corn incubated at 28°C for 30 days contained 39 $\mu g/g$ of deoxynivalenol as detected by GLC, and 35 $\mu g/g$ of 90% pure deoxynivalenol was isolated from this material (Vesonder et al., 1981b). The latter authors also detected unspecified levels of zearalenone in this culture material.

The deoxynivalenol produced by this strain (Vesonder et al., 1977b, 1981b) was presumably responsible for its emetic activity in pigeons (Prentice & Dickson, 1968; Prentice et al., 1959) as well as its feed refusal activity in pigs (Vesonder et al., 1977b). The negative results for emetic activity obtained by Ellison & Kotsonis (1973) were presumably due to the loss of deoxynivalenol-producing ability in the particular subculture as *F. culmorum* NRRL 3288 used by these authors.

A lyophilized culture of this strain was received on 23 October 1979 from NRRL as "*F. culmorum* NRRL 3288." This culture is badly degenerated and pionnotal, but still produced some typical *F. culmorum* macroconidia on CLA.

Strain 8.16 *F. culmorum*, Washington (= R-5251; MRC 1623)

This strain was isolated from feedstuffs associated with outbreaks of hyperestrogenism in pigs in Washington, USA (Mirocha et al., 1976). Cultures of this strain on rice incubated at 24 to 27°C for 1 week followed by 4 weeks at 14°C contained unspecified levels of deoxynivalenol and zearalenone as determined by GC-MS (Mirocha et al., 1976).

A soil culture of this strain was received on 6 March 1978 from C.J. Mirocha as "*F. culmorum*, Washington."

Strain 8.17 *F. culmorum* ITM-122 (=R-5797; MRC 2064)

This isolate, from an unspecified source, was reported to produce deoxynivalenol in cultures on corn incubated at 27°C for 4 weeks (Palmisano et al., 1981). The levels of deoxynivalenol detected in this culture material were 1.59 $\mu g/g$ by GC and 1.39 $\mu g/g$ by differential-pulse polarography.

A slant culture of this strain was received on 18 September 1980 from A. Bottalico as "*F. culmorum* ITM-122."

Strain 8.18 *F. culmorum* No. 34 = *Calonectria nivalis* CMI 14764 = *Micronectriella nivalis* ATCC 26559 (= R-6353; MRC 2434)

This strain (as *F. culmorum* CMI 14764) was isolated from wheat culms in New Zealand in 1926, according to the Catalogue of the Culture Collection of the CMI, Seventh Ed. (1975). According to the ATCC Catalogue of Strains, Fifteenth Ed. (1982), *Micronectriella nivalis* ATCC 26559 = CMI 14764.

The trichothecenes calonectrin and 15-deacetylcalonectrin were first isolated and characterized from cultures of "*Calonectria nivalis* CMI 14764" in Czapek-Dox medium incubated in stationary culture for 21 days (Gardner et al., 1972). These authors also isolated 50 mg of the sesquiterpene culmorin from 70ℓ of this culture material. It is not clear why Gardner et al. (1972) referred to strain CMI 14764 as *Calonectria nivalis*. Blight & Grove (1974) stated that "the organism described as *Calonectria nivalis* (= *F. nivale*), which produces calonectrin and 15-de-0-acetylcalonectrin, is now considered to be *F. culmorum*." The strain referred to as *F. culmorum* No. 34 by Blight & Grove (1974) is *F. culmorum* CMI 14764, according to J.F. Grove (personal communication, 28 November 1980). This strain was found to produce culmorin (0.9 mg/ℓ), 2-acetylquinazolin-4(3H)-one (0.6 mg/ℓ), and 52 mg/ℓ of acetyldeoxynivalenol (= 3α-acetoxy-12,13-epoxy-7α,15-dihydroxytrichothec-9-en-8-one) in culture in Czapek-Dox medium incubated at 25°C for 10 days (Blight & Grove, 1974). The dermotoxicity of this strain to rat skin (Blight & Grove, 1974) was presumably due to the acetyldeoxynivalenol and possibly also the calonectrins known to be produced (Gardner et al., 1972) by this strain (as *Calonectria nivalis* CMI 14764).

This strain (as *Calonectria nivalis* ATCC 26559) was used in studies on the biological acetyl conjugation of T-2 toxin and its derivatives by Yoshizawa et al. (1980). The latter authors stated that the calonectrins were originally isolated from this strain by Gardner et al. (1972).

A slant culture of this strain was received on 1 June 1981 from ATCC as "*Micronectriella nivalis* ATCC 26559." This culture is red and almost completely mycelial on PDA. A few abnormal macroconidia were produced in cultures irradiated with NUV light for 4 weeks or longer. This strain is definitely not *F. nivale,* but is impossible to identify with absolute certainty in its present condition. However, this strain was originally identified as *F. culmorum* by A.H.S. Onions (Blight & Grove, 1974), and we agree that this is probably correct.

Calonectria nivalis, the incorrect name for this strain introduced by Gardner et al. (1972), has persisted in the literature; the strain is still listed as *Micronectriella nivalis* ATCC 26559 in the ATCC Catalogue of Strains, Fifteenth Ed. (1982). This taxonomic error is probably the main reason for the fact that the calonectrins have only been isolated from this strain, since they have probably been looked for only in isolates of *F. nivale* by other investigators. The examination of other strains of *F. culmorum* for the production of the calonectrins is clearly indicated.

Fusarium graminearum Schwabe

Perfect State: *Gibberella zeae* (Schw.) Petch

Incidence and Distribution

Fusarium graminearum has a world-wide distribution as a soil inhabitant and as a serious pathogen that causes root, foot, crown, stem, and ear rot and head blight (scab) of cereals and of a large number of other hosts (Booth, 1971; CMI Descriptions of Pathogenic Fungi and Bacteria, No. 384, 1973; Doidge, 1938; Domsch et al., 1980; Gordon, 1952; Hacking et al., 1976, 1977; Joffe, 1960a, b, 1971; Marasas et al., 1979a; Meyer & Frank, 1979; Wollenweber & Reinking, 1935; Ylimäki, 1981; Yoshizawa et al., 1979).

In Australia, *F. graminearum* is the predominant *Fusarium* associated with crown rot of wheat (Burgess et al., 1975) and with stalk rot of corn (Francis & Burgess, 1975). Two populations of this fungus have been differentiated and designated as *F. roseum* 'Graminearum' Group 1 and Group 2 (Francis & Burgess, 1977). Members of Group 1 are usually associated with diseases of the crowns of plants and do not form perithecia in culture except in compatible crosses. Members of Group 2 are usually associated with diseases of the aerial parts of plants and readily form perithecia in single-spore cultures (Francis & Burgess, 1977).

Association with Human and/or Animal Diseases

Estrogenic Syndrome (= Hyperestrogenism, = Vulvo-Vaginitis). Sporadic field outbreaks of hyperestrogenism in animals, particularly pigs, are caused by the consumption of cereals, particularly corn and barley, infected by *F. graminearum* (see also *F. culmorum*, estrogenic syndrome) and contaminated with the estrogenic metabolite, zearalenone, in many countries including Australia, Canada, Denmark, Finland, France, Germany, Hungary, Italy, Japan, Mexico, Romania, South Africa, UK, USA, USSR, and Yugoslavia (Christensen, 1979; Kurtz & Mirocha, 1978a; Mirocha & Christensen, 1974; Mirocha et al., 1967a, b, 1971, 1977a, b; Nelson et al., 1973; Pathre & Mirocha, 1980).

The estrogenic syndrome in pigs involves primarily the genital system, and the characteristic clinical signs are hyperemia and edematous swelling of the vulva in prepuberal gilts. In severe cases, prolapse of the vagina and rectum may occur. Enlargement of the mammary glands and hypertrophy of the nipples are also prominent clinical signs in prepuberal gilts. The signs of hyperestrogenism have also been recorded in newborn piglets. Signs of estrus in gilts and prolonged estrus cycles have been reported, but standing estrus does not appear to be a prominent feature of field outbreaks. Pathological changes in the genital tract are characterized by interstitial edema and cellular proliferation and metaplasia of the mucosal epithelium of the vagina and cervix. The vulva, vagina, cervix, and myometrium are all thickened due to edema and a combination of cellular hypertrophy and hyperplasia of the wall elements. The uterine horns are grossly enlarged and the endometrium thickened by edematous fluid. The ovaries are hypoplastic with numerous small follicles but without formation of corpora lutea, and there is evidence of follicular atresia and oocyte degeneration. Other reproductive problems in sows, such as infertility, mummification, fetal resorption, abortions, reduced litter size, weakened piglets, splay legs, etc., have also been attributed to the consumption of *Fusarium*-infected grains contaminated with zearalenone. In male pigs, hypertrophy of the mammary glands, atrophy of the testes, swelling of the prepuce, and decreased libido have been reported (Bristol & Djurickovic, 1971; Kurtz & Mirocha, 1978a; Mirocha & Christensen, 1974; Mirocha et al., 1971, 1977a, b; Nelson et al., 1973).

The characteristic clinical signs of the vulvo-vaginitis were first reproduced experimentally in pigs by McNutt et al. (1928) with moldy corn from a field outbreak in Iowa, USA. The corn had been damaged by early frost and unusually wet weather during the previous year. Subsequently vulvo-vaginitis was reproduced experimentally in two young sows with moldy barley from a field outbreak in Ireland by McErlean (1952). One of the fungi isolated from the moldy barley was *F. graminearum*, but the causative role of this fungus was not proven by McErlean (1952). Vulvar and mammary enlargement in immature female pigs were first induced experimentally with a pure culture of *F. graminearum* by Stob et al. (1962). These authors used cultures on corn incubated at 24°C for 2 to 3 weeks of an unnumbered isolate of *F. graminearum* from moldy corn associated with field outbreaks of vulvo-vaginitis in Indiana, USA, during 1957/58. The culture material used in the successful experimental reproduction of vulvo-vaginitis in gilts also induced uterine hypertrophy in ovariectomized mice and so did alcoholic extracts from which the active metabolite was isolated (Stob et al., 1962). The isolate of *F. graminearum* used in the first experimental reproduction of the estrogenic syndrome by Stob et al. (1962) was unfortunately not numbered, but the isolate specified in United States patents on the estrogenic and anabolic activity of the active metabolite (FES = Fermentation Estrogenic Substance = zearalenone) was "*Gibberella zeae* NRRL 2380" (should read 2830), according to Mirocha & Christensen (1974) and Mirocha et al. (1971). We have confirmed the identity of this strain, which was referred to as "*G. zeae* H331 (ATCC 20272)" by Hidy et al. (1977), as *F. graminearum* (Strain 8.19).

The finding by Stob et al. (1962) that *F. graminearum* produced an estrogenic metabolite in culture on corn was confirmed by Christensen et al. (1965). The latter authors investigated two outbreaks of hyperestrogenism in pigs in Minnesota, USA, during 1963/64. Naturally infected moldy corn from one of these field outbreaks induced enlargement of the vulva, mammary glands, and uterus in an immature gilt and also caused uterine hypertrophy in virgin weanling rats. Cultures on whole corn kernels incubated at 20 to 25°C for 3 weeks followed by 2 weeks at 12°C of 12 of the 40 *Fusarium* isolates tested by Christensen et al. (1965) caused increases of five to eight times in the weight of the uteri of virgin female weanling rats. The uterotrophic *Fusarium* strains were not specifically identified, but Christensen et al. (1965) stated that "all or most of the isolates of *Fusarium* tentatively appear to be *F. culmorum* or *F. graminearum*." We have identified one of these isolates that caused uterine hypertrophy in rats, i.e., *Fusarium* No. 5, as *F. graminearum* (Strain 8.20). This strain apparently also induced vulvar enlargement and prolapse in a young gilt according to Figs. 10 and 11 of Christensen (1979) and Fig. 2 of Mirocha et al. (1971). The unnumbered strain of "*F. roseum*" used in the successful experimental reproduction of the estrogenic syndrome in male as well as female pigs by Christensen et al. (1972b) was apparently "*F. roseum* (strain Mapleton 10)" according to Nelson et al. (1973). We have identified this strain, which was isolated from moldy corn associated with infertility problems in pigs in Minnesota and designated as "*F. roseum* 'Graminearum' Mapleton No. 10" (Mirocha et al., 1971; Wolf & Mirocha, 1973), as *F. graminearum* (Strain 8.21). Cultures of this strain were also shown to have estrogenic effects in rats (Mirocha & Christensen, 1974) and to inhibit spermatogenesis in ganders and turkey cocks (Palyusik et al., 1971).

Christensen et al. (1965) isolated the estrogenic metabolite zearalenone (= F-2) from cultures of *F. graminearum* No. 5 (Strain 8.20) by means of the rat uterine hypertrophy bioassay and demonstrated that it was identical to the uterotrophic metabolite isolated from *G. zeae* by Stob et al. (1962). The structure of zearalenone was determined as one of the enantiomorphs of 6-(10-hydroxy-6-oxo-trans-1-undecenyl)-β-resorcylic acid lactone by Urry et al. (1966).

The characteristic clinical signs and pathological changes of hyperestrogenism, including vulvar and mammary swelling and uterine hypertrophy as well as certain associated reproductive problems such as reduced libido, infertility, nymphomania, pseudopregnancy, small litters, small pigs, deformed pigs, and juvenile hyperestrogenism, have been reproduced experimentally in pigs with pure zearalenone (Berger et al., 1981; Chang et al., 1979; Christensen, 1979; Hidy et al., 1977; Kurtz & Mirocha, 1978a; Kurtz et al., 1969; Mirocha & Christensen, 1974; Mirocha et al., 1967a, b, 1971, 1977a, b). Some of these estrogenic effects have also been induced with pure zearalenone in cattle (Christensen, 1979), dogs (Hidy et al., 1977), poultry (Allen et al., 1981a, c; Christensen, 1979; Chi et al., 1980a, b; Hidy et al., 1977; Meronuck et al., 1970; Mirocha & Christensen, 1974; Mirocha et al., 1971; Willemart & Schricke, 1981), primates (Hidy et al., 1977;

Hobson et al., 1977), rodents (Christensen et al., 1956; Hidy et al., 1977; Knaus, 1978; Mirocha et al., 1968a, b, 1971, 1977a, b; Ruzsas et al., 1978, 1979; Stob et al., 1962; Ueno & Tashiro, 1981; Ueno et al., 1974), and sheep (Hidy et al., 1977).

Zearalenone is not acutely toxic and single oral doses of 20 g/kg in mice and rats (Hidy et al., 1977) or 15 g/kg in chickens (Chi et al., 1980b) failed to cause any mortalities. Zearalenone can be considered to be a non-steroidal fungal hormone rather than a mycotoxin in the sense that low dietary levels are physiologically active but not toxic. Thus the oral administration of 1 mg of zearalenone/day to young female pigs for 8 days (approximately 0.02 mg/kg/day) will result in clinical signs of hyperestrogenism (Mirocha et al., 1971) and so will the consumption of diets containing from 1 to 5 mg/kg of zearalenone (Hidy et al., 1977; Mirocha & Christensen, 1974). Moreover, zearalenone has anabolic properties in cattle, mice, and sheep (Hidy et al., 1977) and also acts as a sex-regulating hormone in fungi in that low levels (0.0001 to 1.0 ppm) stimulate perithecial production in *F. graminearum* as well as in several other species of fungi (Nelson et al., 1968; Wolf & Mirocha, 1973, 1977; Wolf et al., 1972). Zearalenone is weakly teratogenic in pigs (Chang et al., 1979) and rats (Hidy et al., 1977; Ruddick et al., 1976), but not in mice (Arora et al., 1981). Zearalenone has been reported to be mutagenic in a recombination-deficient mutant of *Bacillus subtilis* (Ueno & Kubota, 1976), but not in the Ames test with *Salmonella typhimurium* (Kuczuk et al., 1978; Ueno et al., 1978; Wehner et al., 1978). Although Schoental (1981c, d) has referred to her own unpublished findings that rats exposed to large doses of zearalenone developed pituitary tumors and other unspecified abnormalities and neoplasias in other target organs, no published experimental results are available to indicate that zearalenone is carcinogenic.

Although the characteristic clinical signs and pathological lesions of the estrogenic syndrome have been reproduced experimentally with pure zearalenone in pigs, certain complications seen in field outbreaks associated with moldy corn have not. Thus feed refusal and emesis are now considered to be caused by trichothecenes such as deoxynivalenol produced by *F. graminearum* (see feed refusal and emetic syndromes, below). The situation is less clear with regard to abortions and stillbirths that have been reported in addition to signs of hyperestrogenism in some field outbreaks (Miller et al., 1973; Mirocha et al., 1976; Mitton et al., 1975; Ozegovic, 1970; Shreeve et al., 1975, 1978). Abortions have been induced in pigs (Christensen, 1979; Mirocha & Christensen, 1974; Mirocha et al., 1967a, b, 1971, 1977a, b) and rats (Ruzsas et al., 1979) with *Fusarium* cultures, but not with pure zearalenone (Chang et al., 1979). Abortions have, however, been induced in sows by the intravenous injection of 0.21 to 0.41 mg/kg of pure T-2 toxin (Weaver et al., 1977, 1978b). It has been suggested that the abortion facet of the estrogenic syndrome may be due to trichothecenes that are also produced by *F. graminearum* (Christensen, 1979; Mirocha & Christensen, 1974; Mirocha et al., 1977b). T-2 toxin has been reported to occur naturally in feedstuffs together with zearalenone at levels from 20 to 2,100 µg/kg (Barnikol et al., 1981; Bennett & Shotwell, 1979; Jemmali et al., 1978; Lafont & Lafont, 1980; Mirocha et al., 1976, 1979a). In our opinion, the co-occurence of T-2 toxin and zearalenone in these cases can best be explained by the simultaneous infection of the cereals involved by *F. sporotrichioides* and *F. graminearum* which produced the T-2 toxin and zearalenone, respectively. *Fusarium graminearum* is a weak producer of T-2 toxin, if it produces T-2 toxin at all (see T-2 toxin, below). Thus T-2 toxin may be responsible for the abortions reported in field outbreaks associated with moldy cereals, but probably not for those induced by cultures of *F. graminearum*. It is well established that the trichothecenes nivalenol and deoxynivalenol are produced by *F. graminearum* in culture (see mycotoxins produced, below). One or both of these trichothecenes have also frequently been found to occur together with zearalenone in nature (Bennett & Shotwell, 1979; Bennett et al., 1981; Bottalico et al., 1981; Jemmali et al., 1978; Lew et al., 1979; Marasas et al., 1977, 1979c; Mirocha, 1979; Mirocha et al., 1976, 1977a, 1979a; Pathre & Mirocha, 1979; Trenholm et al., 1981; Vesonder et al., 1979c; Young et al., 1981). Nivalenol and deoxynivalenol have, however, not been reported to be abortifacient. It is clear that the entire question of abortions and stillbirths caused by *F. graminearum* is unresolved. The ability of the naturally occurring trichothecenes nivalenol and deoxynivalenol to cause abortion, the possibility of synergism between these trichothecenes and zearalenone, and the possibility of the involvement of other unidentified mycotoxins require further investigation.

Field outbreaks of hyperestrogenism in animals associated with moldy corn occur sporadically and are dependent upon a complexity of factors that culminate in the formation of physiologically significant levels of zearalenone in the feed consumed by the animals. Among these factors are varietal susceptibility and climatic conditions conducive to the infection of corn ears by *F. graminearum* and the subsequent elaboration of zearalenone in infected ears. In an analysis of the *F. graminearum* ear rot epiphytotics in Indiana, USA, in 1965 and 1972, Tuite et al. (1974) found that optimal conditions for infection were at least 9 days of rain and a mean temperature below 21°C during silking. The frequency of zearalenone contamination of corn in Canada was found to be strongly correlated with rainfall for August, but weakly correlated with temperature (Sutton et al., 1980a). In the midwestern United States zearalenone is usually not present in high concentrations in freshly harvested corn (Caldwell & Tuite, 1970). However, zearalenone has been detected at levels up to 10 μg/g in freshly harvested, severely *F. graminearum*–infected corn kernels in Indiana, USA (Caldwell & Tuite, 1974) and up to 7 μg/g in ears damaged by birds or mechanically in Canada (Sutton et al., 1980b). In some countries in Europe where cold and rainy conditions may occur during the corn growing season, zearalenone has also been detected in preharvest corn at levels as high as 110 μg/g (Bottalico, 1977a, 1978a, 1979; Bottalico et al., 1977, 1981). Zearalenone can also occur in *F. graminearum*–infected corn stalks (Mirocha et al., 1979a; Sutton et al., 1976), but such contaminated stalks have not been implicated in outbreaks of hyperestrogenism in animals. Field outbreaks of the estrogenic syndrome in pigs are typically associated with ear corn harvested during autumns with exceptionally high rainfall and stored on the cob in open cribs; they may also occur when ears are left on plants in the field after the normal harvest period (Christensen, 1979). In these situations, high levels of zearalenone are produced during the storage of high moisture content, *F. graminearum*–infected corn ears under conditions of fluctuating high and low temperatures (Christensen, 1979; Mirocha & Christensen, 1974; Mirocha et al., 1971, 1977a, b).

The fact that the complex conditions for zearalenone production occur during some seasons in many countries is attested to by the natural occurrence of zearalenone at levels ranging from 0.01 to 2909 mg/kg in corn and other feedstuffs associated with field outbreaks of hyperestrogenism in animals (Aucock et al., 1980; Barnikol et al., 1981; Bennett & Shotwell, 1979; Christensen 1979; Funnell, 1979; Harwig & Munro, 1975; Hesseltine, 1974; Jemmali et al., 1978; Knaus, 1978; Mirocha & Christensen, 1974; Mirocha et al., 1971, 1974, 1976, 1977a, b, 1979a; Pathre & Mirocha, 1980; Scott, 1978; Shotwell, 1977). One of the derivatives of zearalenone produced by *F. graminearum*, α-zearalenol, is three to four times more active than zearalenone (Hagler et al., 1979; Mirocha et al., 1978; Ueno & Tashiro, 1981) and has also been found to occur naturally at levels ranging from 0.15 to 4.0 mg/kg in feeds associated with hyperestrogenism in pigs (Mirocha et al., 1979b). The possibility also exists that α- and β-zearalenol can be formed in feeds by the biotransformation of zearalenone by certain yeasts (Palyusik et al., 1980). The possible role of zearalenol in causing field outbreaks of hyperestrogenism merits further investigation.

Zearalenone has been found to occur naturally in foodstuffs intended for human consumption, but has not been proven to be associated with hyperestrogenism or any other disease in humans. Surveys of commercial corn in the United States during various years have revealed the presence of zearalenone in 1 to 17% of samples at levels ranging from 0.4 to 5.0 mg/kg (Eppley et al., 1974; Shotwell, 1977; Shotwell et al., 1970, 1971; Stoloff et al., 1976). Zearalenone has also been detected in 9 of 11 samples of corn meal obtained from retail stores in Washington, DC, at levels from 11 to 69 µg/kg (Ware & Thorpe, 1978), in 8 of 56 samples of breakfast cereals in the UK at levels up to 140 µg/kg (Jarvis, 1981), and in a sample of corn flakes in Canada at a level of 20 µg/kg (Scott et al., 1978). The natural occurrence of zearalenone has also been reported in 19 of 42 wheat samples from the southeastern United States at levels from 0.4 to 11.0 mg/kg (Hesseltine et al., 1978; Shotwell, 1977; Shotwell et al., 1977). Another possible avenue of human exposure to zearalenone is in the form of milk. Zearalenone is known to be transmitted into bovine milk in the form of the parent compound as well as the metabolite zearalenol, which is three to four times more estrogenic than zearalenone (Mirocha et al., 1981). Surveys of milk for the presence of zearalenone and zearalenol have, however, not been conducted. In Africa, zearalenone has been detected in corn intended for human consumption, either as porridge or as beer, at levels (where reported) ranging from 0.06 to 12.8 mg/kg in the Republic of South Africa (Marasas et al., 1977), Swaziland (Martin, 1974; Martin & Gilman, 1976; Martin & Keen, 1978), Transkei (Marasas et al., 1979c), and Zambia (Lovelace & Nyathi, 1977; MacDonald & Raemakers, 1974; Marasas et al., 1977). In addition, zearalenone is apparently not destroyed during the making of home-brewed corn beer in Africa, and levels of 0.3 to 2.0 µg/g have been found in 17 of 140 samples in Lesotho (Martin & Keen, 1978), 8 to 53 µg/g in 6 of 55 samples in Swaziland (Martin & Keen, 1978), and 0.01 to 4.6 mg/ℓ in 23 beer samples in Zambia (Lovelace & Nyathi, 1977). The above reports indicate that humans are being exposed to zearalenone, particularly by the consumption of foodstuffs containing corn. In certain parts of Africa where corn is the main human dietary staple and is also used in the making of home-brewed beer, the exposure levels may be quite high. The effects of zearalenone in humans are for the most part unknown, although zearalenone and certain derivatives apparently have estrogenic effects in humans since they have been patented for use in treatment of the postmenopausal syndrome and as oral contraceptives in humans (Schoental, 1981d; Verdeal & Ryan, 1979). The suggestions that zearalenone may be responsible for the high incidence of human cervical cancer in Africa (Martin, 1974; Martin & Gilman, 1976; Martin & Keen, 1978) and for the variable incidence of neoplasms in the sex organs of human males and females as well as animals (Schoental, 1974, 1975, 1977b, 1978, 1979a, b, 1980b, 1981c, d, e), have not yet been supported by published experimental results on the carcinogencity of zearalenone. The unpublished results that rats exposed to large doses of zearalenone developed pituitary tumors and other neoplasias in other target organs, referred to by Schoental (1981c, d), have still not been published as far as we know.

In conclusion, it has been established beyond doubt that field outbreaks of hyperestrogenism in animals, particularly pigs, are caused by the estrogenic metabolite zearalenone produced in cereals infected by *F. graminearum* (see also *F. culmorum*, estrogenic syndrome). In fact, the estrogenic syndrome is one of the few mycotoxicoses for which the equivalent of Koch's postulates—i.e., the natural occurrence of a fungal metabolite in feeds at levels capable of inducing the characteristic clinical signs and pathological lesions of the disease experimentally—has been established. Although considerable evidence for human exposure to zearalenone has been presented, it has not been proven to be the cause of any known human disease. It should be borne in mind, however, that "zearalenone keeps some very bad and dangerous company" (Christensen, 1979) and that the presence of zearalenone may be considered as an indicator (Lew, 1978) for the presence of other *Fusarium* toxins such as trichothecenes that are much more difficult to detect chemically. Members of this "bad company" are most probably responsible for the signs of toxicity not attributable to zearalenone, such as those seen in field outbreaks of the hemorrhagic syndrome (see *F. sporotrichioides*, hemorrhagic syndrome) and feed refusal and emetic syndromes (see feed refusal and emetic syndrome, below) associated with moldy corn, in cases of poisoning in humans following the consumption of scabby cereals (see akakabi-byo, below), and in animals administered culture material of *F. graminearum* (see toxicity to experimental animals, below).

Feed Refusal and Emetic Syndromes. Sporadic field outbreaks of feed refusal in pigs, sometimes associated with vomiting, are caused by the infection of cereals, particularly corn and barley, by *F. graminearum* in the midwestern United States, Japan, and elsewhere (Curtin & Tuite, 1966; Kurata, 1978; Kurtz & Mirocha, 1978b; Prentice & Dickson, 1968; Smalley & Strong, 1974; Ueno, 1971, 1973a, b, 1977a, 1980; Uraguchi, 1971; Vesonder et al., 1973, 1976). The reduced palatability of the scabby grain is reflected in decreased weight gains and slower growth rates of pigs and is associated with nausea and emesis in animals forced to eat the grain by starvation. A similar syndrome in humans is discussed under akakabi-byo (=Scabby Grain Intoxication, see akakabi-byo, below).

A severe epiphytotic of barley scab caused by *F. graminearum* occurred in the midwestern United States in 1928; the scabby barley was refused by pigs and also caused outbreaks of a toxicosis characterized by nausea and vomiting (Christensen & Kernkamp, 1936; Mains et al., 1930; Mundkur, 1934). Some of this *F. graminearum*–infected barley was exported to Europe, where it also caused numerous feeding problems in pigs to such an extent that American barley had to be tested for toxicity prior to being admitted into Germany (Beller & Wedemann, 1929; Miessner & Schoop, 1929). At that time, several investigators in Germany and the United States proved experimentally that naturally infected scabby barley caused a marked feed refusal and/or emetic response in pigs (Beller & Wedemann, 1929; Christensen & Kernkamp, 1936; Hoyman, 1941; Mains et al., 1930; Miessner & Schoop, 1929; Mundkur, 1934). The causative role of *F. graminearum* was proven by the demonstration that pure cultures of this fungus on autoclaved, scab-free barley or artificial media caused feed refusal (Mains et al., 1930; Miessner & Schoop, 1929) and that barley or corn inoculated artificially with *F. graminearum* caused emesis (Christensen & Kernkamp, 1936; Shands, 1937) in pigs. The production of emetic material in artificial media by *Fusarium* was first reported by Prentice et al. (1959) and Prentice & Dickson (1968), who found that extracts of cultures in Richard's solution of several *Fusarium* strains, including "*F. roseum* 162," caused emesis upon injection in pigeons. The strain used by Prentice & Dickson (1968) is "*F. roseum* NRRL A-15,666," according to Ellison & Kotsonis (1973) and Vesonder et al. (1977b). We have not seen this culture and consequently do not know whether this is *F. graminearum*. Some of the *Fusarium* isolates tested for emetic activity by Prentice & Dickson (1968) had previously been shown to cause "gushing" of beer, which is a problem associated with the barley crops of certain seasons (Gjertsen, 1967; Gjertsen et al., 1966; Prentice & Sloey, 1960; Sloey & Prentice, 1962).

During the fall and winter of 1965, pigs in Indiana, USA, refused to consume corn infected by *F. graminearum,* but emesis was not a prominent clinical sign associated with these outbreaks (Curtin & Tuite, 1966). Naturally infected corn was refused by pigs under experimental conditions and water as well as methanol extracts of the refused corn caused emesis in pigs upon oral, intravenous, or intraperitoneal administration (Curtin & Tuite, 1966). Thus the emetic activity was not due to local irritation of the gastric mucosa. It was not clearly established whether refusal of the *F. graminearum*-infected corn was due to an odor or taste associated with the emetic substance or whether another chemical entity was involved. However, Curtin & Tuite (1966) concluded that emesis was independent of feed refusal in experimental pigs and that different metabolites were responsible for the two effects. The naturally infected corn did not cause uterine hypertrophy in immature female mice or rats, and Curtin & Tuite (1966) concluded that the estrogenic metabolite zearalenone (see estrogenic syndrome, above) was not involved in either the feed refusal or emetic syndromes.

Another *F. graminearum* ear rot epiphytotic of corn occurred in Indiana, USA, in 1972, and an analysis of the climatic conditions associated with the 1965 and 1972 epiphytotics revealed that optimal conditions for infection were at least 9 days of rain and a mean temperature below 21°C during silking of corn (Tuite et al., 1974). During the unusually cool and wet season of 1972, the *F. graminearum* ear rot epiphytotic was again associated with reports that pigs refused to eat the infected corn or vomited after the consumption of small quantities (Vesonder et al., 1973, 1976). Outbreaks of the feed refusal and emetic syndromes in pigs in Austria during 1977/78 were associated with abnormal weather conditions—an early frost at the end of September followed by a relatively mild October—that favored infection of corn ears by *F. graminearum* (Lew et al., 1979). Low temperatures with concomitant high rainfall and humidity during the summer of 1980 in Ontario, Canada, also favored the infection of wheat by *F. graminearum* and resulted in outbreaks of feed refusal, emesis, and death in pigs (Neish & Cohen, 1981; Scott et al., 1981; Trenholm et al., 1981).

The emetic principle, referred to as vomitoxin, was isolated from naturally infected corn that caused vomiting in pigs in Ohio, USA, in 1972 by Vesonder et al. (1973) and tentatively assigned the structure 3,7,15-trihydroxy-12,13-epoxytrichothec-9-en-8-one. The same trichothecene was isolated from *Fusarium*-infected barley in Japan and referred to as Rd toxin by Morooka et al. (1972). This compound was called deoxynivalenol by Yoshizawa & Morooka (1973), who determined it has a structure identical to that proposed for vomitoxin by Vesonder et al. (1973). The emetic principle was also isolated from a sample of corn from Ohio, USA, and shown to be a trichothecene by Ueno et al. (1974a); Ishii et al. (1975) isolated 26 mg of pure deoxynivalenol from 3.5 kg of this corn. The level of deoxynivalenol in the corn from Ohio was determined to be 40 ppm by Vesonder et al. (1976), who also demonstrated that pure deoxynivalenol caused feed refusal as well as emesis in pigs. On the basis of these findings, Vesonder et al. (1976) concluded that deoxynivalenol was responsible for the feed refusal as well as the emetic syndromes and that pigs will refuse corn containing high levels of deoxynivalenol, but will consume sufficient quantities of corn containing lower levels to elicit the emetic response.

The first *Fusarium* strain that was reported to produce deoxynivalenol in culture was *F. roseum* No. 117 (= ATCC 28114), isolated from moldy barley contaminated with deoxynivalenol and nivalenol in Japan (Morooka et al. 1972; Yoshizawa & Morooka, 1973; Yoshizawa et al., 1978). We have identified this isolate as *F. graminearum* (Strain 8.22). Cultures of this strain on rice were refused by rats and contained 220 µg/g of deoxynivalenol (Yoshizawa et al., 1978). We have also confirmed the identity of *F. graminearum* NRRL 5883, which was "isolated from naturally infected corn" in the United States and reported to produce deoxynivalenol in culture by Vesonder et al. (1976), as *F. graminearum* (Strain 8.23). Cultures of this strain on rice were refused by pigs and 60 mg/kg of chromatographically pure deoxynivalenol was isolated from this culture material (Vesonder et al., 1976). We have also confirmed the identity of the Hungarian isolate *F. graminearum* F-59, which has been reported to cause feed refusal and emesis in pigs (Palyusik, 1973) and to produce deoxynivalenol (Pathre & Mirocha, 1978), as *F. graminearum* (Strain 8.38). Other isolates from corn refused by pigs and reported to produce deoxynivalenol in culture that we have identified as *F. graminearum* include *F. graminearum* NRRL 5864 (Strain 8.24), isolated from corn in Minnesota (Vesonder et al., 1981b), and *F. graminearum* NRRL 6450, 6451, and 6452 (Strains 8.25, 8.26, and 8.27), isolated from Austrian corn contaminated with deoxynivalenol (Vesonder & Ciegler, 1979).

Feed refusal as well as emesis have been reproduced experimentally in pigs with pure deoxynivalenol (Forsyth et al., 1977; Vesonder et al., 1976, 1979b; Yoshizawa & Morooka, 1977). Deoxynivalenol also has feed refusal activity in mice (Burmeister et al., 1980a) and rats (Vesonder et al., 1979b; Yoshizawa & Morooka, 1977; Yoshizawa et al., 1978), and emetic activity in cats (Ueno et al., 1974a), dogs (Yoshizawa & Morooka, 1974), and ducklings (Ishii et al., 1975; Ueno, 1977b; Ueno et al., 1974a; Yoshizawa and Morooka, 1974). Other clinical signs and pathological changes reported to be caused by deoxynivalenol include radiomimetic cellular damage in the actively dividing tissues and pericellular edema with petechial hemorrhages in the brain of cats (Ueno et al., 1974a), extensive ecchymotic hemorrhaging throughout the carcass, widespread deposition of urates, neural disturbances, and irritation of the upper gastrointestinal tract in chickens (Huff et al., 1981), and "marked dilation with hemorrhage of the gastrointestinal tract and engorgement of the testes" in mice (Yoshizawa & Morooka, 1974). The lack of mutagenicity of deoxynivalenol to *Salmonella typhimurium* (Wehner et al., 1978) indicates that it is probably not carcinogenic, but the chronic effects in experimental animals are unknown at present.

Deoxynivalenol has been reported to occur naturally at levels ranging from 25 to 50,500 µg/kg in corn and other feedstuffs associated with field outbreaks of the feed refusal and emetic syndromes in pigs in Austria, Canada, France, and the United States (Bottalico et al., 1981; Christensen, 1979; Forsyth et al., 1977; Ishii et al., 1975; Jemmali et al., 1978; Kuroda et al., 1979; Lew et al., 1979; Mirocha, 1979; Mirocha et al., 1976, 1977a; Neish & Cohen, 1981; Pathre & Mirocha, 1977, 1979; Trenholm et al., 1981; Vesonder & Ciegler, 1979; Vesonder et al., 1973, 1976, 1979c). The co-occurrence of deoxynivalenol and zearalenone in corn and other cereals infected by *F. graminearum* has frequently been reported (Bennett & Shotwell, 1979; Bennett et al., 1981; Bottalico et al., 1981; Jemmali et al., 1978; Lew et al., 1979; Marasas et al., 1977, 1979c; Mirocha, 1979; Mirocha et al., 1976, 1977a; Pathre & Mirocha, 1979; Trenholm et al., 1981; Vesonder et al., 1979c; Young et al., 1981). In addition, deoxynivalenol is also known to occur naturally together with other trichothecenes such as diacetoxyscirpenol (Bottalico et al., 1981), nivalenol (Jemmali et al., 1978; Kuroda et al., 1979; Morooka et al., 1972; Yoshizawa & Morooka, 1973, 1977; Yoshizawa et al., 1979), and T-2 toxin (Bennett & Shotwell, 1979; Jemmali et al., 1978; Mirocha et al., 1979a).

The observation that corn naturally infected by *F. graminearum* often has a greater feed refusal effect in pigs than can be accounted for by the amount of deoxynivalenol detected by chemical analysis has led to suggestions that deoxynivalenol may act synergistically with zearalenone, with other known naturally occurring trichothecenes, and/or with other unidentified toxins (Forsyth et al., 1977; Kotsonis et al., 1975a; Lew et al., 1979; Marasas et al., 1977; Trenholm et al., 1981; Vesonder et al., 1981a, b; Yoshizawa et al., 1978; Young et al., 1981). Zearalenone has been reported to have a low degree of feed refusal activity in rats and to enhance the refusal of feed containing high levels of T-2 toxin (Kotsonis et al., 1975a). In contrast to these findings, Burmeister et al. (1980a) reported that zearalenone was not a feed refusal factor in mice and did not enhance the feed refusal activity of either deoxynivalenol or T-2 toxin. Feed refusal and emetic activities have, however, been well established for the trichothecenes that are known to occur naturally together with deoxynivalenol, i.e., diacetoxyscirpenol, nivalenol, and T-2 toxin (Burmeister et al., 1980a; Ellison & Kotsonis, 1973; Kotsonis et al., 1975a; Kurtz & Mirocha, 1978b; Lutsky et al., 1978; Sato & Ueno, 1977; Sato et al., 1975, 1978; Stähelin et al., 1968; Ueno, 1977b, 1980; Ueno et al., 1971c, 1974a; Vesonder et al., 1979b; Weaver et al., 1977, 1978a, b, c, 1981; Yagen et al., 1977; Yoshizawa & Morooka, 1974). Consequently these trichothecenes may contribute to the feed refusal and/or emetic effects of naturally *Fusarium*-infected cereals. Lastly, the possible involvement of other unidentified mycotoxins has to be considered. Two unidentified compounds believed to be trichothecenes have been isolated from corn naturally infected by *F. graminearum* and associated with field outbreaks of porcine feed refusal in the United States (Vesonder et al., 1976). A new trichothecene similar to deoxynivalenol, 3,15-dihydroxy-12,13-epoxytrichothec-9-ene-8-one, has recently been isolated from corn inoculated in the field with *F. graminearum* (Bennett et al., 1981). In our experience, cultures on corn of most isolates of *F. graminearum* are highly toxic to experimental animals and this acute toxicity cannot be attributed to the levels of deoxynivalenol (or zearalenone) detected by chemical analysis. This discrepancy may be due to the underestimation of the actual amounts of deoxynivalenol present (Burmeister et al., 1980a; Vesonder & Ciegler, 1979) or to the presence of butenolide and/or nivalenol, which are known to be produced simultaneously with deoxynivalenol in cultures of *F. graminearum* (Morooka et al., 1972; Yoshizawa et al., 1979). However, the possibility of other unidentified toxic metabolities, such as the unknown trichothecene detected in cultures of *F. graminearum* by Wu et al. (1981), can not be excluded. The production of unidentified mycotoxins by *F. graminearum,* the natural occurrence of these postulated compounds, and their possible involvement in the feed refusal and emetic syndromes require further investigation.

In conclusion, it has been established beyond doubt that deoxynivalenol occurs in cereals infected by *F. graminearum* and associated with field outbreaks of feed refusal and emesis in animals, particularly pigs, at levels capable of inducing the characteristic clinical signs of these syndromes under experimental conditions. However, there is reason to believe that deoxynivalenol alone is not responsible for all of the feed refusal and emetic activity of *F. graminearum*–infected cereals and that other factors may also be involved.

Akakabi-byo (= Scabby Grain Intoxication) of Humans. An outbreak of "Taumelgetriede" (= staggering grain) intoxication of humans in the USSR was described by Woronin (1891). He attributed clinical signs of headache, vertigo, shivering chills, nausea, vomiting, and visual disturbances in humans in eastern Siberia to consumption of rye bread made from scabby rye infected by *F. graminearum* (as *F. roseum* = *G. saubinettii*). Similar clinical signs in humans in the USSR following the consumption of "drunken" or "intoxicating bread" have subsequently been reported during 1912 (Shapovalov, 1919) and 1923 (Dounin, 1926), when wide-spread epiphytotics of *F. graminearum* head blight of cereals occurred in association with abnormally wet and cool seasons. A similar toxic bread in China is known as "Mi-chum," according to Uraguchi (1971). In the southern part of Korea, the consumption of scabby barley infected by *F. graminearum* during 1963 resulted in clinical signs of nausea, vomiting, abdominal pain, and diarrhea in humans (Cho, 1964).

In Japan, sporadic epiphytotics of "akakabi-byo" (= red mold disease or scab) of wheat, barley, oats, rye, and rice caused by *F. graminearum* can affect more than one-third of the national production and are frequently associated with outbreaks of a human mycotoxicosis characterized by anorexia, nausea, vomiting, headache, abdominal pain, diarrhea, chills, giddiness, and convulsions (Ichinoe, 1978b; Kurata, 1978; Kurata et al., 1968; Saito & Ohtsubo, 1974; Saito & Okubo, 1970; Saito & Tatsuno, 1971; Tsunoda, 1970; Udagawa et al., 1970; Ueno, 1971, 1973a, b, 1977a, 1980; Ueno et al., 1971a; Uraguchi, 1971; Yoshizawa & Morooka, 1977).

No reports are available in the literature on the mycotoxins present in samples of scabby grain actually implicated in outbreaks of the human mycotoxicosis with clinical signs as described above. However, the trichothecenes deoxynivalenol and nivalenol have been found to occur naturally in Japan. Deoxynivalenol occurred at levels of 20 to 7,300 μg/kg and nivalenol at 20 to 7,000 μg/kg in scabby barley and wheat infected predominantly by *F. graminearum* (Kuroda et al., 1979; Morooka et al., 1972; Yoshizawa & Morooka, 1977; Yoshizawa et al., 1976). Deoxynivalenol as well as nivalenol are known to cause feed refusal and emesis in animals (see feed refusal and emetic syndromes, above). The production of deoxynivalenol in culture by a *Fusarium* isolated from scabby cereals in Japan was first reported for the strain designated as *F. roseum* No. 117 (Morooka et al., 1972; Yoshizawa & Morooka, 1973). We have identified this strain as *F. graminearum* (Strain 8.22). Subsequently Suzuki et al. (1980, 1981b) reported that 4 of 57 isolates of *F. graminearum* from wheat and barley in Japan produced deoxynivalenol and that 43 of 57 isolates produced nivalenol in culture. In addition Yoshizawa et al. (1979) found that 21 of 61 Japanese isolates of *F. graminearum* produced "deoxynivalenols" in culture, that 20 of 61 isolates produced "nivalenols," and that 2 of 61 strains produced "deoxynivalenols" as well as "nivalenols" in culture. According to Yoshizawa et al. (1979), these results provided conclusive evidence that the natural occurrence of deoxynivalenol and nivalenol in barley and wheat in southern Japan could be attributed to infection of these cereals by strains of *F. graminearum* capable of producing either deoxynivalenol, nivalenol, or both of these trichothecenes. We have not seen any of these Japanese isolates of *F. graminearum* reported to produce deoxynivalenol and/or nivalenol by Suzuki et al. (1980, 1981b) and Yoshizawa et al. (1979) or the strain reported, as "*G. zeae* Ohoita-II," to produce nivalenol by Ueno et al. (1973a). However, in addition to the known deoxynivalenol-producing isolate *F. graminearum* (= *F. roseum*) No. 117 (Strain 8.22), we have also confirmed the identity of the Japanese isolate *F. graminearum* NHL-F-1118 that produces deoxynivalenol, according to H. Kurata (personal communication, 29 April 1980), as *F. graminearum* (Strain 8.34). We have also confirmed the identity of two Japanese isolates, *F. graminearum* NHL-F-1104 and NHL-F-1112, that produce nivalenol in culture (H. Kurata, personal communication, 29 April 1980) as *F. graminearum* (Strains 8.35 and 8.36).

In our opinion, the following Japanese strains that have been reported to produce trichothecenes other than deoxynivalenol and nivalenol in culture are also strains of *F. graminearum*: *F. rigidiusculum* M-1-3, which produces diacetoxyscirpenol, neosolaniol, and T-2 toxin (Ueno et al., 1973a); *F. tricinctum* A-R-5, which produces neosolaniol and

T-2 toxin (Ueno et al., 1975); and *F. tricinctum* Abashiri-2, which produces diacetoxyscirpenol, HT-2 toxin, neosolaniol, and T-2 toxin (Ueno et al., 1973a). We have accordingly identified these three strains as *F. graminearum* (Strains 8.29, 8.32 and 8.33).

In addition to the trichothecenes mentioned above, Japanese isolates of *F. graminearum* have also been reported to produce butenolide (Morooka et al., 1972; Yoshizawa & Morooka, 1973; Yoshizawa et al., 1979) and zearalenone (Ichinoe et al., 1977; Ishii et al., 1974; Suzuki et al., 1978). We have identified *F. roseum* No. 117, reported to produce butenolide (Morooka et al., 1972; Yoshizawa & Morooka, 1973), as *F. graminearum* (Strain 8.22). We have also identified three Japanese isolates reported to produce zearalenone, i.e., *G. zeae* Ishii (Ueno et al., 1971a), *F. roseum* 'Culmorum' M-3-2 (Ishii et al., 1974; Ueno et al., 1974b, 1977a), and *F. roseum* 'Gibbosum' A-0-2 (Ishii et al., 1974), as *F. graminearum* (Strains 8.28, 8.30, and 8.31).

Although *F. graminearum* is the predominant fungus associated with scabby cereals in Japan, a fungus referred to in the literature as *F. nivale* was isolated from 5 to 6% of scabby wheat grains during the severe epiphytotic of *Fusarium* head blight in 1963 (Saito & Okubo, 1970; Tsunoda, 1970). One of these isolates, *F. nivale* Fn-2B, proved to be highly toxic and to produce four trichothecenes—diacetylnivalenol, fusarenon-X, nivalenol, and T-2 toxin—in culture (Fujimoto et al., 1972; Joffe, 1978b; Joffe & Yagen, 1977; Morooka & Tatsuno, 1970; Morooka et al., 1971; Ohta et al., 1977, 1978; Tatsuno, 1968, 1976; Tatsuno et al., 1968, 1969, 1970, 1973; Tsunoda, 1970; Ueno, 1971, 1977a; Ueno et al., 1969a, 1970a, b, c, 1971a, b, c, 1972b, 1975; Yoshizawa & Morooka, 1975a, b; Yoshizawa et al., 1979, 1980). We have identified this strain as *F. sporotrichioides* (Strain 4.9). This finding implies that *F. sporotrichioides* may also be involved in scabby grain intoxication of humans in Japan (see *F. sporotrichioides*, akakabi-byo).

It follows from the above discussion that: 1) Japanese isolates of *F. graminearum* from scabby cereals are known to produce the trichothecenes deoxynivalenol, nivalenol, fusarenon-X, diacetoxyscirpenol, neosolaniol, and T-2 toxin, as well as butenolide and zearalenone, in culture. 2) A Japanese isolate of *F. sporotrichioides* (= *F. nivale* Fn-2B, Strain 4.9) from scabby wheat is known to produce nivalenol, fusarenon-X, diacetylnivalenol, and T-2 toxin in culture. 3) Scabby cereals in Japan that are known to be contaminated with butenolide, deoxynivalenol, and nivalenol may be infected by both *F. graminearum* and *F. sporotrichioides*. In such cases of co-infection, the butenolide and deoxynivalenol may be produced by *F. graminearum* while nivalenol may be produced by both species. In addition, the other trichothecenes and zearalenone known to be produced by these two species may also occur naturally in scabby cereals in Japan. 4) The possibility exists that synergistic interactions between deoxynivalenol and other trichothecenes such as nivalenol, and/or zearalenone, and/or other unidentified toxins (see feed refusal and emetic syndromes, above) may be involved in the human mycotoxicosis caused by the consumption of scabby cereals in Japan.

Deoxynivalenol has also been found to occur naturally in Africa in corn intended for human consumption, but *Fusarium*-infected corn has not been associated with clinical signs comparable to those of akakabi-byo intoxication in humans. Two samples of naturally *F. graminearum*-infected corn, one from South Africa and the other from Zambia, have been found to contain 2.5 and 7.4 mg/kg of deoxynivalenol, respectively (Marasas et al., 1977). Deoxynivalenol has also been reported to occur naturally in 5 of 8 samples of moldy home-grown corn in Transkei at levels ranging from 0.035 to 4.0 mg/kg (Marasas et al., 1979c). Although considerably higher levels of deoxynivalenol were detected in the pooled samples of corn from an area in Transkei with a high esophageal cancer rate than in the corresponding samples from a low-rate area (Marasas et al., 1979c), no additional evidence has been presented that deoxynivalenol may be involved in the etiology of human esophageal cancer. In all of these samples of corn from southern Africa, deoxynivalenol occurred together with zearalenone, but no diacetoxyscirpenol or T-2 toxin could be detected chemically (Marasas et al., 1977, 1979c) and no information was given on the presence or absence of nivalenol, which has been found to occur together with deoxynivalenol in corn in France (Jemmali et al., 1978). In surveys of corn in the United States, deoxynivalenol has been detected in 4 of 9 samples of the 1972 crop at levels of 15 to 28 mg/kg (Vesonder et al., 1979c) and in 24 of 52 samples of 1977 crop at levels ranging from 0.5 to 10.0 mg/kg (Vesonder et al., 1978). Although deoxynivalenol occurred together with zearalenone in some of these corn samples (Vesonder et al., 1979c), none of these contained chemically detectable levels of diacetoxyscirpenol or T-2 toxin (Vesonder et al., 1978, 1979c), and no information was provided on the occurrence of nivalenol. Surveys of Canadian wheat for deoxynivalenol during the scab epiphytotic of 1980 revealed up to 100% contamination of Ontario white winter wheat and Quebec red spring wheat at levels ranging from 0.01 to 8.5 mg/kg (Scott et al., 1981; Trenholm et al., 1981). None of these wheat samples contained chemically detectable levels of either zearalenone (Trenholm et al., 1981) or diacetoxyscirpenol and T-2 toxin (Scott et al., 1981; Trenholm et al., 1981), but nivalenol was not analyzed for because nivalenol was "not carried through our procedures of extraction and cleanup" (Scott et al., 1981). Although the contamination of Canadian wheat with deoxynivalenol is "a problem that requires further research from the viewpoint of potential harm to human health" (Scott et al., 1981), the consumption of this wheat has not been related to any known cases of intoxication in humans.

In conclusion, convincing evidence has been presented that the consumption of cereals infected by *F. graminearum* has resulted in cases of a human mycotoxicosis characterized by emesis in Japan, Korea, and the USSR. Although deoxynivalenol and nivalenol are known to occur naturally in scabby cereals in Japan, these two trichothecenes have not been directly implicated in an actual case of the human mycotoxicosis. Thus it is not known whether deoxynivalenol and/or nivalenol are responsible for the clinical signs of scabby grain intoxication in man or whether other factors are also involved.

Toxicity to Experimental Animals

Cattle. Cattle appear to be much more resistant than pigs to the deleterious effects of scabby cereals naturally infected by *F. graminearum*. Thus scabby barley (Mains et al., 1930) and moldy corn (Noller et al., 1979) refused by pigs in the United States have been fed to cattle without causing any clinical signs of toxicity or estrogenicity. Lactating Holstein cows did not refuse corn severely infected (44%) by *F. graminearum* in the United States, and although these cows gained less weight than cows fed normal corn, milk production was not affected (Noller et al., 1979). This corn contained 500 ppb of zearalenone but it was not analyzed for deoxynivalenol (Noller et al., 1979).

Field outbreaks of hyperestrogenism with overt clinical signs of vulvo-vaginitis attributable to zearalenone in cattle are not well documented in the literature. Zearalenone has, however, been detected at unspecified levels in the feed of diary cattle in Minnesota and 14 ppm in hay in England associated with cases of infertility in dairy cattle (Mirocha et al. 1968b, c). Fertility disturbances in a herd of dairy cows with some external signs of hyperestrogenism in Finland have also been attributed to estrogenic metabolites produced by *Fusarium* spp. in the feed (Roine et al., 1971). Although an isolate of *F. graminearum* from the incriminated feed was shown to produce zearalenone in culture, zearalenone analyses of the feed were "in progress" (Roine et al., 1971) and consequently it is not known whether the feed actually contained zearalenone or not.

Clinical signs of hyperestrogenism including swollen and hyperemic external genitalia have been induced experimentally in cows with pure zearalenone (Christensen, 1979). Zearalenone is also known to be transmitted into bovine milk in the form of the parent compound as well as the metabolite zearalenol (Mirocha et al., 1981). Implants of the derivative zearalanol are used commercially as a growth promoter in cattle (Hidy et al., 1977).

No reports have been found in the literature on the toxicity of pure cultures of *F. graminearum* to cattle. The effects of pure deoxynivalenol in cattle have also not been reported.

Chickens. Scabby cereals naturally infected by *F. graminearum* appear to have little or no adverse effects in chickens. Barley that was heavily infected by *F. graminearum* during the severe scab epiphytotic in the United States in 1928 caused feed refusal and emesis in pigs, but caused no feed refusal or significant reductions in weight gain when fed to chickens (Mains et al., 1980; Mundkur, 1934). Water extracts of scabby barley in Korea caused emesis in pigs, but had no toxic effects upon dosing to day-old chicks (Cho, 1964). Corn infected by *F. graminearum* during the ear rot epiphytotic in Indiana, USA, in 1972 and refused by pigs had no effect on weight gain or feed efficiency of chickens (Featherston, 1973). However, corn severely damaged by *F. graminearum* in the field and containing 0.5 ppm of zearalenone caused reduced feed consumption and egg production in laying hens (Adams & Tuite, 1976). This corn probably also contained toxins other than zearalenone, such as deoxynivalenol and/or other trichothecenes. Adams & Tuite (1976) concluded that corn in commercial poultry rations would rarely be as severely damaged as the corn used in their experiments, and that corn infected by *F. graminearum* would seldom cause field problems in laying hens. However, Lew et al. (1979) reported that corn naturally infected by *F. graminearum* in Austria during 1977 and containing 1 to 20 ppm of deoxynivalenol and 0.05 to 1.6 ppm of zearalenone was lethal to chickens. This corn was refused by pigs, but not by chickens, and the chickens died within a few days after consuming this corn. Wheat infected by *F. graminearum* during the scab epiphytotic in Canada in 1980 and containing 1.0 ppm of deoxynivalenol caused no marked changes in feed consumption, weight gain, and egg production in Leghorn chickens (Trenholm et al., 1981).

Pure cultures on autoclaved barley of a strain of *F. graminearum* isolated from scabby barley in the midwestern United States in 1928 were fed to chickens by Mundkur (1934). The culture material was not acutely toxic to growing chicks, but caused a reduction in weight gain.

Bacon & Marks (1976) fed cultures on corn of an unnumbered strain of *F. graminearum* "isolated from corn" to commercial broiler chicks and Japanese quail. Diets containing this culture material and supplying 1 to 30 ppm of zearalenone had no adverse effects on either broiler chickens or quail during a feeding period of 8 weeks. Diets containing culture material of the same strain of *F. graminearum* and supplying either 25 or 100 ppm of zearalenone also had no adverse effects on the reproductive performance of laying hens (Marks & Bacon, 1976).

The strain used in toxicity experiments in chickens and referred to as *F. roseum* or *F. roseum* "Gibbosum" by Speers et al. (1971, 1972, 1977), is *F. semitectum* (Strain 6.4).

Pure zearalenone is remarkably well tolerated by chickens and a single oral dose of 15 g/kg body weight resulted in no mortalities and no noticeable gross or histopathological lesions (Chi et al., 1980b). Dietary levels of zearalenone up to 800 mg/kg had no significant effects on the growth and feed intake of broiler chickens or the reproductive performance of mature male and female chickens (Allen et al., 1981a, c; Bacon & Marks, 1976; Chi et al., 1980a, b; Christensen, 1979; Hidy et al., 1977; Meronuck et al., 1970; Marks & Bacon, 1976; Speers et al., 1971). However, certain estrogenic effects, such as an increase in the number and size of cysts in the right ovary (Chi et al., 1980a; Speers et al., 1971) and a reduction in testicular weight (Allen et al., 1981a), have been induced in chickens with pure zearalenone.

The acute toxicosis in chickens dosed with pure deoxynivalenol was characterized by extensive ecchymotic hemorrhaging throughout the carcass, widespread deposition of urates, disturbance of the nervous system, and irritation of the upper gastrointestinal tract (Huff et al., 1981). These pathological changes suggest that deoxynivalenol may be involved in the hemorrhagic anemia syndrome (see *F. sporotrichioides,* hemorrhagic syndrome) of chickens caused by moldy feed (Huff et al., 1981).

Ducklings. Cultures on autoclaved corn of *F. graminearum* MRC 120, isolated from Zambian corn, were acutely toxic to day-old Pekin ducklings and caused the death of 4 out of 4 ducklings in a mean time of 6.7 days (Marasas et al., 1978). We have confirmed the identity of this isolate as *F. graminearum* (Strain 8.37). This culture material was found to contain 1 g/kg of zearalenone (Marasas et al., 1978). The toxicity of this material to ducklings could not be attributed to zearalenone, and the chemical nature of the mycotoxin(s) produced by this strain of *F. graminearum* is unknown.

Pure deoxynivalenol is known to cause emesis in ducklings (Ishii et al., 1975; Ueno, 1977b; Ueno et al., 1974a; Yoshizawa & Morooka, 1974).

Geese. Palyusik et al. (1971) reported that cultures on corn of an unnumbered isolate of *F. graminearum* caused inhibition of spermatogenesis and infertility in ganders. According to M. Palyusik (personal communication, 16 December 1977) the strain used in this work was *F. roseum* 'Graminearum' Mapleton No. 10. We have identified this strain, which was isolated in the United States and is known to produce large amounts of zearalenone in culture, as *F. graminearum* (Strain 8.21). Culture material of this strain caused marked decreases in sperm volume of ganders and regressive changes in the testes. Eggs laid by geese mated to a gander fed this culture material were not fertile. The inhibition of spermatogenesis in ganders by zearalenone-containing culture material of *F. graminearum* was confirmed by Palyusik et al. (1974). The strain of *F. graminearum* used in this work was *F. graminearum* F-59 (= ATCC 34909), according to M. Palyusik (personal communication, 16 December 1977). We have confirmed the identity of this strain, which was isolated in Hungary and is known to produce zearalenone in culture, as *F. graminearum* (Strain 8.38). No indication was given of the levels of zearalenone present in the cultures of the two isolates of *F. graminearum* (Strains 8.21 and 8.38) used in these experiments by Palyusik et al. (1971, 1974). It is also not known whether zearalenone alone was responsible for the inhibition of spermatogenesis in ganders or whether other toxic metabolites present in the culture material were also involved.

In addition to ganders and turkey cocks (see turkeys, below), inhibition of spermatogenesis and infertility have also been induced in male guinea fowl fed "grain containing 30–40 mg/kg" of zearalenone (Vanyi & Széky, 1980) and in male pheasants injected with "dihydrogenated zearalenone" (Willemart & Schricke, 1981).

In contrast to the pronounced effects of zearalenone-containing culture material of *F. graminearum* on the fertility of ganders, similar cultures had no effects on laying geese (Palyusik & Koplik-Kovacs, 1975a, b). Diets containing culture material of an unnumbered isolate of *F. culmorum* and supplying 100 ppm of zearalenone had no significant effects on feed consumption, weight gain, egg production, or egg fertility (Palyusik & Koplik-Kovacs, 1975a, b). According to M. Palyusik (personal communication 16 December 1977), the strain used in this work was *F. culmorum* F-79 (= ATCC 34910). We have identified this strain, which was isolated in Hungary and is known to produce zearalenone in culture, as *F. graminearum* (Strain 8.39).

Guinea Pigs. Pelleted feed that caused hyperestrogenism in pigs in Minnesota, USA, caused uterine hypertrophy when fed to guinea pigs (Christensen et al., 1965). Feed refusal and weight loss have been reported in guinea pigs fed scabby barley and corn naturally infected by *F. graminearum* (Curtin & Tuite, 1966; Mains et al., 1930; Mundkur, 1934) or cultures of *F. graminearum* on autoclaved barley (Mundkur, 1934).

Hamsters. Cultures of "*F. roseum* Ohio Isolate C" on autoclaved corn incubated at 20 to 30°C for 12 days followed by 14 days at 12°C proved to be acutely toxic to hamsters and caused the death of 10 out of 10 animals within 21 days (Sharda et al., 1971). Gross pathological changes included icterus and petechial hemorrhages in the liver and lungs. A culture of "*F. roseum* Ohio Isolate C" is no longer available (L.E. Williams, personal communication, 16 June 1981) and consequently it was impossible to determine its identity.

Insects. Harein et al. (1970) reported that culture material on rice of an unnumbered isolate of *F. roseum* 'Graminearum' containing 9,000 ppm of zearalenone decreased egg production and egg hatch of the confused flour beetle (*Tribolium confusum*). Larvae that emerged from these eggs were also abnormally small and sluggish. Diets containing 1,000 ppm of pure zearalenone had no effect on the fecundity or fertility of the beetles. Harein et al. (1970) concluded that this discrepancy may have been due to the difference in dietary zearalenone concentration or the presence of other toxic metabolite(s) in the culture material.

Cultures on rice of an unnumbered isolate of *F. roseum* 'Graminearum' containing up to 10,000 ppm of zearalenone were fed to confused flour beetles and the lesser meal worm (*Alphitobius diaperinus*) by Eugenio et al. (1970b). Zearalenone was ingested from the culture material by both species of insects and persisted through metamorphosis and after death. Consequently these insects may act as mobile sources of zearalenone in feeds and foods (Eugenio et al., 1970b).

Mice. Uterine hypertrophy was induced in mature ovariectomized mice with cultures on corn and alcoholic extracts of culture material of nine isolates of *F. graminearum* from corn associated with field outbreaks of porcine hyperestrogenism in the United States (Stob et al., 1962). These nine isolates of *F. graminearum* and the one from which the uterotrophic metabolite was isolated and partially purified were not specified by Stob et al. (1962). However, the isolate specified in United States patents on the estrogenic and anabolic activity of the active metabolite (FES = Fermentation Estrogenic Substance = zearalenone) was "*Gibberella zeae* NRRL 2380" (should read 2830), according to Mirocha & Christensen (1974) and Mirocha et al. (1971). We have confirmed the identity of this strain as *F. graminearum* (Strain 8.19).

Stob et al. (1962) reported that a water-soluble fraction, toxic for mice, was "occasionally produced concurrently" with the estrogenic metabolite in cultures of *F. graminearum*. Water extracts of scabby barley naturally infected by *F. graminearum* have also been found to be acutely toxic to suckling mice by Cho (1964). It is possible that this water-soluble metabolite, toxic to mice and present in naturally infected barley as well as cultures of F. graminearum, was deoxynivalenol.

Cultures on autoclaved corn of "F. roseum Ohio Isolate C" proved to be acutely toxic to mice and caused the death of 8 of 12 animals within 22 days (Sharda et al., 1971). Gross pathological changes included icterus and petechial hemorrhages in the liver and lungs. A culture of "F. roseum Ohio Isolate C" is no longer available (L.E. Williams, personal communication, 16 June 1981) and consequently it was not possible to determine its identity.

Extracts of cultures in PSC medium of two Japanese isolates of G. zeae (Ohoita-II and Ishii) were found to be toxic to mice upon intraperitoneal injection by Ueno et al. (1973a). The toxic extracts of G. zeae Ohoita-II were found to contain fusarenon-X and nivalenol, but the mycotoxin(s) produced by G. zeae Ishii (Strain 8.28) were not identified. We have identified four other Japanese strains reported to be toxic to mice, i.e., F. roseum No. 117 (Morooka et al., 1972), F. rigidiusculum M-1-3, F. roseum 'Culmorum' M-3-2, and F. tricinctum Abashiri-2 (Ueno et al., 1972a, 1973a), as F. graminearum (Strains 8.22, 8.29, 8.30 and 8.33). Fusarium roseum No. 117 (Strain 8.22) was reported to produce deoxynivalenol (= Rd toxin) and butenolide by Morooka et al. (1972), while toxic extracts of cultures in PSC medium of two of the other strains of F. graminearum, i.e., F. rigidiusculum M-1-3 (Strain 8.29) and F. tricinctum Abashiri-2 (Strain 8.33), have been found to contain diacetoxyscirpenol, neosolaniol, and T-2 toxin (Ueno et al., 1973a).

Cultures on rice of F. graminearum NRRL 2830 were fed on a life-long basis to mice by Saito & Ohtsubo (1974). We have confirmed the identity of this strain, which was originally isolated in the United States and is known to produce zearalenone and deoxynivalenol, as F. graminearum (Strain 8.19). The culture material on rice of this strain caused the death of 21 of 43 mice within 70 days, "mostly showing diarrhea," and one case each of hemangiosarcoma of the liver and lymphocytic leukemia were observed (Saito & Ohtsubo, 1974).

Yoshizawa et al. (1979) reported that 24 of 106 isolates of Fusarium spp. (predominantly F. graminearum) from barley and wheat in Japan were toxic to mice when extracts of cultures in PSC medium were administered intraperitoneally to male mice. Chemical analyses of these toxic extracts of F. graminearum strains revealed the presence of butenolide, zearalenone, deoxynivalenol, and nivalenol, either singly or in various combinations (Yoshizawa et al., 1979).

Pure zearalenone has estrogenic effects in mice and causes dose-dependent increases in uterine weights of immature females and ovariectomized mature mice (Hidy et al., 1977; Knaus, 1978; Mirocha et al., 1968b; Ueno et al., 1974b, 1977a). Zearalenone is not acutely toxic to mice, and a single oral dose of 20 g/kg caused no mortalities (Hidy et al., 1977). Zearalenone is also not teratogenic in mice (Arora et al., 1981). Zearalenone is not a feed refusal factor in mice and the feed refusal activity of deoxynivalenol in mice is not enhanced by the addition of zearalenone (Burmeister et al., 1981a).

Pigeons. Prentice & Dickson (1968) reported that extracts of cultures in Richard's solution of several Fusarium strains, including "F. roseum 162," caused emesis upon injection into the wing vein of pigeons. According to Prentice & Dickson (1968), "F. roseum 162" was isolated from scabby barley in Korea in 1963, and this strain is "F. roseum NRRL A-15,666," according to Ellison & Kotsonis (1973) and Vesonder et al. (1977b). Contrary to the findings of Prentice & Dickson (1968), ethylacetate extracts of cultures of this strain in Richard's solution or on corn exhibited no emetic activity upon force feeding to pigeons (Ellison & Kotsonis, 1973). We have not seen this culture and consequently it was not possible to determine whether it is *F. graminearum* or not.

Nine isolates of *F. graminearum* (as *G. roseum* 'Graminearum' or *F. roseum* 'Graminearum') from corn refused by pigs in Wisconsin, USA, in 1972, were tested for emetic activity in pigeons by Kotsonis et al. (1975a). Cultures of these strains on autoclaved corn were incubated at room temperature for 1 week followed by 4 weeks at 12°C and ethylacetate extracts were force-fed to adult pigeons. With one possible exception, none of these strains of *F. graminearum* caused emesis in pigeons. *Fusarium roseum* 'Graminearum' No. 8 induced a limited, barely detectable vomiting response which was difficult to verify because the pigeons also vigorously refused the extract by expectoration. Six of the nine isolates of *F. graminearum* caused a severe feed refusal response in pigs as well as rats. Chemical analyses of culture material of four of these isolates revealed the presence of zearalenone at levels ranging from 13.8 to 47.8 μg/g, but "careful analysis of crude extracts of our most active refusal isolates failed to detect known emetic trichothecenes such as T-2 toxin, HT-2 toxin, or 4-deoxynivalenol" (Kotsonis et al., 1975a). The analytical method used for deoxynivalenol determination was unfortunately not specified. Kotsonis et al. (1975a) concluded that the pronounced feed refusal activity of these isolates of *F. graminearum* in pigeons and other animals may have been due to "a combination of zearalenone and one or more of the trichothecenes."

Wu et al. (1981) reported that extracts of cultures of *F. graminearum* FG-1, isolated from wheat in China, caused "violent emesis" in test pigeons. On the basis of TLC analyses, Wu et al., (1981) concluded that the emetic principle produced by this strain was not deoxynivalenol but possibly an unknown trichothecene.

Pigs. Diets containing cereals infected by *F. graminearum* are known to cause field outbreaks of three distinct syndromes in pigs, i.e., estrogenic, feed refusal, and emetic syndromes. The occurrence of field outbreaks of these syndromes in pigs and their etiology have already been discussed (see estrogenic syndrome and feed refusal and emetic syndromes, above). The following discussion will be limited to a summary of the main clinical signs and pathological lesions that have been induced experimentally in pigs, either with cereals naturally infected by *F. graminearum*, or with pure cultures of *F. graminearum*, or with crystalline material of two known metabolites of *F. graminearum*, i.e., zearalenone and deoxynivalenol.

The following estrogenic effects have been induced experimentally in pigs fed barley and corn naturally infected by *F. graminearum*: swelling, edema, tumefaction, and hyperemia of the vulva (Christensen et al., 1965; McErlean, 1952; McNutt et al., 1928); prolapse of the vagina (McNutt et al., 1928); enlargement of the uterus (Christensen et al., 1965; Etienne & Jemmali, 1979); enlargement of the mammae (Christensen et al., 1965; McErlean, 1952); reduction in litter size (Sharma et al., 1974); and increase in fetal mummification (Sharma et al., 1974).

Feed refusal as evidenced by a reduction in feed consumption and weight gain has been induced experimentally in pigs fed barley, wheat, and corn naturally infected by *F. graminearum* (Beller & Wedeman, 1929; Christensen & Kernkamp, 1936; Curtin & Tuite, 1966; Forsyth, 1974; Forsyth et al., 1976, 1977; Mains et al., 1930; Mundkur, 1934; Trenholm et al., 1981; Tuite et al., 1974; Vesonder et al., 1976).

Emesis has been induced experimentally in pigs with barley, wheat, and corn naturally infected by *F. graminearum*, either by the feeding of the scabby cereals (Christensen & Kernkamp, 1936; Forsyth, 1974; Mundkur, 1934) or by the administration of extracts of these cereals (Cho, 1964; Christensen & Kernkamp, 1936; Curtin & Tuite, 1966; Forsyth, 1974; Hoyman, 1941; Tuite et al., 1974; Vesonder et al., 1973).

The following estrogenic effects have been induced experimentally in pigs with pure cultures of *F. graminearum:* swelling, edema, tumefaction, and hyperemia of the vulva (Christensen et al., 1972b; Mirocha et al., 1971; Palyusik, 1973; Stob et al., 1962); prolapse of the vagina (Christensen et al., 1972b; Mirocha et al., 1971); enlargement of the uterus (Christensen et al., 1972b; Mirocha et al., 1971; Palyusik, 1973); enlargement of the mammae (Christensen et al., 1972b; Stob et al., 1962); degeneration of oocytes (Palyusik, 1973); increase in embryonal mortality (Sharma et al., 1974); abortion (Mirocha & Christensen, 1974; Mirocha et al., 1971); reduction in litter size (Mirocha et al., 1971); increase in fetal mummification (Sharma et al., 1974; Wilson et al., 1967); and reduction in testicular weight (Christensen et al., 1972b). We have identified the following *Fusarium* strains that have apparently been found to have estrogenic effects in pigs as *F. graminearum* (Strains 8.19, 8.20, 8.21, and 8.38):

1) *F. graminearum* (= *F. roseum* = *G. zeae*) NRRL 2830 (Strain 8.19). The isolate used in the first successful experimental reproduction of vulvar and mammary enlargement in pigs with a pure culture of *F. graminearum* was not specified by Stob et al. (1962). However, the isolate designated in United States patents on the estrogenic and anabolic metabolite (= zearalenone) was "*G. zeae* NRRL 2380" (should read 2830) according to Mirocha & Christensen (1974) and Mirocha et al. (1971). Thus it is reasonable to assume that *G. zeae* NRRL 2830 was one of the nine isolates of *G. zeae* reported to have uterotrophic activity by Stob et al. (1962), but it cannot be stated unequivocally that this is the strain that these authors used in the experimental reproduction of vulvar and mammary enlargement in pigs. Subsequent to the original isolation of zearalenone from this strain by Stob et al. (1962), several other authors have also reported zearalenone production by *F. graminearum* NRRL 2830 (Bolliger & Tamm, 1972; Hesseltine, 1972; Hidy et al., 1977; Ueno et al., 1971a; Vesonder et al., 1981b). This strain is also known to produce deoxynivalenol together with zearalenone in culture (Vesonder et al., 1981b).

2) *Fusarium* No. 5 (Strain 8.20). This strain was isolated from corn associated with an outbreak of the estrogenic syndrome in Minnesota, USA, and cultures on corn incubated at 20 to 25°C for 21 days followed by 14 days at 12°C caused uterine hypertrophy in rats (Christensen et al., 1965). This strain apparently also caused vulvar enlargement and vaginal prolapse in a young gilt, according to Figs. 10 and 11 of Christensen (1979) and Fig. 2 of Mirocha et al. (1971). This strain of *F. graminearum* is known to produce zearalenone (Christensen et al., 1965).

3) *F. roseum* 'Graminearum' Mapleton No. 10 (Strain 8.21). This strain was "originally isolated from feed associated with problems of infertility in swine in southern Minnesota" (Mirocha et al., 1971). Culture material on corn of an unnumbered strain of "*F. roseum*" was found to cause vulvar and mammary enlargement, rectal prolapse, and uterine hypertrophy in sexually immature female pigs and reduced testicular weight in male pigs by Christensen et al. (1972b). The same experiment was referred to by Nelson et al. (1973), who specified the strain used as "*F. roseum* (strain Mapleton No. 10)." This strain of *F. graminearum* is known to produce large amounts of zearalenone in culture (Eugenio et al., 1970a; Mirocha & Christensen, 1974; Mirocha et al., 1967a, b, 1968a, b, 1971; Scott et al., 1970; Sherwood & Peberdy, 1972, 1974a, b; Steele et al., 1974, 1976, 1977a, b; Wolf & Mirocha, 1973, 1977).

4) *F. graminearum* F-59 = ATCC 34909 (Strain 8.38). This strain was isolated from millet in Hungary (Szathmary et al., 1976, 1977). According to M. Palyusik (personal communication, 16 December 1977), this was the unnumbered "home strain of *F.*

graminearum" used in the experimental reproduction of the estrogenic syndrome in pigs by Palyusik (1973). Cultures on corn caused characteristic signs of hyperestrogenism, including uterine hypertrophy and oocyte degeneration in gilts (Palyusik, 1973). This strain is known to produce zearalenone (Szathmary et al., 1976, 1977) and also produces deoxynivalenol (Pathre & Mirocha, 1978).

Feed refusal in pigs has been induced experimentally with pure cultures of *F. graminearum* on a variety of media (Christensen & Kernkamp, 1936; Kotsonis et al., 1975a; Mains et al., 1930; Miessner & Schoop, 1929; Palyusik, 1973; Vesonder et al., 1976, 1977b). We have identified the following *Fusarium* strains that have been shown experimentally to have feed refusal activity in pigs as *F. graminearum* (Strains 8.23 and 8.38):

1) *F. graminearum* NRRL 5883 (Strain 8.23). This strain was isolated from "naturally infected corn" in the USA (Vesonder et al., 1976). Cultures on either rice (Vesonder et al., 1976) or corn (Vesonder et al., 1977b) were refused by pigs. This strain is known to produce deoxynivalenol (Lindenfelser et al., 1978; Vesonder et al., 1976, 1977b, 1981b) and also produced zearalenone together with deoxynivalenol in culture (Vesonder et al., 1981b).

2) *F. graminearum* F-59 = ATCC 34909 (Strain 8.38). This strain was isolated from millet in Hungary (Szathmary et al., 1976, 1977). According to M. Palyusik (personal communication, 16 December 1977), this was the unnumbered "home strain of *F. graminearum*" that was shown to cause hyperestrogenism in pigs by Palyusik (1973). The prolonged cultivation of this strain on corn "made the maize emetic, thence its consumption was eventually refused by pigs" (Palyusik, 1973). As far as we are aware, this is the only strain of *F. graminearum* that has been reported to cause hyperestrogenism, feed refusal, and emesis in pigs under experimental conditions. *F. graminearum* F-59 is known to produce deoxynivalenol (Pathre & Mirocha, 1978) as well as zearalenone (Szathmary et al., 1976, 1977) in culture.

Emesis has been induced experimentally in pigs with barley, wheat, and corn artificially inoculated with *F. graminearum*, either by the feeding of the inoculated corn (Christensen & Kernkamp, 1936) or the administration of water extracts of the scabby cereals by stomach tube (Christensen & Kernkamp, 1936; Shands, 1937). The ability of pure cultures of *F. graminearum*, or extracts of these cultures, to cause emesis in pigs is not well established. Kotsonis et al. (1975a) found that cultures on corn incubated at room temperature for 1 week followed by 4 weeks at 12°C of 8 of 9 isolates of *F. graminearum* caused feed refusal in pigs. Culture material of two of these isolates also caused signs of hyperestrogenism, but none of them caused emesis. Vesonder et al. (1976) reported that cultures on rice incubated at 28°C for 13 days of *F. graminearum* NRRL 5883 (Strain 8.23) were refused by pigs and that only fractions containing deoxynivalenol were refused by pigs. Neither the culture material nor any of the fractions caused emesis. Vesonder et al. (1976) suggested that the levels of deoxynivalenol present may determine whether or not pigs will consume sufficient amounts of material to elicit the emetic response. Thus the fractions of the culture material of *F. graminearum* 5883 (Strain 8.23) containing high levels of deoxynivalenol were refused by pigs and did not cause emesis, while the fractions of naturally infected moldy corn assayed by Vesonder et al. (1973) contained lower levels and caused emesis. According to Palyusik (1973), cultures on corn of *F. graminearum* F-59 (Strain 8.38) caused hyperestrogenism in pigs and the "prolonged growing of the mould made the maize emetic, thence its consumption was eventually refused by the pigs." This strain is known to produce zearalenone (Szathmary et al., 1976, 1977) as well as deoxynivalenol (Pathre & Mirocha, 1978). These findings by Palyusik (1973) suggest the interesting possibility that different levels of zearalenone and deoxynivalenol may be present in cultures of *F. graminearum* at different stages of incubation and that the feeding of culture material to pigs after different incubation periods may result in either hyperestrogenism, emesis, or feed refusal.

The following clinical signs and pathological changes have been induced experimentally in female pigs with pure zearalenone: swelling, edema, tumefaction, and hyperemia of the vulva, prolapse of the vagina, enlargement of the mammae, and enlargement of the uterus (Chang et al., 1979; Christensen, 1979; Kurtz & Mirocha, 1978a; Kurtz et al., 1969; Mirocha & Christensen, 1974; Mirocha et al., 1971); atrophy of the ovaries with follicular atresia, oocyte degeneration, and absence of corpora lutea (Chang et al., 1979; Kurtz & Mirocha, 1978a; Kurtz et al., 1969; Mirocha & Christensen, 1974; Mirocha et al., 1971; Vanyi et al., 1974); reduction in litter size and weight (Chang et al., 1979; Christensen, 1979; Miller et al., 1973; Mirocha & Christensen, 1974; Mirocha et al., 1971, 1977a); stillbirths and piglets with "splayleg" (Miller et al., 1973); and infertility (Chang et al. 1979; Christensen, 1979; Mirocha et al., 1977a). Diets containing 25 to 100 ppm of 95% pure zearalenone have been shown to cause multiple reproductive deficiencies in sows, including abnormal estrous cycles, nymphomania, pseudopregnancy, malformed piglets, juvenile hyperestrogenism, diminished fertility, and complete infertility (Chang et al., 1979). The effects of pure zearalenone on male pigs are not as well documented. Ruhr et al. (1978) reported that dietary zearalenone levels up to 60 ppm had no effects on the reproductive potential of mature boars. However, the consumption of a diet containing 40 ppm of zearalenone resulted in reduced libido scores and depressed plasma testosterone concentrations in prepubertal boars (Berger et al., 1981).

Pure deoxynivalenol has been shown to cause feed refusal (Forsyth et al., 1977; Vesonder et al., 1976, 1979b) as well as emesis (Forsyth et al., 1977; Vesonder et al. 1973, 1976) in pigs. Pigs responded to deoxynivalenol in the diet with decreased feed intake inversely proportional to the concentration added. Thus 3.6 ppm of deoxynivalenol in the diet decreased consumption by 20%, while 40 ppm caused 90% rejection (Forsyth et al., 1977). The minimum emetic dose of deoxynivalenol in pigs was 0.05 mg/kg intraperitoneally and 0.1 to 0.2 mg/kg orally (Forsyth et al., 1977).

Rabbits. Extracts of cultures of one out of three isolates of *F. graminearum* from overwintered cereals in the USSR (see *F. sporotrichioides,* ATA) have been reported to be mildly toxic to rabbit skin (Joffe, 1960a, b, 1971). The mycotoxin(s) produced by this dermotoxic strain have not been identified.

Joffe & Palti (1975) identified *F. tricinctum* NRRL 5908, which was isolated from fescue hay in the United States, as *F. sporotrichioides.* Subsequently Joffe (1978b) and Joffe & Yagen (1977) reported that extracts of cultures on wheat of this strain (as *F. sporotrichioides* NRRL 5908) were mildly toxic to rabbit skin and contained T-2 toxin. We have identified this strain as *F. graminearum* (Strain 8.41).

Takatori et al. (1980) found that *F. graminearum* was the dominant *Fusarium* species (19 of 53 isolates) colonizing hay and rice straw in Hokkaido, Japan. Extracts of these isolates were screened for toxicity by the rabbit skin test. Of these 19 isolates of *F. graminearum,* two caused a "very strong" inflammatory reaction, two were slightly dermotoxic, and three were very slightly dermotoxic. Takatori et al. (1980) concluded that "from the results of chemical and biological assay, the production of T-2 toxin, neosolaniol, and diacetoxyscirpenol was presumed."

Cultures of "*F. roseum* Ohio Isolate C" on autoclaved corn incubated at 20 to 30°C for 12 days followed by 14 days at 12°C caused the death of one out of three rabbits after 13 days (Sharda et al., 1971). Gross pathological changes included icterus and petechial hemorrhages in the liver and lungs. A culture of "*F. roseum* Ohio Isolate C" is no longer available (L.E. Williams, personal communication, 16 June 1981) and consequently it was not possible to determine its identity.

Rats. In this discussion, a summary will be given of the estrogenic (see estrogenic syndrome, above), feed refusal (see feed refusal and emetic syndromes, above), and toxic effects in rats caused by cereals naturally infected by *F. graminearum,* pure cultures of *F. graminearum,* and crystalline material of two known metabolites of *F. graminearum,* i.e., zearalenone and deoxynivalenol.

Estrogenic effects, as evidenced by an increase in the weight of the uterus, were first induced experimentally in weanling virgin rats by Christensen et al. (1965) with corn naturally infected by *F. graminearum* and associated with a field outbreak of the estrogenic syndrome in pigs in the United States. A slight increase in uterine weight of virgin female rats has also been induced by the feeding of a hand-selected sample of corn naturally infected by *F. graminearum* in South Africa and containing 6.4 ppm of zearalenone (Marasas et al., 1977).

Feed refusal, as evidenced by a reduction in feed consumption and weight gain and sometimes mortality, has been induced experimentally in rats fed scabby barley (Mains et al., 1930) or moldy corn naturally infected by *F. graminearum* (DeUriarte et al., 1976; Featherston, 1973; Forsyth, 1974; MacDonald & Raemakers, 1974; Marasas et al., 1977). Hand-selected samples of corn naturally infected by *F. graminearum* in South Africa and Zambia that caused feed refusal in rats (Marasas et al., 1977), contained deoxynivalenol (2.5 and 7.4 ppm, respectively) as well as zearalenone (6.4 and 12.8 ppm, respectively).

Pure cultures on corn of 12 of 40 *Fusarium* isolates from corn in the United States were found to cause uterine hypertrophy in weanling virgin rats by Christensen et al. (1965). These uterotrophic *Fusarium* strains were not specifically identified, but Christensen et al. (1965) stated that "all or most of the isolates of *Fusarium* tentatively appear to be *F. culmorum* or *F. graminearum.*" We have identified one of these strains that caused uterine hypertrophy in rats, i.e., "*Fusarium* No. 5," as *F. graminearum* (Strain 8.20). The estrogenic metabolite zearalenone (= F-2) was isolated from the culture material of this strain by means of the rat uterus bioassay (Christensen et al., 1965). We have identified another strain, i.e., "*Fusarium* YN-13," reported to cause an increase in the weight of the uterus in rats by Christensen et al. (1965), as *F. sporotrichioides* (Strain 4.25). However, we have reason to believe that this strain was originally a mixture of *F. sporotrichioides* and *F. graminearum* (see Strain 4.25). Enlargement of the uterus in rats has also been induced experimentally with pure cultures of several Finnish isolates of *F. graminearum* known to produce zearalenone (Kallela & Korpinen, 1973; Kallela & Saastamoinen, 1981a, b; Korpinen et al., 1972; Roine et al., 1971); with culture material on corn of four isolates of *F. roseum* 'Graminearum' from the United States that contained 13.8 to 47.8 ppm of zearalenone (Kotsonis et al., 1975a); with culture material on corn of two isolates of *F. graminearum* (MRC 121 and MRC 460) from corn in southern Africa that contained 114.0 to 127.0 ppm of zearalenone (Marasas et al., 1977); and with culture material on millet of three Hungarian isolates, including *F. graminearum* F-59 (Strain 8.38) and *F. graminearum* (= *F. culmorum*) F-79 (Strain 8.39), that contained up to 3400 ppm of zearalenone (Szathmary et al., 1976, 1977). In Hungary, zearalenone-containing culture material on corn of an unspecified isolate of *F. graminearum* has been reported to cause multiple reproductive disorders in male and female rats (Mess et al., 1979; Ruzsas et al., 1978, 1979). The effects on the reproductive system of rats induced by this culture material included reduced fertility, complete infertility, abortion, decreased gonadal weights, reduction in the number of corpora lutea, inhibition of spermatogenesis, neonatal mortality, and permanent changes in the reproductive organs associated with decreased fertility in the offspring (Mess et al., 1979; Ruzsas et al., 1978, 1979).

We have identified the following *Fusarium* strains that have been reported to cause increases in the weight of the uterus of weanling virgin rats as *F. graminearum* (Strains 8.38, 8.39, 8.42, 8.43, and 8.45):

1) *F. graminearum* F-59 (Strain 8.38). This strain was isolated from millet in Hungary (Szathmary et al., 1976, 1977). Cultures on millet containing 2% malt extract incubated at 23 to 27°C for 1 week, followed by 2 weeks at 5 to 8°C and 1 week at 23 to 27°C, caused enlargement of the uterus in virgin weanling rats and contained 200 to 730 µg/g of zearalenone (Szathmary et al., 1976, 1977).

2) *F. culmorum* F-79 (Strain 8.39). This strain of *F. graminearum* was isolated from millet in Hungary (Szathmary et al., 1976, 1977). Cultures on millet containing 2% malt extract incubated at 23 to 27°C for 1 week, followed by 2 weeks at 5 to 8°C and 1 week at 23 to 27°C, caused enlargement of the uterus in virgin weanling rats and contained 540 to 3,400 µg/g of zearalenone (Szathmary et al., 1976, 1977).

3) *F. graminearum* MRC 121 (Strain 8.42). This strain was isolated from moldy Zambian corn (Marasas et al., 1977). Cultures on corn incubated at 25°C for 2 weeks followed by 6 weeks at 12°C were acutely toxic to rats and contained 1,270.0 ppm of zearalenone and 6.0 ppm of deoxynivalenol. Because of the high degree of toxicity of the culture material, it was diluted 1:10 with commercial corn meal. This material, containing 127.0 ppm of zearalenone, was fed to weanling virgin female rats for 7 days and caused marked increases in uterine weight (Marasas et al., 1977).

4) *F. graminearum* MRC 460 (Strain 8.43). This strain was isolated from moldy South African corn (Marasas et al., 1977). Cultures on corn incubated at 25°C for 2 weeks followed by 6 weeks at 12°C were acutely toxic to rats and contained 1,140.0 ppm of zearalenone and 1.3 ppm of deoxynivalenol. Because of the high degree of toxicity of the culture material, it was diluted 1:10 with commercial corn meal. This material, containing 114.0 ppm of zearalenone, was fed to weanling female rats for 7 days and caused marked increases in uterine weight (Marasas et al., 1977).

5) *F. graminearum* 7137 (Strain 8.45). This strain (as *F. graminearum* No. 13) was isolated from a commercial feed mix in Finland (Korpinen et al., 1972). According to A. Ylimäki (personal communication, 10 April 1979), the strain referred to as *F. graminearum* No. 13 by Korpinen et al. (1972) is *F. graminearum* 7137 (Strain 8.45). Cultures on a mixture of oats, barley, and wheat were incubated at room temperature for one month. This material was fed to immature female rats for 7 days and caused marked increases in uterine weight (Korpinen et al., 1972). The zearalenone content of this culture material was not determined by Korpinen et al. (1972). Subsequently Kallela & Korpinen (1973) reported that one of the *F. graminearum* strains that was "found to cause clear estrogenic changes in rats" by Korpinen et al. (1972) produced zearalenone in culture. Since *F. graminearum* No. 13 caused the greatest increase in uterine weight of all the strains tested by Korpinen et al. (1972), it is possible that the unnumbered strain of *F. graminearum* found to produce zearalenone by Kallela & Korpinen (1973) may have been isolate No. 13.

Feed refusal and weight loss in rats have been induced experimentally with numerous strains of *F. graminearum* on a variety of media (Korpinen & Uoti, 1974; Korpinen et al., 1972; Kotsonis et al., 1975a; Mains et al., 1930; Marasas et al., 1977, 1978; Mitchell & Beadles, 1940; Roine et al., 1971; Sharda et al., 1971; Yoshizawa et al., 1978). In cases where cultures of *F. graminearum* have been reported to be lethal to rats (Korpinen & Uoti, 1974; Marasas et al., 1977, 1978; Mirocha & Christensen, 1974; Mitchell & Beadles, 1940; Sharda et al., 1971; Wilson et al., 1967), it has not always been clear whether the mortality was due to the toxicity of the culture material or to starvation due to feed refusal. A typical example of this problem is the statement by Mitchell & Beadles (1940) that corn inoculated with *F. graminearum* "proved to be so toxic to rats that too little was eaten to permit a fair test. However, even with this limited consumption of the ration containing the damaged corn, many of the rats died." We have identified the following *Fusarium* strains that have been reported to cause feed refusal and/or mortality in rats, as *F. graminearum* (Strains 8.19, 8.21, 8.22, 8.37, 8.42, 8.43, 8.44, and 8.45):

1) *F. graminearum* NRRL 2830 (Strain 8.19). Cultures of this strain on rice were fed to male rats on a life-long basis by Saito & Ohtsubo (1974). All of the rats (18) fed this culture material survived for one year and died at random intervals thereafter. No mention was made of the occurrence of tumors by Saito & Ohtsubo (1974), but one lymphosarcoma was recorded in these rats according to Ohtsubo & Saito (1977) and Saito et al. (1980). It could be concluded from this experiment that the culture material of this strain, which is known to produce zearalenone and deoxynivalenol in culture (see Strain 8.19), was not carcinogenic in rats.

2) *F. roseum* 'Graminearum' Mapleton No. 10 (Strain 8.21). According to Mirocha & Christensen (1974) this strain "evidently produces a complex of toxins, and although oestrogenic, on some occasions it has been lethal to white rats in feeding tests." Although this strain of *F. graminearum* is known to produce large amounts of zearalenone and certain derivatives of zearalenone in culture (see Strain 8.21), we have not been able to find any reports on the production of other mycotoxins by this strain.

3) *F. roseum* No. 117 = ATCC 28114 (Strain 8.22). This strain was isolated from scabby barley in Japan (Morooka et al., 1972). Cultures of this strain on rice incubated at 25°C for 15 days caused a marked feed refusal response in rats (Yoshizawa et al., 1978). This culture material that was refused by rats contained large amounts of deoxynivalenol (220 µg/g) and 3-acetyldeoxynivalenol (160 µg/g), as well as trace amounts of diacetyldeoxynivalenol, and the greater part of the feed refusal activity was found in the fraction containing these trichothecenes (Yoshizawa et al., 1978).

4) *F. graminearum* MRC 120 (Strain 8.37). This strain was isolated from Zambian corn (Marasas et al., 1978). Culture material on corn caused the death of 2 of 4 rats within 14 days and marked feed refusal and weight loss in the survivors (Marasas et al., 1978). The culture material contained zearalenone (1 g/kg), but was not analyzed for deoxynivalenol or other trichothecenes. Thus the chemical nature of the mycotoxin(s) produced by this strain and responsible for the feed refusal and mortality in rats is unknown.

5) *F. graminearum* MRC 121 (Strain 8.42). Cultures on corn of this strain, which was isolated from Zambian corn, were acutely toxic to rats and caused the death of 4 of 5 rats within 3 days (Marasas et al., 1977). The toxic culture material contained deoxynivalenol and zearalenone at levels of 6.0 and 1,270.0 ppm, respectively (Marasas et al., 1977).

6) *F. graminearum* MRC 460 (Strain 8.43). Cultures on corn of this strain, which was isolated from South African corn, were acutely toxic to rats and caused the death of 2 of 5 rats within 3 days (Marasas et al., 1977). The toxic culture material contained deoxynivalenol and zearalenone at levels of 1.3 and 1,140.0 ppm, respectively (Marasas et al., 1977).

7) *F. graminearum* 72322 (Strain 8.44). Korpinen & Uoti (1974) tested 10 unnumbered isolates of *F. graminearum* from cereal seeds in Finland for toxicity to rats. Cultures on a

mixture of oats, barley, and wheat were fed to rats for 14 days and caused marked reductions in weight gain. Most of the animals died before the tenth day, and hemorrhage and ulceration of the gastric mucosa were prominent post mortem findings (Korpinen & Uoti, 1974). According to A. Ylimäki (personal communication, 10 April 1979) *F. graminearium* 72322 (Strain 8.44) was one of the unnumbered strains of *F. graminearum* found to cause weight reductions in rats by Korpinen & Uoti (1974). The chemical nature of the mycotoxin(s) produced by this strain is unknown.

8) *F. graminearum* 7137 (Strain 8.45). This strain (as *F. graminearum* No. 13) was isolated from a commercial feed mix in Finland (Korpinen et al., 1972). According to A. Ylimäki (personal communication, 10 April 1979), the strain referred to as *F. graminearum* No. 13 by Korpinen et al. (1972) is *F. graminearum* 7137 (Strain 8.45). Cultures of this strain on a mixture of oats, barley, and wheat were fed to immature female rats and caused marked increases in the weight of the uterus. However, the reduced palatability of this culture material also resulted in weight loss of these rats (Korpinen et al., 1972). The mycotoxins present in this culture material were not determined by Korpinen et al. (1972), but *F. graminearum* 7137 has subsequently been shown to produce deoxynivalenol in cultures on rice (Pathre & Mirocha, 1978).

Pure zearalenone is not acutely toxic to rats and a single oral dose of 20 g/kg caused no mortalities (Hidy et al., 1977). Zearalenone has estrogenic effects in rats and causes dose-dependent increases in uterine weight in immature virgin female rats (Christensen et al., 1965; Hidy et al., 1977; Kallela & Saastamoinen, 1981a, b; Mirocha et al., 1967a, b, 1968a, b, 1971; Smith, 1980; Ueno et al., 1974b). Other reproductive disorders that have been induced experimentally in rats with pure zearalenone include reduced numbers of corpora lutea, reduced testicular size, and arrested maturation of spermatocytes (Hidy et al., 1977). The administration of zearalenone to neonatal rats causes permanent reproductive disorders and decreased fertility associated with decreased gonadal size, reduced numbers of corpora lutea, testicular degeneration, and inhibition of spermatogenesis (Mess et al., 1979; Ruzsas et al., 1978, 1979). Zearalenone is also teratogenic in rats and causes fetal skeleton anomalies (Hidy et al., 1977; Ruddick et al., 1976).

Pure deoxynivalenol causes feed refusal in rats (Vesonder et al., 1979b; Yoshizawa & Morooka, 1977; Yoshizawa et al., 1978). Feed intake by rats decreases linearly with increasing dietary deoxynivalenol concentration, and the ED_{50} for rat feed refusal is approximately 100 ppm (Yoshizawa et al., 1978). Zearalenone has also been reported to have some feed-refusal activity in rats (Kotsonis et al., 1975a; Smith, 1980, 1981) and to enhance the refusal of feed containing T-2 toxin (Kotsonis et al., 1975a). Thus zearalenone may act synergistically with deoxynivalenol in causing feed refusal by rats of culture material of *F. graminearum* containing both of these metabolites (see feed refusal and emetic syndromes, above).

Turkeys. An estrogenic response in turkey poults characterized by swollen vents and prolapsed cloacae has been induced experimentally with cultures on corn of an unnumbered isolate of "*F. roseum*" (Meronuck et al., 1970). According to Meronuck et al. (1970), the "isolate of *F. roseum* was one of the higher producers of the estrogenic compound of F-2, as reported by Mirocha et al." In the reference cited, Mirocha et al. (1967b) referred to "two estrogen-producing isolates of *F. graminearum* designated as numbers of 9 and 10." We are assuming here that the strain used in the experimental induction of an estrogenic response in turkeys was *F. roseum* 'Graminearum' Mapleton No. 10, which we have identified as *F. graminearum* (Strain 8.21). This strain is known to produce large amounts of zearalenone in culture (see Strain 8.21), but the zearalenone content of the culture material used by Meronuck et al. (1970) was not indicated. However, in an apparent reference to the feeding trial in turkeys conducted by Meronuck et al. (1970), the zearalenone content of rations containing 10% of "*Fusarium*-invaded corn" was stated to have been 800 ppm by Mirocha et al. (1971). According to these authors, this diet caused swelling of the vents, cystic development of the right genital tract, evertion of the cloaca, and enlargement of the cloacal bursa and oviduct in turkey poults. Mirocha et al. (1971) also stated that a similar estrogenic response had been induced in turkey poults with diets containing 300 ppm of pure zearalenone. Allen et al. (1981a) reported that diets containing 200 mg/kg or more of zearalenone caused swelling of the vent in turkey poults, but levels up to 800 mg/kg did not cause prolapse of the vent area and had no effects on the weights of the bursa and ovaries.

Inhibition of spermatogenesis, regressive changes in the testes, and infertility in turkey cocks have been reported to be caused by culture material on corn of an unnumbered isolate of *F. graminearum* by Palyusik et al. (1971). According to M. Palyusik (personal communication, 16 December 1971), the strain used in this work was *F. roseum* 'Graminearum' Mapleton No. 10. This is *F. graminearum* (Strain 8.21). This strain is known to produce large amounts of zearalenone in culture (see Strain 8.21), but the zearalenone content of the culture material used by Palyusik et al. (1971) was not indicated. However, Allen et al. (1981a) found that zearalenone levels up to 800 mg/kg had no effect on the weight of the testes in male turkey poults and caused no histopathological changes in the testes.

In view of the findings by Allen et al. (1981a) that pure zearalenone caused swelling of the vents but otherwise had minimal effects in turkey poults, it seems that some of the effects reported to be caused by culture material of *F. graminearum* (Strain 8.21) in male (Palyusik et al., 1971) as well as female (Meronuck et al., 1970; Mirocha et al., 1971) turkeys were not caused by zearalenone, but rather by other unidentified mycotoxins produced by this strain.

Mycotoxins Produced

4-Acetamido-2-butenoic acid. This metabolite was first isolated by Vesonder et al. (1977a) from the fraction containing deoxynivalenol of cultures on rice incubated at 28°C for 13 days of *F. graminearum* NRRL 5883. We have confirmed the identity of this strain as *F. graminearum* (Strain 8.23). No toxic, teratogenic, or antibiotic activity was exhibited by 4-acetamido-2-butenoic acid, but the possibility has been suggested that this compound may be an intermediate in the biosynthetic pathway of butenolide (Vesonder et al., 1977a). However *F. graminearum* NRRL 5883 (Strain 8.23) has not been reported to produce butenolide as far as we are aware.

3-Acetyldeoxynivalenol (= Deoxynivalenol monoacetate). This trichothecene was first isolated from the culture broth of "*F. roseum* No. 117" by Yoshizawa & Morooka (1973). We have identified this strain, which was isolated from barley in Japan, as *F. graminearum* (Strain 8.22). The production of 3-acetyldeoxynivalenol by this strain has subsequently been reported by Forsyth et al. (1977), Yoshizawa & Morooka (1974, 1975a, 1977), and Yoshizawa et al. (1978, 1979). In cultures on either PSC medium (Yoshizawa & Morooka, 1975a) or on rice (Yoshizawa & Morooka, 1977; Yoshizawa et al, 1978) incubated at either 15 or 25°C, maximal growth of *F. roseum* No. 117 = ATCC 28114 (Strain 8.22) and maximal levels of 3-acetyldeoxynivalenol occurred after 10 to 14 days, followed by a decrease in the level of the monoacetate and an increase in the concentration of deoxynivalenol. The mycelium of this strain is known to be able to deacetylate 3-acetyldeoxynivalenol as well as diacetyldeoxynivalenol to deoxynivalenol (Yoshizawa & Morooka, 1975a, 1977). The levels of 3-acetyldeoxynivalenol detected in cultures of this strain on rice, some incubated at 15°C for 30 days and others at 25°C for 15 days, were 5 and 160 µg/g, respectively (Yoshizawa et al., 1978). Culture material of this strain on rice caused a marked feed refusal response in rats (Yoshizawa et al., 1978). The yield of crystalline 3-acetyldeoxynivalenol isolated from cultures of *F. graminearum* (= *F. roseum*) No. 117 (Strain 8.22) on rice incubated at 25°C for 10 to 12 days was 400 mg/kg, according to Yoshizawa & Morooka (1977).

The production of 3-acetyldeoxynivalenol by other isolates of *F. graminearum* from wheat and barley in Japan has subsequently been reported by Morooka et al. (1980) and Yoshizawa et al. (1979). According to Morooka et al. (1980), 5 of 20 isolates of *F. graminearum* produced unspecified levels of 3-acetyldeoxynivalenol, and two of these isolates also produced fusarenon-X. Yoshizawa et al. (1979) reported that 23 of 61 isolates of *F. graminearum* produced 3-acetyldeoxynivalenol in cultures in PSC medium incubated at 25°C for 2 weeks. Two of these isolates also produced fusarenon-X. We have not seen any of these Japanese isolates of *F. graminearum* reported by Morooka et al. (1980) and Yoshizawa et al. (1979) to produce 3-acetyldeoxynivalenol. We have, however, confirmed the identify of *F. graminearum* NHL-F-1118, which was also isolated from wheat in Japan and produced 3-acetyldeoxynivalenol, according to H. Kurata (personal communication, 20 April 1980), as *F. graminearum* (Strain 8.34).

Butenolide. Morooka et al. (1972) and Yoshizawa & Morooka (1973) isolated unspecified amounts of butenolide from extracts of cultures of "*F. roseum* No. 117" that were toxic to mice. We have identified this strain, which was isolated from scabby barley in Japan, as *F. graminearum* (Strain 8.22).

The production of butenolide by isolates of *F. graminearum* from wheat and barley in Japan has also been reported by Suzuki et al. (1981a, b) and Yoshizawa et al. (1979). Six isolates of *F. graminearum* produced butenolide in cultures on rice at levels of 0.09 to 4.24 µg/g as detected by GC with an electron capture detector (Suzuki et al., 1981a, b). According to Yoshizawa et al. (1979), 36 of 61 isolates of *F. graminearum* produced unspecified levels of butenolide as detected by TLC in PSC medium incubated at 25°C for 2 weeks. Among these, eight isolates produced butenolide only, four produced butenolide and zearalenone, 21 produced butenolide and trichothecenes, and three produced butenolide, trichothecenes, and zearalenone (Yoshizawa et al., 1979).

Deoxynivalenol (= Vomitoxin). The first *Fusarium* strain that was reported to produce deoxynivalenol in culture was *F. roseum* No. 117 (= ATCC 28114), which was isolated from scabby barley contaminated with deoxynivalenol and nivalenol in Japan (Morooka et al., 1972; Yoshizawa & Morooka, 1973). We have identified this strain as *F. graminearum* (Strain 8.22). The production of deoxynivalenol by this strain has subsequently been confirmed by Yoshizawa & Morooka (1974, 1975a, 1977) and Yoshizawa et al. (1978, 1979). In cultures of this strain on either PSC medium (Yoshizawa & Morooka, 1975a) or rice (Yoshizawa & Morooka, 1977; Yoshizawa et al., 1978), maximal levels of deoxynivalenol occurred subsequent to maximal vegetative growth, and the increase in the concentration of deoxynivalenol was associated with a decrease in the concentration of 3-acetyldeoxynivalenol. The mycelium of this strain is known to be able to deacetylate both 3-acetyldeoxynivalenol and diacetyldeoxynivalenol to deoxynivalenol (Yoshizawa & Morooka, 1975a; 1977). The levels of deoxynivalenol detected in cultures of this strain on rice incubated either at 15°C for 30 days or at 25°C for 15 days were 15 and 220 µg/g, respectively (Yoshizawa et al., 1978). Culture material of this strain on rice caused a marked feed refusal response in rats (Yoshizawa et al., 1978). The yield of crystalline deoxynivalenol isolated from cultures of F. graminearum (= F. roseum) No. 117 (Strain 8.22) on rice incubated at 25°C for 10 to 12 days was 100 mg/kg, according to Yoshizawa & Morooka (1977).

The production of deoxynivalenol by other isolates of F. graminearum from barley and wheat in Japan has subsequently been reported by several investigators (Ichinoe et al., 1980; Morooka et al. 1980; Suzuki et al., 1980, 1981b; Yoshizawa et al., 1979). According to Ichinoe et al. (1980), only 4 of 112 isolates of F. graminearum from different geographical areas in Japan produced deoxynivalenol, and none of these isolates produced both deoxynivalenol and nivalenol. Morooka et al. (1980) reported that 8 of 20 isolates produced deoxynivalenol and that one of these isolates produced fusarenon-X as well. Suzuki et al. (1980, 1981b) found that F. graminearum was the predominant *Fusarium* species (73.7% of 810 isolates) isolated from barley and wheat in Saitama Prefecture. Analyses of cultures of these isolates by GC-MS revealed that 4 of 57 strains of F. graminearum produced deoxynivalenol at levels ranging from 0.77 to 30.56 µg/g. The production of "deoxynivalenols" in PSC medium incubated at 25°C for 21 days by 23 of 61 isolates of F. graminearum was reported by Yoshizawa et al. (1979). These 23 isolates produced unspecified levels of 3-acetyldeoxynivalenol as detected by TLC and GLC. Although Yoshizawa et al. (1979) analyzed these cultures for 3-acetyldeoxynivalenol and not for deoxynivalenol, they apparently assumed that strains of F. graminearum which produce the acetylated metabolite also produce deoxynivalenol. According to Yoshizawa et al. (1979), two of these 23 strains produced both 3-acetyldeoxynivalenol and fusarenon-X. We have not seen any of these Japanese isolates of F. graminearum reported to produce deoxynivalenol or 3-acetyldeoxynivalenol. We have, however, confirmed the identity of F. graminearum NHL-F-1118, which was also isolated from wheat in Japan and produces deoxynivalenol as well as 3-acetyldeoxynivalenol, according to H. Kurata (personal communication, 29 April 1980), as F. graminearum (Strain 8.34).

The first *Fusarium* strain that was shown experimentally to cause feed refusal in pigs (see feed refusal and emetic syndromes, above), was F. graminearum NRRL 5883 (Vesonder et al., 1976). We have confirmed the identity of this strain, which was isolated from moldy corn in the United States, as F. graminearum (Strain 8.23). Cultures of this strain on rice incubated at 28°C for 13 days were refused by pigs, and 60 mg/kg of chromatographically pure deoxynivalenol was isolated from this material (Vesonder et al., 1976). Cultures of this strain on corn incubated under similar conditions were also refused by pigs and contained unspecified levels of deoxynivalenol as detected by TLC and IR (Vesonder et al., 1977b). Cultures on corn incubated at 28°C for 30 days contained 128 µg/g of deoxynivalenol as determined by TLC and GLC-MS, and 90 µg/g of 90% pure deoxynivalenol was isolated from this material (Vesonder et al., 1981b). This strain produced only small amounts (5 µg/g) of deoxynivalenol in cultures on wild rice (*Zizania aquatica*) incubated at 28°C for 3 weeks (Lindenfelser et al., 1978). In addition to F. graminearum NRRL 5883 (Strain 8.23), Vesonder et al. (1981b) also reported the production of deoxynivalenol by 14 other strains of F. graminearum in cultures on corn at levels ranging from 5 to 236 µg/g. We have confirmed the identity of two of these isolates, i.e., F. graminearum NRRL 2830, which produced 41 µg/g, and F. graminearum NRRL 5864, which produced 95 µg/g of deoxynivalenol (Vesonder et al., 1981b), as F. graminearum (Strains 8.19 and 8.24).

The production of deoxynivalenol in cultures on rice by four unnumbered isolates of

"*F. roseum*" from feedstuffs associated with suspected mycotoxicoses was reported by Mirocha et al. (1976). It is not known whether any of these unnumbered strains of "*F. roseum*" were in fact *F. graminearum* or whether any of them are represented in the ITFRC. Subsequently Pathre & Mirocha (1978) reported that three isolates of *F. graminearum* from feedstuffs associated with mycotoxicoses in farm animals produced deoxynivalenol at levels of 125 to 320 µg/g in cultures on rice incubated at 25°C for 7 days followed by 21 days at 12°C. We have confirmed the identity of two of these isolates (*F. graminearum* F-59 and 7137) as *F. graminearum* (Strains 8.38 and 8.45). *Fusarium graminearum* F-59 (Strain 8.38), which was originally isolated in Hungary and has been reported to cause feed refusal and emesis in pigs (Palyusik, 1973), produced 320 µg/g of deoxynivalenol in cultures on rice and 15 mg/ℓ in PSC medium incubated at 21°C for 21 days (Pathre & Mirocha, 1978). *Fusarium graminearum* 7137 (Strain 8.45), which was originally isolated in Finland and has been reported to cause weight loss in rats (Korpinen et al., 1972), produced 125 µg/g of deoxynivalenol in cultures on rice and none in PSC medium (Pathre & Mirocha, 1978).

Two isolates of *F. graminearum* (MRC 121 and MRC 460) from Zambian and South African corn naturally contaminated with deoxynivalenol have been reported to produce deoxynivalenol in cultures on corn incubated at 25°C for 2 weeks followed by 6 weeks at 12°C (Marasas et al., 1977). We have confirmed the identity of these two isolates as *F. graminearum* (Strains 8.42 and 8.43). The culture material of these two strains caused feed refusal and mortality in rats and contained deoxynivalenol (6.0 and 1.3 ppm, respectively) as determined by GC-MS.

Four isolates of *F. graminearum* from Austrian corn naturally contaminated with deoxynivalenol have been reported to produce deoxynivalenol in cultures on corn incubated at 28°C for 13 days (Vesonder & Ciegler, 1979). We have confirmed the identity of three of these isolates (*F. graminearum* NRRL 6450, 6451, and 6452) as *F. graminearum* (Strains 8.25, 8.26, and 8.27). The levels of deoxynivalenol in the cultures of these three isolates as analyzed by GLC and confirmed by GC-MS were 11.0, 16.7, and 3.7 µg/g, respectively (Vesonder & Ciegler, 1979). Four unnumbered isolates of *G. zeae* from Austrian corn naturally contaminated with deoxynivalenol were also reported to produce deoxynivalenol in cultures on corn incubated at 27°C for 4 weeks by Bottalico et al. (1981). The analyses were done by TLC and GC, and levels of deoxynivalenol ranging from 8.8 to 16.8 µg/g were detected in the culture material of these four strains (Bottalico et al., 1981).

Two Italian isolates of *F. graminearum* (ITM 2 and 126) have been reported to produce deoxynivalenol in cultures on corn incubated at 27°C for 4 weeks by Palmisano et al. (1981). We have confirmed the identity of *F. graminearum* IMT 126 as *F. graminearum* (Strain 8.49). The levels of deoxynivalenol in the culture material of this strain were 1.20 and 1.18 µg/g as detected by GC and differential-pulse polarography, respectively (Palmisano et al. 1981).

Nine isolates of *F. graminearum* from Canadian wheat and barley naturally contaminated with deoxynivalenol have been reported to produce deoxynivalenol, together with zearalenone, in cultures on grains (wheat, barley, oats, or rice) incubated either at 25°C for 1 week followed by 2 weeks at 10°C and a further 3 to 4 weeks at 4°C, or at 25°C for 15 days (Neish & Cohen, 1981). Analyses of the culture material of these strains were done by capillary GLC, and levels of deoxynivalenol ranging from 0.1 to 6.8 ppm were found. Five of the isolates from wheat (DAOM 177406 to 177410) were also cultured on rice at 28°C for 14 days and analysed by R.F. Vesonder, who found deoxynivalenol at levels ranging from 4.0 to 16.5 µg/g (Neish & Cohen, 1981). Six isolates of *F. gramine-*

arum from Canadian wheat naturally contaminated with deoxynivalenol have also been reported to produce unspecified levels of deoxynivalenol by Trenholm et al. (1981).

Deoxynivalenol has also been reported to be produced in corn inoculated in the field with cultures of *F. graminearum*. Hart & Stebbins (1981) reported that three unnumbered isolates of *G. zeae* produced deoxynivalenol at levels up to 500 µg/g in 33 commercial lines of sweet corn inoculated artificially in the field. We have confirmed the identity of these three isolates (*G. zeae* B-507, B-601, and W-8) which all produce from 100 to 400 ppm of deoxynivalenol in sweet corn according to L.P. Hart (personal communication, September 1981), as *F. graminearum* (Strains 8.46, 8.47 and 8.48). Corn inoculated with "Fusaria cultures (NRRL 6206 and NRRL 6207)" was found to contain 20 ppm of deoxynivalenol, together with zearalenone and a new trichothecene, 3,15-dihydroxy-12,13-epoxytrichothec-9-ene-8-one, by Bennett et al. (1981). These two "Fusaria cultures" were *F. graminearum* NRRL 6206 and NRRL 6207, which had both been reported to produce deoxynivalenol (Vesonder et al., 1981b) as well as zearalenone in cultures on corn (Shannon et al., 1980; Vesonder et al., 1981b). According to Young et al. (1981), corn inoculated in the field "with *Fusarium*" contained 37.5 ppm of deoxynivalenol and 19.5 to 39.0 ppm of zearalenone. The identity of this unnumbered and unspecified *Fusarium* strain is not known to us.

In view of the extensive literature on deoxynivalenol production by *F. graminearum*, we conclude that *F. graminearum* is the most important *Fusarium* sp. known to produce deoxynivalenol. Approximately 100 isolates of *F. graminearum* have been reported to produce deoxynivalenol in cultures at levels ranging from 0.1 to 320 mg/kg. Sixteen isolates of *F. graminearum* represented in the ITFRC (Table 8.1) have been reported to produce deoxynivalenol in culture at levels ranging from 1.2 to 320 mg/kg. The highest known deoxynivalenol-producing strain in the ITFRC is *F. graminearum* F-59 (Strain 8.38).

Diacetoxyscirpenol. Ueno et al. (1973a) reported that diacetoxyscirpenol was detected by TLC in cultures in PSC medium incubated at 25 to 27°C for 2 weeks of *inter alia* the following three *Fusarium* strains: *F. rigidiusculum* M-1-3, *F. tricinctum* Abashiri-2, and *F. roseum* "Culmorum" 70-K-11. We have identified these three strains as *F. graminearum* (Strains 8.29, 8.33, and 8.40). The production of diacetoxyscirpenol by *F. graminearum* (= *F. roseum*) 70-K-11 (Strain 8.40), which was isolated from corn in New Guinea, has also been reported by Ueno et al. (1972b).

According to Takatori et al. (1980), two isolates of *F. graminearum* from hay and rice straw in Hokkaido, Japan, were dermotoxic to rabbit skin, and on the basis of the analyses of extracts of cultures of these two strains by TLC, "the production of T-2 toxin, neosolaniol and diacetoxyscirpenol was presumed."

The production of diacetoxyscirpenol in cultures on rice by three unnumbered isolates of "*F. roseum*" from feedstuffs associated with suspected mycotoxicoses was reported by Mirocha et al. (1976). It is not known to us whether any of these unnumbered isolates of "*F. roseum*" were in fact *F. graminearum* or whether any of them are represented in the ITFRC.

Negative results for diacetoxyscirpenol production by *F. graminearum* have been reported for four Italian isolates by Bottalico et al. (1981), two isolates from corn in southern Africa (Strains 8.42 and 8.43) by Marasas et al. (1977), and 57 isolates from barley and wheat in Japan by Suzuki et al. (1980, 1981b).

In conclusion, the production of diacetoxyscirpenol by *F. graminearum* is equivocal at present and needs to be confirmed by appropriate analytical methods in authenticated strains of *F. graminearum*.

Diacetyldeoxynivalenol (= Deoxynivalenol diacetate). This trichothecene was first isolated from cultures on rice incubated at 25°C for 10 to 12 days of "*F. roseum* No. 117" (Yoshizawa & Morooka, 1977). We have identified this strain, which was isolated from barley in Japan, as *F. graminearum* (Strain 8.22). The production of diacetyldeoxynivalenol by this strain has subsequently also been confirmed by Yoshizawa et al. (1978, 1979). The yield of crystalline diacetyldeoxynivalenol isolated from cultures of *F. graminearum* (= *F. roseum*) No. 117 (Strain 8.22) on rice incubated at 25°C for 10 to 12 days was 10 mg/kg, according to Yoshizawa & Morooka (1977).

3,15-Dihydroxy-12,13-epoxytrichothec-9-ene-8-one. This new trichothecene was first isolated, from corn inoculated in the field with "Fusaria cultures (NRRL 6206 and 6207)." We have not seen these two "Fusaria cultures," but according to Shannon et al. (1980) and Vesonder et al. (1981b), these two strains are *F. graminearum* NRRL 6206 and 6207. This compound was found to co-migrate with deoxynivalenol on TLC plates and was separated from deoxynivalenol isolated from the inoculated corn by means of HPLC (Bennett et al., 1981).

Fusarenon-X (= 4-Acetylnivalenol). The production of fusarenon-X by *G. zeae* Ohoita II, which was isolated from barley in southern Japan, was reported by Ueno et al. (1971a, 1972b, 1973a). Extracts of cultures of this strain in PSC medium incubated at 25 to 27°C for 2 weeks were toxic to mice, and unspecified levels of fusarenon-X, together with nivalenol, were detected in these extracts by TLC.

The production of fusarenon-X by other isolates of *F. graminearum* from barley or wheat in Japan has subsequently been reported by several investigators (Morooka, et al., 1980; Suzuki et al., 1980, 1981b; Yoshizawa et al., 1979). According to Morooka et al. (1980), 11 of 20 isolates of *F. graminearum* produced fusarenon-X and four of these isolates produced this compound together with either deoxynivalenol or 3-acetyldeoxynivalenol. Suzuki et al. (1980, 1981b) found that *F. graminearum* was the predominant *Fusarium* species (73.3% of 810 isolates) isolated from barley and wheat in Saitama Prefecture. Analyses of cultures of these isolates by GC-MS revealed that 43 of 57 isolates of *F. graminearum* produced fusarenon-X (together with nivalenol) at levels ranging from 0.06 to 10.80 µg/g. The production of fusarenon-X by 22 of 61 isolates of *F. graminearum* in PSC medium incubated at 25°C for 2 weeks was reported by Yoshizawa et al. (1979). These 22 isolates produced unspecified levels of fusarenon-X as detected by TLC and GLC, and two of these isolates also produced 3-acetyldeoxynivalenol.

We have not seen any of the above 77 Japanese isolates of *F. graminearum* reported to produce fusarenon-X by Morooka et al. (1980), Suzuki et al. (1980, 1981b), Ueno et al. (1971a, 1972b, 1977a), and Yoshizawa et al. (1979). We have, however, confirmed the identity of *F. graminearum* NHL-F-1104 and NHL-F-1112, both of which were isolated from barley in Japan and produce fusarenon-X (together with nivalenol) according to H. Kurata (personal communication, 28 April 1980), as *F. graminearum* (Strains 8.35 and 8.36).

At present, *F. graminearum* strains that produce fusarenon-X are known only from Japan. Two isolates from the United States that have been analyzed for fusarenon-X proved to be negative, i.e., *G. zeae* NRRL 2830 (Ueno et al., 1971a, 1973a) and *F. graminearum* NRRL 5883 (Vesonder et al., 1977b). We have confirmed the identity of both of these isolates that do not produce fusarenon-X as *F. graminearum* (Strains 8.19 and 8.23).

HT-2 Toxin. Ueno et al. (1972b, 1977a) detected HT-2 toxin by TLC in cultures of *F. roseum* "Culmorum" 70-K-11 in PSC medium incubated at 25 to 27°C for 2 weeks. We have identified this strain, which was isolated from corn in New Guinea according to Ueno et al. (1972b), as *F. graminearum* (Strain 8.40). We have also identified *F. tricinctum* Abashiri-2, which was isolated in Japan and reported to produce HT-2 toxin by Ueno et al. (1973a), as *F. graminearum* (Strain 8.33).

Negative results for HT-2 toxin production by *F. graminearum* have been reported for three Hungarian isolates, including *F. graminearum* F-59 (Strain 8.38) and *F. graminearum* (= *F. culmorum*) F-79 (Strain 8.39), by Szathmary et al. (1976, 1977); for two Japanese isolates, *G. zeae* Ohoita II and *G. zeae* Ishii (Strain 11.28), by Ueno et al. (1973a); and for two isolates from the United States, i.e., *G. zeae* NRRL 2830 (Strain 8.19) by Ueno et al. (1973a) and *F. graminearum* NRRL 5883 (Strain 8.23) by Vesonder et al. (1977b). No HT-2 toxin could be detected by Kotsonis et al. (1975a) in isolates of *F. graminearum* from corn in the United States that caused feed refusal in pigs and rats.

Monoacetoxyscirpenol. According to Mirocha et al. (1976), an unnumbered isolate of "*F. roseum*" from India produced monoacetoxyscirpenol in cultures on rice. We have not seen this strain and consequently it has not been possible to determine its identity.

Negative results for monoacetoxyscirpenol production by *F. graminearum* have been reported for two isolates from southern Africa, i.e., *F. graminearum* MRC 121 and MRC 460 (Strains 8.42 and 8.43), by Marasas et al. (1977).

Neosolaniol. Ueno et al. (1973a) detected neosolaniol by TLC in cultures in PSC medium incubated at 25 to 27°C for 2 weeks of *inter alia* the following three *Fusarium* strains: *F. rigidiusculum* M-1-3, *F. tricinctum* Abashiri-2, and *F. roseum* "Culmorum" 70-K-11. We have identified these three strains as *F. graminearum* (Strains 8.29, 8.33, and 8.40). The production of neosolaniol by *F. graminearum* (= *F. roseum*) 70-K-11 has also been reported by Ueno et al. (1972b, 1975). We have also identified *F. tricinctum* A-R-5, which was reported to produce neosolaniol in shake culture by Ueno et al. (1975), as *F. graminearum* (Strain 8.32).

According to Takatori et al. (1980), two isolates of *F. graminearum* from hay and rice straw in Hokkaido, Japan, were dermotoxic to rabbit skin, and on the basis of the analyses of extracts of cultures of these two strains by TLC, "the production of T-2 toxin, neosolaniol, and diacetoxyscirpenol was presumed."

Negative results for neosolaniol production by *F. graminearum* have been reported for three Hungarian isolates, including *F. graminearum* F-59 (Strain 8.38) and *F. graminearum* (= *F. culmorum*) F-79 (Strain 8.39), by Szathmary et al. (1976, 1977); for three isolates of *G. zeae*, i.e., Ohoita II, Ishii (Strain 8.28), and NRRL 2830 (Strain 8.19), by Ueno et al. (1973a); and for 57 Japanese isolates of *F. graminearum* by Suzuki et al. (1980, 1981b).

Nivalenol. The production of nivalenol by *G. zeae* Ohoita II, which was isolated from barley in southern Japan, was reported by Ueno et al. (1971a, 1972b, 1973a). Extracts of cultures of this strain in PSC medium incubated at 25 to 27°C for 2 weeks were toxic to mice, and unspecified levels of nivalenol (together with fusarenon-X) were detected in these extracts by TLC.

The production of nivalenol by other isolates of *F. graminearum* from barley and wheat in Japan has subsequently been reported by several investigators (Ichinoe et al., 1980; Suzuki et al., 1980, 1981b; Yoshizawa et al., 1979). According to Ichinoe et al. (1980), 66 of 112 isolates of *F. graminearum* from different geographical areas in Japan produced nivalenol, and none of these isolates produced both nivalenol and deoxynivalenol. Suzuki et al. (1980, 1981b) found that *F. graminearum* was the predominant *Fusarium* species (73.7% of 810 isolates) isolated from barley and wheat in Saitama Prefecture. Analyses of cultures of these isolates by GC-MS revealed that 43 of 57 isolates of *F. graminearum* produced nivalenol (together with fusarenon-X) at levels ranging from 0.09 to 16.85 µg/g. The production of "nivalenols" in PSC medium incubated at 25°C for 21 days by 22 of 61 isolates of *F. graminearum* has been reported by Yoshizawa et al. (1979). These 22 isolates produced unspecified levels of fusarenon-X (= 4-acetylnivalenol) as detected by TLC and GLC. Although Yoshizawa et al. (1979) analyzed these cultures for fusarenon-X and not for nivalenol, they apparently assumed that strains of *F. graminearum* that produce the acetylated metabolite also produce nivalenol. According to Yoshizawa et al. (1979), two of these isolates of *F. graminearum* produced both fusarenon-X and 3-acetyldeoxynivalenol.

We have not seen any of the above 132 Japanese isolates of *F. graminearum* reported to produce nivalenol by Ichinoe et al. (1980), Suzuki et al. (1980, 1981b), Ueno et al. (1971a, 1972b, 1977a), and Yoshizawa et al. (1979). We have, however, confirmed the identity of *F. graminearum* NHL-F-1104 and NHL-F-1112, both of which were isolated from barley in Japan and produce nivalenol (together with fusarenon-X) according to H. Kurata (personal communication, 29 April 1980), as *F. graminearum* (Strains 8.35 and 8.36).

At present, *F. graminearum* strains that produce nivalenol are known only from Japan, where approximately 132 of 232 isolates tested have been reported to produce this compound.

T-2 Toxin. Ueno et al. (1973a) detected T-2 toxin by TLC in cultures in PSC medium incubated at 25 to 27°C for 2 weeks of *inter alia* the following three *Fusarium* strains: *F. rigidiusculum* M-1-3, *F. tricinctum* Abashiri-2, and *F. roseum* "Culmorum" 70-K-11. We have identified these three strains as *F. graminearum* (Strains 8.29, 8.33, and 8.40). The production of T-2 toxin by *F. graminearum* (= *F. roseum*) 70-K-11 has also been reported by Ueno et al. (1972b, 1975). We have also identified *F. tricinctum* A-R-5, which was reported to produce T-2 toxin in shake culture by Ueno et al. (1975), as *F. graminearum* (Strain 8.32).

According to Burmeister et al. (1972), one out of six unnumbered isolates of *G. zeae* from corn in the United States produced unspecified levels of T-2 toxin as detected by TLC in cultures on corn grits incubated at 15°C for 21 days.

Joffe & Palti (1975) identified *F. tricinctum* NRRL 5908, which was isolated from fescue hay in the United States, as *F. sporotrichioides*. This strain, as *F. sporotrichioides* NRRL 5908, has been reported to produce T-2 toxin (0.6 mg/10 g) as determined by GLC in cultures on wheat incubated at 12°C for 21 days by Joffe (1978b) and Joffe & Yagen (1977). We have identified this strain as *F. graminearum* (Strain 8.41).

According to Takatori et al. (1980), two isolates of *F. graminearum* from hay and rice straw in Hokkaido, Japan, were dermotoxic to rabbit skin, and on the basis of the analyses of extracts of cultures of these two strains by TLC, "the production of T-2 toxin, neosolaniol, and diacetoxyscirpenol was presumed."

Hitokoto et al. (1981) reported that "a T-2 toxin–producing strain of *Fusarium graminearum* F-135 isolated from wheat flour" produced T-2 toxin in cultures on various types of beans and peas incubated at 25°C. The highest level of T-2 toxin (30.2 µg/g) as determined by TLC was produced by this strain in cultures on green peas.

Negative results for T-2 toxin production by *F. graminearum* have been reported for three Hungarian isolates, including *F. graminearum* F-59 (Strain 8.38) and *F. graminearum*

(= *F. culmorum*) F-79 (Strain 8.39), by Szathmary et al. (1976, 1977); for six Italian isolates by Bottalico (1977c); for two Japanese isolates, i.e., *G. zeae* Ohoita II and Ishii (Strain 8.28), by Ueno et al. (1973a); for 57 isolates of *F. graminearum* from wheat and barley in Japan by Suzuki et al. (1980, 1981b); for two isolates from southern Africa, i.e., *F. graminearum* MRC 121 and 460 (Strains 8.42 and 8.43), by Marasas et al. (1977); and for eight isolates from the United States by Burmeister et al. (1972), Marasas et al. (1971), Ueno et al. (1973a), and Vesonder et al. (1977b). The isolates from the United States that have been reported not to produce T-2 toxin included *F. graminearum* NRRL 2830 (Ueno et al., 1973a) and NRRL 5883 (Vesonder et al., 1977b). We have confirmed the identity of both of these isolates as *F. graminearum* (Strains 8.19 and 8.23). In addition, no T-2 toxin could be detected by Kotsonis et al. (1975a) in isolates of *F. graminearum* from corn in the United States that caused feed refusal in pigs and rats. Negative results for T-2 toxin production have also been reported for 63 zearalenone-producing isolates of *F. graminearum* from corn in the United States (Cullen, 1981).

In conclusion, the production of T-2 toxin by *F. graminearum* is equivocal at present and needs to be confirmed by appropriate analytical methods in authenticated strains of *F. graminearum*.

Zearalenone (= **FES** = **F-2**). The estrogenic metabolite zearalenone was first isolated from cultures on corn incubated at 24°C for 2 to 3 weeks of an isolate of *G. zeae* from corn associated with field outbreaks of porcine hyperestrogenism (see estrogenic syndrome, above) in the United States (Stob et al., 1962). Culture material of this strain caused vulvar and mammary enlargement in immature pigs and also induced uterine hypertrophy in ovariectomized mice. The above metabolite was isolated and partially characterized from this culture material of an unnumbered isolate of *G. zeae* by Stob et al. (1962). The strain designated in United States patents on the estrogenic and anabolic activity of the active metabolite (FES = Fermentation Estrogenic Substance = zearalenone) was "*Gibberella zeae* NRRL 2380" (should read 2830) according to Mirocha & Christensen (1974) and Mirocha et al. (1971), or *F. roseum* NRRL 2830 according to Vesonder et al. (1981b). We have confirmed the identity of this strain as *F. graminearum* (Strain 8.19). Subsequent to the original isolation of zearalenone from this strain by Stob et al. (1962), several other authors have also reported zearalenone production by *F. graminearum* (= *F. moniliforme* = *F. roseum*) NRRL 2830 (Strain 8.19) (Bolliger & Tamm, 1972; Hesseltine, 1972; Hidy et al., 1977; Ueno et al., 1971a; Vesonder et al., 1981b). Maximal yields of zearalenone by this strain were obtained by Bolliger & Tamm (1972) in cultures on corn incubated at 12°C for 10 weeks. These authors isolated 11.6 g of zearalenone from 7.5 kg of culture material on corn, and also isolated several related metabolites from this material.

In addition to *F. graminearum* NRRL 2830 (Strain 8.19), we have also identified the following *Fusarium* strains that have been reported to have estrogenic effects in pigs and rats and to produce zearalenone as *F. graminearum* (Strains 8.20, 8.21, 8.38, 8.39, 8.42, 8.43, and 8.45):

1) *Fusarium* No. 5 (Strain 8.20). Cultures on corn of this strain incubated at 20 to 25°C for 21 days followed by 14 days at 12°C caused marked increases in the weight of the uterus of virgin weanling rats, and an unspecified amount of the estrogenic metabolite zearalenone (= F-2) was isolated from this culture material by Christensen et al. (1965).

2) *F. roseum* 'Graminearum' Mapleton No. 10 (Strain 8.21). Culture material of this strain on corn caused vulvar and mammary enlargement, rectal prolapse, and uterine hypertrophy in sexually immature female pigs (Christensen et al., 1972b; Nelson et al., 1973). This

strain of *F. graminearum* is known to produce large amounts of zearalenone in culture (Eugenio et al., 1970a; Ishii et al., 1974; Mirocha & Christensen, 1974; Mirocha & Pathre, 1979; Mirocha et al., 1967a, b, 1968a, b, 1971; Scott et al., 1970; Sherwood & Peberdy, 1972, 1974a, b; Steele et al., 1974, 1976, 1977a, b; Wolf & Mirocha, 1973, 1977). Maximal yields of zearalenone reported for this strain (as *F. roseum* Isolate No. 1 from corn, Mapleton) by Eugenio et al. (1970a) were 3,087 and 2,585 µg/g on parboiled polished rice and corn, respectively, incubated for 2 weeks at 24 to 27°C followed by 8 weeks at 12°C. Mirocha et al. (1967a, b) reported yields of 1,600 µg/g of zearalenone in cultures of "*F. graminearum* No. 10" in cultures on corn incubated for 2 weeks at 25 to 28°C followed by 8 weeks at 12°C. According to Mirocha et al. (1971), yields as high as 30,000 to 60,000 ppm have been obtained in cultures of "*F. graminearum* Mapleton No. 10" on rice or corn amended with glucose. According to Mirocha & Christensen (1974) yields as high as 38,000 to 40,000 ppm have been obtained in cultures of "*F. roseum* Mapleton 10" on solid media under optimal incubation conditions. The maximal yield of zearalenone by this strain (as *F. graminearum* IMI 155426) obtained by Sherwood & Peberdy (1972) was 1,068 µg/g in cultures on wheat at 37% moisture content incubated at 25°C for 10 weeks followed by 6 weeks at 12°C. Sherwood & Peberdy (1974a) reported that *F. graminearum* IMI 155426 produced 1976 µg/g of zearalenone in cultures on grains at 31% moisture content incubated at 12°C for 4 months. Wolf & Mirocha (1973, 1977) stated that *F. roseum* 'Graminearum' JRL MAP 10 (ATCC 24689) was a subculture of a mass isolate designated as *F. roseum* 'Graminearum' Mapleton No. 10, and was a high producer (> 15.0 mg/g) of zearalenone in culture on a corn-rice medium. These authors also stated that *F. roseum* 'Graminearum' ascospore No. 55 (ATCC 24688) was a single ascospore isolate from the mass isolate *F. roseum* 'Graminearum' Mapleton No. 10, and was a low producer (< 1.5 mg/g) of zearalenone in culture on a corn-rice medium. Ishii et al. (1974) isolated 50 to 130 mg/kg of crystalline zearalenone from cultures of "*F. roseum* Map. 10 (the strain from C.J. Mirocha)" on rice incubated at 25°C for 14 days followed by 14 days at 12 to 15°C.

3) *F. graminearum* F-59 = ATCC 34909 (Strain 8.38). Culture material on corn of this strain (unnumbered) caused uterine hypertrophy in pigs (Palyusik, 1973). Cultures on millet containing 2% malt extract incubated at 23 to 27°C for 1 week followed by 2 weeks at 5 to 8°C and 1 week at 23 to 27°C caused uterine hypertrophy in rats and contained 200 to 730 µg/g of zearalenone as determined by TLC and GLC (Szathmary et al., 1976, 1977).

4) *F. culmorum* F-79 = ATCC 34910 (Strain 8.39). Cultures of this strain of *F. graminearum* on millet containing 2% malt extract incubated at 23 to 27°C for 1 week followed by 2 weeks at 5 to 8°C and 1 week at 23 to 27°C caused uterine hypertrophy in rats and contained 540 to 3,400 µg/g of zearalenone as determined by TLC and GLC (Szathmary et al., 1976, 1977). This strain has also been reported to produce zearalenone by Kovacs et al. (1975), Palyusik & Koplik-Kovacs (1975a, b), and Sandor et al. (1980).

5) *F. graminearum* MRC 121 (Strain 8.42). Cultures of this strain on corn incubated at 25°C for 2 weeks followed by 6 weeks at 12°C caused uterine hypertrophy in rats and contained 1,270 ppm of zearalenone as determined by TLC (Marasas et al., 1977).

6) *F. graminearum* MRC 460 (Strain 8.43). Cultures of this strain on corn incubated at 25°C for 2 weeks followed by 6 weeks at 12°C caused uterine hypertrophy in rats and contained 1,140 ppm of zearalenone as determined by TLC (Marasas et al., 1977).

7) *F. graminearum* 7137 (Strain 8.45). This strain (as *F. graminearum* No. 13) was isolated from a commercial feed mix in Finland (Korpinen et al., 1972). According to A. Ylimäki (personal communication, 10 April 1979), the strain referred to as *F. graminearum* No. 13 by Korpinen et al. (1972) is *F. graminearum* 7137 (Strain 8.45). Cultures on a

mixture of oats, barley, and wheat were incubated at room temperature for one month. This material was fed to immature female rats for 7 days and caused marked increases in uterine weight (Korpinen et al., 1972). The zearalenone content of this culture material was not determined by Korpinen et al. (1972). Subsequently Kallela & Korpinen (1973) reported that one of the *F. graminearum* strains, which was "found to cause clear estrogenic changes in rats" by Korpinen et al. (1972), produced zearalenone in culture. Since *F. graminearum* No. 13 caused the greatest increase in uterine weight of all the strains tested by Korpinen et al. (1972), it is possible that the unnumbered strain of *F. graminearum* found to produce zearalenone by Kallela & Korpinen (1973) may have been isolate No. 13. The levels of zearalenone detected by Kallela & Korpinen (1973) in cultures of the unnumbered Finnish isolate of *F. graminearum* on either barley grains, oat grains, or wheat bran incubated at room temperature for 5 weeks were 407, 45, and 3.2 µg/g, respectively.

Kotsonis et al. (1975a) reported that four isolates of *F. roseum* 'Graminearum' from corn in the United States had estrogenic effects in pigs and rats and also caused feed refusal in both species. Analyses by TLC of cultures of these four isolates on corn incubated at room temperature for 1 week followed by 4 weeks at 12°C revealed the presence of zearalenone at levels ranging from 13.8 to 47.8 µg/g.

Conflicting results have been reported in the literature concerning the optimal conditions for the production of zearalenone in culture by strains of *F. graminearum*. Eugenio et al. (1970a) found that "*F. roseum*, Isolate No. 1 from corn from Mapleton, Minnesota" was the highest producer (1,900 ppm) among 43 isolates of *F. roseum* tested. According to C.J. Mirocha (personal communication, 6 March 1978), this is the same isolate of "*F. roseum* 'Graminearum' Mapleton No. 10" (Mirocha & Christensen, 1974; Mirocha & Pathre, 1979; Mirocha et al., 1971; Wolf & Mirocha 1973, 1977) that we have identified as *F. graminearum* (Strain 8.21). This strain produced only small amounts of zearalenone in cultures incubated at a constant temperature of 24 to 27°C and maximal yields (3,087 µg/g) were obtained in cultures on parboiled polished rice incubated for 2 weeks at 24 to 27°C followed by 8 weeks at 12°C (Eugenio et al., 1970a). Maximal yields of zearalenone (1,068 µg/g) by this strain (as *F. graminearum* IMI 155426) have been obtained by Sherwood & Peberdy (1972) in cultures on wheat incubated at 25°C for 4 weeks followed by 6 weeks at 12°C. On the basis of these results, which were obtained for the most part with a single strain of *F. graminearum* (Strain 8.21), the routine method for zearalenone production developed at the University of Minnesota involved the incubation of cultures on rice or corn at 24 to 27°C for 1 to 2 weeks followed by 4 to 6 weeks at 12–14°C (Mirocha & Christensen, 1974; Mirocha & Pathre, 1979; Mirocha et al., 1967a, b, 1968a, 1971, 1977a, b). However, much higher levels of zearalenone (8,375 ppm) have been obtained in cultures of *F. graminearum* (= *F. roseum* 'Graminearum') Mapleton No. 10 (Strain 8.21) incubated first at 24 to 27°C, then at 12 to 14°C and again at 27°C (Mirocha et al., 1968a, 1971). An enhancing effect on zearalenone production by incubating the cultures at 25°C for 14 days following the exposure to 12 to 15°C has also been reported by Ishii et al. (1974) for *F. roseum* M-3-2, which we have identified as *F. graminearum* (Strain 8.30). Moreover, incubation at low temperatures has not been found to have an enhancing effect on zearalenone production by certain other isolates of *F. graminearum*. Thus Schroeder & Hein (1975) reported that *F. roseum* S-74-1c, which was isolated from sorghum naturally contaminated with zearalenone in Texas, USA, produced more zearalenone in cultures on corn as well as on sorghum incubated at a constant temperature of 25°C (1,593 and 3,030 mg/kg, respectively), than in cultures incubated at 25°C followed by 10°C (499 and 300 mg/kg, respectively). Similarly Naik et al. (1978) found that four out of five isolates of *F. graminearum* from corn in Ontario, Canada, produced more zearalenone in cultures on corn incubated at a constant temperature of 25°C for 6 weeks (up to 571 µg/g) than in cultures incubated at 25°C for 2 weeks followed by 4 weeks at either 10 (up to 18 µg/g) or 15°C (up to 31 µg/g). Greatly increased yields of zearalenone (up to 60,000 ppm according to Mirocha et al., 1971) have also been obtained in cultures of *F. graminearum* on corn or rice supplemented with either glucose (Mirocha et al., 1968a, 1971) or 1% peptone (Ishii et al., 1974). It seems that the routine method for zearalenone production (Mirocha & Christensen, 1974; Mirocha & Pathre, 1979; Mirocha et al., 1967a, b, 1968a, 1971, 1977a, b), which has been used by numerous investigators in screening *Fusarium* isolates for zearalenone production, does not necessarily provide optimal conditions for zearalenone production by all isolates of *F. graminearum*.

In liquid media, isolates of *F. graminearum* have been reported to produce only trace amounts of zearalenone (Eugenio et al., 1970a; Ishii et al., 1974; Mirocha & Pathre, 1979). However, Bacon et al. (1977) screened seven isolates of *F. graminearum* from corn and sorghum in the United States for zearalenone production in a liquid starch-glutamate medium. The zearalenone yields obtained from six of these isolates in stationary cultures in this medium incubated at 24 to 28°C for 2 weeks were 22 to 86 mg/ℓ. The same strains produced zearalenone at levels ranging from 0.090 to 0.596 mg/g as determined by TLC in cultures on corn incubated at 25°C for 2 weeks followed by 6 weeks at 12°C (Bacon et al., 1977). Chemically produced mutants selected from *G. zeae* ATCC 20028 have been found to produce large amounts of zearalenone in surface as well as submerged fermentations in suitable liquid media (Hidy et al., 1977). In the industrial production of zearalenone in submerged cultures in fermentors at 24°C, yields in excess of 20g/ℓ have been obtained with these mutant strains of *F. graminearum* (Hidy et al., 1977).

The observation that there is an association between the production of perithecia and zearalenone by isolates of *F. graminearum* (Eugenio et al., 1970a, Caldwell et al., 1970) led to the discovery that zearalenone regulates sexual reproduction in *F. graminearum* (Inaba & Mirocha, 1979; Wolf & Mirocha, 1973, 1977; Wolf et al., 1972). Thus low amounts of zearalenone (1 to 10 ng/g of culture) enhance and high amounts (10 to 100 µg/g of culture) inhibit perithecial formation (Wolf & Mirocha, 1973, 1977). It follows that isolates of *F. graminearum* that produce high levels of zearalenone in culture should not produce perithecia, and this appears to be the case (Inaba & Mirocha, 1979; Bacon et al., 1977; Wolf & Mirocha, 1973). The difference in perithecial production by the two populations of *F. graminearum* designated as Groups 1 and 2 by Francis & Burgess (1977) may be due to an inherent difference in zearalenone-producing ability. Members of *F. graminearum* Group 1 may not form perithecia in culture because they are high producers of zearalenone, and conversely the members of Group 2 may form perithecia in culture because they are low producers of zearalenone. This possibility has not yet been tested experimentally.

Caldwell et al. (1970) screened 110 *Fusarium* strains isolated in the United States and belonging to seven species for zearalenone production by incubating cultures on corn at 16°C for 3 weeks. Analyses of these cultures by TLC revealed that 21 of 23 isolates of *F. roseum* 'Graminearum' produced zearalenone at levels ranging from 0.2 to 230 µg/g. The production of zearalenone by 63 of 63 isolates of *F. graminearum* from corn in the United States was reported by Cullen (1981). Zearalenone yields of up to 2,500 µg/plate were obtained in cultures on nutrient broth–moistened vermiculite incubated at 19°C for 3 weeks. Vesonder et al. (1981b) found that 15 of 15 isolates of *F. graminearum* from corn in the United States produced zearalenone, together with deoxynivalenol, in cultures on corn incubated at 28°C for 30 days. Zearalenone was detected by TLC but was not quantitated. However, Vesonder et al. (1981b) isolated 40 µg/g of zearalenone from the culture material of *F. graminearum* NRRL 5883. We have confirmed the identity of this isolate as *F. graminearum* (Strain 8.23). We have also identified two other strains reported to produce zearalenone by Vesonder et al. (1981b), i.e., *F. graminearum* NRRL 2830 and NRRL 5864, as *F. graminearum* (Strains 8.19 and 8.24). The first of these two strains, *F. graminearum* NRRL 2830, is the strain from which zearalenone was first isolated by Stob et al. (1962), and zearalenone production by this strain has subsequently also been reported by several investigators (see Strain 8.19). The second strain, *F. graminearum* NRRL 5864 (Strain 8.24), was isolated in the United States and has also been reported to produce zearalenone in cultures on either wild rice (*Zizania aquatica*) or sorghum incubated at 25°C for 38 days at levels up to 220 and 1500 µg/g, respectively, by Lindenfelser et al. (1978). This strain has also been found to produce zearalenone at levels ranging from 1 to 46 mg/kg as detected by TLC in cultures on grain from different corn hybrids and inbreds incubated at 25°C for 3 weeks followed by 2 weeks at 15°C (Shannon et al., 1980).

Caldwell & Tuite (1970) found that corn inoculated in the field with *F. graminearum* in the United States contained only low levels of zearalenone. Thus four isolates of *F. roseum* 'Graminearum' that produced 50 to 1,100 ppm of zearalenone in cultures on corn incubated at 24°C for 2 weeks followed by 8 weeks at 12°C produced only 0.2 to 5.0 ppm of zearalenone in corn ears inoculated artificially in the field. Hart & Stebbins (1981) found that zearalenone and deoxynivalenol were produced in 33 commercial corn hybrids inoculated in the field with *G. zeae* in the United States. All three isolates of *G. zeae* produced deoxynivalenol in inoculated corn ears, but only *G. zeae* W-8 produced zearalenone. We have confirmed the identity of *G. zeae* W-8, which produced more than 180 µg/g of zearalenone in artificially inoculated corn ears, as *F. graminearum* (Strain 8.48). Hart & Stebbins (1981) as well as Caldwell & Tuite (1970) found no apparent correlation between the severity of infection and zearalenone content. However, a correlation between disease severity and zearalenone content in two corn hybrids and four inbreds inoculated in the field with isolates of *G. zeae* has been reported by Cullen (1981) and Cullen et al. (1981). Zearalenone levels reported in corn inoculated in the field in the United States by other investigators include 30 ppm in corn inoculated with "Fusaria cultures (NRRL 6202 and NRRL 6207)" by Bennett et al. (1981) and 19.5 to 39.0 ppm in corn inoculated "with *Fusarium*" by Young et al. (1981). The "Fusaria cultures" used by Bennett et al. (1981) were *F. graminearum* NRRL 6206 and NRRL 6207, which were both isolated in the United States and had both been reported to produce zearalenone in culture by Shannon et al. (1980) and Vesonder et al. (1981b). The identity of the unnumbered and unspecified *Fusarium* strain used by Young et al. (1981) is unknown to us. The production of zearalenone in cultures on corn kernels of 14 corn inbreds and four single cross hybrids by three isolates of *F. graminearum* (NRRL 5864, NRRL 6206, and NRRL 6207) was screened by Shannon et al. (1980). We have confirmed the identity of *F. graminearum* NRRL 5864 as *F. graminearum* (Strain 8.24). Analyses by TLC of the cultures of these three strains on kernels of different corn inbreds and hybrids incubated at 25°C for 3 weeks followed by 2 weeks at 15°C revealed zearalenone levels ranging from 1 to 210 mg/kg (Shannon et al., 1980). The corn hybrids appeared to have less resistance to zearaleneone formation than most of the inbreds, and there was highly significant variation between corn varieties and fungal isolates. The amount of fungal growth did not correlate with the amount of zearalenone produced (Shannon et al., 1980).

In Canada, zearalenone has been detected by TLC in the stems and ears of corn plants inoculated artificially in either the stems, ears, or both with an unnumbered isolate of *F. graminearum* "from corn near Guelph, Ontario" (Sutton et al., 1976). We have confirmed the identity of this strain, which produced 480 µg/g of zearalenone in cultures on corn incubated at 20°C for 6 weeks (Sutton et al., 1976) and was *F. graminearum* isolate G-73-2 according to J.C. Sutton (personal communication, 5 May 1981), as *F. graminearum* (Strain 8.52). The average levels of zearalenone detected in the stems of corn plants inoculated in the stems was 132 µg/100g compared to 1,236 µg/100g in the ears of plants inoculated in the ears. The finding of an average level of 18.3 µg/100g of zearalenone in the healthy ears of four stem-inoculated plants provided circumstantial evidence for the translocation of zearalenone from the stem to the ear in corn plants, according to Sutton et al. (1976).

Nine isolates of *F. graminearum* from barley and wheat contaminated with deoxynivalenol in Ontario, Canada, have been found to produce zearalenone, together with deoxynivalenol, in cultures on wheat, barley, oats, and rice incubated either at 25°C for 1 week followed by 2 weeks at 10°C and a further 3 to 4 weeks at 4°C, or at 25°C for 15 days (Neish & Cohen, 1981). Analyses of these cultures by HPLC and fluorescence detection revealed the presence of zearalenone in all nine isolates at levels ranging from 0.4 to 90.6 ppm (Neish & Cohen, 1981). Although these nine isolates of *F. graminearum* from Canadian wheat were shown to be able to produce relatively high levels of zearalenone in cultures on wheat by Neish & Cohen (1981), zearalenone has apparently not been found to co-occur with deoxynivalenol in Canadian wheat naturally infected by *F. graminearum* during the scab epiphytotic of 1980 (Neish & Cohen, 1981; Scott et al., 1981; Trenholm et al., 1981).

Extensive screening programs of *Fusarium* strains for zearalenone production have been conducted in Japan. Ishii et al. (1974) analyzed cultures on rice incubated at 25°C for 12 days followed by 14 days at 12 to 15°C of 166 *Fusarium* isolates from cereal grains in Japan for zearalenone by TLC. Unspecified levels of zearalenone were detected in cultures of 32 strains, including 15 of 40 isolates referred to as "*F. roseum*." We have identified the following three isolates, reported to produce zearalenone by Ishii et al. (1974), as *F. graminearum*: *F. roseum* M-3-2 (Strain 8.30), *F. roseum* (Gibbosum) A-O-2 (Strain 8.31), and *F. tricinctum* A-B-2 (Strain 8.33). Ishii et al. (1974) isolated 30 to 40 mg/kg of crystalline zearalenone from the cultures on rice of *F. graminearum* (= *F. roseum*) M-3-2 (Strain 8.30). The yield of zearalenone by this strain was increased to 250 mg/kg by an additional incubation period of 14 days of 25°C following the incubation at 12 to 15°C, and further increased to 407 mg/kg by supplementing the rice with 1% peptone (Ishii et al., 1974). This strain has also been reported to produce zearalenone by Ueno (1973b) and Ueno et al. (1974b, 1977a). We have also confirmed the identity of "*G. zeae* Ishii," which was reported by Ueno et al. (1971a) to produce small amounts of zearalenone, as *F. graminearum* (Strain 8.28).

According to Ichinoe et al. (1977), 11 of 30 isolates of *F. graminearum* from cereals and legumes in Japan produced zearalenone at levels ranging from 0.2 to 3,051 µg/g in cultures on rice. Ichinoe et al. (1980) reported that 56 of 105 isolates of *F. graminearum* from barley and wheat produced in different geographical areas in Japan produced zearalenone at levels ranging from 0.13 to 81.8 µg/g as detected by TLC and GLC in cultures on rice supplemented with 0.5% peptone and incubated at 25°C for 1 week followed by 2 weeks at 15°C. The highest incidence of zearalenone-producing strains (7 of 10) occurred in Hokkaido in northern Japan, and the highest-producing strains (average zearalenone yield 16.38 µg/g) were also isolated from wheat and barley produced in this area (Ichinoe et al., 1980).

Suzuki et al. (1978) detected zearalenone at levels of 0.23 to 0.77 µg/g by GC with a flame ionization detector in cultures on rice of 3 of 6 isolates of *F. graminearum* from barley and wheat in Saitama Prefecture, Japan. The same analytical method revealed the presence of 6 to 3,051 µg/g of zearalenone in cultures of 5 of 5 isolates of *F. graminearum* from corn imported from the United States (Suzuki et al. 1978). According to Suzuki et al. (1981b), 30 of 40 isolates of *F. graminearum* from barley and wheat in Saitama Prefecture produced zearalenone in culture. These authors also found that 36 of 52 isolates of *F. graminearum* from different geographical areas in Japan produced zearalenone; among these, nine produced zearalenone only, 11 produced zearalenone and butenolide, one produced zearalenone and trichothecenes, and 15 produced zearalenone, butenolide, and trichothecenes simultaneously.

Yoshizawa et al. (1979) reported that 17 of 61 isolates of *F. graminearum* from barley and wheat in southern Japan produced zearalenone in cultures on rice incubated for 1 week at 25°C followed by 2 weeks at 15°C. Zearalenone analyses were done by TLC, but the levels produced by these 17 isolates of *F. graminearum* were not stated by Yoshizawa et al. (1979). Among these 17 zearalenone-producing isolates, seven produced zearalenone only, four produced zearalenone and butenolide, three produced zearalenone and the trichothecenes fusarenon-X and 3-acetyldeoxynivalenol, and three produced zearalenone, butenolide, and trichothecenes simultaneously.

We have not seen any of the approximately 150 Japanese isolates of *F. graminearum* reported to produce zearalenone by Ichinoe et al. (1977, 1980), Suzuki et al. (1978, 1981b), and Yoshizawa et al. (1979). We have, however, confirmed the identity of *F. graminearum* NHL-F-1118 and NHL-F-1112, which were both isolated in Japan and produce zearalenone according to H. Kurata (personal communication, 29 April 1980), as *F. graminearum* (Strains 8.34 and 8.36).

Fusarium isolates from barley in the UK were screened for zearalenone production by Hacking et al. (1976, 1977). These authors found that 3 of 44 isolates of *F. graminearum* produced unspecified levels of zearalenone as detected by TLC in cultures on rice incubated at 25°C for 4 weeks followed by 6 weeks at 12°C. The main zearalenone-producing species in barley in the UK identified in the study by Hacking et al. (1976, 1977) was *F. culmorum* (see *F. culmorum*, zearalenone).

The screening of *Fusarium* isolates from cereals in Italy revealed that 10 of 10 isolates of *F. graminearum* produced zearalenone at levels ranging from 20 to 1,590 µg/g as determined by TLC and UV spectrophotometry in cultures on corn incubated at 25°C for 2 weeks followed by 8 weeks at 12°C (Bottalico, 1977b, 1978b, 1979). In addition, one isolate of *F. graminearum* from wheat (Bottalico, 1975) and one from corn (Bottalico, 1977a) have also been reported to produce zearalenone at levels of 61 to 270 µg/g and 136 to 375 µg/g, respectively, in cultures on corn incubated under similar conditions. We have not seen any of these 12 Italian isolates of *F. graminearum* reported to produce zearalenone by Bottalico (1975, 1977a, b, 1978b, 1979). We have, however, confirmed the identity of *F. graminearum* ITM 126, which was also isolated in Italy according to A. Bottalico (personal communication, 18 September 1980), as *F. graminearum* (Strain 8.49), but we do not know whether this strain, which is known to produce deoxynivalenol (Palmisano et al., 1981), also produces zearalenone.

An unnumbered strain of *F. graminearum* isolated from feed suspected of causing infertility in cattle in Finland produced 9.55 mg/kg of zearalenone in cultures on a mixture of oats, barley, and wheat grain incubated at room temperature for one month (Roine et al., 1971). A Finnish isolate of *F. graminearum* (see Strain 8.45) "which in earlier investigations was found to cause clear estrogenic changes in rats" has been reported to produce zearalenone by Kallela & Korpinen (1973). The levels of zearalenone produced by this unnumbered strain of *F. graminearum* in cultures on either barley grain, oats grains, or wheat bran incubated at room temperature for 5 weeks were 407, 45, and 3.2 µg/g, respectively (Kallela & Korpinen, 1973). Another unnumbered isolate of *F. graminearum*, "which in the field had caused the vulvovaginitis syndrome in swine," has been reported to produce zearalenone in cultures on oats or mixtures of oats, wheat, and barley incubated at 20 to 24°C for 2 weeks followed by 2 days at 5°C (Kallela & Saastamoinen, 1981a, b). The total amounts of zearalenone in the cultures were determined by adding the soluble amount of zearalenone determined by liquid chromatography to the insoluble amount remaining in the solid residue after extraction, which was estimated semiquantitatively by the rat uterus bioassay (see rats, above). The total levels of zearalenone determined by these methods in cultures of this strain on oats and a mixture of oats and wheat were 2,524 and 3,999 ppm, respectively (Kallela & Saastamoinen, 1981a, b).

Two isolates of *F. graminearum* (F-107 and F-110) from feedstuffs in Germany have been reported to produce zearalenone in cultures on corn incubated at 25°C for 21 days followed by 14 days at 14°C (Möller et al., 1978). We have confirmed the identity of both of these isolates as *F. graminearum* (Strains 8.50 and 8.51). The levels of zearalenone detected by TLC and UV spectrophotometry in cultures of these two strains were 248 and 324 mg/kg, respectively (Möller et al., 1978). *F. graminearum* F-107 and F-110 (Strains 8.50 and 8.51) have also been found to produce zearalenone at levels up to 232 and 405 mg/kg, respectively, in corn adjusted to different moisture levels and incubated at 20°C for 10 days, 31 days at 10°C, and finally at 20°C for 111 days (Müller & Thaler, 1981). Neither of these two strains produced zearalenone in cultures on corn amended with propionic acid during an incubation period of 4 months (Müller & Thaler, 1981). In cultures on non-sterilized corn with a moisture content of 39%, *F. graminearum* F-107 (Strain 8.50) produced 8.0 mg/kg of zearalenone after incubation at 10°C for 40 days (Thaler et al., 1979). Air-drying of the corn inoculated with this strain did not entirely prevent zearalenone formation, and levels up to 0.87 mg/kg could be detected after 20 days (Thaler et al., 1979).

Two isolates of *F. graminearum* (F-59 and 184a) from millet in Hungary have been reported to produce zearalenone at levels up to 1900 µg/g in cultures on millet containing 2% malt extract incubated at 23 to 27°C for 1 week followed by 2 weeks at 5 to 8°C and another week at 23 to 27°C (Szathmary et al., 1976, 1977). We have confirmed the identity of one of these isolates, i.e., *F. graminearum* F-59, as *F. graminearum* (Strain 8.38). We have also identified *F. culmorum* F-79, which was also isolated from millet in Hungary and reported to produce up to 3,400 µg/g of zearalenone in cultures on millet by Szathmary et al. (1976, 1977), as *F. graminearum* (Strain 8.39). According to M. Palyusik (personal communication, 16 December 1977), the unnumbered Hungarian isolate of *F. culmorum* reported to produce zearalenone by Kovacs et al. (1975) and Palyusik & Koplik-Kovacs (1975a, b) was also *F. graminearum* (= *F. culmorum*) F-79 (Strain 8.39). Sandor et al. (1980) reported that this strain (as *F. culmorum* F-79) produced 23.1 ppm of zearalenone as detected by TLC in cultures on corn incubated at 24 to 28°C for 14 days. The level of zearalenone in these cultures increased to 31.2 ppm following 4 weeks of cold exposure at 12 to 14°C and re-exposure to the growth temperature of 24 to 28°C. The levels of zearalenone in these cultures were further increased (up to 83.3 ppm) by exposure to ionizing gamma radiation (Sandor et al., 1980).

Unspecified numbers of unnumbered Hungarian isolates of *F. graminearum* have also been reported to produce unspecified levels of zearalenone by Lasztity & Wöller (1975a, b, c; 1977) and Lasztity et al. (1977). Corn inoculated with a mixture of unnumbered Hungarian isolates of *F. culmorum* and *F. graminearum* contained 34 µg/g of zearalenone and was used in a study on the effects of zearalenone on the fertility of rats by Mess et al. (1979) and Ruzsas et al. (1978, 1979).

Vesela et al. (1981) reported that 15 of 19 isolates of *F. graminearum* from wheat in Czechoslovakia contaminated with zearalenone produced zearalenone in culture. In Yugoslavia, *F. graminearum* has been isolated from corn contaminated with zearalenone (Brodnik, 1975), and one of these isolates produced 60 ppm of zearalenone in cultures on corn incubated at 25°C for 7 days followed by 90 days at 12°C (Brodnik et al., 1978).

In southern Africa, three isolates of *F. graminearum* from corn contaminated with zearalenone have been reported to produce zearalenone in culture (Marasas et al., 1977, 1978). We have confirmed the identity of these three isolates, i.e., *F. graminearum* MRC 120 and MRC 121, isolated from Zambian corn, and MRC 460, isolated from South African corn, as *F. graminearum* (Strains 8.37, 8.42, and 8.43). The levels of zearalenone determined by TLC and UV spectrophotometry in cultures of these three isolates on corn incubated at 25°C for 2 weeks followed by 6 weeks at 12°C were 1,000, 1,270 and 1,140 ppm, respectively (Marasas et al., 1977, 1978).

In addition to zearalenone, *F. graminearum* is known to produce several metabolites structurally related to zearalenone in culture (Bolliger & Tamm, 1972; Hagler et al., 1979; Hidy et al., 1977, Hurd, 1977; Jackson et al., 1974; Mirocha & Christensen, 1974; Mirocha et al., 1968b, 1971, 1977a, b; Pathre et al., 1980; Steele et al., 1974, 1976, 1977a, b). One of these derivatives of zearalenone produced by *F. graminearum*, α-zearalenol, is three to four times more active than zearalenone (Hagler et al., 1979; Mirocha et al., 1978; Ueno & Tashiro, 1981) and has been found to occur naturally at levels ranging from 0.15 to 4.0 mg/kg in feeds associated with hyperestrogenism in pigs (Mirocha et al., 1979b).

In conclusion, approximately 360 isolates of *F. graminearum* have been reported to produce zearalenone in culture at levels up to 60,000 mg/kg. Twenty-three of these zearalenone-producing isolates of *F. graminearum* are represented in the ITFRC (Table 8.1). The highest known zearalenone-producing strain in the ITFRC is *F. graminearum* Mapleton No. 10 (Strain 8.21), which has been reported to produce up to 60 g/kg of zearalenone. Other strains of *F. graminearum* that have been reported to produce zearalenone in the g/kg range include two isolates from the United States, i.e., *G. zeae* NRRL 2830 and NRRL 5864 (Strains 8.19 and 8.24), the Hungarian isolate of *F. graminearum* F-79 (Strain 8.39), and two isolates from southern Africa, i.e., *F. graminearum* MRC 121 and MRC 460 (Strains 8.42 and 8.43). Together with *F. culmorum* (see *F. culmorum*, zearalenone), *F. graminearum* has to be considered as the most important zearalenone-producing species of *Fusarium*.

Toxigenic Strains in the ITFRC

The *Fusarium* strains listed under *F. graminearum* in Table 8.1 all have been identified as *F. graminearum*, according to Nelson et al. (1983). Toxigenic strains referred to as members of the section Discolor but identified as other species by us are listed in Table 8.2.

Strain 8.19 *Gibberella zeae* NRRL 2830 = ATCC 20272 (= R-5318; MRC 1782)

This strain was isolated from corn implicated in a field outbreak of the estrogenic syndrome (see estrogenic syndrome, above) in pigs in Indiana, USA, during 1957/58 (Stob et al., 1962). Vulvar and mammary enlargement in immature female pigs were first induced experimentally with a pure culture of *F. graminearum* by Stob et al. (1962). These authors used cultures on corn incubated at 24°C for 2 to 3 weeks of an isolate of *F. graminearum* from moldy corn associated with field outbreaks of vulvo-vaginitis in Indiana, USA, during 1957/58. The culture material that was used in the successful experimental reproduction of vulvo-vaginitis in gilts also induced uterine hypertrophy in ovariectomized mice, as did alcoholic extracts from which the active metabolite was isolated (Stob et al., 1962). The isolate of *F. graminearum* used in the first experimental reproduction of the estrogenic syndrome by Stob et al. (1962) was unfortunately not numbered, but the isolate specified in United States patents on the estrogenic and anabolic activity of the active metabolite (FES = Fermentation Estrogenic Substance = zearalenone) was "*Gibberella zeae* NRRL 2380" (should read 2830), according to Mirocha & Christensen (1974) and Mirocha et al. (1972). This strain was referred to as "*F. roseum* NRRL 2830, isolated from corn, in Indiana" by Vesonder et al. (1981b), who stated that zearalenone was produced by this strain according to "F.N. Andrews and M. Stob, U.S. Patent 3,196,019, month 1965." Hidy et al. (1977) referred to this strain as "*G. zeae* H331 (ATCC 20272)." According to the ATCC Catalogue of Strains, Fifteenth Ed. (1982), *G. zeae* ATCC 20272 was obtained from "Commercial Solvents Corp. H331 (NRRL 2830). Production of an anabolic and estrogenic compound (U.S. Pat. 3,196,019)."

Although it seems reasonable to assume that *G. zeae* NRRL 2830 was one of the nine isolates of *G. zeae* reported to have uterotrophic activity by Stob et al. (1962), it cannot be stated unequivocally that this is the strain that these authors used in the experimental reproduction of vulvar and mammary enlargement in pigs. It is clear, however, that zearalenone was first isolated from this strain by Stob et al. (1962).

Subsequent to the original isolation of zearalenone from this strain by Stob et al. (1962), several other authors have also reported zearalenone production by *F. graminearum* (= *F. moniliforme* = *F. roseum*) NRRL 2830 (Bolliger & Tamm, 1972; Hesseltine, 1972; Hidy et al., 1977; Ueno et al., 1971a; Vesonder et al., 1981b). Maximal yields of zearalenone by this strain were obtained by Bolliger & Tamm (1972) in cultures on corn incubated at 12°C for 10 weeks. These authors isolated 11.6 g of zearalenone from 7.5 kg of culture material on corn, and also isolated several metabolites structurally related to zearalenone, i.e., 5-formylzearalenone, 7'-dehydrozearalenone, 8'-hydroxyzearalenone, and 8'-epihydroxyzearalenone.

Stob et al. (1962) reported that a water-soluble fraction, toxic to mice, was "occasionally produced concurrently" with the estrogenic metabolite in cultures of *G. zeae*. It is possible that this water-soluble metabolite toxic to mice may have been deoxynivalenol. *Fusarium graminearum* NRRL 2830 has subsequently been found to produce 41 µg/g of deoxynivalenol, together with zearalenone, in cultures on corn incubated at 28°C for 30 days (Vesonder et al., 1981b).

Cultures of *F. graminearum* NRRL 2830 on rice were fed to male mice and rats on a life-long basis by Saito & Ohtsubo (1974). This culture material caused the death of 21 of 43 mice within 70 days, "mostly showing diarrhea," and one case each of hemangiosarcoma of the liver and lymphocytic leukemia were observed (Saito & Ohtsubo, 1974). All of the rats (18) fed this culture material survived for one year and died at random intervals thereafter. No mention was made of the occurrence of tumors by Saito & Ohtsubo (1974), but one lymphosarcoma was recorded in these rats according to Ohtsubo & Saito (1977) and Saito et al. (1980). It could be concluded from this experiment that the culture material of this strain was not carcinogenic in mice and rats.

A lyophilized culture of this strain was received on 23 October 1979 from NRRL as "*F. graminearum* NRRL 2830." Single-spore isolates obtained from this culture produced mature perithecia and ascospores within 3 weeks on CLA.

Strain 8.20 *Fusarium* **No. 5** (= R-5250; MRC 1621)

This strain was isolated from corn implicated in a field outbreak of the estrogenic syndrome (see estrogenic syndrome, above) in pigs in Minnesota, USA, during 1963/64 (Christensen et al., 1965). Cultures on whole corn kernels incubated at 20 to 25°C for 3 weeks followed by 2 weeks at 12°C of 12 of 40 *Fusarium* isolates, including *Fusarium* No. 5, tested by Christensen et al. (1965) caused increases of five to eight times in the weight of the uteri of virgin female weanling rats. The uterotrophic *Fusarium* strains were not specifically identified, but Christensen et al. (1965) stated that "all or most of the isolates of *Fusarium* tentatively appear to be *F. culmorum* or *F. graminearum*." *Fusarium* No. 5 apparently also induced vulvar enlargement and prolapse in a young gilt, according to Figs. 10 and 11 of Christensen (1979) and Fig. 2 of Mirocha et al. (1971), but details of this experimental reproduction of vulvo-vaginitis with culture material of *F. graminearum* (= *Fusarium*) No. 5 have not been published as far, as we are aware.

Christensen et al. (1965) isolated the estrogenic metabolite zearalenone (= F-2) from cultures of *Fusarium* No. 5 by means of the rat uterine hypertrophy bioassay and demonstrated that it was identical to the uterotrophic metabolite isolated from *G. zeae* (Strain 8.19) by Stob et al. (1962). The amount of zearalenone produced by this strain was not indicated by Christensen et al. (1965).

A soil culture of this strain was received on 6 March 1978 from C.J. Mirocha as "*Fusarium* No. 5." Single-spore isolates obtained from this culture of *F. graminearum* produced mature perithecia and ascospores within 3 weeks on CLA.

Strain 8.21 *F. roseum* 'Graminearum' Mapleton No. 10 (= R-6136; MRC 2329)
=*G. zeae* JRL MAP 10 = ATCC 24689 (= R-5453; MRC 1806)
=*G. zeae* Ascospore No. 55 = ATCC 24688 (= R-5454; MRC 1904)

This strain, as "*F. graminearum* isolate designated Mapleton No. 10," was isolated from "feed associated with problems of infertility in swine in southern Minnesota" (Mirocha et al., 1971). According to Wolf & Mirocha (1973), the mass isolate "*F. roseum* 'Graminearum' Mapleton No. 10" was isolated from "*Fusarium*-infected corn obtained from Mapleton, Minnesota." The strain referred to as "*F. roseum* 'Graminearum' JRL" is a subculture of Mapleton No. 10 according to Wolf & Mirocha (1973) and is "*G. zeae* ATCC 24689" according to Wolf & Mirocha (1977). The strain referred to as "*F. roseum* 'Graminearum' ascospore No. 55" is a single-ascospore isolate of Mapleton No. 10 according to Wolf & Mirocha (1973) and is "*G. zeae* ATCC 24688" according to Wolf & Mirocha (1977).

This important toxigenic strain has become very widely used and has been cited in more papers on *Fusarium* mycotoxicology than any other strain of *F. graminearum* that we are aware of. The literature is very difficult to follow, however, because this strain has been cited under a variety of different names and numbers and sometimes unnumbered. In addition to the names and numbers cited above, this strain has also been referred to by the following names according to C.J. Mirocha (personal communication, December 9, 1977): *F. roseum* Isolate No. 1, from corn from Mapleton, Minnesota (Eugenio et al., 1970a); *F. graminearum* No. 10 (Mirocha et al., 1967a, b); *F. gramine-*

arum Mapleton No. 10 (Mirocha et al., 1971); *F. roseum* Mapleton No. 10 (Mirocha & Christensen, 1974; Nelson et al., 1973); *F. roseum* strain Mapleton 10 (Bullerman et al., 1975); *F. roseum* Map. 10 (Ishii et al., 1974); *F. roseum* JRL (MAP 10) from moldy corn collected near Mapleton, Minnesota (Steele et al., 1976); *F. roseum,* unnumbered, from moldy corn collected near Mapleton, Minnesota (Steele et al., 1974); *F. roseum* 'Graminearum,' unnumbered (Eugenio et al., 1970b; Harein et al., 1970; Wolf et al., 1972); *F. roseum,* unnumbered (Christensen et al., 1972; Meronuck et al., 1980; Pathre et al., 1980); *F. graminearum,* unnumbered (Mirocha et al., 1968b); *F. graminearum* IMI 155426 = ATCC 26557 (Sherwood & Peberdy, 1972; 1974a, b).

Culture material on corn of an unnumbered strain of "*F. roseum*" was found to cause vulvar and mammary enlargement, rectal prolapse, and uterine hypertrophy in sexually immature female pigs and reduced testicular weight in male pigs by Christensen et al. (1972b). The same experiment was referred to by Nelson et al. (1973), who specified the strain used as "*F. roseum* (strain Mapleton No. 10)." This strain of *F. graminearum* is known to produce large amounts of zearalenone in culture (Eugenio et al., 1970a; Ishii et al., 1974; Mirocha & Christensen, 1974; Mirocha & Pathre, 1979; Mirocha et al., 1967a, b, 1968a, b, 1971; Scott et al., 1970; Sherwood & Peberdy, 1972, 1974a, b; Steele et al., 1974, 1976, 1977a, b; Wolf & Mirocha, 1973, 1977). Maximal yields of zearalenone reported for this strain (as *F. roseum* Isolate No. 1 from corn Mapleton, Minnesota) by Eugenio et al. (1970a) were 3,087 and 2,585 µg/g on parboiled polished rice and corn, respectively, incubated for 2 weeks at 24 to 27°C followed by 8 weeks at 12°C. Mirocha et al. (1967a, b) reported yields of 1,600 µg/g of zearalenone in cultures of "*F. graminearum* No. 10" in cultures on corn incubated for 2 weeks at 25 to 28°C followed by 8 weeks at 12°C. According to Mirocha et al. (1971), yields as high as 30,000 to 60,000 ppm have been obtained in cultures of "*F. graminearum* Mapleton No. 10" on rice or corn amended with glucose. According to Mirocha & Christensen (1974), yields as high as 38,000 to 40,000 ppm have been obtained in clutures of "*F. roseum* Mapleton 10" on solid media under optimal incubation conditions. The maximal yield of zearalenone by this strain (as *F. graminearum* IMI 155426) obtained by Sherwood & Peberdy (1972) was 1,068 µg/g in cultures on wheat at 37% moisture content incubated at 25°C for 10 weeks followed by 6 weeks at 12°C. Sherwood & Peberdy (1974a) reported that *F. graminearum* IMI 155426 produced 1,976 µg/g of zearalenone in cultures on grains at 31% moisture content incubated at 12°C for 4 months. Wolf & Mirocha (1973, 1977) stated that *F. roseum* 'Graminearum' JRL MAP 10 (ATCC 24689) was a subculture of a mass isolate designated as *F. roseum* 'Graminearum' Mapleton No. 10, and was a high producer (>15.0 mg/g) of zearalenone in cultures on a corn-rice medium. These authors also stated that *F. roseum* 'Graminearum' ascospore No. 55 (ATCC 24688) was a single-ascospore isolate from the mass isolate *F. roseum* 'Graminearum' Mapleton No. 10, and was a low producer (<1.5 mg/g) of zearalenone in culture on a corn-rice medium. Ishii et al. (1974) isolated 50 to 130 mg/kg of crystalline zearalenone from cultures of "*F. roseum* Map. 10 (the strain from C.J. Mirocha)" on rice incubated at 25°C for 14 days followed by 14 days at 12 to 15°C.

An estrogenic response in turkey poults characterized by swollen vents and prolapsed cloacae has been induced experimentally with cultures on corn of an unnumbered isolate of "*F. roseum*" (Meronuck et al., 1970). According to Meronuck et al. (1970), the "isolate of *F. roseum* was one of the higher producers of the estrogenic compound F-2, as reported by Mirocha et al." In the reference cited, Mirocha et al. (1967b) referred to "two estrogen-producing isolates of *F. graminearum* designated as numbers of 9 and 10." Although *F. roseum* 'Graminearum' Mapleton No. 10 is known to produce large amounts of zearalenone in culture, the zearalenone content of the culture material used by Meronuck et al. (1970) was not indicated. However, in an apparent reference to the feeding trial in turkeys conducted by Meronuck et al. (1970), the zearalenone content of rations containing 10% of "*Fusarium*-invaded corn" was stated to have been 800 ppm by Mirocha et al. (1971). According to these authors this diet caused swelling of the vents, cystic development of the right genital tract, evertion of the cloaca, and enlargement of the cloacal bursa and oviduct in turkey poults.

Inhibition of spermatogenesis, regressive changes in the testes, and infertility in turkey cocks have been reported to be caused by culture material on corn of an unnumbered isolate of *F. graminearum* by Palyusik et al. (1971). According to M. Palyusik (personal communication, 16 December 1971), the strain used in this work was *F. roseum* 'Graminearum' Mapleton No. 10. This strain is known to produce large amounts of zearalenone in culture, but the zearalenone content of the culture material used by Palyusik et al. (1971) was not indicated.

Culture material of this strain (as *F. roseum* 'Graminearum,' unnumbered) on rice has been fed to insects by Eugenio et al. (1970b) and Harein et al. (1970). Culture material containing up to 10,000 ppm of zearalenone was fed to confused flour beetles (*Tribolium confusum*) and the lesser meal worm (*Alphitobius diaperinus*) by Eugenio et al. (1970b). Zearalenone was ingested from the culture material by both species of insects and persisted through metamorphosis and after death. Harein et al. (1970) reported that culture material containing 9,000 ppm of zearalenone decreased egg production and egg hatch of the confused flour beetle. Larvae that emerged from these eggs were also abnormally small and sluggish.

According to Mirocha & Christensen (1974), *F. roseum* 'Graminearum' Mapleton No. 10 "evidently produces a complex of toxins, and although oestrogenic, on some occasions it has been lethal to white rats in feedings tests." Although this strain of *F. graminearum* is known to produce large amounts of zearalenone and certain derivatives of zearalenone in culture, we have not been able to find any reports on the production of other mycotoxins by this strain.

According to Inaba & Mirocha (1979) and Wolf & Mirocha (1973), an inverse relationship between perithecial formation and zearalenone-producing potential among isolates of *F. graminearum* was reported by C.P. Eugenio in a Ph.D. thesis at the University of Minnesota. However, the highest zearalenone-producing strain among 43 isolates of "*F. roseum*" tested by Eugenio et al. (1970a) was also a producer of perithecia, because these authors obtained "90 single-ascospore lines of isolate No. 1 of *F. roseum*." According to C.J. Mirocha (personal communication, 9 December 1971), the strain used by Eugenio et al. (1970a) was *F. graminearum* Mapleton No. 10. The levels of zearalenone produced by these 90 single-ascospore isolates in cultures on corn ranged from 70 to 1,623 ppm (Eugenio et al., 1970a). One of these single-ascospore isolates, designated as *F. roseum* 'Graminearum' ascospore No. 55 (= ATCC 24688), and a subculture of the original isolate *F. graminearum* Mapleton No. 10, designated *F. roseum* 'Graminearum' JRL, have been used in studies on the mechanism of regulation of sexual reproduction in *F. graminearum* (Inaba & Mirocha, 1979; Wolf & Mirocha, 1973, 1977; Wolf et al., 1972). The subculture *F. roseum* 'Graminearum' JRL was found to be a high producer of zearalenone and produced more than 15.0 mg/g on a corn-rice medium and 0.15 µg/g wet weight on Coon's agar. The single-ascospore isolate No. 55 was found to be a low producer that produced less than 1.5 mg/g of zearalenone on the corn-rice medium and only traces in agar cultures (Wolf & Mirocha, 1973).

This strain of *F. graminearum* has also been used in a number of studies on the biosynthesis and metabolism of zearalenone (Mirocha & Pathre, 1979; Steele et al., 1974, 1976, 1977a, b). Several metabolites structurally related to zearalenone have been isolated from cultures of *F. graminearum* Mapleton No. 10. (Mirocha & Christensen, 1974; Mirocha et al., 1968b, 1971). Some of these zearalenone derivatives that have been isolated from cultures of this strain and chemically characterized include 8'-hydroxyzearalenones (Jackson et al., 1974), 3'-hydroxyzearalenones (Pathre et al., 1980), and 6',8'-dihydroxyzearalene (Steele et al., 1976).

We have examined the following cultures of this strain of *F. graminearum* that have been deposited in the ITFRC:

1) *F. roseum* 'Graminearum' Mapleton No. 10. A soil culture of this strain was received on 4 October 1981 from C.J. Mirocha as "Fus. Map-10, JRL-HP (10/8/79)." This is a badly degenerated culture, but it still produced some macroconidia on CLA that allowed it to be identified as *F. graminearum*. Subcultures of this culture have been lyophilized and deposited in the ITFRC as *F. graminearum* R-6136 and MRC 2329. It is probable that the cultures deposited in CMI and ATCC as *F. graminearum* IMI 155426 and ATCC 26557, respectively, are also subcultures of this culture obtained from C.J. Mirocha.

2) *G. zeae* ATCC 24689. A slant culture of this strain was received on 12 December 1979 from ATCC as "*G. zeae* ATCC 24689." This culture is listed in the ATCC Catalogue of Strains, Fifteenth Ed. (1982) as "J.C. Wolf JRL, J.R. Lieberman, Mapleton 10. Corn." According to Wolf & Mirocha (1973), this culture is a subculture of *F. roseum* 'Graminearum' Mapleton No. 10 and is a high producer of zearalenone. This culture of *F. graminearum* is in better condition than the one discussed above and has been lyophilized and deposited in the ITFRC as *F. graminearum* R-5453 and MRC 1806.

3) *G. zeae* Ascospore No. 55. A silica gel culture of this strain was received on 17 March 1980 from C.J. Mirocha as "*F. roseum* Map 10 Asco 55." According to Wolf & Mirocha (1973), this culture is a single-ascospore isolate of *F. roseum* 'Graminearum' Mapleton No. 10 and is a low producer of zearalenone. This is a typical culture of *F. graminearum*, and single-spore isolates obtained from this culture produced mature perithecia and ascospores within 3 weeks on CLA. Subcultures of this culture have been lyophilized and deposited in the ITFRC as *F. graminearum* R-5454 and MRC 1904.

Strain 8.22 *F. roseum* No. 117 = ATCC 28114 (= R-5796; MRC 1946)

This strain (as *F. roseum* No. 117) was isolated from scabby barley contaminated with deoxynivalenol and nivalenol (see akakabi-byo, above) in Japan (Morooka et al., 1972; Yoshizawa & Morooka, 1973). This strain has also been referred to as *F. roseum* 117 (ATCC 28114) by Yoshizawa et al. (1978, 1979), *F. graminearum* ATCC 28114 by Yoshizawa et al. (1980), and *F. roseum* 117 (ATCC 28117) by Forsyth et al. (1977).

Cultures of this strain on rice incubated at 25°C for 15 days caused a marked feed refusal response in rats (Yoshizawa et al., 1978). This culture material that was refused by rats contained large amounts of deoxynivalenol (220 μg/g) and 3-acetyldeoxynivalenol (160 μg/g), as well as trace amounts of diacetyldeoxynivalenol. The greater part of the feed refusal activity was found in the fraction containing these trichothecenes (Yoshizawa et al., 1978).

3-Acetyldeoxynivalenol was first isolated from the culture broth of *F. roseum* No. 117 by Yoshizawa & Morooka (1973). The production of 3-acetyldeoxynivalenol by this strain has subsequently also been reported by Forsyth et al. (1977), Yoshizawa & Morooka (1974, 1975a, 1977), and Yoshizawa et al. (1978, 1979). The levels of 3-acetyldeoxynivalenol detected in cultures of this strain on rice incubated either at 15°C for 30 days or at 25°C for 15 days were 5 and 160 μg/g, respectively (Yoshizawa et al., 1978). The yield of crystalline 3-acetyldeoxynivalenol isolated from cultures of this strain on rice incubated at 25°C for 10 to 12 days was 400 mg/kg according to Yoshizawa & Morooka (1977).

Butenolide has been isolated at unspecified levels from cultures of this strain in PSC medium (Morooka et al., 1972; Yoshizawa & Morooka, 1973).

This strain (as *F. roseum* No. 117) was the first *Fusarium* strain that was reported to produce deoxynivalenol (= Rd toxin) in culture (Morooka et al., 1972; Yoshizawa & Morooka, 1973). The production of deoxynivalenol by this strain has subsequently been confirmed by Yoshizawa & Morooka (1974, 1975a, 1977) and Yoshizawa et al. (1978, 1979). In cultures of this strain on either PSC medium (Yoshizawa & Morooka, 1975a) or on rice (Yoshizawa & Morooka, 1977; Yoshizawa et al., 1979), maximal levels of deoxynivalenol occurred subsequent to maximal vegetative growth, and the increase in the concentration of deoxynivalenol was associated with a decrease in the concentration of 3-acetyldeoxynivalenol. The mycelium of this strain is known to be able to deacetylate both 3-acetyldeoxynivalenol and diacetyldeoxynivalenol to deoxynivalenol (Yoshizawa & Morooka, 1975a; 1977). The levels of deoxynivalenol detected in cultures of this strain on rice incubated either at 15°C for 30 days or at 25°C for 15 days were 15 and 220 μg/g, respectively (Yoshizawa et al., 1978). The yield of crystalline deoxynivalenol isolated from cultures of this strain on rice incubated at 25°C for 10 to 12 days was 100 mg/kg, according to Yoshizawa & Morooka (1977).

Diacetyldeoxynivalenol was first isolated from cultures on rice incubated at 25°C for 10 to 12 days of *F. roseum* No. 117 (Yoshizawa & Morooka, 1977). The production of diacetyldeoxynivalenol by this strain has subsequently also been confirmed by Yoshizawa et al. (1978, 1979). The yield of crystalline diacetyldeoxynivalenol isolated from cultures of this strain on rice incubated at 25°C for 10 to 12 days was 10 mg/kg according to Yoshizawa & Morooka (1977).

This strain is listed in the ATCC Catalogue of Strains, Fifteenth Ed. (1982) as "*F. roseum* ATCC 28114, K. Ishii, KU-117, S. Morooka, Barley. Production of zearalenone (Appl. Microbiol. 27: 625–628, 1974)." On the basis of this statement, we have to assume that *F. graminearum* No. 117 was one of the unnumbered isolates of *F. roseum* that was found to produce zearalenone in cultures on rice incubated at 25°C for 14 days followed by 14 days at 12 to 15°C by Ishii et al. (1974).

A slant culture of this strain was received on 17 March 1980 from ATCC as "*F. roseum* ATCC 28114." This is a typical culture of *F. graminearum*, and single-spore isolates obtained from this culture produced mature perithecia and ascospores within 3 weeks on CLA.

Strain 8.23 *F. graminearum* NRRL 5883 (= R-5320; MRC 1785)

This strain was isolated from corn contaminated with deoxynivalenol and implicated in a field outbreak of the feed refusal syndrome (see feed refusal and emetic syndromes, above) in pigs in Ohio, USA, in 1973 (Vesonder et al., 1976, 1981b). Cultures of this strain on rice incubated at 28°C for 13 days were refused by pigs, and 60 mg/kg of chromatographically pure deoxynivalenol was isolated from this material (Vesonder et al., 1976). Cultures of this strain on corn incubated under similar conditions were also refused by pigs and contained unspecified levels of deoxynivalenol as detected by TLC and IR (Vesonder et al., 1977b). Cultures on corn incubated at 28°C for 30 days contained 128 μg/g of deoxynivalenol as determined by TLC and GLC-MS, and 90 μg/g of 90% pure deoxynivalenol was isolated from this material (Vesonder et al., 1981b). In addition, 40 μg/g of zearalenone was also isolated from this culture material on corn by Vesonder et al. (1981b). This strain produced only small amounts (5 μg/g) of deoxynivalenol in cultures on wild rice (*Zizania aquatica*) incubated at 28°C for 3 weeks (Lindenfelser et al., 1978).

4-Acetamido-2-butenoic acid was first isolated by Vesonder et al. (1977a) from the fraction containing deoxynivalenol of cultures on rice incubated at 28°C for 13 days of *F. graminearum* NRRL 5883. The possibility has been suggested that this compound may be an intermediate in the biosynthetic pathway of butenolide (Vesonder et al., 1977a). However, this strain has not been reported to produce butenolide as far as we are aware.

A lyophilized culture of this strain was received on 23 October 1979 from NRRL as "*F. graminearum* NRRL 5883." Single-spore isolates obtained from this culture produced mature perithecia and ascospores within 3 weeks on CLA.

Strain 8.24 *F. graminearum* NRRL 5864 (= R-5316; MRC 1693)

This strain was isolated from corn in Minnesota, USA (Vesonder et al., 1981b). Cultures of this strain on corn incubated at 28°C for 30 days contained 95 μg/g of deoxynivalenol (Vesonder et al., 1981b). This culture material also contained unspecified levels of zearalenone (Vesonder et al., 1981b). This strain produced zearalenone in cultures on wild rice and sorghum grain incubated at 25°C for 38 days at maximal levels of 220 and 1,500 μg/g, respectively (Lindenfelser et al., 1978). The production of zearalenone by this strain in cultures on kernels from four single-cross corn hybrids and 14 inbreds was studied by Shannon et al. (1980). The levels of zearalenone were determined in these cultures following incubation at 25°C for 3 weeks and 2 weeks at 15°C. This strain produced zearalenone in cultures on kernels of all four corn hybrids at levels ranging from 8 to 46 mg/kg, but zearalenone could be detected at levels from 1 to 37 mg/kg in only six of the 14 corn inbreds (Shannon et al., 1980).

A slant culture of this strain was received on 5 September 1979 from NRRL as "*F. graminearum* NRRL 5864." Single-spore isolates obtained from this culture produced mature perithecia and ascospores within 3 weeks on CLA.

Strain 8.25 *F. graminearum* NRRL 6450 (= R-6055; MRC 2233)

This strain was isolated from corn contaminated with deoxynivalenol and implicated in a field outbreak of the feed refusal syndrome (see feed refusal and emetic syndromes, above) in pigs in Austria (Vesonder & Ciegler, 1979). Cultures of this strain on corn incubated at 28°C for 13 days contained 11 μg/g of deoxynivalenol (Vesonder & Ciegler, 1979).

A lyophilized culture of this strain was received on 9 March 1981 from NRRL as "*F. graminearum* NRRL 6450." Single-spore isolates obtained from this culture produced mature perithecia and ascospores within 3 weeks on CLA.

Strain 8.26 *F. graminearum* NRRL 6451 (= R-6337; MRC 2436)

This strain was isolated from corn contaminated with deoxynivalenol and implicated in a field outbreak of the feed refusal syndrome (see feed refusal and emetic syndromes, above) in pigs in Austria (Vesonder & Ciegler, 1979). Cultures of this strain on corn incubated at 28°C for 13 days contained 16.7 μg/g of deoxynivalenol (Vesonder & Ciegler, 1979).

A slant culture of this strain was received on 6 July 1981 from NRRL as "*F. graminearum* NRRL 6451." Single-spore isolates obtained from this culture produced mature perithecia and ascospores within 3 weeks on CLA.

Strain 8.27 F. graminearum NRRL 6452 (= R-6056; MRC 2234)
This strain was isolated from corn contaminated with deoxynivalenol and implicated in a field outbreak of the feed refusal syndrome (see feed refusal and emetic syndromes, above) in pigs in Austria (Vesonder & Ciegler, 1979). Cultures of this strain on corn incubated at 28°C for 13 days contained 3.7 µg/g of deoxynivalenol (Vesonder & Ciegler, 1979).

A lyophilized culture of this strain was received on 9 March 1981 from NRRL as "F. graminearum NRRL 6452."

Strain 8.28 Gibberella zeae Ishii (= R-6796; MRC 2581)
This strain was isolated from barley in Kagawa, Japan, in 1963 (Saito & Okubo, 1970; Ueno et al., 1971a). Extracts of cultures of this strain on rice or in PSC medium incubated at 25 to 27°C for 2 weeks were lethal to mice (Ueno et al., 1971a, 1973a). These extracts did not inhibit protein synthesis in rabbit reticulocytes or give a positive reaction in the rabbit skin test, and no known trichothecenes could be detected chemically in these extracts (Ueno et al., 1971a, 1973a). Cultures of this strain on rice incubated at 25°C for 2 weeks contained a "minor amount" of zearalenone as detected by TLC (Ueno et al., 1971a).

A slant culture of this strain was received during November 1981 from Y. Ueno as "G. zeae Ishii." This culture proved to be almost completely mycelial, but still produced a few macroconidia on CLA that allowed it to be identified as F. graminearum.

Strain 8.29 F. rigidiusculum M-1-3 (= R-6795; MRC 2580)
This strain (as Fusarium sp. M-1-3) was isolated from moldy soybean hulls associated with field outbreaks of bean hulls poisoning (see F. sporotrichioides, bean hulls poisoning) of horses in Hokkaido, Japan, in 1970 (Ueno et al., 1972a). Extracts of cultures of this strain in PSC medium incubated at 27°C for 12 days were acutely toxic to mice upon intraperitoneal injection (Ueno et al., 1972a, 1973a). The radiomimetic pathological changes in mice dosed with these extracts were characterized by cellular degradation and karyorrhexis in the actively dividing cells of the bone-marrow, thymus, spleen, testes, and the mucosal epithelium of the small intestine (Ueno et al., 1972a). These extracts also caused a marked inhibition of protein synthesis in rabbit reticulocytes (Ueno et al., 1972a; 1973a). Ueno et al. (1973a) detected diacetoxyscirpenol, neosolaniol, and T-2 toxin by TLC in extracts of this strain (as F. rigidiusculum M-1-3) that had been shown to be toxic to mice.

A slant culture of this strain was received during November 1981 from Y. Ueno as "F. rigidiusculum M-1-3." This culture proved to be almost completely mycelial, but still produced a few macroconidia on CLA that allowed it to be identified as F. graminearum.

Strain 8.30 F. roseum 'Culmorum' M-3-2 = ATCC 28112 (= R-6798; MRC 2583)
This strain (as Fusarium sp. M-3-2) was isolated from moldy soybean hulls associated with field outbreaks of bean hulls poisoning (see F. sporotrichioides, bean hulls poisoning) of horses in Hokkaido, Japan, in 1970 (Ueno et al., 1972a). Extracts of cultures of this strain in PSC medium incubated at 27°C for 12 days were acutely toxic to mice upon intraperitoneal injection and caused radiomimetic cellular damage in the actively dividing cells of the hematopoietic and lymphopoietic tissues (Ueno et al., 1972a, 1973a). These extracts also inhibited protein synthesis in rabbit reticulocytes. No known trichothecenes could be detected by TLC in extracts of this strain (as F. roseum 'Culmorum' M-3-2) by Ueno et al. (1973a). The statement by Ishii et al. (1974) that this strain (as F. roseum M-3-2) had previously been "shown capable of producing trichothecenes" is not in agreement with the results of Ueno et al. (1973a).

The production of zearalenone by this strain (as *F. roseum* M-3-2) in cultures on rice incubated at 25°C for 14 days followed by 14 days at 12 to 15°C has been reported by Ishii et al. (1974), Ueno (1973b), and Ueno et al. (1974b, 1977a). The yield of crystalline zearalenone isolated from the culture material on rice by Ishii et al. (1974) was 30 to 40 mg/kg. The zearalenone yield was increased to 250 mg/kg by an additional incubation of 14 days at 25°C following the incubation at 12 to 15°C, and further increased to 407 mg/kg by supplementing the rice with 1% peptone (Ishii et al., 1974).

A slant culture of this strain was received during November 1981 from Y. Ueno as "*F. roseum* 'Culmorum' M-3-2." This culture proved to be almost completely mycelial, but still produced a few macroconidia and protoperithecia on CLA that allowed it to be identified as *F. graminearum*.

Strain 8.31 *F. roseum* 'Gibbosum' A-O-2 (= R-6802; MRC 2606)

This strain was isolated in Japan (Ishii et al., 1974), but the exact source is unknown to us. This strain produced unspecified levels of zearalenone in cultures on rice incubated at 25°C for 14 days followed by 14 days at 12 to 15°C (Ishii et al., 1974). A "very small amount" of zearalenone was detected by TLC in cultures of this strain in liquid media (Ishii et al., 1974). According to Ishii et al. (1974), this strain has previously been "shown capable of producing trichothecenes" by Ueno et al. (1972a, 1973a), but no reference to this strain could be found in these two publications.

A slant culture of this strain was received during November 1981 from Y. Ueno as "*F. roseum* 'Gibbosum' A-O-2." This strain proved to be almost completely mycelial, but still produced a few macroconidia on CLA that allowed it to be identified as *F. graminearum*.

Strain 8.32 *F. tricinctum* A-R-5 (= R-6801; MRC 2605)

This strain was isolated from "blackened rice grains" in Hokkaido, Japan, in 1973 (Ueno et al., 1975). This strain was found to produce unspecified levels of T-2 toxin in stationary cultures in PSC medium by Ueno et al. (1975). In shake cultures incubated at 27°C for 5 days, this strain produced T-2 toxin and neosolaniol in a medium containing glucose, peptone, and yeast extract, but not in PSC medium (Ueno et al., 1975).

A slant culture of this strain was received during November 1981 from Y. Ueno as "*F. tricinctum* A-R-5." This strain proved to be almost completely mycelial, but still produced a few macroconidia on CLA that allowed it to be identified as *F. graminearum*.

Strain 8.33 *F. tricinctum* A-B-2 = Abashiri-2 (= R-6777; MRC 2563)

This strain was isolated in Japan (Ishii et al., 1974; Ueno et al., 1973a), but the exact source is unknown to us. Extracts of cultures of this strain (as *F. tricinctum* Abashiri-2) in PSC medium incubated at 25 to 27°C for 2 weeks were acutely toxic to mice and inhibited protein synthesis in rabbit reticulocytes (Ueno et al., 1973a). Analyses of these extracts by TLC revealed the presence of diacetoxyscirpenol, neosolaniol, T-2 toxin, and HT-2 toxin (Ueno et al., 1973a). The production of unspecified amounts of zearalenone by this strain (as *F. tricinctum* A-B-2) in cultures on rice incubated at 25°C for 14 days followed by 14 days at 12 to 25°C has been reported by Ishii et al. (1974).

A slant culture of this strain of *F. graminearum* was received during November 1981 from Y. Ueno as "*F. tricinctum* Abashiri-2."

Strain 8.34 *F. graminearum* NHL-F-1118 (= R-5467; MRC 1960)

This strain was isolated from wheat grain in Japan and produces 3-acetyldeoxynivalenol, deoxynivalenol, and zearalenone, according to H. Kurata (personal communication, 29 April 1980). We are not aware of any published reports on the toxigenicity of this strain.

A slant culture of this strain was received on 29 April 1980 from H. Kurata as "*F. graminearum* NHL-F-1118." Single-spore isolates obtained from this typical culture of *F. graminearum* produced mature perithecia and ascospores within 2 weeks on CLA.

Strain 8.35 *F. graminearum* NHL-F-1104 (= R-5469; MRC 1963)

This strain was isolated from barley grain in Japan and produces fusarenon-X and nivalenol, according to H. Kurata (personal communication, 29 April 1980). We are not aware of any published reports on the toxigenicity of this strain.

A slant culture of this strain was received on 29 April 1980 from H. Kurata as "*F. graminearum* NHL-F-1104." Single-spore isolates obtained from this typical culture of *F. graminearum* produced mature perithecia and ascospores within 2 weeks on CLA.

Strain 8.36 *F. graminearum* NHL-F-1112 (= R-5470; MRC 1976)

This strain was isolated from barley grain in Japan and produces fusarenon-X, nivalenol, and zearalenone, according to H. Kurata (personal communication, 29 April 1980). We are not aware of any published reports on the toxigenicity of this strain.

A slant culture of this strain was received on 29 April 1980 from H. Kurata as "*F. graminearum* NHL-F-1112." This culture proved to be pionnotal but produced macroconidia on CLA that allowed it to be identified as *F. graminearum*.

Strain 8.37 *F. graminearum* MRC 120 (= R-6800; MRC 120)

This strain was isolated from corn produced in Zambia during 1974/75 (Marasas et al., 1978). Culture material of this strain on corn incubated at 25°C for 2 weeks followed by 6 weeks at 12°C was acutely toxic to day-old ducklings and caused the death of 4 out of 4 ducklings in a mean time of 6.75 days (Marasas et al., 1978). This culture material also caused the death of 2 of 4 rats within 14 days and marked feed refusal and weight loss in the survivors (Marasas et al., 1978). The culture material contained 1 g/kg of zearalenone but was not analyzed for deoxynivalenol or other trichothecenes. Thus the chemical nature of the mycotoxin(s) produced by this strain and responsible for the toxicity of this strain to ducklings and rats is unknown.

A lyophilized culture of this strain was received on 5 April 1982 from W.F.O. Marasas as "*F. graminearum* MRC 120." Single-spore isolates obtained from this culture produced mature perithecia and ascospores within 3 weeks on CLA.

Strain 8.38 *F. graminearum* F-59 = ATCC 34909 (= R-4838; MRC 1646)

This strain was isolated from millet in Hungary (Szathmary et al., 1976, 1977). According to M. Palyusik (personal communication, December 16, 1977) this was the unnumbered "home strain of *F. graminearum*" used in the experimental reproduction of the estrogenic syndrome in pigs by Palyusik (1973). Cultures on corn caused characteristic signs of hyperestrogenism including uterine hypertrophy and oocyte degeneration in gilts (Palyusik, 1973).

Palyusik et al. (1974) reported that cultures on corn of an unnumbered isolate of *F. graminearum* caused inhibition of spermatogenesis in ganders. According to M. Palyusik (personal communication, 16 December 1977) the strain used in this work was also *F. graminearum* F-59 (= ATCC 34909). Although this strain is known to produce zearalenone in culture, the levels of zearalenone present in the culture material that caused hyperestrogenism in pigs (Palyusik, 1973) and inhibition of spermatogenesis in ganders (Palyusik et al., 1974) were not indicated.

Cultures of this strain on millet containing 2% malt extract incubated at 23 to 27°C for

1 week followed by 2 weeks at 5 to 8°C and one week at 23 to 27°C caused uterine hypertrophy in rats and contained 200 to 730 µg/g of zearalenone as determined by TLC and GLC (Szathmary et al., 1976, 1977).

The production of deoxynivalenol by this strain was reported by Pathre & Mirocha (1978), who detected 320 µg/g of deoxynivalenol in cultures on rice and 15 mg/ℓ in PSC medium incubated at 21°C for 21 days. The level of 320 µg/g of deoxynivalenol detected in rice cultures of *F. graminearum* F-59 by Pathre & Mirocha (1978) is the highest yield of deoxynivalenol that has ever been reported for a culture of *F. graminearum* as far as we are aware.

A slant culture of this strain was received on 16 December 1977 from M. Palyusik as "*F. graminearum* F-59 (= ATCC 34909)."

Strain 8.39 *F. culmorum* F-79 = ATCC 34910 (= R-4613; MRC 2940)

This strain was isolated from millet in Hungary (Szathmary et al., 1976, 1977).

Diets containing culture material of an unnumbered Hungarian isolate of *F. culmorum* and supplying 100 ppm of zearalenone had no significant effects on feed consumption, weight gain, egg production, or egg fertility in laying geese (Palyusik & Koplik-Kovacs, 1975a, b). According to M. Palyusik (personal communication, 16 December 1977), the strain used in this work was *F. culmorum* F-79 (= ATCC 34910).

Cultures of this strain of *F. graminearum* (as *F. culmorum* F-79) on millet containing 2% malt extract incubated at 23 to 27°C for 1 week followed by 2 weeks at 5 to 8°C and 1 week at 23 to 27°C caused uterine hypertrophy in rats and contained 540 to 3,400 µg/g of zearalenone as determined by TLC and GLC (Szathmary et al., 1976, 1977). According to M. Palyusik (personal communication, 16 December 1977), this strain was the unnumbered Hungarian isolate of *F. culmorum* reported to produce zearalenone by Kovacs et al. (1975). Sandor et al. (1980) also reported that this strain (as *F. culmorum* F-79) produced 23.1 ppm of zearalenone as detected by TLC in cultures on corn incubated at 24 to 28°C for 14 days. The level of zearalenone in these cultures increased to 31.2 ppm following 4 weeks of cold exposure at 12 to 14°C and re-exposure to the growth temperature of 24 to 28°C. The levels of zearalenone in these cultures were further increased (up to 83.3 ppm) by exposure to ionizing gamma radiation (Sandor et al., 1980).

A slant culture of this strain was received on 16 December 1977 from M. Palyusik as "*F. culmorum* F-79 (= ATCC 34910)." Single-spore isolates obtained from this typical culture of *F. graminearum* produced mature perithecia and ascospores within 2 weeks on CLA.

Strain 8.40 *F. roseum* 'Culmorum' 70-K-11 (= R-6767; MRC 2560)

This strain (as *F. roseum* 70-K-11) was isolated from corn in New Guinea (Ueno et al., 1972b). Extracts of cultures of this strain (as *F. roseum* 'Culmorum' 70-K-11) in PSC medium incubated at 25 to 27°C for 2 weeks were acutely toxic to mice and also inhibited protein synthesis in rabbit reticulocytes (Ueno et al., 1973a). Analyses of these extracts by TLC revealed the presence of diacetoxyscirpenol, neosolaniol, T-2 toxin, and HT-2 toxin (Ueno et al., 1972b; 1973a). In shake cultures incubated at 27°C for 5 days, this strain produced unspecified levels of T-2 toxin and neosolaniol in a medium containing glucose, peptone, and yeast extract, but not in PSC medium (Ueno et al., 1975).

A slant culture of this strain was received during November 1981 from Y. Ueno as "*F. roseum* 'Culmorum' 70-K-11." Single-spore isolates obtained from this culture produced mature perithecia and ascospores within 3 weeks on CLA.

Strain 8.41 *F. tricinctum* NRRL 5908 (= R-5317; MRC 1781)

This strain was isolated from fescue hay in the United States according to Joffe & Palti (1975), who identified it as *F. sporotrichioides*. This strain (as *F. sporotrichioides* NRRL 5908) has been reported to be slightly dermotoxic to rabbit skin and to produce T-2 toxin (0.6 mg/10 g) as determined by GLC in cultures on wheat incubated at 12°C for 21 days (Joffe, 1978b; Joffe & Yagen, 1977).

A lyophilized culture of this strain was received on 23 October 1979 from NRRL as "*F. tricinctum* NRRL 5908." Single-spore isolates obtained from this typical culture of *F. graminearum* produced mature perithecia and ascospores within 2 weeks on CLA. It is not clear to us how this perithecial strain of *F. graminearum* could have been identified as either *F. tricinctum* or *F. sporotrichioides*. Possibly *F. tricinctum* NRRL 5908 was originally a mixture of *F. sporotrichioides* and *F. graminearum*.

Strain 8.42 *F. graminearum* MRC 121 (= R-4507; MRC 121)

This strain was isolated from corn contaminated with deoxynivalenol and zearalenone produced in Zambia during 1973/74 (Marasas et al., 1977). Cultures of this strain on corn incubated at 25°C for 2 weeks followed by 6 weeks at 12°C were acutely toxic to rats and caused the death of 4 of 5 rats within 3 days (Marasas et al., 1977). This material contained 1,270.0 ppm of zearalenone and 6.0 ppm of deoxynivalenol. Because of the high degree of toxicity of the culture material, it was diluted 1:10 with commercial corn meal. This material, containing 127.0 ppm of zearalenone, was fed to weanling virgin female rats for 7 days and caused marked increases in uterine weight (Marasas et al., 1977).

A lyophilized culture of this strain was received on 17 October 1978 from W.F.O. Marasas as "*F. graminearum* MRC 121."

Strain 8.43 *F. graminearum* MRC 460 (= R-4053; MRC 460)

This strain was isolated from corn contaminated with deoxynivalenol and zearalenone produced in South Africa during 1975 (Marasas et al., 1977). Cultures of this strain on corn incubated at 25°C for 2 weeks followed by 6 weeks at 12°C were acutely toxic to rats and caused the death of 2 of 5 rats within 3 days (Marasas et al., 1977). This material contained 1,140.0 ppm of zearalenone and 1.3 ppm of deoxynivalenol. Because of the high degree of toxicity of the culture material, it was diluted 1:10 with commercial corn meal. This material, containing 114.0 ppm of zearalenone, was fed to weanling female rats for 7 days and caused marked increases in uterine weight (Marasas et al., 1977).

A lyophilized culture of this strain was received on 17 October 1978 from W.F.O. Marasas as "*F. graminearum* MRC 460."

Strain 8.44 *F. graminearum* 72322 (= R-5176; MRC 1391)

This strain was isolated from wheat in Finland during 1972 according to A. Ylimäki (personal communication, 2 August 1978).

Korpinen & Uoti (1974) tested 10 unnumbered isolates of *F. graminearum* from cereal seeds in Finland for toxicity to rats. Cultures on a mixture of oats, barley and wheat were fed to rats for 14 days and caused marked reductions in weight gain. Most of the animals died before the 10th day and hemorrhage and ulceration of the gastric mucosa were prominent *post mortem* findings (Korpinen & Uoti, 1974). According to A. Ylimäki (personal communication, 10 April 1979) *F. graminearum* 72322 was one of the unnumbered strains of *F. graminearum* found to cause weight reductions in rats by Korpinen & Uoti (1974). The chemical nature of the mycotoxin(s) produced by this strain is unknown.

A slant culture of this strain was received on 2 August 1978 from A. Ylimäki as "*F. graminearum* 72322." Single-spore cultures obtained from this typical culture of *F. graminearum* produced mature perithecia and ascospores within 2 weeks on CLA.

Strain 8.45 *F. graminearum* 7137 (= R-5953; MRC 1393)
This strain (as *F. graminearum* No. 13) was isolated from a commercial feed mix in Finland (Korpinen et al., 1972). According to A. Ylimäki (personal communication, 10 April 1979), the strain referred to as *F. graminearum* No. 13 by Korpinen et al. (1972) is *F. graminearum* 7137.

Cultures of this strain on a mixture of oats, barley, and wheat were incubated at room temperature for one month. This material was fed to immature female rats for 7 days and caused marked increases in uterine weight (Korpinen et al., 1972). The zearalenone content of this culture material was not determined by Korpinen et al. (1972). Subsequently Kallela & Korpinen (1973) reported that one of the *F. graminearum* strains that was "found to cause clear estrogenic changes in rats" by Korpinen et al. (1972) produced zearalenone in culture. Since *F. graminearum* No. 13 caused the greatest increase in uterine weight of all the strains tested by Korpinen et al. (1972), it is possible that the unnumbered strain of *F. graminearum* found to produce zearalenone by Kallela & Korpinen (1973) may have been isolate No. 13.

The reduced palatability of the culture material of this strain (as *F. graminearum* No. 13) that caused uterine hypertrophy in rats also resulted in weight loss of these rats (Korpinen et al., 1972). Subsequently this strain (as *F. graminearum* 7137) was shown to produce 125 µg/g of deoxynivalenol in cultures on rice incubated at 21°C for 21 days (Pathre & Mirocha, 1978).

A slant culture of this strain was received on 2 August 1978 from A. Ylimäki as "*F. graminearum* 7137." Single-spore isolates obtained from this typical culture of *F. graminearum* produced mature perithecia and ascospores within 2 weeks on CLA.

Strain 8.46 *Gibberella zeae* B-507 (= R-6574; MRC 2638)
This strain was isolated from corn stalks in Michigan, USA, and was one of the unnumbered isolates of *G. zeae* used by Hart & Stebbins (1981) to inoculate sweet corn hybrids in the field, according to L.P. Hart (personal communication, September 1981). This strain (unnumbered) produced deoxynivalenol, but not zearalenone, at levels ranging from none detected to over 500 µg/g of tissue in ears of 33 commercial sweet corn hybrids inoculated in the field in the United States (Hart & Stebbins, 1981).

A slant culture of this strain was received during September 1981 from L.P. Hart as "*G. zeae* B-507." Single-spore isolates obtained from this typical culture of *F. graminearum* produced mature perithecia and ascospores within 2 weeks on CLA.

Strain 8.47 *Gibberella zeae* B-601 (= R-6575; MRC 2639)
This strain was isolated from corn stalks in Michigan, USA, and was one of the unnumbered isolates of *G. zeae* used by Hart & Stebbins (1981) to inoculate sweet corn hybrids in the field, according to L.P. Hart (personal communication, September 1981). This strain (unnumbered) produced deoxynivalenol, but not zearalenone, at levels ranging from none detected to over 500 µg/g of tissue in ears of 33 commercial sweet corn hybrids inoculated in the field in the United States (Hart & Stebbins, 1981).

A slant culture of this strain was received during September 1981 from L.P. Hart as "*G. zeae* B-601."

Strain 8.48 *Gibberella zeae* W-8 (= R-6576; MRC 2640)

This strain was isolated from scabbed wheat in Michigan, USA, according to L.P. Hart (personal communication, September 1981). This strain produced deoxynivalenol at levels ranging from none detected to over 500 µg/g and zearalenone at levels from none detected to over 180 µg/g of tissue in ears of 33 commercial sweet corn hybrids inoculated in the field in the United States (Hart & Stebbins, 1981).

A slant culture of this strain was received during September 1981 from L.P. Hart as "*G. zeae* W-8." Single-spore isolates obtained from this typical culture of *F. graminearum* produced mature perithecia and ascospores within 2 weeks on CLA.

Strain 8.49 *F. graminearum* ITM 126 (= R-5798; MRC 2065)

This strain was isolated in Italy, according to A. Bottalico (personal communication, 18 September 1980), but the exact source is unknown to us. The production of deoxynivalenol by this strain in cultures on corn incubated at 27°C for 4 weeks was reported by Palmisano et al. (1981). The levels of deoxynivalenol in the culture material of this strain were 1.20 and 1.18 µg/g as detected by GC and differential-pulse polarography, respectively (Palmisano et al. 1981).

A slant culture of this strain was received on 18 September 1980 from A. Bottalico as "*F. graminearum* ITM 126."

Strain 8.50 *F. graminearum* F-107 (= R-4610; MRC 3081)

This strain was isolated from feedstuffs in Germany (Möller et al., 1978). This strain produced 248 mg/kg of zearalenone in cultures on corn incubated at 25°C for 21 days followed by 14 days at 14°C (Möller et al., 1978). This strain has also been found to produce zearalenone at levels up to 232 mg/kg in corn adjusted to different moisture levels and incubated at 20°C for 10 days, at 10°C for 31 days, and finally at 20°C for 111 days (Müller & Thaler, 1981). This strain did not produce zearalenone in cultures on corn amended with propionic acid during an incubation period of 4 months (Müller & Thaler, 1981). In cultures on non-sterilized corn with a moisture content of 39%, *F. graminearum* F-107 produced 8.0 mg/kg of zearalenone after incubation at 10°C for 40 days (Thaler et al., 1979). Air-drying of the corn inoculated with this strain did not entirely prevent zearalenone formation, and levels up to 0.87 mg/kg could be detected after 20 days (Thaler et al., 1979).

A slant culture of this strain was received on 9 December 1977 from J.M. Möller as "*F. graminearum* F-107." Single-spore isolates obtained from this culture produced mature perithecia and ascospores within 3 weeks on CLA. This culture has also been lyophilized at BAF as *F. graminearum* Sp. 898.

Strain 8.51 *F. graminearum* F-110 (= R-4611; MRC 3082)

This strain was isolated from feedstuffs in Germany (Möller et al., 1978). This strain produced 324 mg/kg of zearalenone in cultures on corn incubated at 25°C for 21 days followed by 14 days at 14°C (Möller et al., 1978). This strain has also been found to produce zearalenone at levels up to 405 mg/kg in corn adjusted to different moisture levels and incubated at 20°C for 10 days, at 10°C for 31 days, and finally at 20°C for 111 days (Müller & Thaler, 1981). This strain did not produce zearalenone in cultures on corn amended with propionic acid during an incubation period of 4 months (Müller & Thaler, 1981).

A slant culture of this strain was received on 9 December 1977 from J.M. Möller as "*F. graminearum* F-110." Single-spore isolates obtained from this culture produced mature perithecia and ascospores within 3 weeks on CLA. This culture has also been lyophilized at BAF as *F. graminearum* Sp. 899.

Strain 8.52 *F. graminearum* **G-73-2** (= R-6182; MRC 2393)
This strain (unnumbered) was isolated from corn near Guelph, Ontario, Canada, in 1973 (Sutton et al., 1976). According to J.C. Sutton (personal communication, 5 May 1981), this unnumbered strain was *F. graminearum* G-73-2. This strain produced 480 μg/g of zearalenone in cultures on corn incubated at 20°C for 6 weeks (Sutton et al., 1976). Zearalenone has also been detected in the stems and ears of corn plants inoculated artificially in stems, the ears, or both with this strain of *F. graminearum*. The average levels of zearalenone detected in the stems of corn plants inoculated in the stems was 132 μg/100 g, compared to 1,236 μg/100 g in the ears of plants inoculated in the ears. The finding of an average level of 18.3 μg/100g of zearalenone in the healthy ears of four plants inoculated in the stems with this strain of *F. graminearum* provided circumstantial evidence for the translocation of zearalenone from the stem to the ear in corn plants, according to Sutton et al. (1976).

A slant culture of this strain was received on 5 May 1981 from J.C. Sutton as "*F. graminearum* G-73-2."

9. Section Lateritium

The only species accepted in this section by Nelson et al. (1983), *F. lateritium,* has been reported to be toxigenic.

Fusarium lateritium Nees

Perfect State: *Gibberella baccata* (Wallr.) Sacc.

Incidence and Distribution

Fusarium lateritium occurs worldwide and has a very wide host range, but typically causes dieback, canker, or bark disease of woody trees and shrubs (Booth, 1971; Doidge, 1938; Gerlach & Ershad, 1970; Gordon, 1956, 1960a; Meyer & Frank, 1979; Wollenweber & Reinking, 1935).

Association with Human and/or Animal Diseases

Fusarium lateritium has not been associated with any human or animal toxicoses, although some strains isolated from overwintered cereals associated with ATA (see *F.*

sporotrichioides, ATA) in the USSR and from fescue hay associated with "fescue foot" of cattle (see *F. sporotrichioides*, fescue foot) in the USA have been referred to in the literature as *F. lateritium*.

Toxicity to Experimental Animals

Brine Shrimp. N.D. Davis et al. (1975) reported that *F. lateritium* AUA 553 was toxic to brine shrimp.

Chickens. *Fusarium lateritium* AUA 553 was reported to be toxic to chicken embryos by N.D. Davis et al. (1975). According to Diener et al. (1976), extracts of cultures on nutrient-amended shredded wheat incubated at 25°C for 14 to 21 days of two isolates of *F. lateritium* (844 and 845) from cotton in the United States were toxic to chicken embryos. The chemical nature of the mycotoxin(s) produced by the above strains is unknown.

Guinea Pigs. According to Joffe (1960a, 1965, 1971, 1974c, 1978b), filtrates of liquid cultures and powdered mycelium of unspecified dermotoxic isolates of *F. lateritium* isolated from overwintered cereals associated with ATA in the USSR were lethal to guinea pigs within 5 to 21 days. Autopsies of most animals revealed "haemorrhages in organs and tissues, dilated blood vessels, and haemorrhages in the walls of the intestines" (Joffe, 1960a). The chemical nature of the mycotoxin(s) produced by these dermotoxic strains is unknown.

During the isolation of a metabolite with larvicidal activity against *Lucilia sericata* from culture filtrates of *F. lateritium* IMI 140879 (=ATCC 20227) grown in shake culture at 26°C for 5 days, two assistants developed localized rashes on the face (Cole & Rolinson, 1972). The application of chromatography column fractions of cultures of this strain to the shaved back of a guinea pig resulted in a delayed but marked inflammatory reaction with severe erythema followed by central necrosis and lateral thickening of the skin. The larvicidal and dermotoxic metabolite was identified as diacetoxyscirpenol (Cole & Rolinson, 1972). Cultures of ATCC 20227 obtained from ATCC were not viable and consequently we were unable to confirm the identity of this strain (see diacetoxyscirpenol, below).

Insects. Culture filtrates of *F. lateritium* IMI 140879 (=ATCC 20227) grown in shake culture at 26°C for 5 days were lethal to the larvae of *Lucilia sericata* (Cole & Rolinson, 1972). The larvicidal activity was used as a bioassay to isolate the active metabolite which was identified as diacetoxyscirpenol. Cultures of ATCC 20227 obtained from ATCC were not viable and consequently we were unable to confirm the identity of this strain (see diacetoxyscirpenol, below).

Mice. Yates et al. (1970) reported that the unidentified *Fusarium* strains Nos. 25, 26, and 28, isolated from toxic fescue hay associated with an outbreak of "fescue foot" in cattle, and No. 17, isolated from orchard grass, were toxic to mice upon intraperitoneal injection of extracts of cultures on Sabouraud's agar incubated at 15°C. All of these toxic strains were found to produce butenolide. Burmeister et al. (1971) identified these four strains as *F. lateritium* and confirmed their toxicity to mice as well as the production of butenolide by all these strains and of T-2 toxin by strain No. 17. We were unfortunately not able to examine any of these toxic isolates referred to as *F. lateritium* by Burmeister et al. (1971), but we have reason to believe that they were not correctly identified as *F. lateritium* (see butenolide, below).

Extracts of cultures of three out of four strains of *F. lateritium* isolated from river sediments in Japan were toxic to mice (Ueno et al., 1977b). We have identified two of these toxic strains (Nos. 5013 and 5036) as *F. oxysporum* (Strains 11.6 and 11.7) and these strains are consequently excluded from *F. lateritium*.

Rabbits. Joffe (1960a,b, 1971) reported that 32 isolates of *F. lateritium* from overwintered cereals and their soils associated with outbreaks of ATA (see *F. sporotrichioides*, ATA) in the USSR were tested for dermotoxicity to rabbit skin and that two isolates (Nos. 1421 and 2381) were found to be toxic and three mildly toxic. Upon dosing to rabbits, culture filtrates and powdered dry mycelium of these dermotoxic strains of *F. lateritium* caused death within 8 to 24 days (Joffe 1960a). Autopsies of most animals revealed "haemorrhage in organs and tissues, dilated blood vessels and haemorrhage in the walls of the intestines" (Joffe, 1960a). The chemical nature of the mycotoxin(s) produced by these strains toxic to rabbits is unknown.

Mycotoxins Produced

Butenolide. Yates et al. (1970) reported that four unidentified *Fusarium* isolates, Nos. 25, 26, and 28 from fescue hay associated with an outbreak of "fescue foot" (see *F. sporotrichioides*, fescue foot) in cattle and No. 17 isolated from orchard grass in the United States, produced butenolide (based on TLC and IR evidence) in cultures on Sabouraud's agar incubated at 15°C. Burmeister et al. (1971) identified these four strains as *F. lateritium* and confirmed the production of butenolide by these strains. Burmeister et al. (1971) also stated that a "*Gibberella* stage" was observed in cultures of all four of these butenolide-producing isolates. On the basis of this statement we doubt that these isolates were in fact *F. lateritium* because in *F. lateritium* "both homo- and heterothallic strains exist but even in the homothallic strains mature perithecia are seldom produced on agar" (Booth 1971). We were unfortunately not able to examine any of these isolates referred to as *F. lateritium* by Burmeister et al. (1971) because apparently none of them have been maintained at NRRL (J.J. Ellis, personal communication, 2 June 1981).

Diacetoxyscirpenol. The metabolite with larvicidal and dermotoxic activity in shake cultures of *F. lateritium* IMI 140879 incubated at 26°C for 5 days was identified by elemental analysis, molecular weight, and IR and NMR spectra as diacetoxyscirpenol (Cole & Rolinson, 1972). *Fusarium lateritium* IMI 140879 is not listed in the Catalogue of the Culture Collection of the CMI, Seventh Edition (1975), but is listed in the ATCC Catalogue of Strains, Fifteenth Edition (1982), as ATCC 20227, "*F. lateritium*, Beecham Res. Labs. BRL 886 (IMI 140879)." A lyophilized culture of ATCC 20227 received from the ATCC on 9 March 1981 was not viable, and neither was another lyophilized culture received on 10 June 1981. Consequently we were unable to confirm the identity of this isolate.

The strain reported as *Fusarium* sp. K 5036 (Ishii, 1975) or *F. lateritium* 5036 (Ueno et al., 1977b) to produce diacetoxyscirpenol is *F. oxysporum* (see Strain 11.7).

Diacetylnivalenol. The strain reported as *Fusarium* sp. K. 5036 (Ishii, 1975) or *F. lateritium* 5036 (Ueno et al., 1977b) to produce diacetylnivalenol is *F. oxysporum* (see Strain 11.7).

7,8-Dihydroxydiacetoxyscirpenol. The strain reported as *Fusarium* sp. K. 5036 (Ishii, 1975) or *F. lateritium* 5036 (Ueno et al., 1977b) to produce this new trichothecene is *F. oxysporum* (see Strain 11.7).

Table 9.1. *Fusarium* Strains Published as *F. lateritium* but Identified as Other Species in the International Toxic *Fusarium* Reference Collection (ITFRC)

Strain No.	ITFRC No.	MRC No.	Published Name(s) and Number(s)	Source	Present Identification
11.6	O-1174	2566	*F. lateritium* 5013	Y. Ueno	*F. oxysporum*
11.7	O-1175	2567	*F.* sp. K. 5036 (=*F. lateritium* 5036)	Y. Ueno	*F. oxysporum*

7-Hydroxydiacetoxyscirpenol. The strain reported as *Fusarium* sp. K. 5036 (Ishii, 1975) or *F. lateritium* 5036 (Ueno et al., 1977b) to produce this new trichothecene is *F. oxysporum* (see Strain 11.7).

Neosolaniol. The strain reported as *Fusarium* sp. K. 5036 (Ishii, 1975) or *F. lateritium* 5036 (Ueno et al., 1977b) to produce neosolaniol is *F. oxysporum* (see Strain 11.7).

T-2 Toxin. Yates et al. (1970) reported that *Fusarium* sp. No. 17, isolated from orchard grass in the United States, produced T-2 toxin as detected by TLC in ethyl acetate extracts of cultures on Sabouraud's agar incubated at 15°C. Burmeister et al. (1971) identified isolate No. 17 as *F. lateritium* and confirmed the production of T-2 toxin by this isolate in cultures on Sabouraud's agar incubated at 15°C for 21 days. Burmeister et al. (1972) tested 10 isolates of *F. lateritium* for T-2 toxin production and found that two isolates (Nos. 1 and 2) from fescue hay produced unspecified levels of T-2 toxin as detected by TLC on autoclaved white corn grits incubated at 15°C for 21 days. We were unfortunately not able to examine any of these T-2 toxin–producing isolates referred to as *F. lateritium* by Burmeister et al. (1971, 1972), but we have reason to believe that they were not correctly identified as *F. lateritium* (see butenolide, above).

Zearalenone. Ishii et al. (1974) reported that one unspecified isolate of *F. lateritium* produced zearalenone and this was confirmed by Ueno et al. (1977b), who referred to the producing strain as *F. lateritium* 5013. We have identified this isolate as *F. oxysporum* (Strain 11.6).

Toxigenic Strains in the ITFRC

There are no toxigenic strains of *F. lateritium* represented in the ITFRC. Toxigenic strains that have been referred to as *F. lateritium* in the literature but identified as other species by us are listed in Table 9.1.

10. Section Liseola

The following four species in the section Liseola accepted by Nelson et al. (1983) are considered to be toxigenic: *F. moniliforme, F. proliferatum, F. subglutinans,* and *F. anthophilum.*

Fusarium moniliforme Sheldon

Perfect State: *Gibberella fujikuroi* (Sawada) Wollenw.

Incidence and Distribution

Fusarium moniliforme occurs world-wide on a great variety of plant hosts and is one of the most prevalent fungi associated with corn (*Zea mays* L.) in most corn-producing areas of the world (Booth, 1971; CMI Descriptions of Pathogenic Fungi and Bacteria No. 22, 1964; CMI Distribution Maps of Plant Disease, Map No. 102, Ed. 4, 1972; Doidge, 1938; Gordon, 1944, 1952, 1956, 1960a; Joffe et al., 1973; Kellerman et al., 1972; Meyer & Frank, 1979; Nirenberg, 1976; Palti, 1978; Wollenweber & Reinking, 1935).

Fusarium moniliforme is one of the most prevalent fungi associated with basic human and animal dietary staples such as corn, it has been suspected of being involved in human and animal diseases since its original description in the previous century (see association with human and/or animal diseases, below), and has been proven to be toxic to a variety of experimental animals (see toxicity to experimental animals, below). However, the chemical nature of the mycotoxins produced by this fungus is for the most part unknown (see mycotoxins produced, below).

Association with Human and/or Animal Diseases

Saccardo (1881) described a fungus found on corn kernels in Italy as *Oospora verticillioides* Sacc. This fungus soon became implicated as the possible cause of pellagra because of its prevalence in corn associated with the disease in Italy (Cuboni, 1882; Tiraboschi, 1905) and Russia (Von Deckenbach, 1907). At about the same time, widespread field outbreaks occurred in the United States of a disease in animals associated with the ingestion of moldy corn (Peters, 1904). The hooves of cattle and horses sloughed, pigs shed their bristles, chickens lost their feathers, some animals developed convulsions, and a high-percentage of affected animals died (Peters, 1904). The fungus most commonly associated with the moldy corn and implicated as the cause of this disease was described as *Fusarium moniliforme* Sheldon by Sheldon (1904). Although it was soon pointed out that *F. moniliforme* was identical to *O. verticillioides* (Manns and Adams, 1923; Norton and Chen, 1920; Wineland, 1924), the new combination *Fusarium verticillioides* (Sacc.) Nirenberg was only made by Nirenberg (1976). In this volume, the name *F. moniliforme* Sheldon is, however, used in preference to *F. verticillioides* (Sacc.) Nirenberg because of reasons explained by Nelson et al. (1983).

Equine Leukoencephalomalacia (LEM). The only mycotoxic disease of either humans or animals for which the causative role of *F. moniliforme* has been established beyond doubt is equine leukoencephalomalacia (Bridges, 1978; Buck et al., 1979; Haliburton et al., 1979; Kriek et al., 1981b; Marasas et al., 1976; Pienaar, 1977; Wilson, 1971; Wilson & Maronpot, 1971; Wilson et al., 1973). This neurotoxic disease of horses, donkeys and mules is characterized by liquefactive necrotic lesions in the white matter of the cerebral hemispheres, from which the name equine leukoencephalomalacia (LEM) was derived. Mortalities in horses, preceded by marked nervous symptoms prior to death, have been known since 1850 in the corn-growing regions of the United States (Biester & Schwarte, 1939; Biester et al., 1940; Buckley & MacCallum, 1901; Butler 1902; Graham, 1912, 1935, 1936; Lock, 1974; Schwarte et al., 1937). Thousands of horses died in Maryland around the turn of the century (Buckley & MacCallum, 1901) and in some of the Midwestern states during the 1930s (Biester & Schwarte, 1939; Biester et al., 1940; Graham, 1935, 1936; Schwarte et al., 1937). As recently as 1978–79, the disease has been responsible for the deaths of hundreds of horses in the United States (Buck et al., 1979). The disease is also known to occur in Argentina (Monina et al., 1981; Rodriguez, 1945), China (Iwanoff et al., 1957; Pan, 1981), Egypt (Badiali et al. 1968), and South Africa (Kellerman et al., 1972; Marasas et al., 1976; Pienaar et al., 1981).

As early as 1902 equine LEM was reproduced experimentally in horses by the feeding of naturally contaminated moldy corn in the United States by Butler (1902). His findings were subsequently confirmed by several workers in the United States (Biester et al., 1940; Schwarte et al., 1937), China (Iwanoff et al., 1957), and Egypt (Badiali et al., 1968). Despite these early indications that equine LEM is a mycotoxicosis, the causative fungus was positively identified only in 1971 when the disease was experimentally reproduced in two donkeys with a pure culture of an Egyptian isolate (see Strain 10.1) of *F. moniliforme* (Wilson, 1971; Wilson & Maronpot, 1971; Wilson et al., 1973). This was confirmed in South Africa by the induction of typical lesions in four horses with pure cultures of three local isolates of *F. moniliforme* (see Strains 10.2, 10.3, and 10.4) from moldy corn (Kriek et al., 1981b; Marasas et al., 1976; Pienaar, 1977). These and other toxic South African isolates (see Strains 10.2, 10.3, 10.4, 10.5, and 10.6) were also shown to be capable of causing a fatal hepatic syndrome in horses (Kellerman et al., 1972; Kriek et al., 1981b) as well as a variety of lesions in other experimental animals (Kriek et al., 1981a, b). Recently typical LEM was again experimentally reproduced in one donkey with the original Egyptian isolate (see Strain 10.1) of *F. moniliforme* (Buck et al., 1979; Haliburton et al., 1979). We have confirmed the identity of all the above strains as *F. moniliforme* (Strains 10.2, 10.3, 10.4, 10.5, and 10.6).

The chemical nature of the *F. moniliforme* mycotoxin that causes equine LEM is unknown at present. It is highly unlikely that the mycotoxin monliformin is involved, since culture material of neither the Egyptian (Buck et al., 1979; Haliburton et al., 1979) nor a South African (Kriek et al., 1981b) isolate of *F. moniliforme* (Strains 10.1 and 10.4) that induced typical lesions of the disease contained chemically or biologically detectable levels of moniliformin. Intravenous injection of moniliformin in a donkey resulted in death after 27 days (Buck et al., 1979). No gross lesions of LEM were observed in the brain, but pathological changes included microscopic hemorrhages in the subcortical white matter, satellitosis in the cortex and subcortical areas, and occasional neurophagia. Haliburton et al. (1981) also reported that moniliformin caused overt clinical signs of acute toxicosis in donkeys, but concluded that neither moniliformin nor two other metabolites of *F. moniliforme*, i.e., 5-butylpicolinic acid (= fusaric acid) and 2-methoxy-4-ethylphenol, appear to be involved in the pathogenesis of LEM.

Investigations are in progress in the United States and in South Africa to isolate and characterize the *F. moniliforme* mycotoxin that causes equine LEM. *Fusarium moniliforme* is one of the most prevalent fungi associated with corn, a basic human and animal dietary staple. The natural occurrence in corn of an unidentified mycotoxin capable of causing the remarkable brain lesions of LEM in equine animals is attested to by the occurrence of field outbreaks of the disease in most corn-producing areas of the world. Consequently it is imperative that this mycotoxin produced by *F. moniliforme* should be chemically characterized as soon as possible.

Clinical signs very similar to those of equine LEM have also been described in "bean hulls poisoning," a disease of horses which is very prevalent in Hokkaido, Japan (Konishi & Ichijo, 1970a,b; Ueno et al., 1972a). The main pathological lesions of this disease have been described as "degenerative change of the nerve cells of the cerebral cortex and acute circulatory disturbance" (Ueda et al., 1967), but no specific mention was made of leukoencephalomalacic lesions. Although liquefactive necrotic lesions in the white matter are not found in all cases of LEM (Badiali et al., 1968; Wilson et al., 1973), it is not clear at present whether bean hulls poisoning is identical to LEM or not. Bean hulls poisoning has been reproduced experimentally in Japan by feeding moldy soy bean hulls to horses (Konishi & Ichijo, 1970b). Toxgenic strains of *F. moniliforme* that caused circulatory disturbances and death in mice, but did not produce any known trichothecene toxins, were isolated from the toxic moldy bean hulls (Ueno, 1973b; Ueno et al., 1972a, 1973a). These strains soon lost their toxin-producing ability and were not tested in horses (Ueno, 1973b). We have examined only one of these strains of *F. moniliforme* (No. M-10-2) isolated from bean hulls in Hokkaido (Ueno et al., 1972a, 1973a). We found that this culture is not *F. moniliforme*, but a degenerate strain of *F. avenaceum* (Strain 5.4). It is not known whether the culture which we examined is in fact the same culture originally identified as *F. moniliforme,* or perhaps a contaminant or mislabeled culture, but the available culture has to be excluded from *F. moniliforme.*

Abnormal Bone Development. *Fusarium moniliforme* was one of the predominant fungi isolated from commercial poultry feeds in Arkansas by Sharby et al. (1973). Cultures on autoclaved corn of two of these isolates (ATCC 24088 and 24089) caused a reduction in weight gain and feed efficiency in broiler chicks, and within 12 days the chicks exhibited a severe leg deformity characterized by a bowing outward or a slanting inward of the legs, giving the chicks either a "cowboy" or a "knock-kneed" appearance (Sharby et al., 1973). We have identified both of these isolates as *F. moniliforme* (Strains 10.10 and 10.11). This leg abnormality also occurred in some of the chicks receiving the control diet containing uninoculated autoclaved corn, but the incidence was much lower than in the groups receiving corn inoculated with the two strains of *F. moniliforme.* This finding led Sharby et al. (1973) to conclude that the leg abnormalities in the control chicks were caused by the presence of a metabolite formed during previous mold contamination of the corn used in the control diet.

In France, feeds heavily contaminated by *F. moniliforme* have been implicated in outbreaks of rickets-like diseases characterized by nervousness, paraplegia, and osteodystrophy in pigs and by paralysis, osteomalacia, and 40 to 80% mortality in poultry (Moreau, 1974, 1979). Two hypotheses have been proposed for mechanisms by which metabolites of *F. moniliforme* could cause abnormal bone development. Fusaric acid, a known chelating agent produced by *F. moniliforme* (see fusaric acid, below), may induce a phosphorus deficiency in the animal; because of the importance of phosphorus in bone development—for instance in the synthesis of vitamin D—this deficiency may result in the development of rickets (Moreau, 1974). *Fusarium moniliforme* is also known to produce a thiaminase, and an induced thiamine deficiency may result in locomotion difficulties, paralysis, and death (Moreau, 1974, 1979).

During the autumn and winter of 1974/75, an increased prevalence of rickets-like signs was noted in commercial broilers in Germany (Gedek et al., 1978). The clinical signs and pathological lesions of the disease, characterized by a rapidly developing severe rickets with widening of the growth plates, insufficient cartilage formation, demineralisation of the spongiosa and compacta, and production of fibrous bone marrow with some degree of granulomatous proliferation of osteoclasts, could be reproduced experimentally in broilers with feed contaminated with *F. moniliforme* from an affected flock (Gedek et al., 1978; Köhler et al., 1978). Administration of calcium or phosphorus did not influence the development of the disease, but the addition of vitamin D_3 had a rapid and favorable effect. Analysis of the incriminated feed and of culture material of a strain of *F. moniliforme* isolated from the feed for zearalenone and trichothecenes by TLC yielded negative results. However, in biological tests involving chicken embryos, cell cultures, and the rabbit skin test, positive results were obtained, thus indicating that the feed as well as the *F. moniliforme* culture contained unidentified mycotoxins that could not be detected chemically. On the basis of these findings, Köhler et al. (1978) concluded that *F. moniliforme* influences the availability of vitamin D, not only by the production of specific toxic metabolites, but also by causing changes in the basic steroidal structure.

Although *F. moniliforme* has been implicated as the cause of the rickets-like diseases of poultry and pigs described above because of its prevalence in the associated feeds, the causative role has not been proven in experiments with pure cultures of the fungus and the specific metabolite(s) involved have not yet been identified.

Moldy Sweet Potato Toxicosis. A toxicosis of cattle characterized by severe pulmonary edema has been attributed to the presence of four furanoterpenoids in sweet potatoes (*Ipomoea batatas* L.) infected by *F. solani* (Boyd, 1976; Boyd & Wilson, 1972; Boyd et al., 1973; Doupnik et al., 1971; Peckham et al., 1972; Wilson & Boyd, 1974; Wilson et al., 1970, 1971; Yang et al., 1971). This syndrome is described in detail under *F. solani* (see *F. solani*, moldy sweet potato toxicosis). We have, however, identified an unnumbered isolate of *F. solani* used in the work cited above as *F. moniliforme* (Strain 10.8). Although *F. moniliforme* was also isolated from moldy sweet potatoes implicated in a field outbreak of the disease in Georgia, USA (Doupnik et al., 1971; Peckham et al., 1972), it is unknown whether the culture that we received as *F. solani* and identified as *F. moniliforme* (Strain 10.8) is a contaminant or whether *F. moniliforme* rather than *F. solani* was in fact used in the research on moldy sweet potato toxicosis cited above.

Human Esophageal Cancer. In Africa, the highest human esophageal cancer rate occurs in the southwestern districts of the Transkei, while the rate in the northeastern region of the Transkei is relatively low (Marasas et al., 1979c, 1981). Corn is the main dietary staple in both areas. In a comparative study of the mycoflora of home-grown corn produced in the two areas, the most striking and consistent difference was the significantly higher incidence of *F. moniliforme* (= *F. verticillioides*) in corn produced in the high-rate area (Marasas et al., 1979c, 1981). Several isolates of *F. moniliforme* from corn produced in the high esophageal cancer rate area in Transkei have been found to be acutely toxic to ducklings, but not to produce moniliformin (Kriek et al., 1981a). The acute and long-term effects in experimental animals of two of these isolates, *F. verticillioides* MRC 826 and MRC 602, have been determined, and the lesions induced included cirrhosis and nodular hyperplasia of the liver and intraventricular cardiac thrombosis in rats, leukoencephalomalacia and toxic hepatosis in horses, pulmonary edema in pigs, nephrosis and hepatosis in sheep, and acute congestive heart failure in baboons (Kriek et al., 1981a,b). We have identified both of these isolates as *F. moniliforme* (Strains 10.4 and 10.5). Although these Transkeian isolates of *F. moniliforme* were shown to affect a variety of target organs in different animal species, no experimental evidence has been obtained to date that they can cause esophageal carcinoma. The statement that "leukoencephaly, liver damage, and/or esophagal carcinoma have been observed in mammals fed with corn contaminated with *Fusarium moniliforme*," attributed to W.F.O. Marasas by Scharf & Frauenrath (1980), is incorrect.

In China, *F. moniliforme* is one of the fungi most frequently associated with foodstuffs in Lin Xian county in Henan province, which is one of the highest esophageal cancer risk areas in the world (Li et al., 1979, 1980). According to the reviews by Li et al. (1980), Lin & Tang (1980), and Yang (1980), corn meal inoculated with isolates of *F. moniliforme* from Lin Xian county has been found to induce tumors in several different organs in rats. We have confirmed the identity of an unnumbered Chinese isolate as *F. moniliforme* (Strain 10.7). Although carcinoma of the esophagus was not observed, the *F. moniliforme*-infested corn meal induced epithelial hyperplasia, precancerous changes, and papillomas of the esophagus and stomach of rats and mice. Four nitrosamines have been detected in corn bread inoculated with Chinese isolates of *F. moniliforme* (see Strain 10.7) following nitrosation with sodium nitrite (Li et al., 1979, 1980; Lu et al., 1979, 1980a). At least one of these nitrosamines, N-methyl-N-benzylnitrosamine, is known to selectively produce tumors of the esophagus in rats and has also been shown to be mutagenic in *Salmonella typhimurium* (Lu et al., 1980b).

Several North American and European strains of *F. moniliforme* have also been shown to be mutagenic in *S. typhimurium* (Bjeldanes & Thompson, 1979; Bjeldanes et al., 1978). Four mutagens—fusariogenins A, B, C, and D—have been isolated from one of these strains, and fusariogenin C, the mutagen produced in largest quantity, was reported to be a non-crystalline compound, $C_{23}H_{29}O_7$ (Bjeldanes & Weib, 1980). The major mutagenic compound from *F. moniliforme* was subsequently named fusarin C, and high-resolution mass spectral data and combustion analysis indicated a molecular formula of $C_{23}H_{29}NO_7$ (Wiebe & Bjeldanes, 1981).

At present it is not possible to postulate a causative role for *F. moniliforme* in the etiology of human esophageal cancer because there is no direct experimental evidence of a cause-and-effect relationship. However, the finding that the incidence of *F. moniliforme* in corn is correlated with the esophageal cancer rate in Transkei (Marasas et al., 1981), the frequent association of *F. moniliforme* with foodstuffs in high-rate areas in China (Li et al., 1979, 1980), the presence of mutagenic nitrosamines, some of which are known site-specific esophageal carcinogens, in corn bread inoculated with Chinese isolates of *F. moniliforme* (Li et al., 1979, 1980; Lu et al., 1979, 1980a,b), and the production of mutagenic substances by 21 out of 33 isolates of *F. moniliforme* tested (Bjeldanes & Thompson, 1979; Bjeldanes & Weib, 1980), certainly merit further investigation of the possible role of *F. moniliforme* in the etiology of human esophageal cancer.

Toxicity to Experimental Animals

Baboons. Cultures of *F. moniliforme* (as *F. verticillioides*) MRC 602, isolated from corn produced in a high-incidence area of esophageal cancer (see human esophageal cancer, above) in Transkei, have been fed to three baboons in their rations (Kriek et al., 1981 b). We have confirmed the identity of this strain as *F. moniliforme* (Strain 10.5). Two of the baboons died of acute congestive heart failure after 143 and 248 days, respectively, while the third baboon had cirrhosis of the liver when it was killed after 720 days.

Chickens. Archer (1974) reported that extracts of an isolate of *F. moniliforme* (No. 161 = IMI 173211) from moldy corn was toxic to chick embryos. Extracts of cultures of three isolates of *F. moniliforme* (Auburn University Culture Collection Nos. 573, 587, and 893) from cotton and one (AUA No. 1093) from grain sorghum in the United States were also found to be toxic to chick embryos by Diener et al. (1976, 1981). We have examined one of these isolates from cotton (No. 587) and identified it as *F. proliferatum* (Strain 10.33). This strain is accordingly excluded from *F. moniliforme.*

One isolate of *F. moniliforme* from moldy corn in Wisconsin, USA, was cultured on autoclaved corn at either 10° or 25°C for 21 days and fed to day-old White Leghorn cockerels by Marasas & Smalley (1972). The cultures incubated at 10° and 25°C caused the death of 1 of 10 and 9 of 10 chicks, respectively, within 14 days.

Extracts of cultures of an isolate of *Fusarium* sp. from wheat in India caused the death of one out of three day-old chicks upon oral dosing (Mall et al., 1979). We have identified a culture of this *Fusarium* sp. (PL W-604) as *F. moniliforme* (Strain 10.16).

Diener et al. (1981) reported that an isolate of *F. moniliforme* (AUA 1093) from grain sorghum in the United States caused the death of two out of four day-old White Leghorn chicks intubated with ethanol-chloroform extracts of cultures on grain sorghum incubated at 25°C for 21 days.

The mycotoxin(s) produced by the above strains of *F. moniliforme* toxic to chicks have not been chemically characterized.

Sharby et al. (1973) reported that cultures of two isolates of *F. moniliforme* (ATCC

24088 and 24089) from poultry feed in Arkansas, USA, induced a severe leg abnormality (see abnormal bone development, above) in broiler chicks. We have confirmed the identity of both of these strains as *F. moniliforme* (Strains 10.10 and 10.11).

Fritz et al. (1973) reported that cultures of *F. moniliforme* No. 13, isolated from moldy feed implicated in an outbreak of a hemorrhagic disease in swine in the midwestern United States, caused paralysis with the head retraction typical of polyneuritis in White Plymouth Rock chicks. Four other isolates of *F. moniliforme* did not have this effect. Affected chicks made a dramatic recovery upon injection of thiamine hydrochloride, and addition of thiamine to the diet containing. *F. moniliforme* culture material also prevented the development of polyneuritis. Analysis for thiamine content showed that the negative control diet contained 5.33 mg/kg thiamine, while the *F. moniliforme* culture material that caused polyneuritis contained less that 0.1 mg/kg thiamine. Other experiments indicated that *F. moniliforme* No. 13 produced a thiaminase that was destroyed by autoclaving (Fritz et al., 1973). The implications of thiaminase production by *F. moniliforme* were discussed earlier (see abnormal bone development, above).

Cole et al. (1973) used a cockerel bioassay to purify the highly toxic mycotoxin, moniliformin, from an isolate of *F. moniliforme* (NRRL 5860) from southern leaf blight–damaged corn seed in the United States. We have confirmed the identity of this strain as *F. moniliforme* (Strain 10.23). The oral LD_{50} of the purified toxin in cockerels was determined as 4.0 mg/kg. High doses (40 mg/kg) of moniliformin resulted in death within 45 minutes. Cockerels that died within 2 hours had no detectable lesions in the organs while the lesions in birds that lived more than 2 hours after dosing were ascites with edema of the mesenteries and small hemorrhages in the proventriculus, gizzard, small and large intestine, and skin (Cole et al., 1973). Moniliformin-containing culture material of *F. moniliforme* NRRL 6322 (Strain 10.25) on corn grits was fed to day-old broiler chickens for 21 days by Allen et al. (1981b). We have identified this strain according to Nelson et al. (1983) as the "short-chained" type of *F. moniliforme* (see gibberellins, below). Chicks fed diets containing 64 mg/kg of moniliformin exhibited reduced weight gain and feed consumption. Although the total daily moniliformin consumption by these chicks was nearly twice the reported single oral LD_{50} dose, only 3 of 10 died and no lesions were found upon necropsy. The intravenous LD_{50} of moniliformin isolated from this strain was 1.38 mg/kg body weight in broiler chickens (Allen et al. 1981b).

A strain of *F. moniliforme* (M-1) was reported to produce four substances with mutagenic activity in the Ames assay (Bjeldanes & Weib, 1980; Wiebe & Bjeldanes, 1981). These mutagens were originally referred to as fusariogenins A, B, C, and D by Bjeldanes & Weib (1980), but were subsequently called fusarins A, B, C, and D by Wiebe & Bjeldanes (1981). Cultures on autoclaved corn incubated at 23°C for 2 weeks followed by 2 weeks at 11°C contained fusarins. Feed containing 56% of this culture material did not produce any apparent toxic effects in day-old White Leghorn cockerels fed for 21 days and actually supported better growth than control corn meal (Wiebe & Bjeldanes, 1981). These results suggest that mutagen production by *F. moniliforme* may be independent of the production of compounds that are acutely toxic to or have feed refusal activity in chickens.

Donkeys. Leukoencephalomalacia (see equine leukoencephalomalacia, above) has been induced experimentally in three donkeys with cultures of an unnumbered Egyptian isolate of *F. moniliforme* (Buck et al., 1979; Haliburton et al., 1979; Wilson, 1971; Wilson & Maronpot, 1971; Wilson et al., 1973). We have confirmed the identity of this strain as *F. moniliforme* (Strain 10.1).

Culture material of *F. moniliforme* OP-6, isolated from moldy corn chaff associated with a field outbreak of suspected equine LEM in South Africa, caused a transient pruritis in one donkey and severe liver damage characterized by icterus, edema, and hemorrhages followed by death in another donkey (Kellerman et al., 1972). We have confirmed the identity of this strain as *F. moniliforme* (Strain 10.6).

Ducklings. Richard et al. (1969) reported on the toxicity to day-old Pekin ducklings of seven unidentified isolates of *Fusarium* spp. from visibly moldy shelled corn from storage bins in Iowa, USA. Ether extracts of rice cultures of three of these isolates caused some mortalities (1 of 3 dead in 10 days). We have identified two of these isolates toxic to ducklings, *Fusarium* sp. MC-107 and MC-121, as *F. moniliforme* (Strains 10.17 and 10.18).

Cultures on corn of *F. moniliforme* MRC 137, isolated from corn produced in Zambia during 1974/75, were acutely toxic (2 of 4 dead in 14 days) to day-old Pekin ducklings (Marasas et al., 1978). We have confirmed the identity of this strain as *F. moniliforme* (Strain 10.12).

Scott (1965) found that cultures on autoclaved corn meal of 2 of 10 isolates of *F. moniliforme* from cereal and legume products in South Africa were acutely toxic (3 of 3 dead in 14 days) to day-old Pekin ducklings.

In a study of the toxigenicity of 30 isolates of *F. moniliforme* from commercial South African corn (see Strain 10.9), Marasas et al. (1979a) found that cultures on autoclaved corn of 14 of these isolates (= 46.6%) caused the death of more than two out of four day-old Pekin ducklings within 14 days. None of these acutely toxic strains produced chemically detectable levels of moniliformin in culture. Kriek et al. (1981a) also reported on the acute toxicity to ducklings of six of these 14 non-moniliformin-producing strains of *F. moniliforme* (as *F. verticillioides*), including MRC 548 (see Strain 10.9). In addition, Kriek et al. (1981a) reported that 15 isolates of *F. moniliforme* (as *F. verticillioides*) including MRC 826 and MRC 602 (see Strains 10.4 and 10.5) from home-grown corn in Transkei were acutely toxic to ducklings and that none of these isolates produced moniliformin in culture. We have identified representative cultures of these toxic, non-moniliformin-producing strains from South Africa (MRC 548) and Transkei (MRC 826 and MRC 602) as *F. moniliforme* (Strains 10.9, 10.4, and 10.5).

It can be concluded that the mycotoxin(s) produced by the above strains of *F. moniliforme* toxic to ducklings are unknown at present.

Geese. Palyusik et al. (1974) reported that cultures on corn of unspecified Hungarian isolates of *F. moniliforme* (see Strain 10.13) caused only slight alterations in spermatogenesis in ganders.

Horses. Leukoencephalomalacia has been induced experimentally in four horses with three South African isolates (OP-32, OP-124, and MRC 826, see Strains 10.2, 10.3, and 10.4) of *F. moniliforme* (Kriek et al., 1981a; Marasas et al., 1976). The clinical signs and pathological lesions induced in these animals are described in detail elsewhere in this section (see equine leukoencephalomalacia, above, and Strains 10.2, 10.3, and 10.4).

These three South African isolates of *F. moniliforme* were also shown to be capable of causing a fatal hepatic syndrome in horses (Kellerman et al., 1972; Kriek et al., 1981b, Marasas et al., 1976). Clinical signs in animals that died of hepatosis included subcutaneous edema, and icterus, and the gross pathological lesions consisted of severe cardiac hemorrhages, petechiae and ecchymoses in various organs, edema, icterus, and liver damage. The hepatic lesions were characterized by fatty changes, fibroplasia around the central veins and portal tracts with bile duct proliferation, increased mitotic figures in the hepatocytes, megalocytosis, and biliary stasis (Kellerman et al., 1972).

Leukoencephalomalacia and the hepatoxic syndrome are apparently manifestations of the same toxicosis, since either lesion could be produced by the same culture material, depending on the dosage rate (Marasas et al., 1976). The chemical nature of the causative mycotoxin(s) produced by *F. moniliforme* is, however, unknown at present.

Mice. Richard et al. (1969) reported that ether extracts of rice cultures of four out of seven strains of unidentified *Fusarium* species isolated from moldy shelled corn in storage bins in Iowa, USA, caused some mortalities in female white Swiss-Webster mice. We examined cultures of the four strains toxic to mice (MC-103, MC-127, MC-130, and MC-252) and found them all to be *F. moniliforme* (Strains 10.19, 10.20, 10.21, and 10.22). The most toxic strain, according to Richard et al. (1969), was MC-130 (see Strain 10.21) which killed 3 of 3 mice, while visceral hemorrhages were visible in the one mouse killed by MC-103 (see Strain 10.19).

Fiussello et al. (1970) tested eight isolates of *F. moniliforme* from feeds in Italy for estrogenic activity (uterine hypertrophy) in female white mice. One strain (No. 57) caused a mildly positive response when cultures on corn were incubated at 24°C for 2 weeks followed by 7 weeks at 12°C and ethanol extracts were incorporated into the feed pellets of the mice for 3 days. The increase in uterine weight was about 1/30 of that obtained with cultures of *Gibberella zeae*. It was not determined whether the culture material contained zearalenone or not. Negative results in the mouse bioassay for estrogenic activity were reported for two other Italian isolates of *F. moniliforme* by Ceruti Scurti et al. (1971) and for two isolates of *F. moniliforme* that did not produce zearalenone by Caldwell et al. (1970).

Scott (1965) found that cultures on autoclaved corn meal of *F. moniliforme* M-702 isolated from cereal and legume products in South Africa caused the death of 4 of 10 weanling mice within 28 days. Van Rensburg et al. (1971) also reported on the toxicity of a strain of *F. moniliforme* isolated from cereal and legume products in South Africa to young adult male mice. Only one mouse out of 10 died during the 20-day test period, but the survivors exhibited a severe loss of weight. Pathological changes were present in the kidneys and liver. The renal glomeruli were slightly hyperaemic, and vacuolar degenerative changes were present in the proximal tubules. Occasional cells of the convoluted tubules were necrotic, and in the medullary rays moderate numbers of nuclei were pycnotic. Nuclei in hepatocytes exhibited considerable variation in size and many were binucleate, and occasional single cell necrosis was noted in all mice. The outstanding hepatic alteration was, however, a pronounced bile duct proliferation.

Culture filtrates of *F. moniliforme* PE-8-17, isolated from foodstuffs in Uganda, caused pleomorphism of the liver and the formation of irregular cell cords upon intraperitoneal injection in mice (Itakura & Kinosita, 1975).

Korpinen & Ylimäki (1972) reported on the acute toxicity to mice of four strains of *F. moniliforme* isolated from moldy animal feeds suspected to have been the cause of the outbreaks of disease among domestic animals in Finland. We have confirmed the identity of one of these strains (No. 19) as *F. moniliforme* (Strain 10.14). Cultures of all four strains on autoclaved grain mix were acutely toxic to mice, and the main histopathological findings in most mice were hemorrhages in the mucosa of the gastro-intestinal tract and slight degeneration of the liver and myocardium.

A mouse bioassay in which culture filtrates or extracts of cultures are injected intraperitoneally in male ddS mice has been used extensively in Japan for the detection of toxigenic *Fusarium* strains (Ueno et al., 1971a). Extracts of six Japanese strains of *F. moniliforme* have been found to be lethal to mice in this assay (Ueno et al., 1971a, 1972a, 1973a, 1977b). The finding that the extracts of these strains were toxic to mice but gave negative results in the rabbit reticulocyte assay for inhibition of protein synthesis led Ueno et al. (1971a) to conclude that the mycotoxins produced by these strains are not trichothecenes, and no known trichothecenes or other *Fusarium* toxins have been identified in these strains to date (Ueno et al., 1973a, 1975b). We have examined two of these Japanese strains of *F. moniliforme* reported to be toxic to mice. *Fusarium moniliforme* M-10-2 was originally isolated from moldy bean hulls in Hokkaido (Ueno et al., 1972a, 1973a), and we found that it is not *F. moniliforme* but a degenerate strain of *F. avenaceum* (Strain 5.4). *Fusarium moniliforme* 5003 was originally isolated from river sediments in Japan and shown to be lethal to mice, causing radiomimetic cellular damage, but no known trichothecenes could be identified in the extract (Ueno et al., 1977b); we have identified this culture as a pionnotal strain of *F. moniliforme* (Strain 10.15).

The chemical nature of the mycotoxin(s) produced by the above strains of *F. moniliforme* toxic to mice is unknown at present.

Extracts of *F. moniliforme* CMI-IMI 204057 isolated from moldy sweet corn in India reportedly produced hyperestrogenism in albino mice (Ghosal et al., 1978a). Three mycotoxins, i.e., zearalenone (0.8 mg/liter of culture medium), diacetoxyscirpenol, and T-2 toxin, were detected chemically in these extracts (see mycotoxins produced, below).

Two cytotoxic compounds with antitumor activity have been purified and chemically characterized from *F. moniliforme* strain A-130, which has been found to be acutely toxic to mice in Japan: fusariocin A, with an intraperitoneal LD_{50} of 2.88 mg/kg in mice (Arai & Ito, 1970), and fusariocin C, with an LD_{50} value of 3.97 mg/kg (Ito, 1979). The structure of fusariocin C was recently established by Ito et al. (1981). It is unknown at present whether fusariocins A and C are also responsible for the toxicity of other strains of *F. moniliforme* in mice and other animals.

Pigeons. Prolonged emesis in pigeons was produced by one out of five isolates of *F. moniliforme* (No. 111) tested by Prentice & Dickson (1968) and Prentice et al. (1959). In contrast to these findings, Ellison & Kotsonis (1973) could not detect any emetic activity of this strain (as *F. moniliforme* NRRL 3197) in pigeons, thus indicating that the culture had lost its activity during storage. Vesonder et al. (1981a) re-examined cultures of this strain maintained as lyophilized cultures at NRRL and found "NRRL 3197 to be a variant of the species *F. tricinctum* (Cda.) Sacc. and not a strain of *F. moniliforme*." We also examined a culture of NRRL 3197 and identified it as *F. sporotrichioides* (Strain 4.24). Consequently this strain is excluded from *F. moniliforme*.

According to Ghosal et al. (1978a), extracts of cultures of *F. moniliforme* CMI-IMI 204057 isolated from moldy sweet corn in India caused emetic activity in pigeons, and diacetoxyscirpenol, T-2 toxin, and zearalenone (see mycotoxins produced, below) could be detected chemically in these extracts.

Pigs. During investigations on outbreaks of an unknown disease similar to ergotism among domestic animals in Boyd County, Nebraska, USA, at the turn of the century, corn affected by a "pinkish dry rot" was fed to hogs and within a short time caused the bristles to be shed and the hoofs to slough (Peters, 1904). The same disease was reproduced experimentally in shoats (=young, weanling pigs) fed the "*Fusarium* dry mold" from the toxic corn "cultivated on crackers and corn meal mush" (Peters, 1904). This *Fusarium* was subsequently described as *F. moniliforme* by Sheldon (1904).

Culture material on autoclaved corn of *F. moniliforme* (as *F. verticillioides*) MRC 826 (Strain 10.4) was fed in the ration to three pigs by Kriek et al. (1981b). Two of these pigs died after 6 days and the principal lesion found was pulmonary edema (Kriek et al., 1981b).

Rabbits. A strain of *F. moniliforme* isolated from bean (*Phaseolus vulgaris* L.) hay implicated in an outbreak of a neurotoxic syndrome in horses in South Africa was tested extensively for toxicity to rabbits with negative results (Steyn, 1933, 1950; Van der Walt & Steyn, 1946).

Rabbits either injected with extracts of cultures of *F. moniliforme* isolated from rice in India or fed for 30 days on a diet of rice contaminated with these cultures developed the following signs: "decrease in body weight, increase in heart beat, increase in blood eosinophil, lymphocyte, drooping of ears followed by shot holes in both ears, dragging of hind legs, reduction in stool weight along with deep black colour" (Gangopadhyay & Chakrabarti, 1981). The toxin extracted from the rice cultures of *F. moniliforme* yielded light yellow crystals and was considered to be "very close to zearalenone though not exactly the same."

A skin test involving the application of ether or alcohol extracts of fungal cultures on the shaved skin of rabbits and assessment of the dermotoxic reaction has been used extensively for the screening of *Fusarium*, including *F. moniliforme*, strains for toxicity (Joffe 1960a,b; Joffe & Palti, 1974; Joffe et al., 1973). In a study of the toxicity of *F. moniliforme* isolated from soil and overwintered cereals associated with ATA (see *F. sporotrichioides*, ATA) of humans in the USSR, 37 strains were grown on various media at temperatures ranging from -5 to $8°C$ for 25 to 70 days and the cultures then subjected to freezing and thawing (Joffe, 1960a,b). Extracts of one of these strains were found to be toxic and extracts of four others were found to be mildly toxic to rabbit skin (Joffe, 1960b). The toxic strain (No. 1857) produced edema and hemorrhage, while the mildly toxic strains (Nos. 989, 995, 1863, and 2269) produced either edema only or edema and leucocytorrhea (Joffe, 1960a).

In a study of the toxicity of 82 isolates of *F. moniliforme* from various hosts in Israel and three isolates from other countries, cultures on wheat kernels were incubated at either 12, 18, 24, 30, or $35°C$ for 11 days (Joffe et al., 1973). Extracts of 74 of these 85 strains caused some toxic reaction on rabbit skin, ranging from a slight pinkish inflammation to various degrees of edema, leucocytorrhea, hemorrhage, and necrosis. In general, the most toxic reactions were obtained with cultures incubated at 24 or $30°C$. Histopathological changes in rabbit skin induced by toxic isolates such as No. 45 were characterized by severe inflammatory changes in the epidermis, with subcorneal pustule formation and necrosis of Malpighian layers, and edema and leucocytic infiltration of the dermis (Joffe et al., 1973). According to Joffe & Palti (1974), extracts of 85 of 97 unspecified strains of Liseola Fusaria exhibited some degree of dermotoxicity to rabbit skin and there was a quantitative relationship between the dermotoxicity and phytotoxicity of these isolates.

A strain of *F. moniliforme* isolated from corn associated with an outbreak of aflatoxicosis in humans in India has also been reported to be dermotoxic to rabbit skin (Jain et al., 1980).

The chemical nature of the mycotoxin(s) produced by these dermotoxic strains of *F. moniliforme* is unknown at present.

Rats. Ethyl acetate extracts of cultures of 10 out of 10 strains of *F. moniliforme* isolated from moldy corn associated with outbreaks of toxicoses in animals in the United States caused mild dermotoxic reactions when applied to the shaved skin of rats (Smalley et al., 1970). A strain of *F. moniliforme* isolated from moldy corn in Wisconsin, USA, was cultured on autoclaved corn incubated at either 10 or 25°C for 21 days and fed to female albino Sprague-Dawley rats for 14 days (Marasas & Smalley, 1972). Culture material incubated at 10°C caused no mortalities, but the culture incubated at 25°C caused the death of 2 of 2 rats within 7 days.

Biro et al. (1972) reported a slight inhibitory effect on spermatogenesis in rats by cultures of a Hungarian strain of *F. moniliforme*.

Tuttobello et al. (1974) reported that a strain of *F. moniliforme* "ISS Roma" had uterotrophic effects in virgin weanling female rats when cultured either in liquid shake culture or in still culture on autoclaved corn. The chemical nature of the uterotrophic substance(s) present in the cultures was not determined, but the uterotrophic effect was equivalent to that of about 25 ppm "zearalanol" in the liquid cultures and 15 ppm in the corn cultures.

Cultures of *F. moniliforme* MRC 137 (Strain 10.12), isolated from Zambian corn, caused the death of 1 of 4 weanling male Wistar rats within 14 days and marked reduction in weight gain of the survivors (Marasas et al., 1978).

Scott (1965) found that *F. moniliforme* M-702, isolated from cereal and legume products in South Africa, cultured on autoclaved corn at 26°C for 2 to 3 weeks, caused the death of 2 of 10 weaned rats over a period of 4 weeks.

Kriek et al. (1981a) reported on the toxicity of 21 non-moniliformin-producing strains of *F. moniliforme* (as *F. verticillioides*) isolated from maize in South Africa (including MRC 548, see Strain 10.9) and Transkei (including MRC 826 and MRC 602, see Strains 10.4 and 10.5) to male BD IX rats. Cultures of each strain on autoclaved corn incubated at 25°C for 21 days were incorporated into commercial rat diets at levels of either 8, 16, or 32% and fed to rats for periods up to 77 days. The pathology caused by *F. moniliforme* MRC 602 (Strain 10.5) was fully characterized and compared with that caused by the other strains. This strain caused the death of 20 out of 20 rats in the groups receiving either 16% or 32% of culture material in the diet in mean periods of 78 and 49 days, respectively. The most characteristic histopathological lesions were cirrhosis and nodular hyperplasia of the liver together with acute and proliferative endocardial lesions and concurrent intraventricular thrombosis in the heart. Other lesions included toxic nephrosis, endothelial proliferation in the pulmonary vessels, and thrombosis of the larger vessels in the heart, liver, pancreas, small intestine and lungs. There was a marked variation in toxicity to rats of the other 20 isolates of *F. moniliforme*: 15 caused some mortalities and some of the characteristic hepatic and cardiac lesions, and certain strains, including MRC 826 (Strain 10.4), were noticeably more toxic that MRC 602; the five remaining strains were non-toxic.

Kriek et al. (1981b) confirmed that *F. moniliforme* (as *F. verticillioides*) MRC 826 and MRC 602 (Strains 10.4 and 10.5) cause cirrhosis, intraventricular cardiac thrombosis, and nephrosis in rats.

The chemical nature of the mycotoxin(s) produced by the above strains of *F. moniliforme* toxic to rats is unknown at present.

Kotsonis et al. (1975a) reported that *F. moniliforme* No. 7, isolated from corn refused by swine in Wisconsin, USA, and cultured on corn at room temperature for 1 week followed by 4 weeks at 12°C, caused feed refusal and weight loss, but not an increased fresh weight of the uterus, in rats. The cultures contained 4µg/g zearalenone (see zearalenone, below) as determined by TLC.

Extracts of cultures of *F. moniliforme* CMI-IMI 204057, isolated from moldy corn in

India, were reported to be dermotoxic to albino rats and to contain diacetoxyscirpenol, T-2, and zearalenone (Ghosal et al., 1978a).

According to the reviews by Li et al. (1980), Lin & Tang (1980), and Yang (1980), corn meal inoculated with isolates of *F. moniliforme* from Lin Xian County in the Henan Province of China, which is one of the highest human esophageal cancer (see human esophageal cancer, above) risk areas in the world, has been found to induce tumors in several different organs, but not the esophagus, and to cause epithelial hyperplasia, precancerous changes, and papillomas of the esophagus and stomach in rats. Four nitrosamines have been detected in corn bread inoculated with Chinese isolates of *F. moniliforme* (see Strain 10.7) following nitrosation with sodium nitrite (Li et al., 1979, 1980; Lu et al., 1979), but it is not known whether these nitrosamines were responsible for the reported carcinogenicity to rats of the cultures of Chinese isolates of *F. moniliforme* or not.

Investigations on the use of *F. moniliforme* to produce microbial protein on the aqueous extract of carob (*Ceratonia siliqua*) fermented continuously at 30°C revealed that diets containing the spray-dried mycelium caused no growth depression and no abnormalities in the dissected carcasses of male and female Wistar rats during feeding trials of 21 to 28 days (Drouliscos et al., 1976).

Sheep. A strain of *F. moniliforme* (as *F. verticillioides*) MRC 826 (Strain 10.4) isolated from corn produced in Transkei was cultured on autoclaved corn and dosed to two sheep by rumen fistula (Kriek et al., 1981b). The sheep were dosed at a level of 5 g of culture material/kg/day and both died, one after 10 days and the other after 12 days. The principal lesions were acute nephrosis and hepatosis.

Mycotoxins Produced

Deoxynivalenol. The strain of *F. moniliforme* (NRRL 3197) that was previously reported not to produce deoxynivalenol (Vesonder et al., 1977b), but later found to produce 1.5 µg/g of deoxynivalenol in corn cultures (Vesonder et al., 1981a), is not *F. moniliforme*, but *F. sporotrichioides* (Strain 4.24).

Diacetoxyscirpenol. Ghosal et al. (1978a) reported that *F. moniliforme* CMI-IMI 204057 isolated from moldy sweet corn in India produces diacetoxyscirpenol. According to these authors, the identity of the fungus was confirmed by the CMI. The fungus was grown in Richards solution in still culture at 21°C for 21 days. Extracts of these cultures produced emetic activity in pigeons, dermal toxicity in albino rats, and hyperestrogenism in albino mice. The presence of diacetoxyscirpenol, together with T-2 toxin and zearalenone, was indicated by TLC and the toxins were separated by column chromatography. UV absorption spectra, optical rotation data, and biological assay indicated that the extracts contained 3 mg of diacetoxyscirpenol per liter of culture medium. The levels of diacetoxyscirpenol in the naturally infected sweet corn and in artificially inoculated sweet corn incubated in closed polythene bags were determined as 13 and 17 µg/g, respectively.

Diacetoxyscirpenol has also been found to occur naturally in malformed flowers and shoots of mangoes (*Mangifera indica* L.) in India (Ghosal et al., 1978b). Cultures in mango slurry of *F. moniliforme* isolated from such infected plants were reported to have emetic activity (Ghosal et al., 1976 b), but the emetic principle(s) was not identified. On the basis of the reference to *F. moniliforme* var. *subglutinans* CMI-IMI 225231 in connection with malformation disease of mango by Ghosal et al. (1979), we are assuming that the causal organism of this desease is *F. subglutinans* (see *F. subglutinans*, mycotoxins produced) rather than *F. moniliforme*, although these names are apparently used interchangeably by the above authors.

Several authors have reported the non-production of trichothecenes, including diacetoxyscirpenol, by isolates of *F. moniliforme:* none of 7 by Ichinoe (1976); none of 3 by Suzuki et al. (1980, 1981b); none of 8, including M-10-2 (Strain 5.4) and NRRL 3197 (Strain 4.24), which are not *F. moniliforme*, by Ueno et al. (1973a); and none of 3, including *Fusarium* sp. 5003, which we have identified as *F. moniliforme* (Strain 10.15), by Ueno et al. (1977b).

In conclusion, there is only one unconfirmed report (Ghosal et al., 1978a) of diacetoxyscirpenol production by a single isolate (CMI-IMI 204057) of *F. moniliforme*. We have not seen this isolate and there are no known diacetoxyscirpenol-producing strains of *F. moniliforme* represented in the ITFRC.

Fusaric Acid. Fusaric acid is a well-known phytotoxin (Gaümann, 1957), but is not generally regarded as a mycotoxin. However, in view of the fact that fusaric acid is a chelating agent and may be involved in certain diseases of abnormal bone development in animals in which *F. moniliforme* has been implicated (Moreau, 1974; see abnormal bone development, above), some consideration has to be given here to fusaric acid production by *F. moniliforme*.

Fusaric acid was first isolated from *F. moniliforme* (erroneously referred to as *F. heterosporum* Nees) as a plant-growth inhibitor by Yabuta et al. (1934) during a study of the plant-growth promoters (see gibberellins, below) produced by this fungus isolated from rice with Bakanae disease in Japan. The production of fusaric acid by *F. moniliforme* (as *G. fujikuroi*) was confirmed by Gaümann et al. (1952) and Gaümann (1957). In addition to fusaric acid, *F. fujikuroi* ETH No. M82, isolated from rice in Japan, was also found to produce dehydrofusaric acid (Stoll, 1954; Stoll & Renz, 1957).

Stodola et al. (1955) isolated 15.5 g of fusaric acid from 160 gal. of liquid medium fermented with *F. moniliforme* NRRL 2284, which was originally isolated in Japan. We have identified this culture according to Nelson et al. (1983) as the "short-chained" type of *F. moniliforme* (Strain 10.27). The fusaric acid used in experiments by Burmeister et al. (1980a) was some of the material originally isolated from NRRL 2284 by Stodola et al. (1955) (H.R. Burmeister, personal communication, April 1981). Grove et al. (1958) confirmed the production of fusaric acid by this strain and also isolated dehydrofusaric acid from the cultures. Recently Mutert et al. (1981) detected up to 2,550 mg/ℓ of fusaric acid by high performance liquid chromatography in cultures of this strain (as *G. fujikuroi* CBS 186.56, see Strain 10.27).

Matuo et al. (1976) reported that 50 of 61 isolates of *F. moniliforme* from various sources produced fusaric acid in Richards solution after 15 days' incubation at 25°C. We have examined five of these isolates, i.e., Nos. 374, 881, 883, 975, and 1000, and identified all of them according to Nelson et al. (1983) as the "short-chained" type of *F. moniliforme* (Strains 10.28, 10.29, 10.30, 10.31, and 10.32). Three of these strains, i.e., 1000, 975, and 374 (Strains 10.29, 10.31, and 10.32), were the highest producers of fusaric acid (>3 mg/100 ml) in the study by Matuo et al. (1976).

All of the above fusaric acid–producing strains have also been reported to produce gibberellins (gibberellic acid and gibberellin A, see gibberellins, below) by the respective authors who reported fusaric acid production. Nirenberg (1976) established the new species *F. fujikuroi* Nirenberg to accommodate the causal organism of Bakanae disease of rice. She separated *F. fujikuroi* from *F. moniliforme* on morphological grounds and claimed that isolates of *F. fujikuroi* from rice were characteristically high gibberellin producers. The separation of *F. fujikuroi* from *F. moniliforme* was not accepted by Nelson et al. (1983) and we have consequently placed the six fusaric acid– (and gibberellin-) producing strains examined (Strains 10.27, 10.28, 10.29, 10.30, 10.31, and 10.32) in *F. moniliforme*. However, all these isolates are characterized by having microconidia in short, collapsing chains on monophialides and they can be referred to as the "short-chained" type of *F. moniliforme* (Nelson et al., 1983).

Fusarins (=Fusariogenins). Four compounds with mutagenic activity in *Salmonella typhimurium* have been isolated and partially characterized from cultures on autoclaved corn of *F. moniliforme* isolate M-1 from corn, incubated at 23°C for 2 weeks followed by 2 weeks at 11°C (Wiebe & Bjeldanes, 1981). These compounds were originally referred to as fusariogenins A, B, C, and D, and the molecular formula of the major mutagen fusariogenin C given as $C_{23}H_{29}O_7$ by Bjeldanes & Weib (1980). Subsequently the mutagens were referred to as fusarins A, B, C, and D and the molecular formula of fusarin C given as $C_{23}H_{29}NO_7$ by Wiebe & Bjeldanes (1981). The latter authors obtained approximately 1 g of fusarin C, and lesser yields of fusarins A, B, and D, from 2.5 kg of molded corn.

Culture material containing these mutagens produced no apparent toxic effects when fed to day-old cockerels for 21 days (Bjeldanes & Weib, 1980; Wiebe & Bjeldanes, 1981). These results suggest that the fusarins are not acutely toxic to chickens, at least not at the levels produced in culture by this strain of *F. moniliforme*.

Fusariocins. Two mycotoxins with cytotoxic and antitumor activity have been purified from cultures of *F. moniliforme* A-130 in modified Adye's medium incubated on a gyratory shaker at 27°C for 7 days: fusariocin A, $C_{23}H_{24}O_5 \cdot \frac{1}{2}H_2O$, molecular weight 380, acute intraperitoneal LD_{50} in mice 2.88 mg/kg (Arai & Ito, 1970), and fusariocin C, $C_{27}H_{28}O_6$, molecular weight 448, acute intraperitoneal LD_{50} in mice 3.97 mg/kg (Ito, 1979). The structure of fusariocin A is unknown, but the structure of fusariocin C was recently established by X-ray diffraction to be a pseudodimer containing the tropolone skeleton (Ito et al., 1981).

The description of *F. moniliforme* by Arai & Ito (1970) makes no mention of microconidial chains and it is impossible to determine from their photographs whether this isolate belongs in *F. moniliforme* or *F. subglutinans*.

Gibberellins. The gibberellins were originally isolated as plant-growth promoters from cultures of *F. moniliforme* (as *Gibberella fujikuroi*) by Yabuta & Hayashi (1939) during a study of the Bakanae disease of rice in Japan. The plant-growth-promoting properties of these compounds led to intensive investigations and a voluminous literature on a large variety of compounds based on the gibbane and kaurane skeletons produced by *F. moniliforme*, as well as those occuring as endogenous plant hormones (Turner, 1971). Since the gibberellins are not generally considered to be mycotoxins (Peck et al. 1957), a detailed review of all the compounds of this group produced by *F. moniliforme* is not within the scope of this book. However, in view of the reported estrogenic and uterotrophic effects of gibberellic acid (Maillet & Bouton, 1969; Mirocha et al., 1971) and the possible involvement of this and/or related compounds in the uterine hypertrophy caused by cultures of *F. moniliforme* in mice (Fiussello et al., 1970) and rats (Tuttobello et al., 1974), some consideration has to be given here to the production of gibberellins by *F. moniliforme*.

The strains of *F. moniliforme* from which the gibberellins were originally obtained were isolated from rice in Japan (Yabuta & Hayashi, 1939), as were the strains used in most of the subsequent work on gibberellic acid and other gibberellins produced by this fungus, i.e., isolate 917 (=CBS 183.29, IMI 58290, ATCC 12616), used by the group at the Akers Research Laboratory in England (Borrow et al., 1954), and ETH M82, used by Gaümann's group in Switzerland (Gaümann, 1957; Stoll, 1954). The high-yielding strain NRRL 2284 (=CBS 186.56, IMI 112801, ATCC 14164, see Strain 10.27) used by Stodola et al. (1955), Grove et al. (1958), and Mutert et al. (1981) was originally obtained from Japan, where it was presumably isolated from rice. Gordon (1960b) used a seedling bioassay and found that 23 isolates of *F. moniliforme* from a wide variety of hosts produced gibberellin-like substances, but he did not indicate the relative amounts produced by isolates from different hosts. In other studies comparing the gibberellin-producing ability of isolates from different hosts (Borrow et al. 1955; El-Bahrawi, 1977; Matuo et al., 1976), isolates from rice were consistently found to be the highest producers; for example, the three highest producers of gibberellin-like substances (Nos. 881, 1000, and 883, see Strains 10.28, 10.29, and 10.30) in the study by Matuo et al. (1976) were all isolated from rice in Japan. However, the ability to produce gibberellins is not restricted to isolates of *F. moniliforme* from rice, as stated by Curtis (1957), since isolates that produce gibberellin-like substances have been obtained from sugar cane, corn, cotton, sorghum, and soil (Borrow et al., 1955; El-Bahrawi, 1977; Gordon, 1960b; Matuo et al., 1976). Two of the isolates studied by Matuo et al. (1976), No. 975 from soil (Strain 10.31) and No. 374 from sorghum (Strain 10.32), are examples of gibberellin-producing isolates of *F. moniliforme* from hosts other than rice.

Nirenberg (1976) established the new species *F. fujikuroi* to accommodate the causal fungus of Bakanae disease of rice. She separated *F. fujikuroi* from *F. moniliforme* on morphological grounds and claimed that isolates of *F. fujikuroi* from rice were characteristically high gibberellin producers. The separation of *F. fujikuroi* from *F. moniliforme* was not accepted by Nelson et al. (1983). We have consequently placed the six gibberellin-producing strains that we have examined (Strains 10.27, 10.28, 10.29, 10.30, 10.31, and 10.32), including four high-producing strains from rice (Strains 10.27, 10.28, 10.29, and 10.30) and two lower producers from sources other than rice (Strains 10.31 and 10.32), in *F. moniliforme*. All of these isolates are characterized by having microconidia in short collapsing chains on monophialides, however, and they can be referred to as the "short-chained" type of *F. moniliforme* (Nelson et al., 1983).

Moniliformin. Moniliformin was first isolated by Cole et al. (1973) from a strain of *F. moniliforme* (NRRL 5860) isolated from corn seed damaged by southern leaf blight in the United States. We have confirmed the identity of this strain as *F. moniliforme* (Strain 10.23). According to Springer et al. (1974), the original moniliformin-producing isolate stopped producing the toxin during the course of their study. Consequently another strain, *Gibberella fujikuroi* ATCC 12763, isolated by W.C. Gordon from an unspecified source, was used for the production of moniliformin for chemical characterization. We have identified this strain according to Nelson et al. (1983) as the "short-chained" type of *F. moniliforme* (Strain 10.24, see gibberellins, above). According to Springer et al. (1974) the moniliformin produced by the original strain NRRL 5860 (Strain 10.23) was a sodium salt, while ATCC 12763 (Strain 10.24) produced a potassium salt. They determined the structure of the moniliformin isolated from the latter strain by X-ray diffraction as the potassium salt of 1-hydroxycyclobut-1-ene-3,4-dione.

Two additional strains of *F. moniliforme* have subsequently been reported to produce moniliformin: NRRL 6322 from an unknown source which was found to produce high levels (7–15 g/kg) of moniliformin on corn grits incubated at 28°C for 16 to 20 days (Allen et al., 1981b; Burmeister et al., 1979); and MRC 515, which was isolated from millet in Namibia (M. Steyn et al., 1978). We have identified both of these strains according to Nelson et al. (1983) as the "short-chained" type of *F. moniliforme* (Strains 10.25 and 10.26, see gibberellins, above).

No moniliformin could be detected either chemically or by bioassay (Buck et al., 1979; Haliburton et al., 1979) in culture material of the isolate of *F. moniliforme* from moldy corn in Egypt (Strain 10.1) that was used in the successful experimental induction of LEM (see equine leukoencephalomalacia, above) in donkeys (Buck et al., 1979 Haliburton et al., 1979; Wilson, 1971; Wilson & Maronpot, 1971; Wilson et al., 1973). One of the isolates of *F. moniliforme* from moldy corn in South Africa (Strain 10.4) that induced LEM in horses (Kriek et al., 1981b) has also been reported to be a non-producer of moniliformin (Kriek et al., 1981a,b).

In a study of the toxigenic potential of *F. moniliforme* (as *F. verticillioides*) isolates from South African corn, none of 14 isolates toxic to ducklings produced moniliformin in culture (Marasas et al., 1979a). Similarly, none of 15 toxic isolates of *F. moniliforme* (as *F. verticillioides*) from corn produced in the Transkei produced chemically detectable levels of moniliformin in culture on autoclaved corn incubated at 25°C for 21 days (Kriek et al., 1981a). We have examined representative isolates of these toxic, non-moniliformin-producing strains from South African (MRC 548) and Transkeian corn (MRC 826 and MRC 602) and identified them as *F. moniliforme* (Strains 10.9, 10.4, and 10.5). To date no moniliformin-producing isolates of *F. moniliforme* have been obtained from corn in southern Africa, but several high-yielding strains of *F. subglutinans* have been reported (Kriek et al., 1977; Marasas et al., 1979a) (see *F. subglutinans,* moniliformin, and Strains 10.36 and 10.37).

In conclusion, only four strains of *F. moniliforme* have been reported to produce moniliformin and all of these strains are represented in the ITFRC (Strains 10.23, 10.24, 10.25, and 10.26). One of these strains, *F. moniliforme* NRRL 5860 (Strain 10.23), has reportedly lost its ability to produce moniliformin (Springer et al., 1974). The other three strains are all representatives of the "short-chained" type of *F. moniliforme* (see gibberellins, above). The highest moniliformin-producing strain of *F. moniliforme* in the ITFRC is probably NRRL 6322 (Strain 10.25), which reportedly produces 7–15 g/kg of moniliformin in culture on corn grits (Allen et al., 1981b; Burmeister et al., 1979).

T-2 Toxin. Ghosal et al. (1978a) reported that *F. moniliforme* CMI-IMI 204057, isolated from moldy sweet corn in India, produced T-2 toxin in still culture in Richard's solution incubated at 21°C for 21 days. According to these authors, the identity of the fungus was confirmed by the CMI. Extracts of cultures produced emetic activity in pigeons, dermal toxicity in albino rats, and hyperestrogenism in albino mice. The presence of T-2 toxin, together with diacetoxyscirpenol and zearalenone, was indicated by TLC and the toxins were separated by column chromatography. UV absorption spectra, optical rotation data, and biological assay indicated that the extracts contained 5 mg of T-2 toxin per liter of culture medium. The levels of T-2 toxin in the naturally infected sweet corn and in artificially inoculated sweet corn incubated in closed polythene bags were determined as 16 and 48 μg/g, respectively.

T-2 toxin has also been reported to occur naturally in malformed flowers and shoots of mangoes in India (Ghosal et al., 1978b). Cultures in mango slurry of *F. moniliforme* from such infected plants were reported to have emetic activity (Ghosal et al., 1976b), but the emetic principle(s) was not identified. On the basis of the reference to *F. moniliforme* var. *subglutinans* CMI-IMI 225231 in connection with malformation disease of mango by

Ghosal et al. (1979), we are assuming that the causal organism of this disease is *F. subglutinans* (see *F. subglutinans,* mycotoxins produced) rather than *F. moniliforme,* although these names are apparently used interchangeably by the above authors.

The strain of *F. moniliforme* (NRRL 3197) that was previously reported not to produce T-2 toxin (Ellison & Kotsonis, 1973; Ueno et al., 1972b, 1973a; Vesonder et al., 1977b), but subsequently found to produce 33 µg/g T-2 toxin in corn cultures (Vesonder et al., 1981a), is not *F. moniliforme,* but *F. sporotrichioides* (Strain 4.24).

Several authors have reported the non-production of trichothecenes, including T-2 toxin, by isolates of *F. moniliforme:* none of 7 isolates by Ichinoe (1976); none of 3 by Suzuki et al. (1980; 1981b); none of 8, including M-10-2 (Strain 5.4) and NRRL 3197 (Strain 4.24), which are not *F. moniliforme,* by Ueno et al. (1973a); and none of 3, including *Fusarium* sp. 5003, which we have identified as *F. moniliforme* (Strain 10.15), by Ueno et al. (1977b).

In conclusion, there is only one unconfirmed report (Ghosal et al., 1978a) of T-2 toxin production by a single isolate (CMI-IMI 204057) of *F. moniliforme.* We have not seen this strain and there are no known T-2 toxin-producing strains of *F. moniliforme* represented in the ITFRC.

Zearalenone. The first report that *F. moniliforme* produced the estrogenic metabolite zearalenone (=F-2) was by Mirocha et al. (1969). These authors reported that two unnumbered isolates of *F. moniliforme,* one from shelled corn and the other from feed, produced zearalenone in cultures on autoclaved corn incubated at 22–26°C for 2 weeks followed by 6 weeks at 10–12°C. These two isolates produced zearalenone and a chemically related metabolite, F-3, which was also suspected of having estrogenic activity. Mirocha et al. (1969) did not indicate the levels of zearalenone detected in the cultures of these two strains of *F. moniliforme,* but Mirocha & Christensen (1974) and Mirocha et al. (1977a) stated that in their laboratory *F. moniliforme* had been found to produce only trace amounts (1 to 19 ppm) of zearalenone.

Hacking et al. (1976, 1977) found that 2 of 8 isolates of *F. moniliforme* from barley in the United Kingdom produced zearalenone at a level of 1.5 mg/100 g of rice after incubation at 25°C for 4 weeks followed by 6 weeks at 12°C. One out of 32 tested isolates of *F. moniliforme* from corn in Italy produced low levels (4.5–7.7 ppm) of zearalenone in cultures on autoclaved corn incubated at 25°C for 2 weeks followed by 8 weeks at 12°C (Bottalico, 1976, 1977b, 1979; Bottalico & Piglionica, 1977; Piglionica et al., 1976). Martin & Keen (1978) reported that one of the strains of *F. moniliforme* isolated from sorghum malt and moldy corn in Swaziland and Lesotho that contained zearalenone produced unspecified levels of zearalenone in culture on autoclaved corn incubated at 25°C for 7 days.

Ghosal et al. (1978a) reported that *F. moniliforme* CMI-IMI 204057, isolated from moldy sweet corn in India, produced zearalenone in still culture in Richard's solution incubated at 21°C for 21 days. According to these authors the identity of the fungus was confirmed by the CMI. Extracts of cultures produced emetic activity in pigeons, dermal toxicity in albino rats, and hyperestrogenism in albino mice. The presence of zearalenone, together with diacetoxyscirpenol and T-2 toxin, was indicated by TLC and the toxins were separated by column chromatography. UV absorption spectra, optical rotation data, and biological assay indicated that the extracts contained 0.8 mg of zearalenone per liter of culture medium. The levels of zearalenone in the naturally infected sweet corn and in artificially inoculated sweet corn incubated in closed polythene bags were determined as 4 and 5 µg/g, respectively.

Zearalenone has also been found to occur naturally in malformed flowers and shoots of mangoes in India (Ghosal et al., 1978b). Cultures in mango slurry of *F. moniliforme* from such infected plants were reported to have emetic activity (Ghosal et al., 1976b), but the emetic principle(s) were not identified. On the basis of the reference to *F. moniliforme* var. *subglutinans* CMI-IMI 225231 in connection with malformation disease of mango by Ghosal et al. (1979), we are assuming that the causal organism of this disease is *F. subglutinans* (see *F. subglutinans*, mycotoxins produced) rather than *F. moniliforme*, although these names are apparently used interchangeably by the above authors.

Cultures on rice of *F. moniliforme* isolated from rice in India were found to be toxic to rabbits by Gangopadhyay & Chakrabarti (1981). The toxin extracted from these cultures with hexane and diethyl-ether yielded light yellow crystals which were considered to be "very close to zearalenone though not exactly the same."

The strain of *F. moniliforme* (NRRL 2830) reported to produce zearalenone in cracked corn in both still and continuously agitated culture by Hesseltine (1972) is *F. graminearum* (Strain 8.19) and is consequently excluded from *F. moniliforme*.

The production of zearalenone by Hungarian isolates of *F. moniliforme* on autoclaved corn incubated at 16°C for 3 weeks has been reported by Lasztity & Wöller (1975a,b) and Lasztity et al. (1977), but no indication was given of the number of isolates tested or the levels produced.

Kotsonis et al. (1975b) reported that *F. moniliforme* isolate No. 7 from moldy corn in Wisconsin, USA, produced 4.0 μg/g of zearalenone in cultures on autoclaved corn incubated for 1 week at room temperature followed by 4 weeks at 12°C. This culture material did not, however, induce uterine hypertrophy in rats. Caldwell et al. (1970) also obtained negative results for estrogenic activity in a mouse bioassay with two isolates of *F. moniliforme* that did not produce chemically detectable levels of zearalenone. These authors calculated that a level of 8 μg of zearalenone/g of corn culture material would have caused a positive uterotrophic response. Ceruti Scurti et al. (1971) also reported negative results for estrogenic activity in mice of two isolates (N5 and N136429) of *F. moniliforme*. On the other hand, positive uterotrophic responses were reported in mice with 1 of 8 isolates of *F. moniliforme* by Fiussello et al. (1970) and in rats with *F. moniliforme* isolate "ISS Roma" by Tuttobello et al. (1974). The estrogenic metabolite present in these two positive uterotrophic strains of *F. moniliforme* was not chemically identified, but Tuttobello et al. (1974) calculated that the activity of liquid cultures of isolate "ISS Roma" was equivalent to approximately 25 ppm and of corn cultures to approximately 15 ppm of "zearalanol." The possibility has to be considered, however, that the uterotrophic activity of these two strains may have been due to gibberellic acid (see gibberellins, above), which has been reported to be estrogenically active (Maillet & Bouton, 1969; Mirocha et al., 1971).

F-3, a compound which is chemically related to zearalenone (=F-2) and is also suspected of having estrogenic activity, was found to occur naturally in feed samples implicated in cases of infertility in dairy cattle in Minnesota, USA, by Mirocha et al. (1968a,b). Isolates of *F. moniliforme* from these feeds produced copious amounts of F-3, but not zearalenone, in culture.

Negative results for the production of zearalenone by isolates of *F. moniliforme* have also been reported by several other authors: none of 11 by Caldwell et al. (1970); none of 5 by Eugenio et al. (1970a); none of 11 by Ichinoe et al. (1977); none of 3 by Ishii et al. (1974); none of 10 by Sutton et al. (1976); and none of 3 by Suzuki et al. (1978, 1981b).

In conclusion, approximately seven out of 100 isolates of *F. moniliforme* tested have been reported to produce low levels (1 to 19 ppm) of zearalenone. We have not seen any of these isolates and there are no known zearalenone-producing strains of *F. moniliforme* represented in the ITFRC.

Toxigenic Strains in the ITFRC

The *Fusarium* strains listed under *F. moniliforme* in Table 10.1 all have been identified as *F. moniliforme* according to Nelson et al. (1983). Some of these strains produce microconidia in short collapsing chains on branched monophialides and are considered to be the "short-chained" type of *F. moniliforme* according to Nelson et al. (1983). These latter strains are specifically indicated in the following discussion. Toxigenic strains referred to as *F. moniliforme* in the literature but identified as species in sections other than Liseola by us are listed in Table 10.2.

Strain 10.1 *F. moniliforme* Cairo No. 1 (=M-1225; MRC 2079)
=NRRL 6442 (= M-1245; MRC 2228)

Wilson (1971) first reported that a strain of *F. moniliforme* (unnumbered) isolated from corn associated with a field outbreak of equine leukoencephalomalacia (LEM) in the vicinity of Cairo, Egypt, where flooding of the Nile Delta had occurred, was the causative fungus of this mycotoxicosis (see equine leukoencephalomalacia, above). He found that whole grain corn on which this organism was cultured for 2 weeks at 25°C became toxic to a lethal degree. The feeding of freeze-dried corn culture material (22 kg) to a male Egyptian donkey for 11 days resulted in signs of neurotoxicity as evidenced by walking in circles on the 12th and 13th days. On the 14th day the lower lip drooped and the head was held facing downward into one corner of the pen. The right eyelid was insensitive to stimulation and after another 2 hours of walking in circles the animal fell down and struggled for about an hour before expiring. Upon post-mortem examination, a large necrotic lesion was found in the white matter of the right cerebral hemisphere. This was the first successful experimental reproduction of the pathognomonic cerebral lesion of equine LEM, a disease which had been known in various parts of the world for many decades, with a pure culture of a fungus. This work was reported on in more detail by Wilson & Maronpot (1971), who stated that two out of three Egyptian donkeys that received from 22 to 66 kg of freeze-dried corn inoculated with an unspecified Egyptian isolate of *F. moniliforme* developed characteristic clinical signs and gross brain lesions, but that two Egyptian donkeys given 66 kg of oven-dried culture material exhibited no signs of disease. The culture material that caused LEM in the two donkeys caused no signs of neurotoxicity in goats, pigs, monkeys, chickens, rats, mice, guinea pigs, hamsters, or rabbits. These findings led Wilson & Maronpot (1971) to conclude that certain equines are uniquely susceptible to the toxic metabolite produced in corn by *F. moniliforme,* but the chemical nature of this mycotoxin is still unknown. Wilson et al. (1973) stated that more recent experiments in horses with the Egyptian strain of *F. moniliforme* (unnumbered) have indicated that "the fungus may have temporarily lost its toxigenicity."

According to B.J. Wilson (personal communication, 30 September 1980), the Egyptian strain used in the experimental reproduction of LEM in donkeys (Wilson, 1971; Wilson & Maronpot, 1971; Wilson et al., 1973) was *F. moniliforme* Cairo No. 1. A culture of this strain on corn mash was received from B.J. Wilson on 30 September 1980 with a note that this strain "has been repeatedly transferred on corn and held at about 10°C. May no

Table 10.1. *Fusarium* strains of the Section Liseola in the International Toxic *Fusarium* Reference Collection (ITFRC)

Strain No.	ITFRC No.	MRC No.	Published Name(s) & Number(s)	Source	Toxigenicity
Fusarium moniliforme					
10.1	M-1225	2079	F. moniliforme Cairo No. 1	B.J. Wilson	a, b
10.1	M-1245	2228	F. moniliforme NRRL 6442	NRRL	a, b, c
10.2	M-1253	42	F. moniliforme OP-32	W.F.O. Marasas	a, b
10.3	M-1037	930	F. moniliforme OP-124	W.F.O. Marasas	a, b
10.4	M-1325	826	F. verticillioides MRC 826	W.F.O. Marasas	a, b, c
10.5	M-1167	602	F. verticillioides MRC 602	W.F.O. Marasas	a, b, c
10.6	M-1036	929	F. moniliforme OP-6	W.F.O. Marasas	a, b
10.7	M-1251	2327	F. moniliforme China	R.M. Eppley	a
10.8	M-1246	2326	F. solani (= F. javanicum)	B.J. Wilson	a, b
10.9	M-1353	548	F. verticillioides MRC 548	W.F.O. Marasas	b, c
10.10	M-1247	2316	F. moniliforme F-1 (=ATCC 24088)	ATCC	b
10.11	M-1248	2317	F. moniliforme F-2 (= ATCC 24089)	ATCC	b
10.12	M-1352	137	F. moniliforme MRC 137	W.F.O Marasas	b
10.13	M-1118	1439	F. moniliforme F-76	M. Palyusik	b
10.14	M-1300	2677	F. moniliforme No. 19	A. Ylimäki	b
10.15	M-1327	2535	F. moniliforme 5003	Y. Ueno	b
10.16	M-1293	2633	Fusarium sp.	V.C. Vora	b
10.17	M-1295	2628	Fusarium sp. MC-107	J.L. Richard	b
10.18	M-1296	2629	Fusarium sp. MC-121	J.L. Richard	b
10.19	M-1294	2627	Fusarium sp. MC-103	J.L. Richard	b
10.20	M-1297	2630	Fusarium sp. MC-127	J.L. Richard	b
10.21	M-1298	2631	Fusarium sp. MC-130	J.L. Richard	b
10.22	M-1299	2632	Fusarium sp. MC-252	J.L. Richard	b
10.23	M-1033	1411	F. moniliforme NRRL 5860	NRRL	b, d
10.24	M-1324	69	G. fujikuroi ATCC 12763	ATCC	d
10.24	M-1034	1412	G. fujikuroi ATCC 12763 (= NRRL 6022)	NRRL	d
10.25	M-1138	1784	F. moniliforme NRRL 6322	NRRL	d
10.26	M-1166	515	F. moniliforme MRC 515	C.J. Rabie	d
10.27	M-1250	2322	F. moniliforme NRRL 2284; CBS 186.56	NRRL	e
10.28	M-1268	2387	F. moniliforme 881	T. Mauto	e
10.29	M-1271	2390	F. moniliforme 1000	T. Mauto	e
10.30	M-1269	2388	F. moniliforme 883	T. Matuo	e
10.31	M-1270	2389	F. moniliforme 975	T. Matuo	e
10.32	M-1267	2386	F. moniliforme 374	T. Matuo	e
Fusarium proliferatum					
10.33	M-1243	2324	F. moniliforme 587	U.L. Diener	b
Fusarium subglutinans					
10.34	M-1351	134	F. moniliforme var. subglutinans MRC 134	W.F.O. Marasas	b
10.35	M-1354	940	F. moniliforme var. subglutinans 8549	W. Gerlach	b
10.36	M-811	115	F. moniliforme var. subglutinans MRC 115	W.F.O. Marasas	a, b, d
10.37	M-1168	756	F. moniliforme var. subglutinans MRC 756	W.F.O. Marasas	b, d
Fusarium anthophilum					
10.38	M-1355	941	F. moniliforme var. anthophilum 8998	W. Gerlach	b

a = Associated with human and/or animal disease
b = Culture toxic to experimental animals
c = Non-producer of moniliformin
d = Produces moniliformin
e = Produces fusaric acid and gibberellins

TABLE 10.2. *Fusarium* Strains Published as *F. moniliforme* but Identified as Other Species in the International Toxic *Fusarium* Reference Collection (ITFRC)

Strain No.	ITFRC No.	MRC No.	Published Name(s) & Number(s)	Source	Present Identification
5.4	R-6750	2532	*F. moniliforme* M-10-2	Y. Ueno	*F. avenaceum*
4.24	T-494	2323	*F. moniliforme* 111 (= NRRL 3197)	NRRL	*F. sporotrichioides*
8.19	R-5318	1782	*F. moniliforme* NRRL 2830	NRRL	*F. graminearum*

longer be toxigenic." Strain Cairo No. 1 is being maintained in the ITFRC as M-1225 (=MRC 2079).

According to Haliburton (1979), "a strain of *F. moniliforme* Sheldon was isolated from natural occurring field cases of ELEM along the Nile Delta by Dr. B.J. Wilson, Vanderbilt University, Nashville, TN, submitted to the Northern Regional Research Center (NRRC) in 1967, and maintained as culture collection no. 6442." On the basis of this statement, it is being assumed here that *F. moniliforme* NRRL 6442 is the original Egyptian strain Cairo No. 1.

Typical clinical signs and lesions of equine LEM were induced in one donkey that received a total of 19.85 kg of corn cultured with *F. moniliforme* NRRL 6442 over a period of 39 days (Buck et al., 1979; Haliburton et al., 1979). Another donkey and three rabbits fed the cultured corn did not develop signs of the toxicosis.

No moniliformin was detected by chick bioassay, TLC, and atomic absorption spectroscopy in the culture material of *F. moniliforme* NRRL 6442 that caused a classical case of LEM in one donkey (Buck et al., 1979; Haliburton et al., 1979).

A lyophilized culture of *F. moniliforme* NRRL 6442 was received from NRRL on 9 March 1981. This culture is being maintained in the ITFRC as M-1245 (=MRC 2228). An examination of this culture together with a culture of the original *F. moniliforme* strain Cairo No. 1 (=M-1225; MRC 2079) has revealed that the cultures are identical.

It is concluded that *F. moniliforme* NRRL 6442 is in fact the original strain Cairo No. 1 that has been maintained by lyophilization at NRRL, and that this culture still produces the equine LEM mycotoxin as evidenced by the recent successful experimental induction of the syndrome in a donkey with culture material of NRRL 6442 (Buck et al., 1979; Haliburton et al., 1979).

Strain 10.2 *F. moniliforme* OP-32 (=M-1253; MRC 42)

Kellerman et al. (1972) isolated this strain from corn associated with an unknown fatal disease of horses in South Africa during 1971. Dried culture material on corn was dosed to a horse that died on the 12th day. The gross lesions included severe hemorrhages in the heart, petechiae and ecchymoses in various other organs, edema of the subcutis, and liver damage, but no gross lesions in the brain. A similar hepatic syndrome, together with edema and petechial hemorrhages in the brain, was subsequently induced by Marasas et al. (1976) in another horse with culture material (33.7 kg dosed in 22 days) of the same strain (as *F. moniliforme* OP-32B) incubated on autoclaved corn for 21 days at alternating temperatures of 10°C and 25°C for 12 h/day. The same culture material when administered to a third horse at lower dosage rates (68.3 kg over a period of 144 days) caused the typical clinical signs and pathological lesions of equine LEM characterized by focal areas of liquefactive necrosis in the cerebral white matter (Marasas et al., 1976).

These authors concluded that the hepatotoxic syndrome as well as the LEM are manifestations of the same toxicosis since either lesion could be produced by the same culture material. The two syndromes appeared to be dosage related: a dose equivalent to 0.67–1.94 kg culture material/day for 11–21 days produced a fatal hepatosis, while LEM was precipitated by the equivalent of 0.33–0.44 kg/day for 90–144 days (Marasas et al., 1976).

At present the chemical nature of the causative mycotoxin(s) produced by *F. moniliforme* is unknown and it is not certain whether the liver and brain lesions in horses are caused by different dosage levels of the same toxic metabolite or by different mycotoxins.

A lyophilized culture of this strain was received from W.F.O. Marasas on 2 February 1978 as "*F. moniliforme* OP-32B."

Strain 10.3   *F. moniliforme* OP-124 (=M-1037; MRC 930)

This strain was isolated from ground corn associated with a field outbreak of equine LEM in horses in South Africa during 1973 (Marasas et al., 1976). Cultures of this fungus on autoclaved corn were incubated for 21 days at alternating temperatures of 10°C and 25°C for 12 h/day. Dried culture material (19.6 kg) was dosed to a horse over a period of 90 days. The horse developed typical clinical signs of LEM preceded by mild icterus and petechial hemorrhages in the conjunctiva. In addition to the brain lesions of LEM, in this case represented as focal areas of microcavitation, this horse also had lesions in the liver characteristic of the hepatotoxicosis described by Kellerman et al. (1972). This finding led Marasas et al. (1976) to conclude that the hepatotoxic syndrome and LEM are manifestations of the same toxicosis induced by *F. moniliforme*. Since the chemical nature of the causative mycotoxin(s) has not yet been resolved, it is not known whether the liver and brain lesions in horses are caused by the same toxic metabolite or by different mycotoxins.

A lyophilized culture was received from W.F.O. Marasas on 2 February 1978 as "*F. moniliforme* OP-124."

Strain 10.4   *F. verticillioides* MRC 826 (=M-1325; MRC 826)

This strain was isolated from moldy home-grown corn produced during 1975 in the Kentani district of the Transkei, in an area with a high rate of human esophageal cancer (Kriek et al., 1981a).

Cultures of this strain on autoclaved corn incubated at 25°C for 21 days were found to be extremely toxic to ducklings (Kriek et al., 1981a) rats (Kriek et al., 1981a,b), horses, pigs, and sheep (Kriek et al., 1981b). Diets containing 32% of culture material caused 100% mortality in male BD IX rats after a mean period of 44 days, while five out of six rats receiving 16% in the diet were dead when the experiment was terminated after 70 days (Kriek et al., 1981a). The lesions in both groups of rats were characterized by cirrhosis and nodular hyperplasia of the liver and by acute and proliferative endocardial lesions and intraventricular cardiac thrombosis. These lesions were identical to those caused in rats by *F. moniliforme* (= *F. verticillioides*) MRC 602 (Strain 10.5) as described in detail by Kriek et al. (1981a), except that MRC 826 appeared to be noticeably more toxic.

The same culture material of *F. moniliforme* (= *F. verticillioides*) MRC 826 that was used in the toxicity trial in rats by Kriek et al. (1981a) was dosed by stomach tube to two horses, fed in the ration to three pigs, and administered by rumen fistula to two sheep by Kriek et al. (1981b). The one horse died on the eighth day after receiving a total of

4.287 kg of culture material and the principal lesions were brain edema, early leukoencephalomalacia, toxic hepatosis, and nephrosis. The second horse died of LEM and nephrosis on the 14th day after receiving a total of 3.819 kg of culture material in six doses of 2.5 g/kg each. The amounts of culture material of strain MRC 826 required to induce experimental LEM in horses are much lower than the quantities previously reported (19.85 to 68.3 kg) for one Egyptian (see Strain 10.1) and two South African (see Strains 10.2 and 10.3) strains of *F. moniliforme* (Buck et al., 1979; Haliburton et al., 1979 Marasas et al., 1976; Wilson & Maronpot, 1971).

In pigs, the principal lesion induced by culture material of this strain was pulmonary edema, and in sheep the characteristic lesions were acute nephrosis and hepatosis (Kriek et al., 1981b).

Thus there was a marked variation in the localization of lesions and target organs affected by culture material of *F. moniliforme* MRC 826 in the various animal species examined, namely the brain in horses, the liver and heart in rats, the kidney and liver in sheep, and the lung in pigs. At present it is not known whether the variety of lesions produced in different animal species is caused by a single toxic metabolite or by different mycotoxins produced by *F. moniliforme* MRC 826. This strain does not produce moniliformin (Kriek et al., 1981a), and the chemical nature of the mycotoxin(s) responsible for the lesions caused in experimental animals is unknown.

A lyophilized culture of this strain of *F. moniliforme* was received from W.F.O. Marasas in February 1982 as "*F. verticillioides* MRC 826."

Strain 10.5 *F. verticillioides* MRC 602 (=M-1167; MRC 602; ATCC 42809)

This strain was isolated from moldy home-grown corn produced during 1975 in the Butterworth district of the Transkei, in an area with a high rate of human esophageal cancer (Kriek et al., 1981a).

Culture material on autoclaved corn incubated at 25°C for 21 days was found to be highly toxic to ducklings (Kriek et al., 1981a), rats (Kriek et al., 1981a,b), and baboons (Kriek et al., 1981b). Diets containing either 16 or 32% of culture material of strain MRC 602 caused 100% mortality in male BD IX rats after mean periods of 78 and 49 days, respectively (Kriek et al, 1981a). The most striking pathological lesions in these rats were found in the liver and heart and were characterized by cirrhosis and nodular hyperplasia of the liver and by the occurrence of acute and proliferative endocardial lesions and concurrent intraventricular cardiac thrombosis (Kriek et al., 1981a). The relatively slow development of these remarkable lesions deserves attention because similar hepato- and cardiotoxic strains of *F. moniliforme* will not be detected in acute, short-term toxicity trials in rats. Two baboons fed diets containing culture material died within 143 to 248 days of acute congestive heart failure, while a third baboon that was killed after 720 days had cirrhosis of the liver (Kriek et al., 1981b).

Fusarium moniliforme (= *F. verticillioides*) MRC 602 does not produce moniliformin (Kriek et al., 1981a), and the chemical nature of the mycotoxin(s) responsible for the lesions caused in experimental animals by this strain is unknown. Four new pigments, i.e., the 8-0-methyl derivatives of bostrycoidin, javanicin, solaniol, and fusarubin, have been isolated from cultures of this strain by Steyn et al. (1979), but these pigments are probably not involved in the toxicity of the material.

A lyophilized culture of this strain of *F. moniliforme* was received from W.F.O. Marasas on 22 April 1980 as "*F. verticillioides* MRC 602." A subculture has also been deposited in ATCC as ATCC 42809.

Strain 10.6 *F. moniliforme* OP-6 (=M-1036; MRC 929)

This strain was originally isolated from moldy corn chaff associated with a field outbreak of suspected LEM in horses in South Africa during 1970 (Kellerman et al., 1972). Culture material caused a transient pruritus in one donkey and severe liver damage, icterus, edema, and hemorrhages followed by death in another donkey upon dosing by means of a stomach tube.

A lyophilized culture was received from W.F.O. Marasas on 2 February 1978 as "*F. moniliforme* OP-6."

Strain 10.7 *F. moniliforme,* China (=M-1251; MRC 2327)

A slant culture of this strain was received from R.M. Eppley on 21 April 1981 as "*F. moniliforme,* China." The United States Food and Drug Administration obtained this culture from the Frederick Cancer Research Center and it was reportedly originally isolated by Dr. C. Ji in Lin Xian county in Henan province in northern China, an area with one of the highest human esophageal cancer rates in the world (Li et al., 1980; Lin & Tang, 1980; Yang, 1980). Corn meal inoculated with unspecified strains of *F. moniliforme* from this area has been reported to induce tumors in several different organs, excluding the esophagus, in rats, and epithelial hyperplasia, precancerous changes, and papillomas of the esophagus and stomach of rats and mice (Li et al., 1980; Lin & Tang, 1980; Yang, 1980). Corn meal inoculated with unspecified Chinese strains of *F. moniliforme* and incubated in the presence of sodium nitrite has also been reported to contain three nitrosamines, N-dimethylnitrosamine, N-diethylnitrosamine, and N-methyl-N-benzylnitrosamine, together with a new N-nitroso compound, N-3-methylbutyl-N-1-acetonylnitrosamine (Li et al., 1979, 1980; Lu et al., 1979; 1980a,b). At least one of these nitrosamines, N-methyl-N-benzylnitrosamine, is known to be a site-specific esophageal carcinogen in rats, and this compound, as well as the new nitrosamine, are also mutagenic in *S. typhimurium* (Lu et al., 1980b).

Because of the fact that specific numbers were not assigned to the strains of *F. moniliforme* referred to in the above publications, it is not possible to say whether Strain 10.7 specifically was used in any of the experiments described in these publications. It can only be assumed that this strain was used in the work on the formation of nitrosamines in corn meal inoculated with *F. moniliforme* because it was originally obtained from Dr. C. Ji, who is a co-author of the papers by Li et al. (1979, 1980) and Lu et al. (1979, 1980b).

Strain 10.8 *F. solani* (=M-1246; MRC 2326)

A toxicosis of cattle characterized by severe pulmonary edema has been attributed to the presence of four furanoterpenoids in sweet potatoes (*Ipomoea batatas* L.) infected by *F. solani,* originally referred to as *F. javanicum* (Boyd, 1976; Boyd & Wilson, 1972; Boyd et al., 1973; Burka et al., 1974 a, b, 1977; Wilson, 1971, 1973; Wilson & Boyd, 1974; Wilson et al., 1970, 1971; Yang et al., 1971). This syndrome is discussed in detail under *F. solani* (see *F. solani,* moldy sweet potato toxicosis).

A slant culture of this strain, reportedly capable of catabolizing certain stress metabolites in sweet potatoes to the pulmonary toxins 4-ipomeanol, 1-ipomeanol, 1,4-ipomeadiol, and ipomeanine, was received from B.J. Wilson in December 1980 as "*F. solani.*" This culture produces microconidia in long chains on monophialides and was accordingly identified as *F. moniliforme.* In view of this finding, a second culture was requested from B.J. Wilson, and the culture of *F. solani* received from him on 2 April 1981 also proved to be *F. moniliforme.*

A soil culture of *F. solani* isolated from moldy sweet potatoes associated with a field outbreak of pulmonary edema in cattle in Georgia, USA, that had also caused the disease experimentally when inoculated sweet potatoes were fed to cattle (Doupnik et al. 1971; Peckham et al., 1972) was received from B. Doupnik on 9 September 1981, but this culture was not viable. It is notable that *F. moniliforme* was also isolated from the moldy sweet potatoes associated with the field outbreak (Doupnik et al., 1971; Peckham et al., 1972).

Since we are not aware of any other sources of the culture of *F. solani* used in the original work on moldy sweet potato toxicosis, it is not possible to determine whether the strain that we received as *F. solani* from B.J. Wilson and identified as *F. moniliforme* is a contaminant or was in fact used in the research on this disease cited above.

Strain 10.9   *F. verticillioides* MRC 548 (=M-1353; MRC 548)

This strain was isolated from commercial corn produced in Piet Retief, eastern Transvaal, South Africa, during 1975 (Kriek et al., 1981a). Cultures on autoclaved corn incubated at 25°C for 21 days caused the death of 4 of 4 day-old Pekin ducklings in a mean time of 5.7 days. According to Kriek et al. (1981a), this strain was one of 14 isolates of *F. moniliforme* (as *F. verticillioides*) that Marasas et al. (1979a) previously reported to be toxic to ducklings but non-producers of moniliformin. Culture material of this strain was also toxic to male BD IX rats and caused lesions of toxic hepatosis, cirrhosis, and nodular hyperplasia in the liver and intraventricular thrombi in the heart (Kriek et al., 1981a).

A lyophilized culture of this strain of *F. moniliforme* was received in February 1982 from W.F.O. Marasas as "*F. verticillioides* MRC 548."

Strain 10.10   *F. moniliforme* F-1, ATCC 24088 (=M-1247; MRC 2316)

Sharby et al. (1973) isolated this strain, together with *F. moniliforme* F-2, ATCC 24089 (see Strain 10.11) from commercial poultry feed in Arkansas, USA. Cultures on autoclaved corn inoculated with this strain were incubated at room temperature for periods of 2 to 8 weeks. Diets containing this culture material caused a significant reduction in weight gain and feed efficiency in broiler chicks, and within 12 days the chicks exhibited a severe leg deformity characterized by a bowing outward or a slanting inward of the legs, giving the chicks either a "cowboy" or a "knock-kneed" appearance (see abnormal bone development, above). Although some of the control chicks also developed this leg abnormality, the incidence was much higher in the chicks that received the *F. moniliforme*-inoculated corn.

A lyophilized culture of this strain was received from ATCC on 9 March 1981 as "*F. moniliforme* ATCC 24088."

Strain 10.11   *F. moniliforme* F-2, ATCC 24089 (= M-1248; MRC 2317)

This strain was isolated from poultry feed in Arkansas, USA, by Sharby et al. (1973) and caused the same clinical signs in broiler chickens as those described for *F. moniliforme* F-1, ATCC 24088 (see Strain 10.10).

A lyophilized culture of this strain was received from the ATCC on 9 March 1981 as "*F. moniliforme* ATCC 24089."

Strain 10.12   *F. moniliforme* MRC 137 (= M-1352; MRC 137)

This strain was isolated from corn kernels produced in Zambia during 1974/75 (Marasas et al., 1978). Cultures on autoclaved corn incubated at 25°C for 21 days caused the death of 2 of 4 day-old Pekin ducklings and 1 of 4 weanling male Wistar rats within 14 days, with marked reductions in weight gain of the survivors.

A lyophilized culture was received in February 1982 from W.F.O. Marasas as "*F. moniliforme* MRC 137."

Strain 10.13 *F. moniliforme* F-76 (= M-1118; MRC 1439)

According to M. Palyusik (personal communication, 23 March 1973), this strain was originally isolated from mixed poultry feed suspected of causing disease in broilers in Hungary. This strain is considered here to be representative of the unspecified Hungarian strains of *F. moniliforme* that caused only slight pathological alterations in spermatogenesis in ganders (Palyusik et al., 1974).

A slant culture was received on 23 March 1978 from M. Palyusik as *"F. moniliforme* F-76."

Strain 10.14 *F. moniliforme* No. 19 (M-1300; MRC 2677)

Korpinen & Ylimäki (1972) isolated this strain from commercial poultry feed suspected of causing retarded growth of chickens in Finland. Cultures on an autoclaved mixture of wheat, oats, and barley grains were acutely toxic to mice (2 of 2 dead within 14 days). Different incubation temperatures (22°C for 6 weeks, 22°C for 2 weeks followed by 8°C for 4 weeks, and 22°C for 2 weeks followed by 0°C for 4 weeks) did not have much effect on the toxicity of this strain and cultures incubated at the different temperature regimes were all acutely toxic to mice. The main histopathological findings in most mice were hemorrhages in the gastro-intestinal tract and slight degeneration of the liver and myocardium.

A slant culture of this strain was received on 1 October 1981 from A. Ylimäki as *"F. moniliforme* 7085 II-19."

Strain 10.15 *F. moniliforme* 5003 (M-1327; MRC 2535)

This strain was isolated from river sediments in Japan by Ueno et al. (1977b). Extracts of cultures in liquid medium incubated at 25–27°C for 2 weeks were toxic upon intraperitoneal injection in mice. The pathological changes were characterized as "radiomimetic cellular injury" with karyorrhexis in the spleen and lymph nodes. The rabbit reticulocyte bioassay for trichothecenes was negative and no known trichothecenes could be detected chemically in the cultures.

A slant culture was received from Y. Ueno in November 1981 as *"F. moniliforme* 5003." The culture is pionnotal but produces some chains of microconidia on monophialides.

Strain 10.16 *Fusarium* sp., wheat, India (= M-1293; MRC 2633)

This strain was isolated from wheat in Uttar Pradesh, India (Mall et al., 1979). The chloroform-methanol extract of cultures on wheat incubated at 27–28°C for 15 days was mildly toxic (1 of 3 dead) upon oral dosing to day-old chicks.

A slant culture of this isolate of *F. moniliforme* was received on 24 August 1981 from V.C. Vora as *"Fusarium* sp. PL W-604, from wheat, toxic to day-old chicks."

Strain 10.17 *Fusarium* sp. MC-107 (= M-1295; MRC 2628)

This strain was isolated from visibly moldy shelled corn from a storage bin in Iowa, USA, during 1966 (Richard et al., 1969). The ether extract of cultures on autoclaved rice incubated in shake culture at 28°C caused the death of 1 of 3 day-old Pekin ducklings, but none of 3 female white Swiss-Webster mice, within 10 days.

A slant culture of this strain of *F. moniliforme* was received on 9 September 1981 from J.L. Richard as *"Fusarium* sp. MC-107."

Strain 10.18 *Fusarium* sp. MC-121 (= M-1296; MRC 2629)

This strain was isolated from visibly moldy shelled corn from a storage bin in Iowa, USA, during 1966 (Richard et al., 1969). The ether extract of cultures on autoclaved rice incubated in shake culture at 28°C caused the death of 1 of 3 day-old Pekin ducklings, but none of 3 female white Swiss-Webster mice, within 10 days.

A slant culture of this strain of F. moniliforme was received on 9 September 1981 from J.L. Richard as "*Fusarium* sp. MC-121."

Strain 10.19   *Fusarium* sp. MC-103 (= M-1294; MRC 2627)
This strain was isolated from visibly moldy shelled corn from a storage bin in Iowa, USA, during 1966 (Richard et al., 1969). The ether extract of cultures on autoclaved rice incubated in shake culture at 28°C caused the death with visceral hemorrhages of 1 of 3 female white Swiss-Webster mice, but none of 3 day-old Pekin ducklings, within 10 days.

A slant culture of this strain of F. moniliforme was received on 9 September 1981 from J.L. Richard as "*Fusarium* sp. MC-103."

Strain 10.20   *Fusarium* sp. MC-127 (=M-1297; MRC 2630)
This strain was isolated from visibly moldy shelled corn from a storage bin in Iowa, USA, during 1966 (Richard et al., 1969). The ether extract of cultures on autoclaved rice incubated in shake culture at 28°C caused the death of 2 of 3 female white Swiss-Webster mice, but none of 3 Pekin ducklings, within 10 days.

A slant culture of this strain of F. moniliforme was received on 9 September 1981 from J.L. Richard as "*Fusarium* sp. MC-127." This culture is almost completely mycelial but produces some chains of microconidia on monophialides.

Strain 10.21   *Fusarium* sp. MC-130 (= M-1298; MRC 2631)
This strain was isolated from visibly moldy shelled corn from a storage bin in Iowa, USA, during 1966 (Richard et al., 1969). The ether extract of cultures on autoclaved rice incubated in shake culture at 28°C caused the death of 3 of 3 female white Swiss-Webster mice, but none of 3 Pekin ducklings, within 10 days.

A slant culture of this strain of F. moniliforme was received on 9 September 1981 from J.L. Richard as "*Fusarium* sp. MC-130." The culture is pionnotal but produces some chains of microconidia on monophialides.

Strain 10.22   *Fusarium* sp. MC-252 (= M-1299; MRC 2632)
This strain was isolated from visibly moldy shelled corn from a storage bin in Iowa, USA, during 1966 (Richard et al., 1969). The ether extract of cultures on autoclaved rice incubated in shake culture at 28°C caused the death of 1 of 3 female white Swiss-Webster mice, but none of 3 Pekin ducklings, within 10 days.

A slant culture of this strain of F. moniliforme was received on 9 September 1981 from J.L. Richard as "*Fusarium* sp. MC-252."

Strain 10.23   *F. moniliforme* NRRL 5860 (= M-1033; MRC 1411; ATCC 26263)
This strain (unnumbered) was originally isolated from naturally infected corn damaged by southern leaf blight in the United States in 1970 (Cole et al., 1973). Cultures grown on cracked corn at room temperature were found to be toxic to cockerels. A toxic metabolite, moniliformin, was isolated from the toxic culture material by means of a cockerel bioassay. Moniliformin was shown to be extremely toxic to cockerels (oral LD_{50} 4.0 mg/kg) and also to have phytotoxic effects on corn and tobacco (*Nicotiana tabacum* L.) seedlings.

According to Springer et al. (1974), the original moniliformin-producing isolate of *F. moniliforme* stopped producing the toxin during the course of their study. Consequently a new source, *Gibberella fujikuroi* ATCC 12763 (see Strain 10.24), was used for the production of moniliformin for chemical characterization. The only detectable difference between strains was that the moniliformin produced by the original strain was a sodium salt, while ATCC 12763 produced a potassium salt of moniliformin.

According to R.J. Cole (personal communcation, 15 December 1977), the original moniliformin-producing strain of *F. moniliforme* was deposited in NRRL as *F. moniliforme* NRRL 5860. A lyophilized culture of this strain was received from NRRL on 22 December 1977 as "*F. moniliforme* NRRL 5860." This strain is also deposited in ATCC as *F. moniliforme* ATCC 26263.

Strain 10.24 *Gibberella fujikuroi* **ATCC 12763** (= M-1324; MRC 69)
=**NRRL 6022** (= M-1034; MRC 1412)

Springer et al. (1974) isolated crystalline moniliformin from *Gibberella fujikuroi* ATCC 12763 and determined the structure by X-ray diffraction as the potassium salt of 1-hydroxycyclobut-1-ene-3,4-dione. The origin of this strain is listed in the ATTC Catalogue of Strains, Fifteenth Edition (1982), as "*Gibberella fujikuroi*, ATCC 12763—W.L. Gordon 4150 (*Fusarium moniliforme*)," from an unspecified source. M. Steyn et al. (1978) also reported moniliformin production by this strain (as *F. moniliforme* ATCC 12763).

A lyophilized culture of *G. fujikuroi* ATCC 12763 was obtained from ATCC on 19 June 1975 and is being maintained in ITFRC as M-1324 (= MRC 69). According to R.J. Cole (personal communication, 25 December 1977), this strain is also deposited in NRRL. A lyophilized culture of NRRL 6022 was obtained on 22 December 1977 and is being maintained in ITFRC as M-1034 (= MRC 1412). The cultures of this strain from both sources were found to be identical and to produce microconidia in short collapsing chains on branched monophialides. This strain was identified according to Nelson et al. (1983) as the "short-chained" type of *F. moniliforme* (see gibberellins, above).

Strain 10.25 *Fusarium moniliforme* **NRRL 6322** (= M-1138; MRC 1784)

Burmeister et al. (1979) reported that *F. moniliforme* NRRL 6322 (origin unknown) produced between 7 and 10 g of moniliformin (as determined by TLC) per kg of culture medium consisting of moist, sterile corn grits incubated in static culture at 28°C for 16 to 20 days. Allen et al. (1981b) also cultured strain NRRL 6322 on this medium and determined the moniliformin content by TLC as 15 mg/g of culture medium. About 600 mg of the sodium salt of moniliformin could be recovered per kg of culture medium by extraction with methanol, purification by preparative TLC, and crystallization from ether (Burmeister et al., 1979). Strain NRRL 6322 has been used by Allen et al. (1981b) and Burmeister et al. (1980a, b) as a producer of moniliformin for use in toxicological studies.

Culture material of *F. moniliforme* NRRL 6322 on corn grits was incorporated into diets to provide 8, 16, and 64 mg of moniliformin/kg of diet and fed to broiler chicks from 1 to 21 days of age (Allen et al., 1981b). Diets containing up to 16 mg/kg moniliformin were without effect on chick weight gain, feed consumption, and mortality. The diet containing 64 mg/kg moniliformin caused reduced weight gain and feed consumption. Although the daily moniliformin consumption by the chicks in this group was nearly twice the reported single oral LD_{50} dose, only 3 of 10 died and no lesions were found upon necropsy. Allen et al. (1981b) concluded from these results that chicks can tolerate considerably more moniliformin when consumed in their feed over an extended period than when administered in a single acute dosage.

A slant culture of F. moniliforme NRRL 6322 was received from NRRL on 23 October 1979. This culture was found to produce microconidia in short collapsing chains on branched monophialides and was identified according to Nelson et al. (1983) as the "short-chained" type of F. moniliforme (see gibberellins, above).

Strain 10.26 *Fusarium moniliforme* **MRC 515** (= M-1166; MRC 515)

M. Steyn et al. (1978) reported that cultures of this strain on autoclaved corn kernels incubated at 25°C for 21 days produced the sodium salt of moniliformin. The moniliformin could be purified by methanol extraction, aqueous extraction of the methanol-free residue, ion exchange chromatography with a NaCl concentration gradient, desalination, and crystallization. M. Steyn et al. (1978) did not give the origin of strain MRC 515, but according to C.J. Rabie (personal communication, 22 April 1980), this strain was isolated from millet (*Pennisetum typhoides* (Burm.) Stapf. et C.E. Hubbard) from a household affected by onyalai disease (see *F. chlamydosporum*, onyalai) in Ovambo, Namibia, in June 1974.

A lyophilized culture of F. moniliforme MRC 515 was received from C.J. Rabie on 22 April 1980. This culture was found to produce microconidia in short collapsing chains on branched monophialides and was identified according to Nelson et al. (1983) as the "short-chained" type of F. moniliforme (see gibberellins, above).

Strain 10.27 *F. moniliforme* **NRRL 2284** (= M-1250; MRC 2322)

According to Stodola et al. (1955), this strain was supplied by Dr. J.E. Mitchell, Camp Detrick, who originally obtained it from Dr. Y. Nisikado, Japan. Stodola et al. (1955) did not indicate the host, but it was presumably isolated from Bakanae-diseased rice. According to the CBS List of Cultures, 28th Ed. (1972), F. moniliforme NRRL 2284 = CBS 186.56 = IMI 112801 = ATCC 14164.

Stodola et al. (1955) reported that F. moniliforme NRRL 2284 consistently produced about 12 g of crystalline gibberellin from each 160 gal. of medium in submerged fermentations. The crude crystalline product was composed of gibberellin A and a new gibberellin designated as gibberellin X. Stodola et al. (1955) also isolated 15.5 g of fusarinic acid (= fusaric acid; 5-butylpicolinic acid) from 160 gal. of liquid medium. The fusaric acid used in experiments by Burmeister et al. (1980a) was some of this material originally isolated by Stodola et al. (1955) from NRRL 2284 (H.R. Burmeister, personal communication, April 1981). The production of gibberellin A_1, gibberellic acid, and fusaric acid by this strain (as *G. fujikuroi* NRRL 2284) was confirmed by Grove et al. (1958), who also isolated dehydrofusaric acid from the cultures. Recently Mutert et al. (1981) found that this strain (as *G. fujikuroi* CBS 186.56) still produces up to 2,550 mg/ℓ of fusaric acid in shake culture.

A lyophilized culture of this strain was received from NRRL on 9 March 1981 as "*F. moniliforme* NRRL 2284." This culture produces microconidia in short collapsing chains on branched monophialides and was identified as the "short-chained" type of *F. moniliforme* according to Nelson et al. (1983). The use of the name *F. fujikuroi* for high gibberellin-producing isolates from rice by Nirenberg (1976) has been discussed earlier (see gibberellins, above).

Strain 10.28 *F. moniliforme* **881** (= M-1268; MRC 2387)

According to Matuo et al. (1976), this strain was isolated from adult rice plants in Japan, produces microconidia in chains, and was the highest producer of gibberellin-like substances in their study, but a low producer of fusaric acid.

A slant culture of this strain was received from T. Matuo on 3 June 1981 as "*F. moniliforme* SUF 881." The culture produces microconidia in short collapsing chains on branched monophialides and was identified as the "short-chained" type of *F. moniliforme* according to Nelson et al. (1983). The use of the name *F. fujikuroi* for high gibberellin-producing isolates from rice by Nirenberg (1976) has been discussed earlier (see gibberellins, above).

Strain 10.29 *F. moniliforme* 1000 (= M-1271; MRC 2390)

According to Matuo et al. (1976), this strain was isolated from rice seed in Japan, produces microconidia in chains, was the second highest producer of gibberellin-like substances, and the highest producer of fusaric acid (> 4 mg/100 ml) in their study.

A slant culture of this strain was received from T. Matuo on 3 June 1981 as "*F. moniliforme* SUF 1000." This culture produces microconidia in short collapsing chains on branched monophialides and was identified as the "short-chained" type of *F. moniliforme* according to Nelson et al. (1983). The use of the name *F. fujikuroi* for high gibberellin-producing isolates from rice by Nirenberg (1976) has been discussed earlier (see gibberellins, above).

Strain 10.30 *F. moniliforme* 883 (= M-1269; MRC 2388)

According to Matuo et al. (1976), this strain was isolated from adult rice plants in Japan, produces microconidia in chains, and was the third highest producer of gibberellin-like substances in their study, but a low producer of fusaric acid.

A slant culture of this strain was received from T. Matuo on 3 June 1981 as "*F. moniliforme* SUF 883." This culture produces microconidia in short collapsing chains on branched monophialides and was identified as the "short-chained" type of *F. moniliforme* according to Nelson et al. (1983). The use of the name *F. fujikuroi* for high gibberellin-producing isolates from rice by Nirenberg (1976) has been discussed earlier (see gibberellins, above).

Strain 10.31 *F. moniliforme* 975 (= M-1270; MRC 2389)

According to Matuo et al. (1976), this strain was isolated from soil in a carnation field in Japan, produces microconidia in chains, and was a high producer of gibberellin-like substances as well as fusaric acid.

A slant culture of this strain was received from T. Matuo on 3 June 1981 as "*F. moniliforme* SUF 975." This culture produces microconidia in short collapsing chains on branched monophialides and was identified as the "short-chained" type of *F. moniliforme* according to Nelson et al. (1983). The use of the name *F. fujikuroi* by Nirenberg (1976) has been discussed earlier (see gibberellins, above).

Strain 10.32 *F. moniliforme* 374 (= M-1267; MRC 2386)

According to Matuo et al. (1976), *F. moniliforme* 374 was isolated from *Sorghum* roots in Japan, produces microconidia in chains, and was the third highest producer of fusaric acid (> 3 mg/100 ml) but a lower producer of gibberellin-like substances in their study.

A slant culture of this strain was received from T. Matuo on 3 June 1981 as "*F. moniliforme* SUF 374." The culture produces microconidia in short collapsing chains on branched monophialides and was identified as the "short-chained" type of *F. moniliforme* according to Nelson et al. (1983). The use of the name *F. fujikuroi* by Nirenberg (1976) has been discussed earlier (see gibberellins, above).

Fusarium proliferatum (Matsushima) Nirenberg

Perfect State: Unknown.

Incidence and Distribution

Very little is known about the incidence and distributon of *F. proliferatum*, probably because of mis-identification as *F. moniliforme*. Nirenberg (1976) recorded *F. proliferatum* on a wide vareity of hosts from Germany and also from Egypt, India, Japan, Phillipines, South Korea, and Turkey. Neish & Leggett (1981) examined isolates (as *F. moniliforme* var. *intermedium*) from soil in Canada and from *Phoenix dactylifera* L. in Iraq.

Association with Human and/or Animal Diseases

Fusarium proliferatum has not been associated with human or animal toxicoses.

Toxicity to Experimental Animals

According to Diener et al. (1976), extracts of cultures of *F. moniliforme* (Auburn University Culture Collection No. 587), isolated from cotton bolls and seed in the United States, were assayed for toxicity in brine shrimp, chick embryos, and rats. We have identified this isolate as *F. proliferatum* (Strain 10.33). This strain proved to be "highly toxic to two of the three bioassay systems" (Diener et al., 1976), but it is not entirely clear to which two. The chemical nature of the mycotoxin(s) produced by this strain is unknown.

Mycotoxins Produced

The isolate of *F. moniliforme* (No. 999) described and illustrated by Matuo et al. (1976) as having "pyriform microconidia formed in chains" is most probably *F. proliferatum*. This strain was reported to be a high producer of fusaric acid as well as gibberellin-like substances. Gibberellin production by *F. moniliforme* and the use of the name *F. fujikuroi* by Nirenberg (1976) are discussed elsewhere in this volume (see *F. moniliforme,* gibberellins).

Toxigenic Strains in the ITFRC

The following strain listed under *F. proliferatum* in Table 10.1 has been identified as *F. proliferatum* according to Nelson et al. (1983).

Strain 10.33 *F. moniliforme* Auburn University Culture Collection No. 587 (= M-1243; MRC 2324)

This strain was isolated from cotton balls and seed in the United States (Diener et al., 1976). The fungus was grown on autoclaved nutrient-amended shredded wheat at 25°C for 14–21 days; it was then extracted and assayed for toxicity to brine shrimp, chick embryos, and rats. Extracts of isolate 587 were reported to be "highly toxic to two of the three bioassay systems," but it is not entirely clear to which two.

A lyophilized culture was received on 10 March 1981 from U.L. Diener as "*F. moniliforme* No. 587."

Fusarium subglutinans (Wollenw. & Reink.) Nelson, Toussoun & Marasas

Perfect State: *Gibberella subglutinans* (Edwards) Nelson, Toussoun & Marasas

Incidence and Distribution

Relatively little information is available on the incidence and distribution of *F. subglutinans* because of frequent misidentification as *F. moniliforme* (Booth, 1971). There is little doubt that *F. subglutinans* occurs on a wide range of hosts, particularly Gramineae such as corn and sugar cane, in many areas of the world (Booth, 1971; CMI Descriptions of Pathogenic Fungi and Bacteria No. 23, 1964; Doidge, 1938; Edwards, 1940, 1941; Gordon, 1959, 1960a; Marasas et al., 1979a, 1981; Meyer & Frank, 1979; Nirenberg, 1976; Ullstrup, 1936; Wollenweber & Reinking, 1935). It seems that *F. subglutinans* has a lower optimum temperature for growth and predominates in cooler, more temperate areas than *F. moniliforme* (Booth, 1971; Francis & Burgess, 1975; Marasas et al., 1979a; Ullstrup, 1936).

Association with Human and/or Animal Diseases

Human Esophageal Cancer. *Fusarium subglutinans* is one of the most prevalent fungi associated with home-grown corn produced in high incidence areas of human esophageal cancer (see *F. moniliforme,* human esophageal cancer) in Transkei (Marasas et al., 1979b, 1981). One of these isolates of *F. subglutinans* from Transkeian corn (MRC 115, see Strain 10.36) proved to be extremely toxic to ducklings and rats and to produce large quantities (up to 11.3 g/kg) of moniliformin in culture (Kriek et al., 1977). There is, however, no evidence that either *F. subglutinans* or moniliformin is in any way involved in the etiology of human esophageal cancer.

Toxicity to Experimental Animals

Ducklings. Cultures of *F. moniliforme* var. *subglutinans* MRC 134, isolated from Zambian corn, were found to be acutely toxic (4 of 4 dead) to day-old Pekin ducklings by Marasas et al. (1978). We have identified this isolate as *F. subglutinans* (Strain 10.34).

Culture material of *F. moniliforme* var. *subglutinans* MRC 115, isolated from corn produced in a high-incidence area of human esophageal cancer (see human esophageal cancer, above) in Transkei was found to be extremely toxic to day-old Pekin ducklings (100% mortality in mean times of 90 to 136 minutes) and to produce large quantities (up to 11.3 g/kg) of moniliformin (see moniliformin, below) in culture (Kriek et al., 1977). We have identified this strain as *F. subglutinans* (Strain 10.36). Although ducklings consumed only small amounts of diets containing culture material of this strain before refusing feed, the moniliformin content was so high that the amounts of toxic meal ingested at the first feeding were sufficient to kill them very rapidly. The oral LD_{50} of pure moniliformin isolated from this strain was 3.68 mg/kg in ducklings, and the clinical signs of moniliformin intoxication were progressive muscular weakness, respiratory distress, cyanosis, coma, and death.

In a study of the incidence, geographic distribution, and toxigenicity of *Fusarium* species in South African corn, *F. subglutinans* (as *F. sacchari* var. *subglutinans*) was found to predominate in an area with a relatively cool and humid climate (Marasas et al.,

1979a). The highest percentage of isolates of *F. subglutinans* toxic to ducklings was obtained from the area in which this *Fusarium* species occurred most frequently. In total, cultures on corn incubated at 25°C for 21 days of 23 out of 36 (63.9%) isolates tested were toxic to ducklings (= more than two out of four ducklings dead within 14 days). Moniliformin was produced in culture by 16 out of 23 (69.6%) of the isolates toxic to ducklings at levels ranging from 120 to 1,170 mg/kg (see moniliformin, below).

Rabbits. Two isolates *F. subglutinans* (as *F. moniliforme* var. *subglutinans*) were reported to be dermotoxic to rabbit skin by Joffe et al. (1973). One isolate was only slightly toxic, but extracts of *F. subglutinans* No. 8549 (Strain 10.35) cultured on wheat kernels at 18°C caused a toxic reaction, with leucocytic infiltrations of the dermis accompanied by edema and necrosis of the skin. The chemical nature of the dermotoxic mycotoxin(s) produced by these strains of *F. subglutinans* is unknown.

Rats. Cultures of *F. moniliforme* var. *subglutinans* MRC 134, isolated from Zambian corn, did not cause any mortalities in weanling male Wistar rats fed for 14 days, but caused marked reductions in weight gain (Marasas et al., 1978). We have identified this isolate as *F. subglutinans* (Strain 10.34).

Culture material of *F. moniliforme* var. *subglutinans* MRC 115, isolated from corn produced in a high-incidence area of human esophageal cancer (see human esophageal cancer, above) in Transkei, was reported to be acutely toxic to inbred BD IX rats by Kriek et al. (1977). We have identified this isolate as *F. subglutinans* (Strain 10.36). This strain was found to produce large quantities of moniliformin (up to 11.3 g/kg) in culture (see moniliformin, below). Male rats were more sensitive than females to the toxic effects of low dietary levels of culture material. Diets containing 8% or more of culture material caused the death of 4 of 4 male and 3 of 4 female rats, while 4% moldy meal caused the death of 3 of 4 males but no females. A dietary level of 2% moldy meal caused no mortalities in either males or females and the cumulative amount of moniliformin consumed over a 12-week period by rats in this group was calculated as 286.8 mg and 210.2 mg/rat for males and females, respectively. Kriek et al. (1977) concluded from these results that "in order to have tolerated these remarkably large amounts of moniliformin, these rats must have been able to detoxify the relatively low levels of moniliformin in the feed effectively." The oral LD_{50} of pure moniliformin isolated from this strain was 41.57 and 50.00 mg/kg in female and male rats, respectively. The clinical signs of moniliformin intoxication in rats were progressive muscular weakness, respiratory distress, cyanosis, coma, and death. Deaths occurred within 3 hours of moniliformin administration and rats that did not die within this period recovered completely within 12 hours. The myocardium was the site of the major histopathological lesions in the rats fed either culture material or pure moniliformin and the characteristic lesions were focal myocardial degeneration and necrosis (Kriek et al., 1977).

Mycotoxins Produced

Fusaric Acid. Matuo et al. (1976) reported that *F. subglutinans* (as *F. moniliforme* var. *subglutinans* No. 953 = IMI 131473) produces fusaric acid in Richard's solution incubated at 25°C for 15 days. According to Ghosal et al. (1979), *F. subglutinans* (as *F. moniliforme* var. *subglutinans* CMI-IMI 225231) causes the malformation disease of mango (*Mangifera indica* L.) in India. This strain produces fusaric acid in Richard's solution incubated at 21°C for 21 days, but not in artificially infected mango shoots and inflorescences. In previous papers on mycotoxins produced by the causal fungus of malformation disease (Ghosal et al., 1976b, 1978b), this fungus was referred to as *F. moniliforme* (see *F. moniliforme,* mycotoxins produced).

Gibberellins. Gordon (1960b) used a seedling bioassay to screen 17 isolates of *F. subglutinans* (as *F. moniliforme* var. *subglutinans*) from a variety of hosts from different geographical areas for the production of substances with gibberellin-like biological activity. He reported that 11 of these isolates produced "gibberellins." Matuo et al. (1976) also reported that *F. subgutinans* (as *F. moniliforme* var. *subglutinans* No. 953 = IMI 131473) produces gibberellin-like substances.

On the other hand, Borrow et al. (1955) found that none of four isolates of *F. subglutinans* (as *Gibberella fujikuroi* var. *subglutinans*) produced plant-growth promoting substances or gibberellic acid in culture.

Gibberellin production by *F. moniliforme* and the use of the name *F. fujikuroi* by Nirenberg (1976) are discussed elsewhere in this volume (see *F. moniliforme,* gibberellins).

Moniliformin. The first report of moniliformin production by *F. subglutinans* was by Kriek et al. (1977), who reported that *F. moniliforme* var. *subglutinans* MRC 115 produced large quantities (up to 11.3) g/kg of moniliformin in cultures on autoclaved corn incubated at 25°C for 21 days. We have identified this isolate as *F. subglutinans* (Strain 10.36). These moniliformin-containing cultures were extremely toxic to ducklings (see ducklings, above) and rats (see rats, above). Moniliformin, as a mixture of the sodium and potassium salts of 1-hydroxycylobut-1-ene-3,4-dione, was also isolated from this strain by M. Steyn et al. (1978) and Thiel (1978).

In a study of the incidence, geographic distribution, and toxigenicity of *Fusarium* species in South African corn, *F. subglutinans* (as *F. moniliforme* var. *subglutinans*) was found to predominate in an area with a relatively cool and humid climate (Marasas et al., 1979a). Cultures on autoclaved corn incubated at 25°C for 21 days of 36 isolates of *F. subglutinans* were tested for toxicity to day-old Pekin ducklings (see ducklings, above) and analysed chemically for the presence of moniliformin. A total of 23 out of 36 (= 63.9%) of the isolates tested were toxic to ducklings, and 16 out of the 23 (69.6%) toxic isolates produced moniliformin at levels ranging from 120 to 1,170 mg/kg. A higher proportion of isolates from corn produced in the area where *F. subglutinans* occurred with maximal frequency were toxic to ducklings and produced moniliformin than isolates from the other two areas in the study. The highest moniliformin-producing isolate of *F. subglutinans* (MRC 756, Strain 10.37) also originated in this area. These data suggest that toxic, high moniliformin-producing strains of *F. subglutinans* may be more prevalent in certain geographical areas than in others (Marasas et al., 1979a).

Trichothecenes. According to Ghosal et al. (1979), *F. subglutinans* (as *F. moniliforme* var. *subglutinans* CMI-IMI 225231) causes the malformation disease of mango in India. This strain reportedly produces unspecified 12,13-epoxytrichothecenes *in vitro* and in artificially infected mango shoots and inflorescences. In previous papers on mycotoxins produced by the causal fungus of malformation disease (Ghosal et al., 1976b, 1978b), this fungus was referred to as *F. moniliforme* and was reported to produce diacetoxyscirpenol and T-2 toxin (see *F. moniliforme,* mycotoxins produced). We have not seen any of these isolates and there are no known trichothecene-producing strains of *F. subglutinans* represented in the ITFRC.

Toxigenic Strains in the ITFRC

The following strains listed under *F. subglutinans* in Table 10.1 all have been identified as *F. subglutinans* according to Nelson et al. (1983).

Strain 10.34 *F. moniliforme* var. *subglutinans* MRC 134 (=M-1351; MRC 134)

This strain was isolated from corn produced in Zambia during 1974/75 (Marasas et al., 1978). Cultures on autoclaved corn incubated at 25°C for 3 weeks proved to be acutely toxic to day-old Pekin ducklings (4 of 4 dead in a mean time of 6.25 days). The same culture material did not cause any mortalities in weanling male Wistar rats fed for 14 days, but caused marked reductions in weight gain.

A lyophilized culture of this strain of *F. subglutinans* was received in March 1982 from W.F.O. Marasas as "*F. moniliforme* var. *subglutinans* MRC 134."

Strain 10.35 *F. moniliforme* var. *subglutinans* 8549 (= M-1354; MRC 940)

According to Joffe et al. (1973), this strain was originally isolated from *Haemanthus* sp. in Germany. Extracts of cultures on wheat kernels incubated at 18°C for 11 days caused a dermotoxic leucocytorrhic-edematous reaction with necrosis on rabbit skin. Cultures incubated at 24°C were only mildly dermotoxic.

A culture of this strain of *F. subglutinans* was received on 18 April 1979 from W. Gerlach as "*F. sacchari* var. *elongatum* 8549" (see Nirenberg, 1976).

Strain 10.36 *F. moniliforme* var. *subglutinans* MRC 115 (= M-811; MRC 115; ATCC 38016)

This strain was isolated from home-grown corn produced in a high-incidence area of human esophageal cancer in Transkei during 1975 (Kriek et al., 1977). Culture material on autoclaved corn incubated at 25°C for 21 days proved to be extremely toxic to day-old Pekin ducklings and inbred BD IX rats. In ducklings the toxic culture material resulted in 100% mortality within 90 to 136 minutes and the clinical signs in ducklings as well as rats were progressive muscular weakness, respiratory distress, cyanosis, coma, and death. The characteristic histopathological lesions in rats were focal myocardial degeneraion and necrosis. The toxic principle in these cultures was found to be moniliformin, and this strain produces large quantities (up to 11.3 g/kg) of moniliformin in culture (Kriek et al., 1977). Moniliformin (as a mixture of the sodium and potassium salts) was also isolated from this strain by M. Steyn et al. (1978) and Thiel (1978).

A lyophilized culture of this strain of *F. subglutinans* was received on 17, October 1978 from W.F.O Marasas as "*F. moniliforme* var. *subglutinans* MRC 115."

Strain 10.37 *F. moniliforme* var. *subglutinans* MRC 756 (= M-1168; MRC 756)

This strain was isolated from corn produced in a relatively cool and humid area in the Eastern Transvaal, South Africa. It was obtained during a study of the incidence, geographic distribution, and toxigenicity of *Fusarium* species in South African corn (Marasas et al., 1979a). The maximal incidence of *F. subglutinans* determined in this study occurred in corn from this area.

Culture material of MRC 756 on autoclaved corn incubated at 25°C for 21 days proved to be acutely toxic to day-old Pekin ducklings, and this strain was the highest producer of moniliformin (1,170 mg/kg) detected in the study by Marasas et al. (1979a).

A lyophilized culture of this strain was received on 22 April 1980 from W.F.O Marasas as "*F. moniliforme* var *subglutinans* MRC 756."

Fusarium anthophilum (A. Braun) Wollenw.

Perfect State: Unknown.

Incidence and Distribution

Very little is known about the incidence and distribution of *F. anthophilum*, probably because of mis-identification as *F. moniliforme* or *F. moniliforme* var. *subglutinans*. The species has, however, been recorded on a variety of hosts in a number of countries including Germany, Iran, Kenya, Madagascar, Nigeria, and Trinidad (Gordon, 1956, 1960a, b; Nirenberg, 1976).

Association with Human and/or Animal Diseases

Fusarium anthophilum has not been associated with human or animal toxicoses.

Toxicity to Experimental Animals

Rabbits. Two isolates of *F. anthophilum* (as *F. moniliforme* var. *anthophilum*) were reported to be dermotoxic to rabbit skin by Joffe et al. (1973). One isolate was only slightly toxic, but extracts of isolate 8998 (see Strain 10.38) cultured on wheat kernels at 18°C caused a mildly toxic reaction. The chemical nature of the dermotoxic mycotoxin(s) produced by these strains of *F. anthophilum* is unknown.

Mycotoxins Produced

Gibberellins. Gordon (1960b) detected substances with gibberellin-like biological activity by means of a seedling bioassay in one out of four strains of *F. anthophilum* (as *F. moniliforme* var. *anthophilum*).

Toxigenic Strains in the ITFRC

The following strain listed under *F. anthophilum* in Table 10.1 has been identified as *F. anthophilum* according to Nelson et al. (1983).

Strain 10.38 *F. moniliforme* var. *anthophilum* **8998** (= M-1355; MRC 941)
According to Joffe et al. (1973), this strain was originally isolated from *Hippeastrum* in Germany. Extracts of cultures on wheat kernels incubated at 18°C for 11 days caused a mildly dermotoxic reaction on rabbit skin.

A culture of this strain of *F. anthophilum* was received on 18 April 1979 from W. Gerlach as "*F. anthophilum* 8998" (see Nirenberg, 1976).

11. Section Elegans

The only species accepted in this section by Nelson et al. (1983), *F. oxysporum,* has been reported to be toxigenic.

Fusarium oxysporum Schlecht. emend. Snyd. & Hans.
    Perfect State: Unknown.

Incidence and Distribution

Fusarium oxysporum is a cosmopolitan soil saprophyte, and specialized pathogenic strains cause vascular wilt and damping-off diseases of a great variety of host plants (Booth, 1971; CMI Descriptions Nos. 27 and 28, 1964, and Nos. 211–220, 1970, Doidge, 1938; Domsch et al., 1980; Gordon, 1956, 1960a; Meyer & Frank, 1979; Wollenweber & Reinking, 1935).

    Many pathogenic strains of *F. oxysporum* that are pathogenic only to specific hosts have been designated as formae speciales, e.g., *F. oxysporum* Schlecht. f. sp. *carthami* Klisiewicz & Houston, which causes wilt disease of safflower (*Carthamus tinctorius* L.).

Association with Human and/or Animal Diseases

Moldy Sweet Potato Toxicosis. A toxicosis of cattle characterized by severe pulmonary edema has been attributed to the presence of four furanoterpenoids in sweet potatoes (*Ipomoea batatas* L.) infected by *F. solani* (see both *F. solani* and *F. moniliforme,* moldy sweet potato toxicosis). The lung edemagenic agents 1-ipomeanol, 4-ipomeanol, 1,4-ipomeadiol, and ipomeanine are formed when the furanosesquiterpenoid stress metabolite 4-hydroxymyoporone, formed by sweet potato tissue in response to non-specific stimuli, is catabolized by the pathogenic fungus *F. solani* (Burka et al., 1977). Although *F. oxysporum* is also a pathogen of sweet potato tubers and has been shown *in vitro* to be capable of converting 4-hydroxymyoporone to the lung toxic furanoterpenoids, it was much less effective (Burka et al., 1977). Consequently it is not known whether *F. oxysporum* plays any role in field outbreaks of atypical interstitial pneumonia in cattle consuming moldy sweet potatoes.

Another toxic substance known to occur in moldy sweet potatoes is the phytoalexin ipomeamarone, which is hepatotoxic and has an intraperitoneal LD_{50} of 230 mg/kg in mice (Wilson, 1973). *Fusarium oxysporum* causes surface rot of sweet potato tubers and has been found to be one of the agents that induces high concentrations (350 to 9,480 µg/g) of ipomeamarone in infected sweet potato tissue, but not in surrounding healthy tissue (Martin et al., 1976). These authors concluded that this finding may account for the lack of human poisoning by eating sweet potatoes "because visibly diseased portions are discarded when sweet potatoes are being prepared for consumption." It is evident, however, that the consumption of moldy sweet potatoes is potentially dangerous to human health and that culled sweet potatoes should not be fed to animals.

Toxicity to Experimental Animals

Brine Shrimp. Extracts of cultures of *F. oxysporum* AUA 597 isolated from a carrot obtained from a supermarket in Alabama, USA, were reported to be toxic to brine shrimp by N.D. Davis et al. (1975). Extracts of potatoes inoculated with *F. oxysporum* isolated from potatoes in Germany were also toxic to brine shrimp (Siegfried & Langerfeld, 1978). We have confirmed the identity of both of these isolates as *F. oxysporum* (Strains 11.1 and 11.2). The mycotoxin(s) produced by these strains of *F. oxysporum* toxic to brine shrimp have not been chemically characterized.

Chickens. N.D. Davis et al. (1975) reported that extracts of cultures of *F. oxysporum* AUA 597 (Strain 11.1) caused 100% mortality of chick embryos. Extracts of cultures on nutrient amended shredded wheat incubated at 25°C for 21 days of *F. oxysporum* AUA 1078, isolated from grain sorghum in Alabama, USA, proved to be slightly toxic (1 of 7 dead) to day-old White Leghorn cockerels (Diener et al., 1981). The mycotoxin(s) produced by these two strains of *F. oxysporum* toxic to chick embryos and chickens have not been chemically characterized.

Cultures of *F. oxysporum* Sp. 1028, isolated from barley grains in Germany, were highly toxic to day-old White Leghorn chickens (Marasas et al., 1979b). We have confirmed the identity of this strain as *F. oxysporum* (Strain 11.3). The toxicity of the culture material of this strain was due to the presence of high levels of moniliformin (see moniliformin, below).

According to Meronuck et al. (1970), cultures of an unspecified strain of *F. oxysporum* isolated from corn suspected of being toxic to cattle in the United States were highly toxic to chickens. Speers et al. (1972) reported that cultures of an unspecified isolate of

F. oxysporum caused marked decreases in feed intake, body weight loss, and a complete cessation of egg production in laying hens. According to C.J. Mirocha (personal communication, 6 March 1978), the isolates referred to in the above two publications are the same isolate which was referred to as "*F. roseum* (isolate oxyrose)" by Mirocha & Christensen (1974). We have identified this isolate as *F. semitectum* (Strain 6.4) and it is consequently excluded from *F. oxysporum*.

Ducklings. Martin et al. (1971) reported that 32 of 63 isolates of *F. oxysporum* from foodstuffs in Swaziland were toxic to ducklings. However, in their lists of fungi isolated from different foodstuffs, *F. oxysporum* is not mentioned at all, while *F. moniliforme* is recorded as a "species regularly present in all foodstuffs," whereas in their list of fungi tested for toxicity, *F. moniliforme* is not mentioned, while 32 of 63 isolates of *F. oxysporum* are recorded as being toxic to ducklings. We conclude that the latter statement is a misprint for *F. moniliforme,* and these isolates are consequently excluded from *F. oxysporum*.

Mice. According to Itakura & Kinosita (1975), the culture filtrate of one unspecified strain of *F. oxysporum* isolated from foodstuffs in Uganda was toxic to mice upon intraperitoneal injection.

A mouse bioassay in which culture filtrates or extracts of cultures are injected intraperitioneally in male dds mice has been used extensively in Japan for the detection of toxigenic *Fusarium* strains (Ueno et al., 1971a). Extracts of the following isolates of *F. oxysporum* have been found to be lethal to mice in this assay: unspecified isolate from paddy in Akita, Japan (Ueno et al., 1971a); isolate Abashiri-3 (Ueno et al., 1973a); NRRL 1943 (Strain 11.4) (Ueno et al., 1972b, 1973a); and five out of 24 isolates of *F. oxysporum,* including 5026 (Strain 11.5), from river sediments in Japan (Ueno et al., 1977b). Although some of these isolates of *F. oxysporum* toxic to mice, including 5026 (Strain 11.5) identified as *F. oxysporum* by us, caused radiomimetic pathological changes characteristic of trichothecenes and also inhibited protein synthesis in rabbit reticulocytes, no known trichothecenes have been detected in the lethal extracts and the chemical nature of the mycotoxin(s) produced by these strains is unknown. Two other isolates toxic to mice and referred to as *F. oxysporum* "niveum" Melon-1 and Melon-2 by Ueno et al. (1973a) were, however, reported to produce the trichothecenes fusarenon-X and diacetylnivalenol (see mycotoxins produced, below). We have identified *F. oxysporum* "niveum" Melon-1 as a degenerate strain of *F. sporotrichioides* (Strain 4.11) and this strain is consequently excluded from *F. oxysporum*.

Extracts of *F. lateritium* 5013 and K.5036, isolated from river sediments in Japan, proved to be lethal to mice and to cause radiomimetic pathological changes (Ueno et al., 1977b). We have identified both of these isolates as *F. oxysporum* (Strains 11.6 and 11.7). No known trichothecenes could be detected in the toxic extracts of *F. oxysporum* (= *F. lateritium*) 5013 (Strain 11.6), but this strain was found to produce zearalenone on autoclaved rice (Ishii et al., 1974; Ueno et al., 1977b). Isolate K.5036 (Strain 11.7) was found to produce diacetoxyscirpenol, diacetylnivalenol, neosolaniol, and two new trichothecenes, 7-hydroxydiacetoxyscirpenol and 7,8-dihydroxydiacetoxyscirpenol (Ishii, 1975; Ueno and Shimada, 1974; Ueno et al., 1977b).

Pigeons. Chloroform extracts of cultures in Richard's solution incubated at 21°C for 21 days of a virulent (IMI 186539) and a weakly parasitic strain (IMI 166917) of *F. oxysporum* f. sp. *carthami* from wilt disease of safflower (*Carthamus tinctorius* L.) in India caused prolonged emesis in pigeons at non-lethal concentrations upon oral or intravenous administration (Ghosal et al., 1976a). The extracts were found by TLC to contain about six trichothecene derivatives, of which two were identified as diacetoxyscirpenol and T-2 toxin (see mycotoxins produced, below).

Rabbits. Extracts of cultures of two isolates (*F. oxysporum* 2341 and *F. redolens* [= *F. oxysporum*] 2317) from overwintered cereals associated with outbreaks of ATA (see *F. sporotrichioides,* ATA) in the USSR were dermotoxic to rabbit skin (Joffe 1960a, 1960b, 1971). Extracts of two out of five isolates (2131 and 3163) of *F. oxysporum* from goundnuts in Israel were also reported to be moderately dermotoxic to rabbit skin by Joffe (1973b). According to Joffe & Palti (1974), 149 of 155 unspecified isolates of Elegans Fusaria exhibited some degree of dermotoxicity to rabbit skin. The toxic reactions ranged from a slight reddening of the skin to severe leucocytorrhea, edema, and necrosis. A quantitative relationship was also found between the dermotoxicity of these isolates and their phytotoxicity. The mycotoxin(s) produced by these dermotoxic isolates of *F. oxysporum* have not been chemically characterized.

Rats. According to Diener et al. (1976), extracts of *F. oxysporum* Auburn University No. 973, isolated from cotton in the United States and incubated on nutrient-amended shredded wheat at 25°C for 14 to 21 days, were toxic to rats upon dosing.

Tuttobello et al. (1974) reported that *F. oxysporum* Gaümann 1536/9 had uterotrophic effects in virgin weanling female rats when cultured either in liquid shake culture or in still culture on autoclaved corn. The chemical nature of the uterotrophic substance(s) was not determined, but the uterotrophic effect was equivalent to that of approximately 25 ppm "zearalanol" in the liquid and 12 ppm in the corn cultures.

A virulent (IMI 186539) and a weakly parasitic (IMI 166917) strain of *F. oxysporum* f. sp. *carthami* isolated from wilt disease of safflower in India were reported to be dermotoxic to rat skin by Ghosal et al. (1976a). When chloroform extracts of cultures in Richard's solution incubated at 21°C for 21 days were applied to the skin of rats, edema developed on the second day and the skin lesion became progressively severe, developing into hemorrhage and a heavy scab on the fourth day. When high doses were applied topically, the animals died within 6 days. These dermotoxic extracts were found by TLC to contain about six trichothecene derivatives, of which two were identified as diacetoxyscirpenol and T-2 toxin (see mycotoxins produced, below).

Turkeys. The strain of *F. oxysporum* reported to be highly toxic to turkey poults (Meronuck et al. 1970) and referred to as "*F. roseum* (isolate oxyrose)" by Mirocha & Christensen (1974), is *F. semitectum* (Strain 6.4) and is consequently excluded from *F. oxysporum*.

Mycotoxins Produced

Diacetoxyscirpenol. Ishii (1975) reported that *Fusarium* sp. K.5036, isolated from river sediments in Japan, produced diacetoxyscirpenol as well as four other trichothecenes (see Strain 11.7) and this finding was confirmed (as *F. lateritium* 5036) by Ueno et al. (1977b). We have identified this strain as *F. oxysporum* (Strain 11.7).

According to Chakrabarti et al. (1976), a weakly parasitic strain (IMI 166917) of *F. oxysporum* f. sp. *carthami* isolated from safflower in India produced diacetoxyscirpenol that could be detected in roots and was also translocated to stems, leaves, and seeds of safflower plants grown in sterilized soil inoculated with this fungus. Diacetoxyscirpenol was also produced in Richard's medium incubated in still culture at 21°C for 21 days. The production of diacetoxyscirpenol (together with T-2 toxin; see T-2 toxin, below) in Richard's solution by this strain, as well as by a virulent strain (IMI 186539) of *F. oxysporum* f. sp. *carthami,* was confirmed by Ghosal et al. (1976a). The latter authors isolated 58 mg of crystals from 5 ℓ of culture filtrate of isolate IMI 166917 and identified these crystals as diacetoxyscirpenol by melting point, optical rotation, and spectral properties. Ghosal et al. (1977a) detected diacetoxyscirpenol in seeds of safflower plants naturally infected by *F. oxysporum* f. sp. *carthami* in India. In addition to diacetoxyscirpenol (and T-2 toxin; see T-2 toxin, below), the above two Indian isolates of *F. oxysporum* f. sp. *carthami* have also been reported to produce several other unidentified trichothecene derivatives in Richard's solution and in inoculated safflower plants (Chakrabarti et al., 1976; Chakrabarti & Basu-Chaudhary, 1980; Ghosal et al., 1976a).

Several authors have reported the non-production of trichothecenes, including diacetoxyscirpenol, by isolates of *F. oxysporum*: none of 11 isolates, including NRRL 1943 which we have identified as *F. oxysporum* (Strain 11.4) and excluding *F. oxysporum* "niveum" Melon-1 which we have identified as *F. sporotrichioides* (Strain 4.11), by Ueno et al. (1973a); none of 24, including 5026 which we have identified as *F. oxysporum* (Strain 11.5), by Ueno et al. (1977b); and none of 21 Suzuki et al. (1980, 1981b).

Diacetylnivalenol. Ueno et al. (1973a) reported that *F. oxysporum* "niveum" Melon-1 and Melon-2 produced diacetylnivalenol. We have identified isolate Melon-1 as a degenerate strain of *F. sporotrichioides* (Strain 4.11), and cultures of Melon-2 are no longer available; consequently, these two strains are excluded from *F. oxysporum*.

The production of diacetylnivalenol by *Fusarium* sp. K.5036 was reported by Ishii (1975) and confirmed by Ueno et al. (1977b), who referred to this strain as *F. lateritium* 5036. We have identified this strain as *F. oxysporum* (Strain 11.7).

7,8-Dihydroxydiacetoxyscirpenol. This new trichothecene was isolated from *Fusarium* sp. K.5036 by Ishii (1975). The producing strain was referred to as *F. lateritium* 5036 by Ueno et al. (1977b). We have identified this strain as *F. oxysporum* (Strain 11.7).

Fusarenon-X. Ueno et al. (1973a) reported that *F. oxysporum* "niveum" Melon-1 (see Strain 4.11) and Melon-2 produced fusarenon-X. The former strain (as *F. oxysporum* T-M-1) was also found to produce small amounts of fusarenon-X by Ueno et al. (1975). We have identified isolate Melon-1 as *F. sporotrichioides* (Strain 4.11), and cultures of Melon-2 are no longer available; consequently, these two strains are excluded from *F. oxysporum*.

According to Morooka et al. (1980), *F. oxysporum* No. 55-1, isolated from cereals in Japan, produced fusarenon-X (but not deoxynivalenol or 3-acetyldeoxynivalenol) in liquid medium.

Fusaric Acid. Fusaric acid is a well-known phytotoxin that is produced by several *Fusarium* species, particularly pathogenic strains of *F. oxysporum* causing wilt diseases of a great variety of plants (Gaümann, 1957; Kern, 1972). Although fusaric acid is not generally regarded as a mycotoxin, some attention will be given here to fusaric acid production by *F. oxysporum* because fusaric acid as well as certain other phytotoxins such as lycomarasmin and lycomarasmic acid produced by *F. oxysporum* (Gaümann & Naef-Roth, 1950; Kern, 1972) are chelating agents and may be involved in certain diseases of abnormal bone development in animals (see *F. moniliforme,* abnormal bone development). In addition, fusaric acid is toxic to mice (intraperitoneal LD_{50} 80 mg/kg) and death caused by the lethal dose has been attributed to its hypotensive effect (Hidaka et al., 1969). The ability of fusaric acid to cause significant decreases of blood pressure has also been observed in cats, dogs, rabbits, and rats and has been attributed to the inhibition of dopamin-β-hydroxylase (Bilai et al., 1975; Hidaka, 1971; Hidaka et al., 1969). Fusaric acid has been administered to humans in clinical trials as an antihypertensive agent (Matta & Wooten 1973), in the treatment of Parkinson's disease (Hidaka, 1971; Matta & Wooten, 1973), and at dosage rates up to 1200 mg/day in the treatment of drug addiction (Pozuelo, 1976).

A positive correlation between pathogenicity to plants and the amount of fusaric acid produced has been found for many strains of *F. oxysporum* (Kern, 1972). The production of fusaric acid in Richard's medium incubated at 21°C for 21 days has been reported for a weakly pathogenic (IMI 166917) as well as a virulent (IMI 186539) strain of *F. oxysporum* f. sp. *carthami* by Chakrabarti et al. (1976) and Ghosal et al. (1977b), respectively. However, Chakrabarti & Basu-Chaudhary (1980) found that an unspecified virulent strain of this fungus produced three times more fusaric acid (60–80 mg/ℓ) than an unspecified "mild" strain (20–30 mg/ℓ).

Surico & Graniti (1977) reported that unspecified virulent isolates of *F. oxysporum* f. sp. *albedinis* isolated from Bayoud diseased date palms (*Phoenix dactylifera* L.) in Algeria produced an average of 415 mg/ℓ of fusaric acid and small amounts of anhydro-aspergillomarasmin B. We have identified one of these strains as *F. oxysporum* (Strain 11.8). Recently Mutert et al. (1981) detected moderate amounts (12–290 mg/ℓ) of fusaric acid by HPLC in culture fluid of *F. oxysporum* f. sp. *apii* CBS 184.38 and an isolate of *F. oxysporum* f. sp. *pisi*.

7-Hydroxydiacetoxyscirpenol. This new trichothecene was isolated from *Fusarium* sp. K.5036 by Ishii (1975). The producing strain was referred to as *F. lateritium* 5036 by Ueno et al. (1977b). We have identified this strain as *F. oxysporum* (Strain 11.7).

Moniliformin. *Fusarium oxysporum* Sp. 1028, isolated from barley grain in Germany, produced 1,150 mg/kg of moniliformin in cultures on autoclaved corn incubated at 25°C for 2 weeks (Marasas et al., 1979b). We have confirmed the identity of this strain as *F. oxysporum* (Strain 11.3).

Neosolaniol. The production of neosolaniol by *Fusarium* sp. K.5036 was reported by Ishii (1975) and confirmed by Ueno et al. (1977b), who referred to the producing strain as *F. lateritium* 5036. We have identified this strain as *F. oxysporum* (Strain 11.7).

T-2 Toxin. According to Chakrabarti et al. (1976), a weakly parasitic strain (IMI 166917) of *F. oxysporum* f. sp. *carthami* isolated from safflower in India produced T-2 toxin that could be detected in roots and was also translocated to stems and leaves of safflower plants grown in sterilized soil inoculated with this fungus. T-2 toxin was also produced in Richard's solution incubated in still culture at 21°C for 21 days. The production of T-2 toxin (together with diacetoxyscirpenol; see diacetoxyscirpenol, above) in Richard's solution by this strain, as well as by a virulent strain (IMI 186539) of *F. oxysporum* f. sp. *carthami*, was confirmed by Ghosal et al. (1976a). The latter authors isolated 33 mg of crystals from 5 ℓ culture filtrate of isolate IMI 166917 and identified these crystals as T-2 toxin by melting point, optical rotation, and spectral properties. Ghosal et al. (1977a) detected T-2 toxin in seeds of safflower plants naturally infected by *F. oxysporum* f. sp. *carthami* in India.

According to Nusrath (1979), two Indian isolates of *F. oxysporum* from wilt diseased chick peas and pigeon peas, respectively, produced unspecified levels of T-2 toxin (as well as zearalenone; see zearalenone, below) as determined by TLC in ethyl acetate extracts of cultures in Richard's medium incubated at 25°C for 14 days.

Several authors have reported the non-production of trichothecenes, including T-2 toxin, by isolates of *F. oxysporum:* none of 11, including NRRL 1943 which we have identified as *F. oxysporum* (Strain 11.4) and excluding *F. oxysporum* "niveum" Melon-1 which we have identified as *F. sporotrichioides* (Strain 4.11), by Ueno et al. (1973a); none of 24, including 5026 which we have identified as *F. oxysporum* (Strain 11.5), by Ueno et al. (1977b); and none of 21 by Suzuki et al. (1980, 1981b).

In conclusion, only four Indian isolates of *F. oxysporum* out of at least 56 tested have been reported to produce T-2 toxin. We have not seen any of these Indian isolates, and there are no known T-2 toxin-producing strains of *F. oxysporum* represented in the ITFRC.

Zearalenone. Mirocha & Christensen (1974) stated that *F. oxysporum* had been found to produce zearalenone in their laboratory, but gave no indication of the number of strains that had been found to be positive or of the levels produced. The strain reported as "*F. roseum* isolate oxyrose" to produce "copious amounts" of zearalenone by Mirocha & Christensen (1974), and previously referred to as *F. oxysporum* by Meronuck et al. (1970) and Speers et al. (1972), is *F. semitectum* (Strain 6.4).

Ishii et al. (1974) reported that one unspecified isolate of *F. lateritium* produced zearalenone and this was confirmed by Ueno et al. (1977b), who referred to the producing strain as *F. lateritium* 5013. We have identified this strain as *F. oxysporum* (Strain 11.6).

Fusarium sp. 5029, isolated from river sedimens in Japan, was reported to produce unspecified levels of zearalenone by Ueno et al. (1977b). We have identified this strain as *F. oxysporum* (Strain 11.9).

According to Nusrath (1979), two Indian isolates of *F. oxysporum* from chick pea and pigeon pea, respectively, produced unspecified levels of zearalenone (together with T-2 toxin; see T-2 toxin, above) in Richard's solution incubated at 25°C for 14 days.

Several authors have reported on the non-production of zearalenone by isolates of *F. oxysporum:* none of 15 by Caldwell et al. (1970); neither of 2 by Eugenio et al. (1970a); none of 11 by Ichinoe et al. (1977); none of 12 by Ishii et al. (1974a); none of 5 by Suzuki et al. (1978, 1981b); and none of 24, including isolate 5026 which we have identified as *F. oxysporum* (Strain 11.5), by Ueno et al. (1977b).

In conclusion, only two Indian isolates out of at least 70 isolates of *F. oxysporum* tested have been reported to produce unspecified levels of zearalenone. In addition, we have identified two zearalenone-producing isolates, *F. lateritium* 5013 (Ishii et al., 1974a; Ueno et al., 1977b) and *Fusarium* sp. 5029 (Ueno et al., 1977b), as *F. oxysporum* (Strains 11.6 and 11.9).

Toxigenic Strains in the ITFRC

The toxigenic *Fusarium* strains listed in Table 11.1 all have been identified as *F. oxysporum* according to Nelson et al. (1983). Strains referred to as *F. oxysporum* in the literature but identified as other species by us are listed in Table 11.2.

Strain 11.1 *F. oxysporum* AUA 597 (= O-1055; MRC 2325)

This strain was isolated from a visibly moldy carrot obtained from a supermarket in Alabama, USA (N.D. Davis et al., 1975). The fungus was cultured on autoclaved, nutrient-amended shredded wheat at 25°C for 14 to 21 days. Chloroform-ethanol extracts of this culture material proved to be toxic (20–59% mortality) to brine shrimp and to cause 100% mortality of chicken embryos (N.D. Davis et al., 1975). The chemical nature of the mycotoxin(s) produced by this strain is unknown.

TABLE 11.1 *Fusarium* Strains Identified as *F. oxysporum* in the International Toxic *Fusarium* Reference Collection (ITFRC)

Strain No.	ITFRC No.	MRC No.	Published Name(s) and Number(s)	Source	Toxigenicity
11.1	0-1055	2325	*F. oxysporum* AUA 597	U.L. Diener	a
11.2	0-1071	2199	*F. oxysporum*—Potato (= Berlin 62286)	E. Langerfeld	a
11.3	0-916	1414	*F. oxysporum* Sp. 1028	W.F.O. Marasas	a, b
11.4	0-1011	1694	*F. oxysporum* NRRL 1943	NRRL	a
11.5	0-1173	2564	*F. oxysporum* 5026	Y. Ueno	a
11.6	0-1174	2566	*F. lateritium* 5013	Y. Ueno	a, c
11.7	0-1175	2567	*F.* sp. K 5036 (= *F. lateritium* 5036)	Y. Ueno	a, d
11.8	0-1042	2066	*F. oxysporum* ITM-150	A. Bottalico	e
11.9	0-1166	2536	*F.* sp. 5029	Y. Ueno	c

a = Culture toxic to experimental animals
b = Produces moniliformin
c = Produces zearalenone
d = Produces trichothecenes
e = Produces fusaric acid

A lyophilized culture of this strain was received on 10 March 1981 from U.L. Diener as "*F. oxysporum* No. 597."

Strain 11.2 *F. oxysporum* Berlin 62286 (= O-1071; MRC 2199)

This strain was isolated from rotten potato tubers in Germany (Siegfried & Langerfeld, 1978). Potato tubers of two different cultivars were inoculated with this strain and incubated at 20°C until approximately one-third of each tuber was infected. The infected tissues were then separated and lyophilized, and purified ethyl-acetate extracts of them were assayed for toxicity to brine shrimp. Extracts of both cultivars proved to be toxic (50-80% mortality within 4 hours) to brine shrimp (Siegried & Langerfeld, 1978). The chemical nature of the mycotoxin(s) produced in inoculated potato tubers by this strain is unknown.

A slant culture of this strain was received on 20 January 1981 from E. Langerfeld as "*F. oxysporum* 62286 (Berlin)."

Strain 11.3 *F. oxysporum* Sp. 1028 ((O-916; MRC 1414)

This strain was isolated from barley grains from Bayreuth, Germany (Marasas et al., 1979b). Cultures on autoclaved corn incubated at 25°C for 2 weeks followed by 4 weeks at 10°C were toxic to day-old White Leghorn chickens and caused the death of 4 of 4 chickens in a mean time of 7.5 days after the consumption of an amount of feed equal to only 0.5% of that consumed by the controls (Marasas et al., 1979b). The toxicity of this culture material was due to moniliformin, which was detected in the meal at a level of 1,150 mg/kg.

A lyophilized culture of this strain was received on 17 October 1978 from W.F.O. Marasas as "*F. oxysporum* Sp. 1028."

Strain 11.4 *F. oxysporum* NRRL 1943 (= O-1011; MRC 1694)

According to Ueno et al. (1972b), this strain was obtained from NRRL as *F. oxysporum* NRRL 1943 and was identified by W.L. Gordon, but the exact source is unknown to us. Extracts of cultures of this strain in PSC medium incubated in stationary culture at 26°C for 12 days were lethal to white mice upon intraperitoneal injection (Ueno et al., 1972b, 1973a). These extracts were negative in the rabbit reticulocyte assay for inhibition of

TABLE 11.2. *Fusarium* Strains Published as *F. oxysporum* but Identified as Other Species in the International Toxic *Fusarium* Reference Collection (ITFRC)

Strain No.	ITFRC No.	MRC No.	Published Name(s) and Number(s)	Source	Present Identification
6.4	R-4485	1642	*F. oxysporum*; *F. roseum* (strain oxyrose)	C.J. Mirocha	*F. semitectum*
4.11	T-566	2534	*F. oxysporum* "niveum" Melon-1	Y. Ueno	*F. sporotrichioides*

protein synthesis, and no trichothecenes were detected chemically. Consequently, the chemical nature of the mycotoxin(s) produced by this strain is unknown.

A lyophilized culture of this strain was received on 5 September 1979 from the NRRL as "*F. oxysporum* NRRL 1943."

Strain 11.5 *F. oxysporum* 5026 (O-1173; MRC 2564)

This strain was isolated from river sediments in Japan (Ueno et al., 1977b). Extracts of cultures in PSC medium incubated at 25 to 27°C for 2 weeks were lethal upon intraperitoneal injection in male dd YS mice (Ueno *et al.*, 1977b). The pathological lesions in the mice that died were radiomimetic changes characteristic of trichothecenes, and the extracts also inhibited protein synthesis in rabbit reticulocytes. However, no known trichothecenes could be detected chemically in these extracts by Ueno et al. (1977b).

A slant culture of this pionnotal strain of *F. oxysporum* was received in November 1981 from Y. Ueno as "*F. oxysporum* 5026."

Strain 11.6 *F. lateritium* 5013 (= O-1174; MRC 2566)

This strain was originally isolated from river sediments in Japan (Ueno et al., 1977b). Extracts of cultures in PSC medium incubated at 25–27°C for 2 weeks were lethal upon intraperitoneal injection in mice and caused radiomimetic pathological changes characteristic of trichothecenes (Ueno et al., 1977b). However, no known trichothecenes could be detected in these extracts. This strain produced unspecified levels of zearalenone in cultures on autoclaved rice incubated at 24°C for 2 weeks followed by 1 week at 10–15°C (Ueno et al., 1977b). According to Ueno et al. (1977b) this strain was previously also shown (as an unspecified strain of *F. lateritium*) to produce zearalenone on rice by Ishii et al. (1974).

A slant culture of this pionnotal strain of *F. oxysporum* was received in November 1981 from Y. Ueno as "*F. lateritium* 5013."

Strain 11.7 *F. lateritium* K.5036 (= O-1175; MRC 2567)

This strain (as *Fusarium* sp. K. 5036) was isolated from a sample of river water in Japan, according to Ishii (1975); Ueno et al. (1977b) referred to this strain as *F. lateritium* 5036 and stated that it was isolated from river sediments in Japan.

Extracts of cultures in PSC medium incubated in stationary culture at 25°C for 14 days were lethal to mice and caused the radiomimetic injury characteristic of trichothecenes (Ueno et al., 1977b). These extracts also inhibited protein synthesis in rabbit reticulocytes. This strain was shown to produce five trichothecenes, i.e., diacetoxyscirpenol, diacetylnivalenol, neosolaniol, and two new trichothecenes, 7-hydroxydiacetoxyscirpenol and 7,8-dihydroxydiacetoxyscirpenol (Ishii, 1975; Ueno & Shimada, 1974; Ueno et al., 1977b).

In their description of the morphological characters of "*F. lateritium* 5036," Ueno et al.

(1977b) stated that microconidia are absent, but in their drawings of this isolate they illustrated microconidia, macroconidia, unbranched and branched monophialides, and the terminal as well as intercalary chlamydospores quite typical of F. oxysporum.

A slant culture of this pionnotal strain of F. oxysporum was received in November 1981 from Y. Ueno as "F. lateritium 5036."

Stain 11.8 F. oxysporum ITM-150 (O-1042; MRC 2066)

According to Surico & Graniti (1977), virulent isolates (unspecified) of F. oxysporum f. sp. albedinis obtained from date palms (Phoenix dactylifera L.) with Bayoud disease in Algeria during 1972, produced an average of 415 mg/ℓ of fusaric acid and small amounts of anhydro-aspergillomarasmin B in zinc-amended Czapek-Dox medium incubated in shake culture at 25° for 8 days.

A slant culture was received on 18 September 1980 from A. Bottalico as "F. oxysporum f. sp. albedinis, ITM-150, fusaric acid producer." It is being assumed here that this pionnotal strain of F. oxysporum was one of the unspecified isolates used by Surico & Graniti (1977). This strain (as F. oxysporum ITM 150) did not produce deoxynivalenol in cultures on autoclaved corn incubated at 27°C for 4 weeks (Palmisano et al., 1981).

Strain 11.9 Fusarium sp. 5029 (O-1166; MRC 2536)

This strain was isolated from river sediments in Japan by Ueno et al. (1977b). Extracts of cultures in PSC medium were not toxic to mice and did not contain chemically detectable trichothecenes (Ueno et al., 1977b). Cultures on autoclaved rice incubated at 24°C followed by 1 week at 10–15°C contained unspecified levels of zearalenone (Ueno et al., 1977b).

A slant culture of this pionnotal strain of F. oxysporum was received in November 1981 from Y. Ueno as "Fusarium sp. 5029."

12. Sections Martiella and Ventricosum

Fusarium solani, the only species in the sections Martiella and Ventricosum recognized by Nelson et al. (1983), has been reported in the literature to be toxigenic.

Fusarium solani (Mart.) Appel & Wollenw. emend Snyd. & Hans.

Perfect State: *Nectria haematococca* Berk. & Br.

Incidence and Distribution

Fusarium solani is a cosmopolitan soil saprophyte and facultative parasite associated with wounds and other infections that causes root rots, stem cankers, and storage rots of a great variety of host plants and also keratitis and opportunistic infections of humans and animals (Booth, 1971; CMI Descriptions of Pathogenic Fungi and Bacteria No. 29, 1964; Cuero, 1980; Doidge, 1938; Domsch et al., 1980; Gordon, 1956; Joffe & Palti, 1972; Meyer & Frank, 1979; Wollenweber & Reinking, 1935).

Association with Human and/or Animal Diseases

Moldy Sweet Potato Toxicosis (= Atypical Interstitial Pneumonia). Field outbreaks of a fatal respiratory disease of cattle in Japan and the United States have long been attributed to the ingestion of moldy sweet potatoes (*Ipomoea batatas* L.). This toxicosis is now considered to be caused by four lung-edemagenic furanoterpenoids (4-ipomeanol, 1-ipomeanol, 1,4-ipomeadiol, and ipomeanine) present in sweet potato tubers infected by *F. solani* (Boyd, 1976; Boyd & Wilson, 1972; Boyd et al., 1973; Burka et al., 1974b, 1977; Doster et al., 1978; Peckham et al., 1972; Wilson, 1971, 1973; Wilson & Boyd, 1974; Wilson et al., 1970, 1971). Affected cattle exhibit severe respiratory distress, a rapid respiratory rate, typical extension of the head and neck associated with dyspnea, and frothy exudate around the mouth before death. Pathological changes are limited to the respiratory tract: the lungs are enlarged, wet, and firm, with hemorrhagic foci and marked edema and interstitial emphysema (Doster et al., 1978; Peckham et al., 1972; Wilson, 1971, 1973; Wilson & Boyd, 1974; Wilson et al., 1970, 1971).

The disease has been reproduced experimentally in cattle (Doupnik et al., 1971; Peckham et al., 1972) and mice (Boyd & Wilson, 1972; Boyd et al., 1973; Wilson et al., 1970, 1971) with sweet potatoes artificially inoculated with *F. solani* (= *F. javanicum*) isolated from moldy sweet potatoes associated with a field outbreak of the disease in Georgia, USA. The characteristic pulmonary lesions have also been reproduced experimentally with 4-ipomeanol in cattle (Doster et al., 1978) and with 4-ipomeanol as well as the other three lung-edemagenic furanoterpenoids isolated from *F. solani*–infected sweet potatoes in mice and other laboratory animals (Boyd, 1976; Boyd & Wilson, 1972; Boyd et al., 1973; Wilson, 1973; Wilson & Boyd, 1974; Wilson et al., 1971).

The pulmonary toxins present in moldy sweet potatoes are not mycotoxins in the normal sense of the word because they are not formed by *F. solani* directly; their production is dependent upon a host-pathogen interaction. The pulmonary toxins are formed when certain phytoalexins (= stress metabolites), such as 4-hydroxymyoporone formed by sweet potato tissue in response to non-specific stress factors, are metabolized by the pathogen *F. solani* (and to a lesser extent of *F. oxysporum;* see *F. oxysporum,* moldy sweet potato toxicosis) to the furanoterpenoids 4-ipomeanol, 1-ipomeanol, 1,4-ipomeadiol, and ipomeanine (Burka et al., 1974a, b, 1977; Wilson & Burka, 1979). Recently, Clark et al. (1981) confirmed the accumulation of furanoterpenoids, including 4-ipomeanol, 1,4-ipomeadiol, and ipomeamarone, in living sweet potato tissue inoculated with six isolates of *F. solani* from sweet potatoes. However, these authors also found that several other fungi pathogenic to sweet potatoes were also high-level inducers of the pulmonary toxins 4-ipomeanol and 1,4-ipomeadiol and concluded that these compounds were not specific degradation products of *F. solani* infection.

We have identified a strain received from B.J. Wilson under the name of *F. solani* that induces the accumulation of lung-edemagenic furanoterpenoids in sweet potato tissue as *F. moniliforme* (Strain 10.8). This strain is consequently excluded from *F. solani*. A soil culture of *F. solani* used by Doupnik et al. (1971) and Peckham et al. (1972) in the successful experimental reproduction of pulmonary edema in cattle with artificially inoculated sweet potatoes was received from B. Doupnik, but this culture was not viable. Consequently it is unresolved at present whether the fungus used in the above research on moldy sweet potato toxicosis was *F. solani* or *F. moniliforme.*

Although there is uncertainty at present whether *F. solani* (see Strain 10.8) is in fact involved in the etiology of the bovine atypical interstitial pneumonia and whether the pulmonary toxins present in moldy sweet potatoes are specific degradation products of

F. solani (Burka et al., 1977) or not (Clark et al., 1981), the fact remains that moldy sweet potatoes cause field outbreaks of respiratory disease in cattle (Peckham et al., 1972). Moreover, these potent lung toxins have been found to occur naturally in sweet potatoes offered for sale in supermarkets in the United States and they are not destroyed by normal cooking procedures (Boyd et al., 1973; Wilson, 1971, 1973; Wilson & Boyd, 1974; Wilson et al., 1970, 1971). It is evident that the consumption of moldy sweet potatoes is potentially dangerous to human and animal health and that continued vigilance is necessary.

Androgenic Syndrome. During 1968 in Zimbabwe (Rhodesia), certain flocks of female chicks developed a male-like appearance due to an exaggerated development of their combs and wattles, and this androgenic syndrome was attributed to the ingestion of litter composed of wood shavings of the African tree *Funtumia latifolia* (Stapf.) Stapf. ex Schltr. infected by *F. solani* (Smith & Wells, 1978). A direct comparison between apparently fungus-free and infected samples of wood from a *Funtumia latifolia* tree from Kenya confirmed that only the *F. solani*–infected wood had androgenic activity. The syndrome was also reproduced experimentally in chicks with autoclaved, ground, inactive samples of wood, bark, or leaves of *Funtumia latifolia* inoculated with *F. solani* IMI 187207, which was isolated from the *Funtumia* wood from Kenya and identified by C. Booth, and incubated at 27°C for 3 to 7 days. Inactive material of *Funtumia latifolia* could also be rendered androgenic when incubated with another strain of *F. solani* (IMI 144799) obtained from CMI, but not with cultures of *F. semitectum* isolated from the infected Kenyan wood (Smith & Wells, 1978).

The androgenic principle was tentatively identified as a hydroxyandrosta-1,4-dienedione, $C_{19}H_{24}O_3$, which is formed by the metabolism of the steroidal alkaloids in *Funtumia latifolia* by *F. solani* (Smith & Wells, 1978). In bioassays with chicks, this steroid hormone appeared to be a more potent androgen than testosterone.

The *F. solani* strains (IMI 187207 and IMI 144799) used in the above research have not been preserved at CMI (A.H.S. Onions, personal communication to A.J. Smith, 27 September 1979) and consequently these strains are not represented in the ITFRC.

Bean Hulls Poisoning. One of the most toxic fungi isolated from moldy soybean hulls associated with "bean hulls poisoning" (see *F. sporotrichioides,* bean hulls poisoning) of horses in Hokkaido, Japan, has been designated in the literature as *F. solani* M-1-1 (Ishii et al., 1971; Masuko et al., 1977; Tatsuno et al., 1973; Ueno et al., 1972a, b, 1973a, b, 1975; Yoshizawa et al., 1980). We have identified this strain as *F. sporotrichioides* (Strain 4.12) and it is consequently excluded from *F. solani*.

Toxicity to Experimental Animals

Brine Shrimp. Extracts of potatoes inoculated with one isolate each of *F. solani* and *F. coeruleum* from potatoes in Germany were toxic to brine shrimp (Siegfried & Langerfeld, 1978). We have identified both of these isolates as *F. solani* (Strains 12.1 and 12.2). The chemical nature of the mycotoxin(s) produced by these isolates of *F. solani* toxic to brine shrimp has not been determined.

Cattle. A field outbreak of atypical interstitial pneumonia (see moldy sweet potato toxicosis, above) in cattle in Georgia, USA was associated with the ingestion of culled, moldy sweet potatoes (Doupnik et al., 1971; Peckham et al., 1972; Wilson, 1971, 1973; Wilson et al., 1970). A total of 69 head of cattle out of a herd of 275 died.

A large number of fungi, including *F. solani* (= *F. javanicum*), were isolated from the moldy sweet potatoes (Doupnik et al., 1971; Peckham et al., 1972). The characteristic clinical signs and pathological lesions (see moldy sweet potato toxicosis, above) seen in the natural outbreak of the disease were reproduced experimentally in 3 of 16 head of cattle upon oral administration of living sweet potatoes inoculated with *F. solani*. An ether extract of the culture of *F. solani* on sweet potato tubers also induced the disease in 1 of 12 orally dosed head of cattle (Peckham et al., 1972). A culture of this strain of *F. solani* (see Strain 10.8) received from B. Doupnik was not viable and consequently we could not confirm its identity.

A respiratory disease clinically and pathologically indistinguishable from atypical interstitial pneumonia has also been reproduced experimentally in cattle by the intraruminal administration of 9 to 14 mg/kg of 4-ipomeanol (Doster et al., 1978), one of the furanoterpenoids known to accumulate in moldy sweet potatoes (see moldy sweet potato toxicosis, above, and furanoterpenoids, below).

Chickens. Cultures on corn of *F. solani* isolated from moldy sweet potatoes associated with a field outbreak of atypical interstitial pneumonia (see moldy sweet potato toxicosis, above) of cattle in Georgia, USA, were toxic to day-old chicks (Doupnik et al., 1971; Peckham et al., 1972). A soil culture of this isolate of *F. solani* (see Strain 10.8) received from B. Doupnik was not viable and consequently we could not confirm its identity.

Extracts of cultures of *F. solani* AUA 625, isolated from a cabbage obtained from a supermarket in Alabama, USA, and cultured on autoclaved nutrient-amended shredded wheat at 25°C for 14 to 21 days, caused 100% mortality of chicken embryos (N.D. Davis et al., 1975).

The chemical nature of the mycotoxins produced by these two strains of *F. solani* toxic to chickens and chicken embryos, respectively, is unknown.

Fusarium solani IMI 187207 and IMI 144799 were cultured on autoclaved, ground samples of wood, bark, or leaves of *Funtumia latifolia* covered with Sabouraud-dextrose agar and incubated at 27°C for 3 to 7 days. The cultures of both strains caused an androgenic syndrome (see androgenic syndrome, above), manifested by the enlargement of combs, when fed to one-week-old chicks for seven to eight days (Smith & Wells, 1978). The androgenic principle was tentatively identified as a steroidal hormone (see steroidal hormones, below) that is formed by the metabolism of the steroidal alkaloids in *Funtumia latifolia* by *F. solani* (Smith & Wells, 1978).

Insects. Solvent extracts of cultures in PSC medium of two strains of *F. solani* (CMI 172215 and CMI 197459) showed insecticidal activity on injection in the blowfly, *Calliphora erythrocephala* (Claydon et al., 1977a, b). The insecticidal activity was accounted for by the formation of the naphthazarin pigments fusarubin and anhydrofusarubin by both strains, together with javanicin and fusaric acid (see both pigments and fusaric acid, below), by *F. solani* CMI 197459 (Claydon et al., 1977a, b).

Mice. Ether extracts of naturally moldy sweet potatoes associated with a field outbreak of atypical interstitial pneumonia in cattle in Georgia, USA (see both moldy sweet potato toxicosis and cattle, above) caused pulmonary edema and massive pleural effusion in mice within a few hours (Wilson, 1971, 1973; Wilson & Boyd, 1974; Wilson & Burka, 1979). An identical respiratory disease was produced in mice by the oral or intraperitoneal administration of extracts of living sweet potato slices inoculated with *F. solani* (= *F. javanicum*) and incubated at 20°C for 6 days (Boyd & Wilson, 1972; Boyd et al., 1973; Wilson & Boyd, 1974; Wilson et al., 1970, 1971). We have identified a culture of the strain of *F. solani* used in this work as *F. moniliforme* (Strain 10.8). Consequently it is unresolved at present whether an isolate of F. solani or F. moniliforme was used in the experimental reproduction of this respiratory disease in mice.

The four furanoterpenoids, 4-ipomeanol, 1-ipomeanol, 1,4-ipomeadiol, and ipomeanine, known to accumulate in moldy sweet potatoes (see moldy sweet potato toxicosis, above, and furanoterpenoids, below), have also been shown to cause pulmonary edema and pleural effusion in mice (Boyd, 1976; Boyd & Wilson, 1972; Boyd et al., 1973).

According to Itakura & Kinosita (1975), culture filtrates of one strain of F. solani isolated from foodstuffs in Uganda were toxic to mice upon intraperitoneal injection.

Ueno et al. (1977b) tested six isolates of F. solani from river sediments in Japan for toxicity by injecting mice intraperitoneally with extracts of cultures in PSC medium. Two isolates, i.e., F. solani 5027 and 5034, were lethal to mice and caused radiomimetic pathological changes characteristic of trichothecene toxins. These isolates also caused a marked inhibiton of protein synthesis in the rabbit reticulocyte assay, but no known trichothecenes could be identified in the toxic extracts. We have confirmed the identity of isolate 5027 as F. solani (Strain 12.3).

Mall et al. (1979) reported that 4 of 10 isolates of Fusarium sp. from wheat in India were midly toxic to mice (= 1 of 3 dead within 7 days) when extracts of cultures were injected intraperitoneally. We have identified two of these isolates as F. solani (Strains 12.4 and 12.5). In a continuation of this study, Gupta et al. (1981) reported that three isolates of Fusarium sp. from corn and two from wheat in India were mildly toxic to mice, i.e., did not cause mortality within 7 days, but histopathological lesions were observed in the liver, spleen, and/or kidneys. We have identified one of these isolates from corn (Strain 12.6) and one from wheat (Strain 12.7) as F. solani.

With the exception of the strain that induces the accumulation of lung-edemagenic furanoterpenoids in sweet potatoes, the chemical nature of the mycotoxin(s) produced by the above strains as F. solani toxic to mice is unknown.

The strain reported as F. solani M-1-1 to be toxic to mice (Ishii et al., 1971; Ueno et al., 1972a, 1973a), is F. sporotrichioides (Strain 4.12), and this strain is consequently excluded from F. solani.

Rabbits. In a study of the dermotoxicity of F. solani isolated from soil and overwintered cereals associated with ATA (see F. sporotrichioides, ATA) of humans in the USSR, 32 isolates (including eight of F. javanicum) were grown on various media at temperatures of −5 to 8°C for 25 to 70 days and the cultures then subjected to freezing and thawing (Joffe, 1960a, b, 1971). Extracts of three of these isolates (Nos. 965, 990, and 1664) were found to be mildly toxic to rabbit skin and to cause leukocytorrhea and/or edema (Joffe 1960a, b).

In a study of the dermotoxicity of 109 isolates of F. solani (and F. javanicum) from various hosts in Israel and nine isolates from other countries (Joffe & Palti, 1972), cultures on autoclaved wheat kernels were incubated at either 4, 18, 24, 30, or 35°C for 11 days. Extracts of 66 of these 118 isolates (= 56%) caused some degree of toxic reaction on rabbit skin, ranging from a slight, pinkish inflammation to various degrees of edema, leukocytorrhea, hemorrhage, and necrosis. In general, the most toxic reactions were obtained with cultures incubated at 24 or 30°C. Two of the most toxic isolates of F. solani were 112/2 from banana and 562/2 from groundnut in Israel (Joffe, 1973b; Joffe & Palti, 1972). According to Joffe & Palti (1974), extracts of 75 of 138 unspecified isolates of Martiella Fusaria exhibited some degree of dermotoxicity to rabbit skin. The chemical nature of the mycotoxin(s) produced by these dermotoxic strains of F. solani is unknown at present.

Cuero (1980) used a rabbit eye test to detect active metabolites in two unspecified strains of *F. solani* isolated from patients with mycotic keratitis and from field soil, respectively, in Cali, Colombia. The two strains of *F. solani* were cultured either on Sabouraud agar at 25°C for 7 days or in liquid "asparagine medium" for 10 days and spore suspensions, culture filtrates, and purified chloroform extracts of cultures were instilled in rabbit eyes. The clinical reaction of irritation and erythema was graded subjectively and the purified extract produced the most severe reaction. The rabbit eye test was used as a bioassay in the isolation and preliminary chemical identification of the active metabolite, but the chemical nature is still unknown.

Mycotoxins Produced

Diacetoxyscirpenol. Ripperger et al. (1975) reported the isolation of 286 mg of crystalline diacetoxyscirpenol from 45 ℓ of culture filtrate of an unnumbered strain of *F. solani* var. *coeruleum* obtained from Dr. I. Focke, Bernburg-Strenzfeld, German Democratic Republic.

The strain reported as *F. solani* M-1-1 (Ishii et al., 1971; Tatsuno et al., 1973; Ueno et al., 1972a, b, 1973a, 1975) or *F. solani* sp. 900 (Schmidt et al., 1981a) to produce diacetoxyscirpenol, is *F. sporotrichioides* (Strain 4.12). Consequently this strain is excluded from *F. solani*.

Furanoterpenoids. Four lung-toxic furantoerpenoids, i.e., 4-ipomeanol, 1-ipomeanol, 1,4-ipomeadiol, and ipomeanine, are formed in moldy sweet potatoes (see moldy sweet potato toxicosis, above) as a result of the metabolism by *F. solani,* and to a lesser extent *F. oxysporum,* of certain phytoalexins (= stress metabolites) such as 4-hydroxy-myoporone that are formed by living sweet potato tissue in response to non-specific exogenous stimuli (Burka et al., 1974a, b, 1977). These pulmonary toxins are not mycotoxins in the normal sense of the word because their production is dependent upon a host-pathogen interaction and not upon the fungus alone. Thus the pulmonary toxins are not produced when *F. solani* is grown on autoclaved sweet potato slurry or when living sweet potato tubers are inoculated with fungi other than *F. solani* or *F. oxysporum* (Burka et al., 1974a, b, 1977; Wilson, 1973). Bioproduction of the lung-edemagenic furanoterpenoids, together with the hepatotoxic compounds ipomeamarone and ipomeamaronol, is achieved by inoculating surface-sterilized, fresh sweet potato slices with *F. solani* and incubating in moist chambers at 20 to 24°C for 6 days (Boyd & Wilson, 1972; Boyd et al., 1973; Burka et al., 1974a, b, 1977; Wilson, 1971, 1973; Wilson et al., 1970, 1971). According to Boyd et al. (1973), the relative amounts of lung-toxic furanoterpenoids produced in sweet potatoes inoculated with *F. solani* are as follows: 4-ipomeanol predominates; the level of 1-ipomeanol is about one half of that of 4-ipomeanol; 1,4-ipomeadiol "occurs in amounts comparable to the ipomeanols"; and only traces of ipomeanine are present.

We have identified a culture of *F. solani* (originally referred to as *F. javanicum*) that reportedly causes the accumulation of the four lung-edemagenic furanoterpenoids in inoculated sweet potatoes as *F. moniliforme* (Strain 10.8). Consequently it is unresolved at present whether an isolate of *F. solani* or *F. moniliforme* was used in the research on the bioproduction of the lung-toxic furanoterpenoids cited above.

Recently Clark et al. (1981) reported that 4-ipomeanol and 1,4-ipomeadiol are not specific degradation products of *F. solani* and that several other fungal pathogens of sweet potatoes can also induce the accumulation of relatively high levels of these pulmonary toxins in inoculated sweet potato tubers.

Fusaric Acid. A strain of *F. solani* (D.V. Lightner culture C166 = CMI 197459 identified by C. Booth) isolated from the lobster *Homarus americanus* in New York, USA, produced fusaric acid (up to 464 mg/ℓ) in Czapek-Dox medium incubated at 25°C for 21 days (Claydon et al., 1977a, b).

HT-2 Toxin. The strain reported as *F. solani* M-1-1 (Ueno et al., 1972b, 1973a), or *F. solani* sp. 900 (Schmidt et al., 1981a), or *F. sporotrichioides* ATCC 26553 (Yoshizawa et al., 1980) to produce HT-2 toxin, is *F. sporotrichioides* (Strain 4.12). Consequently this strain is excluded from *F. solani*

Neosolaniol. The strain reported as *F. solani* M-1-1 (Ishii et al., 1971; Tatsuno et al., 1973; Ueno et al., 1972a, b, 1973a, b) or *F. sporotrichioides* ATCC 26554 (Yoshizawa et al., 1980) to produce neosolaniol (originally referred to as solaniol by Ishii et al., 1971), is *F. sporotrichioides* (Strain 4.12). Consequently this strain is excluded from *F. solani*.

Pigments. *Fusarium solani* produces a number of naphthaquinone-type pigments, including anhydrofusarubin, bostrycoidin, dihydrofusarubin, O-ethylfusarubin, O-ethylhydroxydihydrofusarubin, fusarubin, hydroxydihydrofusarubin, isomarticin, javanicin, marticin, norjavanicin, and solaniol, some of which have antimicrobial, phytotoxic, insecticidal, and antitumor properties (Ammar et al., 1979; Baker et al., 1981; Claydon et al., 1977a, b; Gerber & Ammar, 1979; Kern, 1972; Kurobane et al., 1980; Lacey, 1950; Turner, 1971). Several of these pigments of the fusarubin complex have been isolated from PP96, the cholesterol-metabolizing strain of *F. solani* that was isolated from soil in New Jersey, USA, and identified by P.E. Nelson and R.A. Samson (Ammar et al., 1979).

Steroidal Hormones. The ability of *F. solani* to metabolize steroidal compounds is well established, e.g., sapogenins (Kondo & Mitsugi, 1966), 17α-hydroxyprogesterone (Hafez-Zedan & Plourde, 1973), and cholesterol (Ammar et al., 1979; Gerber & Ammar, 1979). The fact that some of the metabolites that arise in this way are steroidal hormones that can cause field outbreaks of an androgenic syndrome (see androgenic syndrome, above) was demonstrated in Zimbabwe (Rhodesia) in 1968 when female chicks developed a male-like appearance (Smith & Wells, 1978). This outbreak was attributed to the ingestion of wood shavings of *Funtumia latifolia* infected by *F. solani*. The androgenic principle was presumably a steroidal hormone formed by the metabolism by *F. solani* of the steroidal alkaloids present in *Funtumia latifolia*. Bioproduction of the androgen was achieved by incubating autoclaved, ground samples of wood, bark, or leaves of *Funtumia latifolia,* or a crude alkaloid fraction from the wood extract, with *F. solani* IMI 187207 or IMI 144799 at 27°C for 3 to 7 days. The androgen was tentatively identified as a hydroxyandrosta-1,4-dienedione, $C_{19} H_{24} O_3$, and appeared to be six to 20 times more potent than testosterone in bioassays with chicks (Smith & Wells, 1978).

T-2 Toxin. The strain reported as *F. solani* M-1-1 (Ishii et al., 1971; Tatsuno et al., 1973; Ueno et al., 1972a, b, 1973a, b, 1975), or *F. solani* NHL-F-111 (Sato & Amano, 1976), or *F. solani* sp. 900 (Schmidt et al., 1981a), or *F. sporotrichioides* ATCC 26553 (Yoshizawa et al., 1980) to produce T-2 toxin, is *F. sporotrichioides* (Strain 4.12). Consequently this strain is excluded from *F. solani*.

Zearalenone. The only available report on zearalenone production by *F. solani* is the statement by Lasztity & Wöller (1975a, b) and Lasztity et al. (1977) that zearalenone is produced by unspecified isolates of *F. solani* from cereals and mixed feeds in Hungary. These authors did not give any indication of the number of strains involved or the levels of zearalenone detected. The validity of this statement is also questionable because in other publications the same authors (Lasztity & Wöller, 1975c, 1977) reported that no zearalenone was produced by Hungarian isolates of *F. solani.*

Several authors have reported on the non-production of zearalenone by isolates of *F. solani:* none of 9 by Caldwell et al. (1970); one nonproducer by Eugenio et al. (1970a); none of 7 by Ichinoe et al. (1977); none of 4 by Ishii et al. (1974); one nonproducer by Mirocha et al. (1969); and none of 6, including isolate 5027 which we have identified as *F. solani* (Strain 12.3), by Ueno et al. (1977b).

In conclusion, there are no confirmed reports of zearalenone production by *F. solani* and there are no known zearalenone-producing strains of *F. solani* represented in the ITFRC.

Toxigenic Strains in the ITFRC

The following toxigenic *Fusarium* strains listed in Table 12.1 all have been identified as *F. solani* according to Nelson et al. (1983). Strains referred to as *F. solani* in the literature but identified as other species by us are listed in Table 12.2.

Strain 12.1   *F. solani*—Berlin 62409 (= S-679; MRC 2198)

This strain was isolated from rotten potato tubers in Germany (Siegfried & Langerfeld, 1978). Potato tubers of the cultivar Maritta were inoculated with this strain and incubated at 20°C until approximately one third of each tuber was infected. The infected tissues were then separated and lyophilized, and purified ethyl-acetate extracts were assayed for toxicity to brine shrimp. These extracts proved to be highly toxic and caused more than 80% mortality within 4 hours (Siegfried & Langerfeld, 1978). The chemical nature of the mycotoxin(s) produced in inoculated potato tubers by this strain is unknown.

A slant culture of this strain was received on 20 January 1981 from E. Langerfeld as "*F. solani* 62409 (Berlin)."

Strain 12.2   *F. coeruleum* 903 (= S-682; MRC 2194)

This strain was isolated from rotten potato tubers in Germany (Siegfried & Langerfeld, 1978). Potato tubers of two different cultivars were inoculated with this strain and incubated at either 6, 15, or 20°C until approximately one third of each tuber was infected. The infected tissues were then separated and lyophilized, and purified ethyl-acetate extracts were assayed for toxicity to brine shrimp. Extracts of both cultivars proved to be toxic and tubers incubated at 20°C appeared to be more toxic (50–80% mortality) than those incubated at lower temperatures (Siegfried & Langerfeld, 1978). The chemical nature of the mycotoxin(s) produced in inoculated potato tubers by this strain is unknown.

A slant culture of this strain was received on 20 January 1981 from E. Langerfeld as "*F. coeruleum* 903." Cultures of this strain have the deep violet-blue color characteristic of the well-known storage rot organism of potatoes, *F. coeruleum* (Lib.) Sacc. in the sense of Wollenweber & Reinking (1935) or *F. solani* var. *coeruleum* (Sacc.) Booth in the sense of Booth (1971), but we have placed this strain in *F. solani* following Nelson et al. (1983).

Strain 12.3   *F. solani* 5027 (= S-735; MRC 2565)

This strain was isolated from river sediments in Japan during 1972 (Ueno et al., 1977b). Extracts of cultures in PSC medium incubated at 25 to 27°C for 2 weeks were lethal upon intraperitoneal injection in mice and caused radiomimetic pathological changes

Table 12.1 *Fusarium* Strains Identified as *F. solani* in the International Toxic *Fusarium* Reference Collection (ITFRC)

Strain No.	ITFRC No.	MRC No.	Published Name(s) and Number(s)	Source	Toxigenicity
12.1	S-679	2198	*F. solani*—Potato (= Berlin 62409)	E. Langerfeld	a
12.2	S-682	2194	*F. coeruleum*—Potato (= 903)	E. Langerfeld	a
12.3	S-735	2565	*F. solani* 5027	Y. Ueno	a
12.4	S-693	2634	*F.* sp. PL W-76	V.C. Vora	a
12.5	S-694	2635	*F.* sp. PL W-168	V.C. Vora	a
12.6	S-787	2803	*F.* sp. PL Z-65	V.C. Vora	a
12.7	S-788	2805	*F.* sp. PL W-771	V.C. Vora	a

a = Culture toxic to experimental animals, mycotoxin(s) unknown

characteristic of trichothecene toxins. This strain also caused a marked inhibition of protein synthesis in the rabbit reticulocyte assay, but neither any known trichothecenes nor zearalenone could be identified in the toxic extracts (Ueno et al., 1977b). Consequently the chemical nature of the mycotoxin(s) produced by this strain is unknown.

A slant culture of this isolate was received in November 1981 from Y. Ueno as "*F. solani* 5027."

Strain 12.4 *Fusarium* sp. PL W-76 (= S-693; MRC 2634)

This strain was isolated from wheat in India (Mall et al., 1979). Extracts of cultures on autoclaved wheat incubated at 27–28°C for 15 days were mildly toxic to mice, causing the death of 1 of 3 animals and/or visible lesions within 7 days (Mall et al., 1979). The chemical nature of the mycotoxin(s) produced by this strain is unknown.

A slant culture of this strain of *F. solani* was received on 24 August 1981 from V.C. Vora as "*Fusarium* sp. PL W-76, from wheat, toxic to mice."

Strain 12.5 *Fusarium* sp. PL W-168 (= S-694; MRC 2635)

This strain was isolated from wheat in India (Mall et al., 1979). Extracts of cultures on autoclaved wheat incubated at 27–28°C for 15 days were mildly toxic to mice, causing the death of 1 of 3 animals and/or visible lesions with 7 days (Mall et al., 1979). The chemical nature of the mycotoxin(s) produced by this strain is unknown.

A slant culture of this strain of *F. solani* was received on 24 August 1981 from V.C. Vora as "*Fusarium* sp. PL W-168, from wheat, toxic to mice."

Strain 12.6 *Fusarium* sp. PL Z-65 (= S-787; MRC 2803)

This strain was isolated from corn in India (Gupta et al., 1981). Chloroform extracts of cultures on autoclaved corn incubated at 28°C for 15 days were mildly toxic to Swiss albino mice upon intraperitoneal injection. These extracts did not cause mortality of the three treated mice within 7 days, but histopathological lesions were observed in the liver, spleen, and/or kidneys (Gupta et al., 1981). The chemical nature of the mycotoxin(s) produced by this strain is unknown.

A slant culture of this strain of *F. solani* was received on 4 June 1982 from V.C. Vora as "*Fusarium* sp. PL Z-65, from maize, toxic to mice."

Table 12.2 *Fusarium* Strains Published as *F. solani* but Identified as Other Species in the International Toxic *Fusarium* Reference Collection (ITFRC)

Strain No.	ITFRC No.	MRC No.	Published Name(s) and Number(s)	Source	Present Identification
10.8	M-1246	2326	*F. solani* (= *F. javanicum*) – Sweet Potato	B.J. Wilson	*F. moniliforme*
4.12	T-422	935	*F. solani* M-1-1 (= Sp. 900)	Y. Ueno (1973)	*F. sporotrichioides*
4.12	T-568	2557	= *F. solani* M-1-1	Y. Ueno (1981)	*F. sporotrichioides*
4.12	T-490	2034	= *F. solani* NHL-F-111 (= Sp. 934 and Sp. 980)	M. Ichinoe	*F. sporotrichioides*
4.12	T-567	2478	= *F. sporotrichioides* ATCC 26553	T. Yoshizawa	*F. sporotrichioides*
4.12	T-492	2319	= *F. sporotrichioides* ATCC 26533	ATCC	*F. sporotrichioides*

Strain 12.7 *Fusarium* sp. PL W-771 (= S-788; MRC 2805)

This strain was isolated from wheat in India (Gupta et al., 1981). Choloroform extracts of cultures on autoclaved wheat incubated at 28°C for 15 days were mildly toxic to Swiss albino mice upon intraperitoneal injection. These extracts did not cause mortality of the three treated mice within 7 days, but histopathological lesions were observed in the liver, spleen, and/or kidneys (Gupta et al., 1981). The chemical nature of the mycotoxin(s) produced by this strain is unknown.

A slant culture of this strain of *F. solani* was received on 4 June 1982 from V.C. Vora as "*Fusarium* sp. PL W-771, from wheat, toxic to mice."

Literature Cited

1. Adams, R.L., and J. Tuite. 1976. Feeding *Gibberella zeae* damaged corn to laying hens. Poult. Sci. 55: 1991–1993.
2. Agarwal, S., and S. Chauhan. 1980. Toxic effects of *Fusarium semitectum* to albino rats. Ind. J. Mycol. Plant Pathol. 10: LXXI (Abstr.).
3. Akhmeteli, M.A. 1977. Epidemiological features of the mycotoxicoses. Ann. Nutr. Alim. 31: 957–976.
4. Akhmeteli, M.A., A.B. Linnik, K.S. Cernov, V.M. Voronin, and L.M. Sabad. 1972a. Study of extracts of barley grain infected with *Fusarium sporotrichioides* No. 63. Bull. World Health Org. 47: 123–124.
5. Akhmeteli, M.A., A.B. Linnik, and K.S. Cernov. 1972b. Hepatocarcinogenesis and the appearance of serum alpha-fetoprotein in mice treated with extracts of barley grain infected with *Fusarium sporotrichioides*. Bull. World Health Org. 47: 663–664.
6. Akhmeteli, M.A., A.B. Linnik, K.S. Cernov, V.M. Voronin, A.J. Hesina, N.A. Guseva, and L.M. Sabad. 1973. Study of toxins isolated from grain infected with *Fusarium sporotrichioides*, p. 209–215. *In* P. Krogh (ed.), Control of Mycotoxins, Butterworths, London.
7. Albright, J.L., S.D. Aust, J.H. Byers, T.E. Fritz, B.O. Brodie, R.E. Olsen, R.P. Link, J. Simon, H.E. Rhoades, and R.L. Brewer. 1964. Moldy corn toxicosis in cattle. J. Amer. Vet. Med. Assoc. 144: 1013–1019.
8. Allen, N.K., C.J. Mirocha, G. Weaver, S. Aakhus-Allen, and F. Bates. 1981a. Effects of dietary zearalenone on finishing broiler chickens and young turkey poults. Poult. Sci. 60: 124–131.

9. Allen, N.K., H.R. Burmeister, G.A. Weaver, and C.J. Mirocha. 1981b. Toxicity of dietary and intravenously administered moniliformin to broiler chickens. Poult. Sci. 60: 1415–1417.
10. Allen, N.K., C.J. Mirocha, S. Aakhus-Allen, J.J. Bitgood, G. Weaver, and F. Bates. 1981c. Effect of dietary zearalenone on reproduction of chickens. Poult. Sci. 60: 1165–1174.
11. Ammar, M.S., N.N. Gerber, and L.E. McDaniel. 1979. New antibiotic pigments related to fusarubin from *Fusarium solani* (Mart.) Sacc. I. Fermentation, isolation, and antimicrobial activities. J. of Antibiotics 32: 679–684.
12. Arai, T., and T. Ito. 1970. Cytotoxicity and antitumor activity of fusariocins, mycotoxins from *Fusarium moniliforme*, p. 87–92. In H. Umezawa (ed.), Progress in antimicrobial and anticancer chemotherapy, Vol. I, Univ. Tokyo Press.
13. Archer, M. 1974. Detection of mycotoxins in foodstuffs by use of chick embryos. Mycopathol. Mycol. Appl. 54: 453–467.
14. Armolik, N. J.G. Dickson, and A.D. Dickson, 1956. Deterioration of barley in storage by microorganisms. Phytopathology 46: 457–461.
15. Arora, R.G., H. Frölén, and A. Nilsson. 1981. Interference of mycotoxins with prenatal development of the mouse. I. Influence of aflatoxin B_1, ochratoxin A and zearalenone. Acta Vet. Scand. 22: 524–534.
16. Aucock, H.W., W.F.O. Marasas, C.J. Meyer, and P. Chalmers. 1980. Field outbreaks of hyperoestrogenism (vulvo-vaginitis) in pigs consuming maize infected by *Fusarium graminearum* and contaminated with zearalenone. J.S. Afr. Vet. Assoc. 51: 163–166.
17. Bacon, C.W., and H.L. Marks. 1976. Growth of broilers and quail fed *Fusarium* (*Gibberella zeae*)-infected corn and zearalenone (F-2). Poult. Sci. 55: 1531–1535.
18. Bacon, C.W., J.D. Robbins, and J.K. Porter. 1977. Media for identification of *Gibberella zeae* and production of (F-2) zearalenone. Appl. Environ. Microbiol. 33: 445–449.
19. Badiali, L., M.H. Abou-Youssef, A.L. Radwin, F.M. Hamdy, and P.K. Hildebrandt. 1968. Moldy corn poisoning as the major cause of an encephalomalacia syndrome in Egyptian Equidae. Amer. J. Vet. Res. 29: 2029–2035.
20. Bailey, W.S., and A.H. Groth. 1959. The relationship of hepatitis x of dogs and moldy corn poisoning of swine. J. Amer. Vet. Med. Assoc. 134: 514–516.
21. Baker, R.A., J.H. Tatum, and S. Nemec, Jr. 1981. Toxin production by *Fusarium solani* from fibrous roots of blight-diseased citrus. Phytopathology 71: 951–954.
22. Bamburg, J.R. 1968. Mycotoxins of the trichothecane family produced by cereal molds. Ph.D. Thesis, University of Wisconsin, Madison, Wisconsin. 161 p.
23. Bamburg, J.R., and F.M. Strong. 1969. Mycotoxins of the trichothecane family produced by *Fusarium tricinctum* and *Trichoderma lignorum*. Phytochemistry 8: 2405–2410.
24. Bamburg, J.R., and F.M. Strong. 1971. 12,13-Epoxytrichothecenes, p. 207–292. In S. Kadis, A. Ciegler, and S.J. Ajl (ed.), Microbial toxins, Vol. VII, Academic Press, New York.
25. Bamburg, J.R., N.V. Riggs, and F.M. Strong. 1968a. The structures of toxins from two strains of *Fusarium tricinctum*. Tetrahedron 24: 3329–3336.
26. Bamburg, J.R., W.F.O. Marasas, N.V. Riggs, E.B. Smalley, and F.M. Strong. 1968b. Toxic spiro-epoxy compounds from Fusaria and other Hyphomycetes. Biotechnol. Bioeng. 10: 445–455.
27. Bamburg, J.R., F.M. Strong, and E.B. Smalley. 1969. Toxins from moldy cereals. J. Agric. Food Chem. 17: 443–450.
28. Barnikol, H., S. Gruber, and A. Thalmann. 1981. Hyperöstrogenismus durch Fusarientoxine bei neugeborenen und abgesetzten Ferkeln. Tierärtz. Umschau 36: 94–105.
29. Beller, K., and W. Wedemann. 1929. Untersuchung über die Schadwirkung amerikanischer Futtergerste (sog. Barley Federal Nr. 11). Z. Infektionskrankh. Parasit. Krankh. Hyg. Haustiere 36: 103–129.
30. Belt, R.J., C.D. Haas, U. Joseph, W. Goodwin, D. Moore, and B. Hoogstraten. 1979. Phase 1 study of anguidine administered weekly. Cancer Treat. Rep. 63: 1993–1995.
31. Bennett, G.A., and O.L. Shotwell. 1979. Zearalenone in cereal grains. J. Amer. Oil Chem. Soc. 56: 812–819.

32. Bennett, G.A., R.E. Peterson, R.D. Plattner, and O.L. Shotwell. 1981. Isolation and purification of deoxynivalenol and a new trichothecene by high pressure liquid chromatography. J. Assoc. Off. Anal. Chem. 58: 1002A–1005A.
33. Berger, T., K.L. Esbenshade, M.A. Diekman, T. Hoagland, and J. Tuite. 1981. Influence of prepubertal consumption of zearalenone on sexual development of boars. J. Anim. Sci. 53: 1559–1564.
34. Bertin, G., K. Chakor, P. Lafont, and C. Frayssinet. 1978. Transmission à la descendance de contamination du regime maternel par les mycotoxines. Collection Med. Leg. Toxicol. Med. 107: 95–100.
35. Bhat, R.V., and C. Rukmini. 1981. Mycotoxins in sorghum: Toxigenic fungi during storage and natural occurrence of T-2 toxin, p. 141–143. In G.D. Bengtson (ed.), Sorghum Diseases: A World Review, ICRISAT, Andra Pradesh, India.
36. Bhat, R.V., V. Nagarajan, and P.G. Tulpule. 1978. Health hazards of mycotoxins in India. National Institute of Nutrition, Hyderabad, Indian Council of Medical Research, New Delhi. U.S. Dept. of Commerce, National Technical Information Service, Springfield, Virginia. PB82-170820. 58p.
37. Biester, H.E., and L.H. Schwarte. 1939. Moldy corn poisoning (Leucoencephalomalacia) in horses with history of previous attack as well as recovery from virus encephalomyelitis. North Amer. Vet. 20: 17–19.
38. Beister, H.E., L.H. Schwarte, and C.H. Reddy. 1940. Further studies on moldy corn poisoning (Leucoencephalomalacia) in horses. Vet. Med. 35: 636–639.
39. Bilai, V.I. 1960. Mikotoksikozy Chelovska i Sel'skokhozyaystvannykh Zhivotnykh, Kiev, 167 p. English translation: Mycotoxicoses of man and agricultural animals, JPRS 7434, U.S. Joint Publications Research Service, U.S. Dept. of Commerce, Washington, D.C.
40. Bilai, V.I. 1970. Phytopathological and hygienic significance of representatives of the Section *Sporotrichiella* in the genus *Fusarium* Lk. Ann. Acad. Sci. Fenn., A IV, Biologica 168: 19–24.
41. Bilai, V.I., O.I. Cherkes, L.O. Bogomolova, and C.B. Frantsozova. 1975. Toxico-biological properties of fusaric acid. Mikrobiologichnyi Zhurnal 37: 325–328 (In Russian).
42. Biro, M., F. Gyuru, and G. Feher. 1972. Study on toxic effect for white rats of *Fusarium* species damaging maize. Mag. Allat. Lapja 27: 597–604 (In Hungarian).
43. Bjeldanes, L.F., and S.V. Thomson. 1979. Mutagenic activity of *Fusarium moniliforme* isolates in the *Salmonella typhimurium* assay. Appl. Environ. Microbiol. 37: 1118–1121.
44. Bjeldanes, L.F., and L.A. Weib. 1980. Mutagenic mycotoxins from *Fusarium moniliforme*. Environ. Mut. 2: 240–241 (Abstr.).
45. Bjeldanes, L.F., G.W. Chang, and S.V. Thomson. 1978. Detection of mutagens produced by fungi with the *Salmonella typhimurium* assay. Appl. Environ. Microbiol. 35: 1150–1154.
46. Blight, M.M., and J.F. Grove. 1974. New metabolic products of *Fusarium culmorum*: Toxic trichothec-9-en-8-ones and 2-acetylquinazolin-4(3H)-one. J. Chem. Soc. Perkin I, 14: 1691–1693.
47. Bolliger, G., and Ch. Tamm. 1972. Vier neue Metabolite von *Giberella zeae:* 5-Formyl-zearalenon, 7'-Dehydrozearalenon, 8'-Hydroxy- und 8'-epi-Hydroxy-zearalenon. Helv. Chim. Acta 55: 3030–3048.
48. Bolton, A.T., and V.W. Nuttall. 1968. Pathogenicity studies with *Fusarium poae*. Can. J. Plant Sci. 48: 161–166.
49. Boonchuvit, B., P.B. Hamilton, and H.R. Burmeister. 1975. Interaction of T-2 toxin with *Salmonella* infections of chickens. Poult. Sci. 54: 1693–1696.
50. Booth, C. 1971. The genus *Fusarium*. Commonwealth Mycol. Inst. Kew, Surrey, England. 237 p.
51. Borrow, A., P.W. Brian, V.E. Chester, P.J. Curtis, H.G. Hemming, C. Henehan, E.G. Jeffreys, P.B. Lloyd, I.S. Nixon, G.L.F. Norris, and M. Radley. 1955. Gibberellic acid, a metabolic product of the fungus *Gibberella fujikuroi:* some observations on its production and isolation. J. Sci. Food Agric. 6: 340–348.

52. Bottalico, A. 1975. Produzione di zearalenone de parte di isolati di *Fusarium* agenti del "mal del piede" del frumento, in Italia. Phytopath. Medit. 14: 134–135.
53. Bottalico, A. 1976. La presenza di *Fusarium moniliforme* Sheld. nelle cariossidi del Granturco (*Zea mays* L.) quale problema fitopatologico e micotossicologico in Italia. II. Aspetti micotossicologici. Phytopath. Medit. 15: 54–58.
54. Bottalico, A. 1977a. Presenza di zearalenone nelle spighe di Granturco (*Zea mays* L.) infette da *Gibberella zeae* (Schw.) Petch, nel Metapontino. Phytopath. Medit. 16: 14–17.
55. Bottalico, A. 1977b. Production of zearalenone by *Fusarium* spp. from cereals, in Italy. Phytopath. Medit. 16: 75–78.
56. Bottalico, A. 1977c. Production of T-2 toxin by *Fusarium* spp. from cereals, in Italy. Phytopath. Medit. 16: 147–148.
57. Bottalico, A. 1978a. Survey of freshly harvested maize from some European localities for zearalenone, in 1977. Phytopath. Medit. 17: 191–192.
58. Bottalico, A. 1978b. On the occurrence of mycotoxins (aflatoxins and zearalenone) in some foods and feeds, in Italy. Proc. Int. Meet. Food Microbiol. and Technol., Tabiano B. (Parma) Italy, April 20–23, 1976: 337–346.
59. Bottalico, A. 1979. On the occurrence of zearalenone in Italy. Mycopathologia 67: 119–121.
60. Bottalico, A., and V. Piglionica. 1977. La sanita delle cariossidi di frumento e di mais con particolare riferimento alla presenza in esse di specie patogene di *Fusarium*. Riv. Agron. 11: 146–152.
61. Bottalico, A., S. Frisullo, and V. Piglionica. 1977. Survey of freshly harvested maize from some European localities for *Fusarium* spp. and zearalenone, in 1976. Phytopath. Medit. 16: 142–144.
62. Bottalico, A., P. Lerario, and A. Visconti. 1981. Occurrence of trichothecenes and zearalenone in preharvest *Fusarium*-infected ears of maize from some Austrian localities. Phytopath. Medit. 20: 1–6.
63. Boyd, M.R. 1976. Role of metabolic activation in the pathogenesis of chemically induced pulmonary disease: mechanism of action of the lung-toxic furan, 4-ipomeanol. Environ. Health Perspect. 16: 127–138.
64. Boyd, M.R., and B.J. Wilson. 1972. Isolation and characterization of 4-ipomeanol, a lung toxic furanoterpenoid produced by sweet potatoes (*Ipomoea batatas*). J. Agr. Food Chem. 20: 428–430.
65. Boyd. M.R., L.T. Burka, T.M. Harris, and B.J. Wilson. 1973. Lung-toxic furanoterpenoids produced by sweet potatoes (*Ipomoea batatas*) following microbial infection. Biochim. Biophys. Acta 337: 184–195.
66. Brian, P.W., A.W. Dawkins, J.F. Grove, H.G. Hemming, D. Lowe, and G.L.F. Norris. 1961. Phytotoxic compounds produced by *Fusarium equiseti*. J. Exp. Bot. 12: 1–12.
67. Bridges, C.H. 1978. Mycotoxicoses in horses, p. 173–181. *In* T.D. Wyllie and L.G. Morehouse (ed.), Mycotoxic fungi, mycotoxins, mycotoxicoses. An encyclopedic handbook, Vol. 2, Marcel Dekker, New York.
68. Bristol, F.M., and S. Djurickovic. 1971. Hyperestrogenism in female swine as the result of feeding mouldy corn. Can. Vet. J. 12: 132–135.
69. Brodnik, T. 1975. Influence of toxic products of *Fusarium graminearum* and *Fusarium moniliforme* on maize seed germination and embryo growth. Seed Sci. Technol. 3: 691–696.
70. Brodnik, T., N. Klemenc, P. Vospernik, and J. Zust. 1978. Influence of toxins from maize infected by *Aspergillus flavus, Penicillium rubrum* and *Fusarium graminearum* and of aflatoxin B_1, rubratoxin A and toxin F-2 on maize embryo growth. Seed Sci. Technol. 6: 965–970.
71. Buck, W.B., J.C. Haliburton, J.P. Thilsted, T.F. Lock, and R.F. Vesonder. 1979. Equine leucoencephalomalacia: comparative pathology of naturally occurring and experimental cases. Amer. Assoc. Vet. Lab. Diagnosticians, 22nd Ann. Proc.: 239–258.
72. Buckley, S.S., and W.G. MacCullum. 1901. Acute haemorrhagic encephalitis prevalent among horses in Maryland. Amer. Vet. Rev. 25: 99–101.

73. Bullerman, L.B., J.M. Baca, and W.T. Stott. 1975. An evaluation of potential mycotoxin-producing molds in corn meal. Cereal Foods World 20: 248-253.
74. Burgess, L.W., A.H. Wearing, and T.A. Toussoun. 1975. Surveys of Fusaria associated with crown rot of wheat in eastern Australia. Aust. J. Agric. Res. 26: 791-799.
75. Burka, L.T., R.M. Bowen, B.J. Wilson, and T.M. Harris. 1974a. 7-Hydroxymyoporone, a new toxic furanosesquiterpene from mold-damaged sweet potatoes. J. Org. Chem 39: 3241-3244.
76. Burka, L.T., L. Kuhnert, B.J. Wilson, and T.M. Harris. 1974b. 4-Hydroxymyoporone, a key intermediate in the biosynthesis of pulmonary toxins produced by *Fusarium solani* infected sweet potatoes. Tetrahedron Letters No. 46: 4017-4020.
77. Burka L.T., L. Kuhnert, B.J. Wilson, and T.M. Harris. 1977. Biogenesis of lung-toxic furans produced during microbial infection of sweet potatoes (*Ipomoea batatas*). J. Amer. Chem. Soc. 99: 2302-2305.
78. Burkhardt, H.J., R.E. Lundin, and W.H. McFadden. 1968. Mycotoxins produced by *Fusarium nivale* (Fries) Cesati isolated from tall fescue (*Festuca arundinacea* Schreb.). Synthesis of 4-acetamido-4-hydroxy-2-butenoic acid-γ-lactone. Tetrahedron 24: 1225-1229.
79. Burmeister, J.R. 1971. T-2 toxin production by *Fusarium tricinctum* on solid substrate. Appl. Microbiol. 21: 739-742.
80. Burmeister, H.R., and C.W. Hesseltine. 1970. Biological assays for two mycotoxins produced by *Fusarium tricinctum*. Appl. Microbiol. 20: 437-440.
81. Burmeister, H.R., J.J. Ellis, and S.G. Yates. 1971. Correlation of biological to chromatographic data for two mycotoxins elaborated by *Fusarium*. Appl. Microbiol. 21: 673-675.
82. Burmeister, H.R., J.J. Ellis, and C.W. Hesseltine. 1972. Survey for Fusaria that elaborate T-2 toxin. Appl. Microbiol. 23: 1165-1166.
83. Burmeister, H.R., G.A. Bennett, R.F. Vesonder, and C.W. Hesseltine. 1974. Antibiotic produced by *Fusarium equiseti* NRRL 5537. Antimicrobial Agents and Chemotherapy 5: 634-639.
84. Burmeister, H.R., R.F. Vesonder, and C.W. Hesseltine. 1977. Swelling of *Penicillium digitatum* conidia by a *Fusarium acuminatum* NRRL 6227 metabolite. Mycopathologia 62: 53-56.
85. Burmeister, H.R., A. Ciegler, and R.F. Vesonder. 1979. Moniliformin, a metabolite of *Fusarium moniliforme* NRRL 6322: purification and toxicity. Appl Environ. Microbiol. 37: 11-13.
86. Burmeister, H.R., R.F. Vesonder, and W.F. Kwolek. 1980a. Mouse bioassay for *Fusarium* metabolites: rejection or acceptance when dissolved in drinking water. Appl. Environ. Microbiol. 39: 957-961.
87. Burmeister, H.R., M.D. Grove, and W.F. Kwolek. 1980b. Moniliformin and butenolide: effect on mice of high-level, long-term oral intake. Appl. Environ. Microbiol. 40: 1142-1144.
88. Burmeister, H.R. J.J. Ellis, and R.F. Vesonder. 1981. Survey for Fusaria that produce an antibiotic that causes conidia of *Penicillium digitatum* to swell. Mycopathologia 74: 29-33.
89. Burnside, J.E., W.L. Sippel, J. Forgacs, W.T. Carll, B. Atwood, and E.R. Poll, 1957. A disease of swine and cattle caused by eating moldy corn. II. Experimental production with pure cultures of molds. Amer. J. Vet. Res. 18: 817-824.
90. Butler, T. 1902. Notes on a feeding experiment to produce leucoencephalitis in a horse with positive results. Amer. Vet. Rev. 26: 748-751.
91. Caldwell, R.W., and J. Tuite. 1970. Zearalenone production in field corn in Indiana. Phytopathology 60: 1696-1697.
92. Caldwell, R.W., and J. Tuite. 1974. Zearalenone in freshly harvested corn. Phytopathology 64: 752-753.
93. Caldwell, R.W., J. Tuite, M. Stob, and R. Baldwin. 1970. Zearalenone production by *Fusarium* species. Appl. Microbiol. 20: 31-34.
94. Ceruti Scurti, J., N. Fiusselo, and G. Cantini. 1971. Metaboliti ad azione estrogena prodotti de miceti. Allionia 17: 55-58.
95. Chakrabarti, D.K., and K.C. Basu Chaudhary. 1980. Correlation between virulence and fusaric acid production in *Fusarium oxysporum* f. sp. *carthami*. Phytopath. Z. 99: 43-46.

96. Chakrabarti, D.K., K.C. Basu Chaudhary, and S. Ghosal. 1976. Toxic substances produced by *Fusarium*. III. Production and screening of phytotoxic substances of *F. oxysporum* f. sp. *carthami* responsible for the wilt disease of safflower *Carthamus tinctorius* Linn. Experientia 32: 608–609.
97. Chang, K., H.J. Kurtz, and C.J. Mirocha. 1979. Effects of the mycotoxin zearalenone on swine reproduction. Amer. J. Vet. Res. 40: 1260–1267.
98. Chi, M.S., and C.J. Mirocha. 1978. Necrotic oral lesions in chickens fed diacetoxyscirpenol, T-2 toxin and crotocin. Poult. Sci. 57: 807–808.
99. Chi, M.S., C.J. Mirocha, H.J. Kurtz, G. Weaver, F. Bates, W. Shimoda, and H.R. Burmeister. 1977a. Acute toxicity of T-2 toxin in broiler chicks and laying hens. Poult. Sci. 56: 103–116.
100. Chi, M.S., C.J. Mirocha, H.J. Kurtz, G. Weaver, F. Bates, and W. Shimoda. 1977b. Subacute toxicity of T-2 toxin in broiler chicks. Poult. Sci. 56: 306–313.
101. Chi, M.S., C.J., Mirocha, H.J. Kurtz, G. Weaver, F. Bates, and W. Shimoda. 1977c. Effects of T-2 toxin on reproductive performance and health of laying hens. Poult. Sci. 56: 628–637.
102. Chi, M.S., T.S. Robison, C.J. Mirocha, and K.R. Reddy. 1978. Acute toxicity of 12,13-epoxytrichothecenes in one-day-old broiler chicks. Appl. Environ. Microbiol. 35: 636–640.
103. Chi, M.S., C.J. Mirocha, H.J. Kurtz, G.A. Weaver, F. Bates, T. Robison, and W. Shimoda. 1980a. Effect of dietary zearalenone on growing broiler chicks. Poult. Sci 59: 531–536.
104. Chi, M.S., C.J. Mirocha, G.A. Weaver, and H.J. Kurtz. 1980b. Effect of zearalenone on female white leghorn chickens. Appl. Environ. Microbiol. 39: 1026–1030.
105. Chi, M.S., M.E. El-Halawani, P.E. Waibel, and C.J. Mirocha. 1981. Effects of T-2 toxin on brain catecholamines and selected blood components in growing chickens. Poult. Sci. 60: 137–141.
106. Cho, B.R. 1964. Toxicity of water extracts of scabby barley to suckling mice. Amer. J. Vet. Res. 25: 1267–1270.
107. Christensen, C.M. 1979. Zearalenone, p. 1–79. *In* W. Shimoda (ed.), Conference on mycotoxins in animal feeds and grains related to animal health, Rockville, Maryland, June 8, 1979. PB-300 300. Report no. FDA/BVM-79/139. U.S. Dept. of Commerce, National Technical Information Service, Springfield, Virginia.
108. Christensen, J.J., and H.C.H. Kernkamp. 1936. Studies on the toxicity of blighted barley to swine. Minnesota Agr. Expt. Stn. Tech. Bull. 113. 28 p.
109. Christensen, C.M., G.H. Nelson, and C.J. Mirocha. 1965. Effect on the white rat uterus of a toxic substance isolated from *Fusarium*. Appl. Microbiol. 13: 653–659.
110. Christensen, C.M., R.A. Meronuck, G.H. Nelson, and J.C. Behrens. 1972a. Effects on turkey poults of rations containing corn invaded by *Fusarium tricinctum* (Cda.) Sny. & Hans. Appl. Microbiol. 23: 177–179.
111. Christensen, C.M., C.J. Mirocha, G.H. Nelson, and J.F. Quast. 1972b. Effect on young swine of consumption of rations containing corn invaded by *Fusarium roseum*. Appl. Microbiol. 23: 202.
112. Chung, C.W., M.W. Trucksess, A.L. Giles Jr., and L. Friedman. 1974. Rabbit skin test for estimation of T-2 toxin and other skin-irritating toxins in contaminated corn. J. Assoc. Off. Anal. Chem. 57: 1121–1127.
113. Claridge, C.A., and H. Schmitz. 1979. Production of 3-acetoxyscirpene-4,15-diol from anguidine (4,15-diacetoxyscirpene-3-ol) by *Fusarium oxysporum* f. sp. *vasinfectum*. Appl. Environ. Microbiol. 37: 693–696.
114. Clark, C.A., A. Lawrence, and F.A. Martin. 1981. Accumulation of furanoterpenoids in sweet potato tissue following inoculation with different pathogens. Phytopathology 71: 708–711.
115. Claydon, N., J.F. Grove, and M. Pople. 1977a. Insecticidal secondary metabolic products from the entomogenous fungus *Fusarium solani*. J. Invertebr. Pathol. 30: 216–223.

116. Claydon, N., J.F. Grove, and M. Pople. 1977b. Fusaric acid from *Fusarium solani.* Phytochemistry 16: 603.
117. Claydon, N., J.F. Grove, and M. Pople. 1979. Insecticidal secondary metabolic products from the entomogenous fungus *Fusarium larvarum.* J. Invertebr. Pathol. 33: 364–367.
118. Coffin, J.L., and G.F. Combs, Jr. 1981. Impaired Vitamin E status of chicks fed T-2 Toxin. Poult. Sci. 60: 385–392.
119. Cole, R.J., and R.H. Cox. 1981. Handbook of toxic fungal metabolites. Academic Press, New York. 937 p.
120. Cole, M., and G.N. Rolinson. 1972. Microbial metabolites with insecticidal properties. Appl. Microbiol. 24: 660–662.
121. Cole, R.J., J.W. Kirksey, H.G. Cutler, B.L. Doupnik, and J.C. Peckham. 1973. Toxin from *Fusarium moniliforme:* effects on plants and animals. Science 179: 1324–1326.
122. Cole, R.J., J.W. Dorner, R.H. Cox, B.M. Cunfer, H.G. Cutler, and B.P. Stuart. 1981. The isolation and identification of several trichothecene mycotoxins from *Fusarium heterosporum.* J. Nat. Prod. 44: 324–330.
123. Collins, G.J., and J.D. Rosen. 1981. Distribution of T-2 toxin in wet-milled corn products. J. Food Sci. 46: 877–879.
124. Connole, M.D., B.J. Blaney, and T. McEwan. 1981. Mycotoxins in animal feeds and toxic fungi in Queensland 1971-80. Aust. Vet. J. 57: 314–318.
125. Coulter, D.B., R.D. Wyatt, and R.G. Stewart. 1977. Electro-retinograms from broilers fed aflatoxin and T-2 toxin. Poult. Sci. 56: 1435–1439.
126. Cuboni, G. 1882. Micromiceti della cariossidi di grano turco in rapporto colla pellagra. Arch. Psichiat. Sci. Penali Antrop. Crim. 3: 353.
127. Cuero, R.G. 1980. Ecological distribution of *Fusarium solani* and its opportunistic action related to mycotic keratitis in Cali, Colombia. J. Clin. Microbiol. 12: 455–461.
128. Cullen, D. 1981. Mycotoxic Fusaria: Pathogenicity, cultural characteristics, taxonomy, and genetics. Ph.D. Thesis, Univ. of Wisconsin, Madison, Wisconsin. 114 p.
129. Cullen, D., and E.B. Smalley. 1981. New process for T-2 toxin production. Phytopathology 71: 212 (Abstr.).
130. Cullen, D., R.W. Caldwell, and E.B. Smalley. 1981. Effect of host genotype on zearalenone contamination of corn. Phytopathology 71: 211–212 (Abstr.).
131. Curtin, T.M., and J. Tuite. 1966. Emesis and refusal of feed in swine associated with *Gibberella zeae*-infected corn. Life Sci. 5: 1937–1944.
132. Curtis, R.W. 1957. Survey of fungi and actinomycetes for compounds possessing gibberellin-like activity. Science 125: 646.
133. Dahlgren, R.R., and D.E. Williams. 1972. Hemorrhagic syndrome in feedlot cattle. The Bovine Practitioner November 1972: 52–53.
134. Davis, G.R.F., and J.D. Smith. 1977. Effect of temperature on production of fungal metabolites toxic to larvae of *Tenebrio molitor.* J. Invertebr. Pathol. 30: 325–329.
135. Davis, G.R.F., and J.D. Smith. 1981. Effect of light and incubation temperature on production by species of *Fusarium* of metabolites toxic to larvae of *Tenebrio molitor* L. Arch. Int. Physiol. Biochim. 89: 81–84.
136. Davis. G.R.F., J.D. Smith, B. Schiefer, and F.M. Loew. 1975. Screening for mycotoxins with larvae of *Tenebrio molitor.* J. Invertebr. Pathol. 26: 299–303.
137. Davis, N.D., R.E. Wagener, D.K. Dalby, G. Morgan-Jones, and U.L. Diener. 1975. Toxigenic fungi in food. Appl. Microbiol. 30: 159–161.
138. Dawkins, A.W. 1966. Phytotoxic compounds produced by *Fusarium equiseti.* Part II. The chemistry of diacetoxyscirpenol. J. Chem. Soc. (C) 1966: 116–123.
139. Dawkins, A.W., J.F. Grove, and B.K. Tidd. 1965. Diacetoxyscirpenol and some related compounds. Chem. Comm. 27: 27–28.
140. DeNicola, D.B., A.H. Rebar, W.W. Carlton, and B. Yagen. 1978. T-2 toxin mycotoxicosis in the guinea pig. Food Cosmet. Toxicol. 16: 601–609.

141. DeSimone, P.A., F.A. Greco, and H.F. Lessner. 1979. Phase 1 evaluation of a weekly schedule of anguidine. Cancer Treat. Rep. 63: 2015-2017.
142. DeUriarte, L.A., D.M. Forsyth, and J. Tuite. 1976. Improved acceptance by rats of *Gibberella zeae*-damaged corn after washing. J. Anim. Sci. 42: 1196-1201.
143. Diener, U.L., R.E. Wagener, G. Morgan-Jones, and N.D. Davis. 1976. Toxigenic fungi from cotton. Phytopathology 66: 514-516.
144. Diener, U.L., G. Morgan-Jones, R.E. Wagener, and N.D. Davis. 1981. Toxigenicity of fungi from grain sorghum. Mycopathologia 75: 23-26.
145. Diggs, C.H., M.J. Scoltock, and P.H. Wiernik. 1978. Phase II evaluation of anguidine (NSC-141537) for adenocarcinoma of the colon or rectum. Cancer Clin. Trials Winter 1978: 297-299.
146. Doerr. J.A., W.E. Huff, H.T. Tung, R.D. Wyatt, and P.B. Hamilton. 1974. A survey of T-2 toxin, ochratoxin, and aflatoxin for their effects on the coagulation of blood in young broiler chickens. Poult. Sci. 53: 1728-1734.
147. Doerr, J.A., P.B. Hamilton, and H.R. Burmeister. 1981. T-2 toxicosis and blood coagulation in young chickens. Toxicol. Appl. Pharmacol. 60: 157-162.
148. Doidge, E.M. 1938. Some South African Fusaria. Bothalia 3: 331-483.
149. Dominik, T., and A. Ihnatowicz. 1975. Soil fungi from Eloka near Abidjan in equitorial West Africa. Zesz. Nauk. Akad. Roln. Szczec. 50: 13-27.
150. Domsch, K.H., W. Gams, and T.H. Anderson. 1980. Compendium of soil fungi. Volume 1. Academic Press, London. 859 p.
151. Doster, A.R., F.E. Mitchell, R.L. Farrell, and B.J. Wilson. 1978. Effects of 4-ipomeanol, a product from mold-damaged sweet potatoes, on the bovine lung. Vet. Pathol. 15: 367-375.
152. Dounin, M. 1926. The fusariosis of cereal crops in European Russia in 1923. Phytopathology 16: 305-308.
153. Doupnik, B., Jr., O.H. Jones, Jr., and J.C. Peckham. 1971. Toxic Fusaria isolated from moldy sweet potatoes involved in an epizootic of atypical interstitial pneumonia in cattle. Phytopathology 61: 890 (Abstr.).
154. Drouliscos. N.J., B.J. Macris, and R. Kokke. 1976. Growth of *Fusarium moniliforme* on carob aqueous extract and nutritional evaluation of its biomass. Appl. Environ. Microbiol. 31: 691-694.
155. Durackova, Z., V. Betina, B. Hornikova, and P. Nemec. 1977. Toxicity of mycotoxins and other fungal metabolites to *Artemia salina* larvae. Zbl. Bakt. Abt. II, 132: 294-299.
156. Edwards E.T. 1940. Internal grain infection and kernel rot in the 1938 American maize crop. J. Aust. Inst. Agric. Sci. 6: 25-31.
157. Edwards, E.T. 1941. Internal grain infection in maize due to *Gibberella fujikuroi* and *Gibberella fujikuroi* var. *subglutinans*. J. Aust. Inst. Agric. Sci. 7: 74-82.
158. El-Bahrawi, S. 1977. Survey of some *Fusarium moniliforme* strains from different host plants for compounds possessing gibberellin-like activity. Zbl. Bakt. Abt. II, 132: 178-183.
159. El-Gholl, N.E., J.J. McRitchie, C.L. Schoulties, and W.H. Ridings. 1978. The identification, induction of perithecia, and pathogenicity of *Gibberella* (*Fusarium*) *tricincta* n. sp. Can. J. Bot. 56: 2203-2206.
160. Ellis, J.J., and S.G. Yates. 1971. Mycotoxins of fungi from fescue. Econ. Bot. 25: 1-5.
161. Ellison, R.A., and F.N. Kotsonis. 1973. T-2 toxin as an emetic factor in moldy corn. Appl. Microbiol. 26: 540-543.
162. Elpidina, O.K. 1959. The antibiotic and antiblastic properties of poin. Antibiotiki (USSR) 4: 431-434.
163. Elpidina, O.K. 1960. Toxic and antibiotic properties of poin. (The toxin of *Fusarium sporotrichiella* var. *poae*), p. 73-81. In V.I. Bilai (ed.), Mycotoxicoses of man and agricultural animals, Kiev. 167 p. English translation, JPRS 7434, U.S. Joint Publications Research Service;, U.S. Dept. of Commerce, Washington, D.C.
164. Enari, T.-M., T. Ilus, M.-L. Niku-Paavola, M. Nummi, A. Ylimäki, and H. Koponen. 1981.

Formation of *Fusarium* metabolites in barley grain. European J. Appl. Microbiol. Biotechnol. 11: 241–243.

165. Eppley, R.M. 1974. Sensitivity of brine shrimp (*Artemia salina*) to trichothecenes. J. Assoc. Off. Anal. Chem. 57: 618–620.
166. Eppley, R.M. 1979. Trichothecenes and their analysis. J. Amer. Oil Chem. Soc. 56: 824–829.
167. Eppley, R.M., L. Stoloff, M.W. Trucksess, and C.W. Chung. 1974. Survey of corn for *Fusarium* toxins. J. Assoc. Off. Anal. Chem. 57: 632–635.
168. Ermakov, V.V., N.A. Kostyunina, and I.A. Kurmanov. 1978. Isolation and identification of the T-2 mycotoxin produced by *Fusarium sporotrichiella*. Dokl. Vses. Akad. S-kh. Nauk. 1: 36–38 (In Russian). English translation in Sov. Agric. Sci. 3: 47–49, 1978.
169. Etienne, M., and M. Jemmali. 1979. Conséquences de l'ingestion de maïs fusarié par la truie reproductrice. C.R. Acad. Sci. Paris, (D) 288: 779–782.
170. Eugenio, C.P., C.M. Christensen, and C.J. Mirocha. 1970a. Factors affecting production of the mycotoxin F-2 by *Fusarium roseum*. Phytopathology 60: 1055–1057.
171. Eugenio, C., E. De las Casas, P.K. Harein, and C.J. Mirocha. 1970b. Detection of the mycotoxin F-2 in the confused flour beetle and the lesser mealworm. J. Econ. Entomol. 63: 412–415.
172. Featherston, W.R. 1973. Utilization of *Gibberella*-infected corn by chicks and rats. Poult. Sci. 52: 2334–2335.
173. Fisher, E.E., A.W. Kellock, and N.A.M. Wellington. 1967. Toxic strain of *Fusarium culmorum* (W.G. Sm.) Sacc. from *Zea mays* L., associated with sickness in dairy cattle. Nature 215: 322.
174. Fiussello, N., J. Ceruti Scurti, and G. Cantini. 1970. Metaboliti ad azione estrogena di miceti isolati da mangimi. Allionia 16: 43–47.
175. Flannigan, B. 1970. Comparison of seed-borne mycofloras of barley, oats and wheat. Trans. Br. Mycol. Soc. 55: 267–276.
176. Flury, E., R. Mauli, and H.P. Sigg. 1965. The constitution of diacetoxyscirpenol. Chem. Comm. 27: 26–27.
177. Forgacs, J. 1965. Stachybotryotoxicosis and moldy corn toxicosis, p. 87–104. *In* G.N. Wogan (ed.), Mycotoxins in Foodstuffs, MIT Press, Cambridge, Massachusetts.
178. Forgacs, J., and W.T. Carll. 1962. Mycotoxicoses. Adv. Vet. Sci. 7: 273–282.
179. Forgacs, J., H. Koch, W.T. Carll, and R.H. White-Stevens. 1962. Mycotoxicoses. I. Relationship of toxic fungi to moldy feed toxicosis in poultry. Avian Dis. 6: 363–380.
180. Forsyth, D.M. 1974. Studies on *Gibberella zeae*–infected corn in diets of rats and swine. J. Anim. Sci. 39: 1092–1098.
181. Forsyth, D.M., L.A. DeUriarte, and J. Tuite. 1976. Improvement for swine of *Gibberella zeae*–damaged corn by washing. J. Anim. Sci. 42: 1202–1206.
182. Forsyth, D.M., T. Yoshizawa, N. Morooka, and J. Tuite. 1977. Emetic and refusal activity of deoxynivalenol to swine. Appl. Environ. Microbiol. 34: 547–552.
183. Francis, R.G., and L.W. Burgess. 1975. Surveys of Fusaria and other fungi associated with stalk rot in maize in eastern Australia. Aust. J. Agric. Res. 26: 801–807.
184. Francis, R.G., and L.W. Burgess. 1977. Characteristics of two populations of *Fusarium roseum* 'Graminearum' in eastern Australia. Trans. Br. Mycol. Soc. 68: 421–427.
185. Frayssinet, C., and P. Lafont. 1976. Les toxines de fusarium responsables d'accidents leucopéniques. Cahiers Nutr. Diet. 11: 77–81.
186. Fritz, J.C., P.B. Mislivec, G.W. Pla, B.N. Harrison, C.E. Weeks, and J.G. Dantzman. 1973. Toxicogenicity of moldy feed for young chicks. Poult. Sci. 52: 1523–1530.
187. Fromentin, H., S. Salazar-Mejicanos, and F. Mariat. 1980. Pouvoir pathogène de *Candida albicans* pour la souris normale ou déprimée par une mycotoxine: le diacétoxyscirpénol. Ann. Microbiol. (Inst. Pasteur) 131B: 39–46.
188. Fromentin, H., S. Salazar-Mejicanos, and F. Mariat. 1981. Experimental cryptococcosis in mice treated with diacetoxyscirpenol, a mycotoxin of *Fusarium*. Sabouraudia 19: 311–313.
189. Fujimoto, Y., Y. Morita, and T. Tatsuno. 1972. Recherches toxicologiques sur les substances

toxiques de *Fusarium nivale:* Etude chimique des toxins principales, nivalenol, fusarenon-X et nivalenol-4, 15-di-O-acétate. Chem. Pharm. Bull. 20: 1194–1203.
190. Funnell, H.S. 1979. Mycotoxins in animal feedstuffs in Ontario, 1972–1977. Can. J. Comp. Med. 43: 243–246.
191. Gajdusek, D.C. 1953. Acute infectious hemorrhagic fevers and mycotoxicoses in the Union of Soviet Socialist Republics, p. 82–106. *In* Med. Sci. Publ. No. 2, Walter Reed Army Med. Center, Washington, D.C.
192. Gams, W., and E. Müller. 1980. Conidiogenesis of *Fusarium nivale* and *Rhynchosporium oryzae* and its taxonomic implications. Neth. J. Pl. Path. 86: 45–53.
193. Gangopadhyay, S., and N.K. Chakrabarti. 1981. Mycotoxins in stored rice. Current Sci. 50: 272–275.
194. Gardner, D., A.T. Glen, and W.B. Turner. 1972. Calonectrin and 15-deacetylcalonectrin, new trichothecanes from *Calonectria nivalis*. J. Chem. Soc. Perkin I, 18: 2576–2578.
195. Garner, G.B., and C.N. Cornell. 1978. Fescue foot in cattle, p. 45–62. *In* T.D. Wyllie and L.G. Morehouse (ed.), Mycotoxic fungi, mycotoxins, mycotoxicoses. An encyclopedic handbook, Vol. 2, Marcel Dekker, New York.
196. Gäumann, E. 1957. Fusaric acid as a wilt toxin. Phytopathology 47: 342–357.
197. Gäumann, E., and St. Naef-Roth. 1959. Uber Lycomarasminsäure, ein Umwandlungsprodukt des Lycomarasmins. Phytopath. Z. 34: 426–431.
198. Gäumann, E., St. Naef-Roth, and H. Kobel. 1952. Uber Fusarinsäure, ein zweites Welketoxin des *Fusarium lycopersici* Sacc. Phytopathol. Z. 20: 1–38.
199. Gedek, B., B. Hüttner, D.I. Kahlau, H. Köhler, and E. Vielitz. 1978. Rachitis bei Mastgeflügel durch Kontamination des Futters mit *Fusarium moniliforme* Sheldon. 1. Mitteilung: Feldbeobachtungen, Reproduktion des Krankheitsbildes und Behandlungsversuche. Zbl. Vet. Med. B., 25: 29–44.
200. Gentry. P.A., and M.L. Cooper. 1981. Effect of *Fusarium* T-2 toxin on hematological and biochemical parameters in the rabbit. Can. J. Comp. Med. 45: 400–405.
201. Gerber, N.N., and M.S. Ammar. 1979. New antibiotic pigments related to fusarubin from *Fusarium solani* (Mart.) Sacc. II. Structure elucidations. J. Antibiotics 32: 685–688.
202. Gerlach, W. 1977. Drei neue Varietäten von *Fusarium merismoides, F. larvarum* und *F. chlamydosporum*. Phytopath. Z. 90: 31–42.
203. Gerlach, W., and D. Ershad. 1970. Beitrag zur Kenntnis der *Fusarium*—und *Cylindrocarpon*—Arten in Iran. Nova Hedwigia 20: 725–784.
204. Getsova, G.V. 1960. Material on the experimental study of *Fusarium* toxicosis, p. 110–117. *In* V.I. Bilai (ed.), Mycotoxicoses of man and agricultural animals, Kiev, 167 p. English translation, JPRS 7434, U.S. Joint Publications Research Service, U.S. Dept. of Commerce, Washington, D.C.
205. Ghosal, S., D.K. Chakrabarti, and K.C. Basu-Chaudhary. 1976a. Toxic substances produced by *Fusarium* I: Trichothecene derivatives from two strains of *Fusarium oxysporum* f. sp. *carthami*. J. Pharm. Sci. 65: 160–161.
206. Ghosal, S., K. Biswas, D.K. Chakrabarti, B.K. Chattopadhyay, and S.K. Bhattacharya. 1976b. The changes in chemical characters from the interaction of *Mangifera indica* with two pathogenic fungi—*Aspergillus niger* and *Fusarium moniliformae*. Ind. J. Pharm. 38: 153 (Abstr.).
207. Ghosal, S., D.K. Chakrabarti, and K.C. Basu-Chaudhary. 1977a. The occurrence of 12, 13-epoxytrichothecenes in seeds of safflower infected with *Fusarium oxysporum* f. sp. *carthami*. Experientia 33: 574–575.
208. Ghosal, S., K. Biswas, D.K. Chakrabarti, and K.C. Basu-Chaudhary. 1977b. Control of *Fusarium* wilt of safflower by mangiferin. Phytopathology 67: 548–550.
209. Ghosal, S., K. Biswas, R.S. Srivastava, D.K. Chakrabarti, and K.C. Basu-Chaudhary. 1978a. Toxic substances produced by *Fusarium*. V: Occurrence of zearalenone, diacetoxyscirpenol, and T-2 toxin in moldy corn infected with *Fusarium moniliforme* Sheld. J. Pharm. Sci. 67: 1768–1769.

210. Ghosal, S., K. Biswas, and B.K. Chattopadhyay. 1978b. Differences in the chemical constituents of *Mangifera indica*, infected with *Aspergillus niger* and *Fusarium moniliformae*. Phytochemistry 17: 689–694.
211. Ghosal, S., D.K. Chakrabarti, K. Biswas, and Y. Kumar. 1979. Toxic substances produced by *Fusarium*. X. Concerning the malformation disease of mango. Experientia 35: 1633–1634.
212. Gilgan, M.W., E.B. Smalley, and F.M. Strong. 1966. Isolation and partial characterization of a toxin from *Fusarium tricinctum* on moldy corn. Arch. Biochem. Biophys. 114: 1–3.
213. Gjertsen, P. 1967. Gushing in beer, its nature, cause and prevention. The Brewers Digest 42: 80–84.
214. Gjertsen, P., B. Trolle, and K. Andersen. 1966. Studies on gushing. II. Gushing caused by microorganisms, specially *Fusarium* species, p. 428–438. In European Brewery Convention, Proc. 10th Congr., Stockholm 1965, Elsevier, Amsterdam.
215. Goodwin, W., C.D. Haas, C. Fabian, I. Heller-Bettinger, and B. Hoogstraten. 1978. Phase I evaluation of anguidine (diacetoxyscirpenol, NSC-141537). Cancer 42: 23–26.
216. Gordon, W.L. 1944. The occurrences of *Fusarium* species in Canada. I. Species of *Fusarium* isolated from farm samples of cereal seed in Manitoba. Can. J. Res., C, 22: 282–286.
217. Gordon, W.L. 1952. The occurrence of *Fusarium* species in Canada. II. Prevalence and taxonomy of *Fusarium* species in cereal seed. Can. J. Bot. 30: 209–251.
218. Gordon, W.L. 1956. The taxonomy and habitats of the *Fusarium* species in Trinidad, B. W. I. Can. J. Bot. 34: 847–864.
219. Gordon, W.L. 1959. The occurrence of *Fusarium* species in Canada, VI. Taxonomy and geographic distribution of *Fusarium* species on plants, insects, and fungi. Can. J. Bot. 37: 257–290.
220. Gordon, W.L. 1960a. The taxonomy and habitats of *Fusarium* species from tropical and temperate regions. Can. J. Bot. 38: 643–658.
221. Gordon, W.L. 1960b. Distribution and prevalence of *Fusarium moniliforme* Sheld. (*Gibberella fujikuroi* (Saw.) Wr.) producing substances with gibberellin-like biological properties. Nature 186: 698–700.
222. Graham, R. 1912. Forage poisoning or so-called cerebro-spinal meningitis in horses, cattle and mules. Kentucky Agric. Exp. Stn. Bull. No. 167: 369–383.
223. Graham, R. 1935. Results of inoculating laboratory animals with equine brain-tissue suspensions and equine brain-tissue filtrates from spontaneous cases of so-called cornstalk disease. J. Amer. Vet. Med. Assoc. 86: 778–780.
224. Graham, R. 1936. Toxic encephalitis or non-virus encephalomyelitis of horses. Vet. Med. 31: 46–50.
225. Greenway, J.A., and R. Puls. 1976. Fusariotoxicosis from barley in British Columbia. I. Natural occurrence and diagnosis. Can. J. Comp. Med. 40: 12–15.
226. Gross, V.J., and J. Robb. 1975. Zearalenone production in barley. Ann. Appl. Biol. 80: 211–216.
227. Grove, J.F. 1970a. Phytotoxic compounds produced by *Fusarium equiseti*. Part V. Transformation products of 4β,15-diacetoxy-3α,7α-dihydroxy-12,13-epoxytrichothec-9-en-8-one and the structures of nivalenol and fusarenone. J. Chem. Soc. (C) 1970: 375–378.
228. Grove. J.F. 1970b. Phytotoxic compounds produced by *Fusarium equiseti*. Part VI. 4β,8α,15-triacetoxy-12,13-epoxytrichothec-9-ene-3α,7α-diol. J. Chem. Soc. (C) 1970: 378–379.
229. Grove, J.F., and M. Pople. 1979. Metabolic products of *Fusarium larvarum* Fuckel. The fusarentins and the absolute configuration of monocerin. J. Chem. Soc. Perkin I, 1979: 2048–2051.
230. Grove, J.F., and M. Pople. 1981. The insecticidal activity of some fungal dihydroisocoumarins. Mycopathologia 76: 65–67.
231. Grove, J.F., P.W. Jeffs, and T.P.C. Mulholland. 1958. Gibberellic acid. Part V. The relation between gibberellin A_1 and gibberellic acid. J. Chem. Soc. 1958: 1236–1240.
232. Grove, M.D., S.G. Yates, W.H. Tallent, J.J. Ellis, I.A. Wolff, N.R. Kosuri, and R.E. Nichols.

1970. Mycotoxins produced by *Fusarium tricinctum* as possible causes of cattle disease. J. Agr. Food Chem. 18: 734-736.
233. Gupta, J., B. Pathak, N. Sethi, and V.C. Vora. 1981. Histopathology of mycotoxicosis produced in Swiss Albino mice by metabolites of some fungal isolates. Appl. Environ. Microbiol. 41: 752-757.
234. Hacking, A., W.R. Rosser, and M.T. Dervish. 1976. Zearalenone-producing species of *Fusarium* on barley seed. Ann. Appl. Biol. 84: 7-11.
235. Hacking, A., W.R. Rosser, and M.T. Dervish. 1977. Incidence of zearalenone-producing strains of *Fusarium* in barley seeds. Ann. Nutr. Alim. 31: 557-562.
236. Hafez-Zedan, H., and R. Plourde. 1973. Steroid 1-dehydrogenation and side-chain degradation enzymes in the life cycle of *Fusarium solani*. Biochim. Biophys. Acta 326: 103-115.
237. Hagler, W.M., and C.J. Mirocha. 1980. Biosynthesis of (^{14}C) zealenone from (1-^{14}C) acetate by *Fusarium roseum* 'Gibbosum.' Appl. Environ. Microbiol. 39: 668-670.
238. Hagler, W.M., C.J. Mirocha, S.V. Pathre, and J.C. Behrens. 1979. Identification of the naturally occurring isomer of zearalenol produced by *Fusarium roseum* 'Gibbosum' in rice culture. Appl. Environ. Microbiol. 37: 849-853.
239. Hagler, W.M., C.J. Mirocha, and S.V. Pathre. 1981. Biosynthesis of radiolabeled T-2 toxin by *Fusarium tricinctum*. Appl Environ. Microbiol. 41: 1049-1051.
240. Haliburton, J.C., R.F. Vesonder, T.F. Lock, and W.B. Buck. 1979. Equine leucoencephalomalacia (ELEM): a study of *Fusarium moniliforme* as an etiologic agent. Vet. Human Toxicol. 21: 348-351.
241. Haliburton, J.C., W.B. Buck, T.F. Lock, R.F. Vesonder, and B.J. Wilson. 1981. Equine leukoencephalomalacia: the toxicity of three *Fusarium moniliforme* metabolites in Equidae. J. Amer. Vet. Med. Assoc. 179: 262-263 (Abstr.).
242. Hamilton, P.B., R.D. Wyatt, and H. Burmeister. 1971. Effect of fusariotoxin T-2 in chickens. Poult. Sci. 50: 1583-1584.
243. Hamilton, P.B., R.D. Wyatt, J.A. Doerr, and H.R. Burmeister. 1974. Response of laying hens to T-2 toxin. Poult. Sci. 53: 1931 (Abstr.).
244. Harein, P.K., E. De Las Casas, C.P. Eugenio, and C.J. Mirocha. 1970. Reproduction and survival of confused flour beetles exposed to metabolites produced by *Fusarium roseum* var. *graminearum*. J. Econ. Entomol. 64: 975-976.
245. Hart, L.P., and T.C. Stebbins. 1981. Production of zearalenone and vomitoxin in commerical sweet corn hybrids inoculated with isolates of *Gibberella zeae*. 1981 Ann. Meeting, Amer. Phytopath. Soc., p. 227 (Abstr.).
246. Harwig, J., and I.C. Munro. 1975. Mycotoxins of possible importance in diseases of Canadian farm animals. Can. Vet. J. 16: 125-141.
247. Harwig, J., and P.M. Scott. 1971. Brine shrimp (*Artemia salina* L.) larvae as a screening system for fungal toxins. Appl. Microbiol. 21: 1011-1016.
248. Harwig, J., P.M. Scott, D.R. Stoltz, and B.J. Blanchfield. 1979. Toxins of molds from decaying tomato fruit. Appl. Environ. Microbiol. 38: 267-274.
249. Hayes, M.A., and H.B. Schiefer. 1979. Quantitative and morphological aspects of cutaneous irritation by trichothecene mycotoxins. Food Cosmet. Toxicol. 17: 611-621.
250. Hayes, M.A., and H.B. Schiefer. 1980. Subacute toxicity of dietary T-2 toxin in mice: influence of protein nutrition. Can. J. Comp. Med. 44: 219-228.
251. Hayes, M.A., J.E.C. Bellamy, and H.B. Schiefer. 1980. Subacute toxicity of dietary T-2 toxin in mice: morphological and hematological effects. Can. J. Comp. Med. 44: 203-218.
252. Hesseltine, C.W. 1972. Solid state fermentations. Biotechnol. Bioeng. 14: 517-532.
253. Hesseltine, C.W. 1974. Natural occurrence of mycotoxins in cereals. Mycopathol. Mycol. Appl. 53: 141-153.
254. Hesseltine, C.W., R.F. Rogers, and O. Shotwell. 1978. Fungi, especially *Gibberella zeae,* and zearlaenone occurrence in wheat. Mycologia 70: 14-18.
255. Hewett, P.D. 1967. A survey of seed-borne fungi of wheat. 2. The incidence of common species of *Fusarium*. Trans. Br. Mycol. Soc. 50: 175-182.

256. Hibbs. C.M., G.D. Osweiler, W.B Buck, and G.P. Macfee. 1974. Bovine hemorrhagic syndrome related to T-2 mycotoxin. 17th Ann. Proc. Amer. Assoc. Vet. Lab. Diagnosticians 1974: 305–310.
257. Hidaka, H. 1971. Fusaric (5-Butylpicolinic) acid, an inhibitor of dopamine β-hydroxylase, affects serotonin and noradrenaline. Nature 231: 54–55.
258. Hidaka, H., T. Nagatsu, K. Takeya, T. Taheuchi, H. Suda, K. Kojiri, M. Matsuzaki, and H. Umezawa. 1969. Fusaric acid, a hypotensive agent produced by fungi. J. Antibiot. 22: 228–230.
259. Hidy, P.H., R.S. Baldwin, R.L. Greasham, C.L. Keith, and J.R. McMullen. 1977. Zearalenone and some derivatives: production and biological activities. Adv. Appl. Microbiol. 22: 59–82.
260. Hitokoto, H., S. Morozumi, T. Wauke, S. Sakai, and H. Kurata. 1981. Fungal contamination and mycotoxin-producing potential of dried beans. Mycopathologia 73: 33–38.
261. Hobson, W., J. Bailey, A. Kowalk, and G. Fuller. 1977. The hormonal potency of zearalenone in non-human primates, p. 365–377. In J.V. Rodricks, C.W. Hesseltine, and M.A. Mehlman (ed.), Mycotoxins in human and animal health, Pathotox Publishers, Park Forest South, Illinois.
262. Hoerr, F.J., and W.W. Carlton. 1979. T-2 mycotoxicosis in male broiler chickens. Toxicol. Appl. Pharmacol. 48: A15 (Abstr.).
263. Hoerr, F.J., W.W. Carlton, and B. Yagen. 1981a. The toxicity of T-2 toxin and diacetoxyscirpenol in combination for broiler chickens. Food Cosmet. Toxicol. 19: 185–188.
264. Hoerr, F.J., W.W. Carlton, and B. Yagen. 1981b. Mycotoxicosis caused by a single dose of T-2 toxin or diacetoxyscirpenol in broiler chickens. Vet. Pathol. 18: 652–664.
265. Hoyman, W.G. 1941. Concentration and characterization of the emetic principle present in barley infected with Gibberella saubinetii. Phytopathology 31: 871–885.
266. Hsu, I.C., E.B. Smalley, F.M. Strong, and W.E. Ribelin. 1972. Identification of T-2 toxin in moldy corn associated with a lethal toxicosis in dairy cattle. Appl. Microbiol. 24:684–690.
267. Huff, W.E., J.A. Doerr, P.B. Hamilton, and R.F. Vesonder. 1981. Acute toxicity of vomitoxin (deoxynivalenol) in broiler chickens. Poult. Sci. 60:1412–1414.
268. Hurd, R.N. 1977. Structure activity relationships in zearalenones, p. 379–391. In J.V. Rodricks, C.W. Hesseltine, and M.A. Mehlman (ed.), Mycotoxins in human and animal health, Pathotox Publishers, Park Forest South, Illinois.
269. Ichinoe, M. 1976. Bioproduction of mycotoxins by fungi isolated from foodstuffs. Proc. Jap. Assoc. Mycotoxicol. 3/4: 16–20 (In Japanese).
270. Ichinoe, M. 1978a. Classification of mycotoxin-producing Fusarium species. Proc. Jap. Assoc. Mycotoxicol. 8: 1–5 (In Japanese).
271. Ichinoe, M. 1978b. Mycotoxins produced by Fusarium. Shokubutsu Boeki 32: 417–422 (In Japanese).
272. Ichinoe, M., H. Kurata, and T. Suzuki. 1977. Zearalenone production by Fusarium species in Japan. Proc. Jap. Assoc. Mycotoxicol. 5/6: 1–2 (In Japanese).
273. Ichinoe, M., R. Amano, N. Morooka, T. Yoshizawa, T. Suzuki, and M. Kurisu. 1980. Geographic difference of toxigenic fungi of Fusarium species. Proc. Jap. Assoc. Mycotoxicol. 11: 20–22 (In Japanese).
274. Ikediobi, C.O., I.C. Hsu, J.R. Bamburg, and F.M. Strong. 1971. Gas-liquid chromatography of mycotoxins of the trichothecene group. Anal. Biochem. 43: 327–340.
275. Ilus, T., P.J. Ward, M. Nummi, H. Adlercreutz, and J. Gripenberg. 1977. A new mycotoxin from Fusarium. Phytochemistry 16: 1839–1840.
276. Ilus, T., M.L. Niku-Paavola, and T.M. Enari. 1981. Chromatographic analysis of Fusarium toxins in grain samples. European J. Appl. Microbiol. Biotechnol. 11: 244–247.
277. Inaba, T., and C.J. Mirocha. 1979. Preferential binding of radiolabeled zearalenone to a protein fraction of Fusarium roseum Graminearum. App. Environ. Microbiol. 37: 80–84.
278. Irfan, M. 1971. The clinical picture and pathology of "Deg Nala Disease" in buffaloes and cattle in West Pakistan. Vet. Rec. 88: 422–424.

279. Ishii, K. 1975. Two new trichothecenes produced by *Fusarium* sp. Phytochemistry 14: 2469–2471.
280. Ishii, K., and Y. Ueno. 1981. Isolation and characterization of two new trichothecenes from *Fusarium sporotrichioides* Strain M-1-1. Appl. Environ. Microbiol. 42: 541–543.
281. Ishii, K., K. Sakai, Y. Ueno, H. Tsunoda, and M. Enomoto. 1971. Solaniol, a toxic metabolite of *Fusarium solani*. Appl. Microbiol. 22: 718–720.
282. Ishii, K., M. Sawano, Y. Ueno, and H. Tsunoda. 1974. Distribution of zearalenone-producing *Fusarium* species in Japan. Appl. Microbiol. 27: 625–628.
283. Ishii, K., Y. Ando, and Y. Ueno. 1975. Toxicological approaches to the metabolites of Fusaria. IX. Isolation of vomiting factor from moldy corn infected with *Fusarium* species. Chem. Pharm. Bull. 23: 2162–2164.
284. Ishii, K., S.V. Pathre, and C.J. Mirocha. 1978. Two new trichothecenes produced by *Fusarium roseum*. J. Agric. Food Chem. 26: 649–653.
285. Itakura, H., and R. Kinosita. 1975. Toxic fungi isolated from Uganda foodstuffs. A histopathological study of acute toxicity of fungal culture filtrates. Trop. Med. 17: 73–90.
286. Ito, T. 1979. Fusariocin C, a new cytotoxic substance produced by *Fusarium moniliforme*. Agric. Biol. Chem. 43: 1237–1242.
287. Ito, Y., K. Ohtsubo, and M. Saito. 1980. Effects of fusarenon-X, a trichothecene produced by *Fusarium nivale*, on pregnant mice and their fetuses. Japan. J. Exp. Med. 50: 167–172.
288. Ito, T., T. Arai, Y. Ohashi, and Y. Sasada. 1981. Structure of fusariocin C, a cytotoxic metabolite from *Fusarium moniliforme*. Agric. Biol. Chem. 45: 1689–1692.
289. Iwanoff, X., C. Yuan, and S. Fang. 1957. Uber die toxische Enzephalomalazie (Moldy corn poisoning) der Einhufer in China. Arch. Exp. Vet. Med. 11: 1033–1056.
290. Jackson, R.A., S.W. Fenton, C.J. Mirocha, and G. Davis. 1974. Characterization of two isomers of 8'-hydroxyzearalenone and other derivatives of zearalenone. J. Agric. Food Chem. 22: 1015–1019.
291. Jain, S.C., S.R.S. Dange, and B.S. Siradhana. 1980. Production of toxic substances by four *Fusarium* species on Ganga-5 maize kernels. Indian Phytopath. 33: 30–32.
292. Jamalainen, E.A. 1955. *Fusarium* species causing plant diseases in Finland. Acta Agral. Fenn. 83: 159–172.
293. Jarvis, B. 1981. Occurrence and significance of mycotoxins in U.K. food. Br. Food Manuf. Ind. Res. Assoc. Tech. Circular No. 751. 16 p.
294. Jemmali, M., Y. Ueno, K. Ishii, C. Frayssinet, and M. Etienne. 1978. Natural occurrence of trichothecenes (nivalenol, deoxynivalenol, T_2) and zearalenone in corn. Experientia 34: 1333–1334.
295. Joffe, A.Z. 1960a. Toxicity and antibiotic properties of some Fusaria. Bull. Res. Counc. Israel 8D: 81–95.
296. Joffe, A.Z. 1960b. The mycoflora of overwintered cereals and its toxicity. Bull. Res. Counc. Israel 9D: 101–126.
297. Joffe, A.Z. 1962. Biological properties of some toxic fungi isolated from overwintered cereals. Mycopathol. Mycol. Appl. 16: 201–221.
298. Joffe, A.Z. 1963. Toxicity of overwintered cereals. Plant and Soil 18: 31–44.
299. Joffe, A.Z. 1965. Toxin production by cereal fungi causing toxic alimentary aleukia in man, p. 77–85. *In* G.N. Wogan (ed.), Mycotoxins in foodstuffs, M.I.T. Press, Cambridge, Massachusetts.
300. Joffe, A.Z. 1969. Toxic properties and effects of *Fusarium poae* (Peck) Wr., *F. sporotrichioides* Sherb. and *Aspergillus flavus* Link. J. Stored Prod. Res. 5: 211–218.
301. Joffe, A.Z. 1971. Alimentary toxic aleukia, p. 139–189. *In* S, Kadis, A. Ceigler, and S.J. Ajl (ed.), Microbiol toxins, Vol. VII, Academic Press, New York.
302. Joffe, A.Z. 1973a. *Fusarium* species of the Sporotrichiella section and relations between their toxicity to plants and animals. Z. PflKrankh. PflSch. 80: 92–99.
303. Joffe, A.Z. 1973b. *Fusarium* species on groundnut kernels and in groundnut soil. Plant and Soil 38: 439–446.

304. Joffe, A.Z. 1974a. A modern system of *Fusarium* taxonomy. Mycopathol. Mycol. Appl. 53: 201–228.
305. Joffe. A.Z. 1974b. Growth and toxigenicity of Fusaria of the Sporotrichiella section as related to environmental factors and culture substrates. Mycopathol. Mycol. Appl. 54: 35–46.
306. Joffe, A.Z. 1974c. Toxicity of *Fusarium poae* and *Fusarium sporotrichioides* and its relation to alimentary toxic aleukia, p. 229–262. *In* I.F.H Purchase (ed.), Mycotoxins, Elsevier, Amsterdam.
307. Joffe. A.Z. 1978a. *Fusarium* toxicosis in poultry, p. 309–321. *In* T.D. Wyllie and L.G. Morehouse (ed.), Mycotoxic fungi, mycotoxins, mycotoxicoses. An encyclopedic handbook, Vol. 2, Marcel Dekker, New York.
308. Joffe. A.Z. 1978b. *Fusarium poae* and *Fusarium sporotrichioides* as principal causal agents of alimentary toxic aleukia, p. 21–86. *In* T.D. Wyllie and L.G. Morehouse (ed.), Mycotoxic fungi, mycotoxins, mycotoxicoses. An encyclopedic handbook, Vol. 3, Marcel Dekker, New York.
309. Joffe, A.Z., and J. Palti. 1967. *Fusarium equiseti* (Cda.) Sacc. in Israel. Israel J. Bot. 16: 1–18.
310. Joffe, A.Z., and J. Palti. 1972. *Fusarium* species of the Martiella section in Israel. Phytopath. Z. 73: 123–148.
311. Joffe, A.Z., and J. Palti. 1974. Relations between harmful effects on plants and on animals of toxins produced by species of *Fusarium*. Mycopathol. Mycol. Appl. 52: 209–218.
312. Joffe, A.Z., and J. Palti. 1975. Taxonomic study of Fusaria of the Sporotrichiella Section used in recent toxicological work. Appl. Microbiol. 29: 575–579.
313. Joffe, A.Z., and B. Yagen. 1977. Comparative study of the yield of T-2 toxin produced by *Fusarium poae, F. sporotrichioides* and *F. sporotrichioides* var. *tricinctum* strains from different sources. Mycopathologia 60: 93–97.
314. Joffe, A.Z., and B. Yagen. 1978. Intoxication produced by toxic fungi *Fusarium poae* and *F. sporotrichioides* on chicks. Toxicon 16: 263–273.
315. Joffe, A.Z., J. Palti, and R. Arbel-Sherman. 1973. *Fusarium moniliforme* Sheld. in Israel (*Gibberella fujikuroi* (Saw.) Wollenw.). Mycopathol. Mycol. Appl. 50: 85–107.
316. Kallela, K., and E.L. Korpinen. 1973. The quantity and estrogenic activity of F-2 toxin produced on different substrates. Nord. Vet. Med. 25: 446–450.
317. Kallela, K., and I. Saastamoinen. 1981a. The effect of grain preservatives on the growth of the fungus *Fusarium graminearum* and on the quantity of zearalenone. Acta Vet. Scand. 22: 417–427.
318. Kallela, K., and I. Saastamoinen. 1981b. Decomposition of the *Fusarium graminearum* toxin zearalenone in storage conditions. Nord. Vet. Med. 33: 451–460.
319. Kalra, D.S., K.C. Bhatia, O.P. Gauatm, and H.V.S. Chauhan. 1973. An obscure disease (possibly Degnala disease) in buffaloes and cattle. Studies on its epizootiology, pathology and etiology. Haryana Agric. Univ. J. Res. 2: 256–264.
320. Kalra, D.S., K.C. Bhatia, O.P. Gautam, and H.V.S. Chauhan. 1977a. *Fusarium equiseti* associated mycotoxins as possible cause of Degnala disease. Ann. Nutr. Alim. 31: 745–752.
321. Kalra, D.S., K.C., Bhatia, O.P. Gautam, and H.V.S. Chauhan. 1977b. Pathology of Degnala disease in cattle and buffaloes. Ann. Nutr. Alim. 31: 753–760.
322. Kalra, D.S., K.C. Bhatia, O.P. Gautam, and H.V.S. Chauhan. 1980. Mycotoxins associated with mouldy rice straw, as a cause of Degnala disease. Proc. Int. Symp. Environ. Pollution Toxicol., 1977. Prog. Ecol. 5/7: 327–333.
323. Kamimura, H., M. Nishijima, K. Yasuda, K. Saito, A. Ibe, T. Nagayama, H. Ushiyama, and Y. Naoi. 1981. Simultaneous detection of several *Fusarium* mycotoxins in cereals, grains, and foodstuffs. J. Assoc. Off. Anal. Chem. 64: 1067–1073.
324. Kanshina, N.F. 1957. Pathomorphology of experimental mycotoxicosis in dogs and white rats. Vopr. Pit. 16: 69–74 (In Russian).
325. Kato, T., Y. Asabe, M. Suzuki, and S. Takitani. 1979. Spectrophotometric and fluorimetric determinations of trichothecene mycotoxins with reagents for formaldehyde. Anal. Chim. Acta 106: 59–65.

326. Kellerman, T.S., W.F.O. Marasas, J.G. Pienaar, and T.W. Naudé. 1972. A mycotoxicosis of Equidae caused by *Fusarium moniliforme* Sheldon. A preliminary communication. Onderstepoort J. Vet. Res. 39: 205–208.
327. Kern, H. 1970. Phytotoxins produced by Fusaria, p. 35–45. *In* R.K.S. Wood, A. Ballio, and A. Graniti (ed.), Phytotoxins in plant diseases, Academic Press, New York.
328. Keyl, A.C., J.C. Lewis, J.J. Ellis, S.G. Yates, and H.L. Tookey. 1967. Toxic fungi isolated from tall fescue. Mycopathol. Mycol. Appl. 31: 327–331.
329. Knaus, E. 1978. Untersuchungen zum Nachweis der fruchtbarkeitsvermindernden Wirkung von Mykotoxinen. Veröff. Landwirtsch.—Chem. Bundesversuchsanst. Linz 11: 105–126.
330. Köhler, H., B. Hüttner, E. Vielitz, D.I. Kahlau, and B. Gedek. 1978. Rachitis bei Mastgeflügel durch Kontamination des Futters mit *Fusarium moniliforme* Sheldon. 2. Mitteilung: histologische und mykotoxikologische Untersuchungen. Zbl. Vet. Med. B., 25: 89–109.
331. Kondo, E., and T. Mitsugi. 1966. Microbiological synthesis of 16-keto steroids from steroidal sapogenins. J. Amer. Chem. Soc. 88: 4737–4738.
332. Konishi, T., and S. Ichijo. 1970a. Clinical studies on bean-hulls poisoning of horse. I. Clinical and biochemical observations in spontaneous cases. Res. Bull. Obihiro Univ. 6: 242–257 (In Japanese).
333. Konishi, T., and S. Ichijo. 1970b. Clinical studies on bean-hulls poisoning of horse. II. Clinical and biochemical observations in experimental cases. Res. Bull. Obihiro Univ. 6: 258–273 (In Japanese).
334. Korpinen, E.-L., and J. Uoti. 1974. The variation in toxic effect of five *Fusarium* species on rats. Ann. Agric. Fenn. 13: 34–42.
335. Korpinen, E.-L., and A. Ylimäki. 1972. Toxigenicity of some *Fusarium* strains. Ann. Agric. Fenn. 11: 308–314.
336. Korpinen, E.-L., K. Kallela, and A. Ylimäki. 1972. Estrogenic activity of *Fusarium graminearum* on rats in experimental conditions. Nord. Vet. Med. 24: 62–66.
337. Kosuri, N.R., M.D. Grove, S.G. Yates, W.H. Tallent, J.J. Ellis, I.A. Wolff, and R.E. Nichols. 1970. Response of cattle to mycotoxins of *Fusarium tricinctum* isolated from corn and fescue. J. Amer. Vet. Med. Assoc. 157: 938–940.
338. Kosuri, N.R., E.B. Smalley, and R.E. Nichols. 1971. Toxicologic studies of *Fusarium tricinctum* (Corda) Snyder et Hansen from moldy corn. Amer. J. Vet. Res. 32: 1843–1850.
339. Kotik, A.N., V.T. Chernobay, N.F. Komissarenko, and V.A. Trufanova. 1979. Isolation of mycotoxin in *Fusarium sporotrichiella* and studies of its physicochemical and toxic properties. Mikrobiol. Zh. 41: 636–639 (In Russian).
340. Kotsonis, F.N., and R.A. Ellison. 1975. Assay and relationship of HT-2 toxin and T-2 toxin formation in liquid culture. Appl. Microbiol. 30: 33–37.
341. Kotsonis, F.N., E.B. Smalley, R.A. Ellison, and C.M. Gale. 1975a. Feed refusal factors in pure cultures of *Fusarium roseum "graminearum."* Appl. Microbiol. 30: 362–368.
342. Kotsonis, F.N., R.A. Ellison, and E.B. Smalley. 1975b. Isolation of acetyl T-2 toxin from *Fusarium poae*. Appl. Microbiol. 30: 493–495.
343. Kovacs, F., Cs. Szathmary, and M. Palyusik. 1975. Data on determination of toxin F-2 (zearalenone) by high-pressure liquid, gas and thin-layer chromatography. Acta Vet. Acad. Sci. Hung. 25: 223–230.
344. Kriek, N.P.J., W.F.O. Marasas, P.S. Steyn, S.J. Van Rensberg, and M. Steyn. 1977. Toxicity of a moniliformin-producing strain of *Fusarium moniliforme* var. *subglutinans* isolated from maize. Food Cosmet. Toxicol. 15: 579–587.
345. Kriek, N.P.J., W.F.O. Marasas, and P.G. Thiel. 1981a. Hepato- and cardiotoxicity of *Fusarium verticillioides* (*F. moniliforme*) isolates from southern African maize. Food Cosmet. Toxicol. 19: 447–456.
346. Kriek, N.P.J., T.S. Kellerman, and W.F.O. Marasas. 1981b. A comparative study of the toxicity of *Fusarium verticillioides* (= *F. moniliforme*) to horses, primates, pigs, sheep and rats. Onderstepoort J. Vet. Res. 48: 129–131.

347. Kubota, T. 1977. Experimental studies on *Fusarium* poisoning in goats. Bull. Nippon Vet. Zootech. Coll. 26: 9–24 (In Japanese).
348. Kuczuk, M.H., P.M. Benson, H. Heath, and A.W. Hayes. 1978. Evaluation of the mutagenic potential of mycotoxins using *Salmonella typhimurium* and *Saccharomyces cerevisiae*. Mut. Res. 53: 11–20.
349. Kurata, H. 1978. Current scope of mycotoxin research from the viewpoint of food mycology, p. 13–64. *In* K. Uraguchi and M. Yamazaki (ed.), Toxicology, biochemistry and pathology of mycotoxins, Kodansha, Tokyo.
350. Kurata. H., F. Sakabe, S. Udagawa, M. Ichinoe, M. Suzuki, and N. Takahashi. 1968. A mycological examination for the presence of mycotoxin-producers on the 1954–1967's stored rice grains. Bull. Nat. Inst. Hyg. Sci. No. 86: 183–187 (In Japanese).
351. Kurmanov, I.A. 1978a. Fusariotoxicosis in cattle and sheep, p. 85–110. *In* T.D. Wyllie and L.G. Morehouse (ed.), Mycotoxic fungi, mycotoxins, mycotoxicoses. An encyclopedic handbook, Vol. 2, Marcel Dekker, New York.
352. Kurmanov, I.A. 1978b. Fusariotoxicosis in chickens in the U.S.S.R., p. 322–326. *In* T.D. Wyllie and L.G. Morehouse (ed.), Mycotoxic fungi, mycotoxins, mycotoxicoses. An encyclopedic handbook, Vol. 2, Marcel Dekker, New York.
353. Kurobane, I., L.C. Vining, A.G. McInnes, and N.N. Gerber. 1980. Metabolites of *Fusarium solani* related to dihydrofusarubin. J. Antibiot. 33: 1376–1379.
354. Kuroda, H., T. Mori, C. Nishioka, H. Okasaki, and M. Takagi. 1979. Studies on gas chromatographic determination of trichothecene mycotoxins in food. J. Food Hyg. Soc. Japan. 20: 137–142 (In Japanese).
355. Kurtz, H.J., and C.J. Mirocha. 1978a. Zearalenone (F-2) induced estrogenic syndrome in swine, p. 256–268. *In* T.D. Wyllie and L.G. Morehouse (ed.), Mycotoxic fungi, mycotoxins, mycotoxicoses. An encyclopedic handbook, Vol. 2, Marcel Dekker, New York.
356. Kurtz, H.J., and C.J. Mirocha. 1978b. Refusal factor and emetic syndrome of swine, p. 265–268. *In* T.D. Wyllie and L.G. Morehouse (ed.), Mycotoxic fungi, mycotoxins, mycotoxicoses. An encyclopedic handbook. Vol. 2, Marcel Dekker, New York.
357. Kurtz,, H.J., J.E. Nairn, G.H. Nelson, C.M. Christensen, and C.J. Mirocha. 1969. Histologic changes in the genital tracts of swine fed estrogenic mycotoxin. Amer. J. Vet. Res. 30: 551–556.
358. Kvashnina, E.S. 1976. Physiological and ecological characteristics of *Fusarium* species of the Sporotrichiella section. Mikol. i Fitopatol. 10: 275–282 (In Russian).
359. Kvashnina, E.S. 1978. Physiological and ecological features of *Fusarium* species of the Sporotrichiella section. Bull. Vses. Ordena Lenina Inst. Eksp. Vet. 32: 42–45 (in Russian). English Abstr., Rev. Med. Vet. Mycol. 17: 111, 1982.
360. Kvashnina, E.S. 1979. Morphological and cultural characteristics of species of the genus *Fusarium,* section *Sporotrichiella,* and its distribution in the U.S.S.R. Mikol. i Fitpatol. 13: 3–10 (In Russian).
361. Kwatra, M.S., and A. Singh. 1973. Experimental reproduction of gangrenous syndrome in buffaloes (*Bos bubalus*). Zbl. Vet. Med. B., 20: 481–489.
362. Lacey, M.S. 1950. The antibiotic properties of fifty-two strains of *Fusarium*. J. Gen. Microbiol. 4: 122–131.
363. Lafarge-Frayssinet, C., G. Lespinats, P. Lafont, F. Loisillier, S. Mousset, Y. Rosenstein, and C. Frayssinet. 1979. Immunosuppressive effects of *Fusarium* extracts and trichothecenes: Blastogenic response of murine splenic and thymic cells to mitogens. Proc. Soc. Expt. Biol. Med. 160: 302–311.
364. Lafont, P., and J. Lafont. 1980. Contaminations du maïs par des mycotoxines. Bull. Acad. Vet. France 53: 533–538.
365. Lafont, P., C. Lafarge-Frayssinet, J. Lafont, G. Bertin, and C. Frayssinet. 1977. Metabolites toxiques de *Fusarium,* agents de l'aleucémie toxique alimentaire. Ann. Microbiol. (Inst. Pasteur) 128B: 215–220.

366. Lansden, J.A., R.J. Cole, J.W. Dorner, R.H. Cox, H.G. Cutler, and J.D. Clark. 1978. A new trichothecene mycotoxin isolated from *Fusarium tricinctum*. J. Agric. Food Chem. 26: 246–249.
367. Lasztity, R., and L. Wöller. 1975a. Toxinbildung bei *Fusarium*-Arten und Vorkommen der Toxine in landwirtschaftlichen Produkten. Die Nahrung 19: 537–546.
368. Lasztity, R., and L. Wöller. 1975b. Toxinerzeugung von Fusariumarten und ihr vorkommen in landwirtschaftlichen Produkten. Penodica Polytechnica Chem. Eng. 19: 249–262.
369. Lasztity, R., and L. Wöller. 1975c. Effects of zearalenone and some derivatives on animals fed on contaminated fodder. Acta Alim. 4: 189–197.
370. Lasztity, R., and L. Wöller. 1977. Investigation of the effect of zearalenone toxin produced by *Fusarium* fungi and its derivatives on animal organisms fed with infested fodder. Zeszyty Problemowe Postepow Nauk Rolniczych (Warsaw) 189: 193–200.
371. Lasztity, R., K. Tamas, and L. Wöller. 1977. Occurrence of *Fusarium* mycotoxins in some Hungarian corn crops and the possibilities of detoxication. Ann. Nutr. Alim. 31: 495–498.
372. Lee, S., and F.S. Chu. 1981. Radioimmunoassay of T-2 toxin in corn and wheat. J. Assoc. Off. Anal. Chem. 64: 156–161.
373. Lee, Y.-W., and C.J. Mirocha. 1981. Toxic components of *Fusarium roseum* isolated from overwintered barley. Phytopathology. 71: 235–236 (Abstr.).
374. Leonov, A.N. 1977. Current view of the chemical nature of factors responsible for alimentary toxic aleukia, p. 323–328. *In* J.V. Rodricks, C.W. Hesseltine, and M.A. Mehlman (ed.), Mycotoxins in human and animal health, Pathotox Publishers, Park Forest South, Illinois.
375. Lew, H. 1978. Zearalenon und Trichothecene. Veröff. Landwirtsch.—Chem. Bundesversuchsanst. Linz 11: 127–143.
376. Lew, H., E. Müllner, R. Hager, and M. Gregor. 1979. Fütterungsprobleme bei Mastschweinen verursacht durch fusarientoxinhältigen Mais. Bodenkultur 30: 309–316.
377. Li, M., S. Lu, C. Ji, M. Wang, S. Cheng, and C. Jin. 1979. Formation of carcinogenic N-nitroso compounds in corn-bread inoculated with fungi. Sci. Sin. 22: 471–477.
378. Li, M., S. Lu, C. Ji, Y. Wang, M. Wang, S. Cheng, and G. Tian. 1980. Experimental studies on the carcinogenicity of fungus-contaminated food from Linxian county, p. 139–148. *In* H.V. Gelboin (ed.), Genetic and environmental factors in experimental and human cancer, Japan Sci. Soc. Press, Tokyo.
379. Lin, P., and W. Tang. 1980. Zur Epidemiologie und Ätiologie des Oesophaguscarcinoms in China. J. Cancer Res. Clin. Oncol. 96: 121–170.
380. Lindenfelser, L.A., E.B. Lillehoj, and H.R. Burmeister. 1974. Aflatoxin and trichothecene toxins: skin tumour induction and synergistic acute toxicity in white mice. J. Nat. Cancer Inst. 52: 113–116.
381. Lindenfelser, L.A., A. Ciegler, and C.W. Hesseltine. 1978. Wild rice as fermentation substrate for mycotoxin production. Appl. Environ. Microbiol. 35: 105–108.
382. Linnainmaa, K., M. Sorsa, and T. Ilus. 1979. Epoxytrichothecene mycotoxins as c-mitotic agents in *Allium*. Hereditas 90: 151–156.
383. Lock, T.F. 1974. Leukoencephalomalacia in two quarter horses. Mod. Vet. Prac. 55: 464.
384. Loeffler, W., R. Mauli, M.E. Ruesch, and W. Stähelin. 1967. Monodeacetylanguidin. Ger. 1,233,098 (Abstr.). Chem. Abstr. 66: 7921, 84744u, 1967.
385. Lovelace, C.E.A., and C.B. Nyathi. 1977. Estimation of the fungal toxins, zearalenone and aflatoxin, contaminating opaque maize beer in Zambia. J. Sci. Food Agric. 28: 288–292.
386. Lu, S., M. Li, C. Ji, M. Wang, Y. Wang, and L. Huang. 1979. A new N-nitroso compound, N-3-methylbutyl-N-1-methylacetonylnitrosamine, in corn-bread inoculated with fungi. Sci. Sin. 22: 601–607.
387. Lu, S., Y. Wang, and M. Li. 1980a. Effects of fungi on the formation of carcinogenic nitrosamines and their precursors in food. Acta Acad. Med. Sin. 2: 24–28 (In Chinese).
388. Lu, S.H., A.M. Camus, C. Ji, Y.L. Wang, M.Y. Wang, and H. Bartsch. 1980b. Mutagenicity in *Salmonella typhimurium* of N-3-methylbutyl-N-1-methyl-acetonyl-nitrosamine and N-methyl-N-benzylnitrosamine, N-nitrosation products isolated from corn-bread contami-

nated with commonly occurring moulds in Linshien county, a high incidence area for oesophageal cancer in Northern China. Carcinogenesis 1: 867–870.
389. Lutsky, I., and N. Mor. 1981a. Experimental alimentary toxic aleukia in cats. Lab. Anim. Sci. 31: 43–47.
390. Lutsky, I.I., and N. Mor. 1981b. Alimentary toxic aleukia (septic angina, endemic panmyelotoxicosis, alimentary hemorrhagic aleukia). T-2 toxin-induced intoxication of cats. Amer. J. Pathol. 104: 189–191.
391. Lutsky, I., N. Mor, B. Yagen, and A.Z. Joffe, 1978. The role of T-2 toxin in experimental alimentary toxic aleukia: a toxicity study in cats. Toxicol. Appl. Pharmacol. 43: 111–124.
392. Macdonald, I.A., and R.H. Raemaekers. 1974. Some results of feeding tests with *Fusarium* and *Diplodia* diseased maize. Productive Farming (Zambia) October 1974: 42–44.
393. Maillet, M., and C. Bouton. 1969. Essai de mise en évidence d'un effet de type oestrogénique de l'acide gibberellique où l'uterus de ratte adulte castrée. Therapie 24: 497–508.
394. Mains, E.B., C.M. Vestal, and P.B. Curtis. 1930. Scab of small grains and feeding trouble in Indiana in 1928. Proc. Indiana Acad. Sci. 39: 101–110.
395. Mall, O.P., P. Mehta, S.C. Agrawal, and V.C. Vora. 1979. A survey of fungal contaminants of field crops with a view to evaluate their ability to produce mycotoxins: I. Fungal contaminants of wheat and barley harvested during spring, 1975. Indian J. Microbiol. 19: 118–122.
396. Mannio, M., and T.-M. Enari. 1973. Uber die Wirkung von *Fusarium*-Schimmelpilzen der Gerste bei der Bierherstellung. Brauwissenschaft 26: 134–137.
397. Manns, T.F., and J.F. Adams. 1923. Parasitic fungi internal of seed corn. J. Agric. Res. 23: 495.
398. Marasas, W.F.O. 1969. Moldy corn: Nutritive value, toxicity and mycoflora with special reference to *Fusarium tricinctum* (Corda) Snyder et Hansen. Ph.D. Thesis, Univ. of Wisconsin, Madison, Wisconsin. 228 p.
399. Marasas, W.F.O., and E.B. Smalley. 1972. Mycoflora, toxicity and nutritive value of mouldy maize. Onderstepoort J. Vet. Res. 39:1–10.
400. Marasas, W.F.O., E.B. Smalley, P.E. Degurse, J.R. Bamburg, and R.E. Nichols. 1967. Acute toxicity to rainbow trout (*Salmo gairdnerii*) of a metabolite produced by the fungus *Fusarium tricinctum*. Nature 214: 817–818.
401. Marasas, W.F.O., J.R. Bamburg, E.B. Smalley, F.M. Strong, W.L. Ragland, and P.E. Degurse. 1969. Toxic effects on trout, rats, and mice of T-2 toxin produced by the fungus *Fusarium tricinctum* (Cd.) Snyd. et Hans. Toxicol. Appl. Pharmacol. 15: 471–482.
402. Marasas, W.F.O., E.B. Smalley, J.R. Bamburg, and F.M. Strong. 1971. Phytotoxicity of T-2 toxin produced by *Fusarium tricinctum*. Phytopathology 61: 1488–1491.
403. Marasas, W.F.O., T.S. Kellerman, J.G. Pienaar, and T.W. Naudé. 1976. Leukoencephalomalacia: a mycotoxicosis of equidae caused by *Fusarium moniliforme* Sheldon. Onderstepoort J. Vet. Res. 43: 113–122.
404. Marasas, W.F.O., N.P.J. Kriek, S.J. van Rensburg, M. Steyn, and G.C. van Schalkwyk. 1977. Occurrence of zearalenone and deoxynivalenol, mycotoxins produced by *Fusarium graminearum* Schwabe, in maize in southern Africa. S. Afr. J. Sci. 73: 346–349.
405. Marasas, W.F.O., N.P.J. Kriek, M. Steyn, S.J. van Rensburg, and D.J. van Schalkwyk. 1978. Mycotoxicological investigations on Zambian maize. Food Cosmet. Toxicol. 16: 39–45.
406. Marasas, W.F.O., N.P.J. Kriek, V.M. Wiggins, P.S. Steyn, D.K. Towers, and T.J. Hastie. 1979a. Incidence, geographic distribution, and toxigenicity of *Fusarium* species in South African corn. Phytopathology 69: 1181–1185.
407. Marasas, W.F.O., L. Leistner, G. Hofmann, and C. Eckardt. 1979b. Occurrence of toxigenic strains of *Fusarium* in maize and barley in Germany. European J. Appl. Microbiol. Biotechnol. 7: 289–305.
408. Marasas, W.F.O., S.J. van Rensburg, and C.J. Mirocha. 1979c. Incidence of *Fusarium* species and the mycotoxins, deoxynivalenol and zearalenone, in corn produced in esophageal cancer areas in Transkei. J. Agric. Food Chem. 27: 1108–1112.

409. Marasas, W.F.O., F.C. Wehner, S.J. van Rensburg, and D.J. van Schalkwyk. 1981. Mycoflora of corn produced in human esophageal cancer areas in Transkei, southern Africa. Phytopathology 71: 792–796.
410. Marks, H.L., and C.W. Bacon. 1976. Influence of *Fusarium*-infected corn and F-2 on laying hens. Poult. Sci. 55: 1864–1870.
411. Martin, P.M.D. 1974. Fungi associated with common crops and crop products and their significance. S. Afr. Med. J. 48: 2374–2378.
412. Martin, P.M.D., and G.A. Gilman. 1976. A consideration of the mycotoxin hypothesis with special reference to the mycoflora of maize, sorghum, wheat and groundnuts. Rep. Trop. Prod. Inst. G10. Tropical Products Institute, London. 112 p.
413. Martin, P.M.D., and P. Keen. 1978. The occurrence of zearalenone in raw and fermented products from Swaziland and Lesotho. Sabouraudia 16: 15–22.
414. Martin, P.M.D., G.A. Gilman, and P. Keen. 1971. The incidence of fungi in foodstuffs and their significance, based on a survey in the eastern Transvaal and Swaziland, p. 281–290. *In* I.F.H. Purchase (ed.), Mycotoxins in human health, Macmillan, London.
415. Martin, W.J., V.C. Hasling, and E.A. Catalano. 1976. Ipomeamarone content in diseased and nondiseased tissues of sweet potatoes infected with different pathogens. Phytopathology 66: 678–679.
416. Masuko, M., Y. Ueno, M. Otokawa, and K. Yaginuma. 1977. The enhancing effect of T-2 toxin on delayed hypersensitivity in mice. Japan. J. Med. Sci. Biol. 30: 159–163.
417. Matsumoto, H., T. Ito, and Y. Ueno. 1978. Toxicological approaches to the metabolites of Fusaria. XII. Fate and distribution of T-2 toxin in mice. Japan. J. Exp. Med. 48: 393–399.
418. Matsuoka, Y., and K. Kubota. 1981. Studies on mechanisms of diarrhea induced by fusarenon-X, a trichothecene mycotoxin from *Fusarium* species. Toxicol. Appl. Pharmacol. 57: 293–301.
419. Matsuoka, Y., K. Kubota, and Y. Ueno. 1979. General pharmacological studies of fusarenon-X, a trichothecene mycotoxin from *Fusarium* species. Toxicol. Appl. Pharmacol. 50: 87–94.
420. Matta, R.J., and G.F. Wooten. 1973. Pharmacology of fusaric acid in man. Clin. Pharmacol. Therap. 14: 541–546.
421. Matthews, J.G., D.S.P. Patterson, B.A. Roberts, and B.J. Shreeve. 1977. T-2 toxin and haemorrhagic syndromes of cattle. Vet. Rec. 101: 391.
422. Matuo, T., T. Endo, and Y. Yoshii. 1976. Morphological and physiological characters of Japanese isolates of *Fusarium moniliforme* Sheldon. Trans. Mycol. Soc. Japan 17: 295–305 (In Japanese).
423. Mayer, C.F. 1953a. Endemic panmyelotoxicosis in the Russian grain belt. Part I. The clinical aspects of alimentary toxic aleukia (ATA). A comprehensive review. Milit. Surg. 113: 173–189.
424. Mayer, C.F. 1953b. Endemic panmyelotoxicosis in the Russian grain belt. Part II. The botany, phytopathology, and toxicology of Russian cereal food. Milit. Surg. 113: 295–315.
425. McErlean, B.A. 1952. Vulvovaginitis of swine. Vet. Rec. 64: 539–540.
426. McNutt, S.H., P. Purwin, and C. Murray. 1928. Vulvovaginitis in swine. Preliminary report. J. Amer. Vet. Med. Assoc. 73: 484–492.
427. Meronuck, R.A., K.H. Garren, C.M. Christensen, G.H. Nelson, and F. Bates. 1970. Effects on turkey poults and chicks of rations containing corn invaded by *Penicillium* and *Fusarium* species. Amer. J. Vet. Res. 31: 551–555.
428. Mess, B., Cs. Ruzsas, L. Wöller, and M. Biro-Gosztonyi. 1979. Alterations in reproductive functions of albino rats treated with a fungous toxin, zearalenone (F_2), in the adult age or during the neonatal period. Neuroendocrinol. Lett. 1: 1–15.
429. Messiaen, C.M., and R. Cassini. 1968. Recherches sur les fusarioses. IV. La systématique des *Fusarium*. Ann. Epiphyt. 19: 387–454.
430. Meyer, H., and H.K. Frank. 1979. Bibliographie der *Fusarium*-Toxine. Berichte der Bundesforschungsanstalt für Ernährung, Karlsruhe. 1979/2. 310 p.

431. Miessner, H., and G. Schoop. 1929. Uber den Pilzbefall amerikanischer "Giftgerste." Deutsch. Tierärztl. Wochenschr. 37: 167–170.
432. Miller, J.K., A. Hacking, J. Harrison, and V.J. Gross. 1973. Stillbirths, neonatal mortality and small litters in pigs associated with the ingestion of *Fusarium* toxin by pregnant sows. Vet. Rec. 93: 555–559.
433. Mills, J.T., and C.C. Frydman. 1980. Mycoflora and condition of grains from overwintered fields in Manitoba, 1977–78. Can. Plant Dis. Surv. 60: 1–7.
434. Mirocha, C.J. 1979. Trichothecene toxins produced by *Fusarium*, p. 289–373. *In* W. Shimoda (ed.), Conference on mycotoxins in animal feeds and grains related to animal health, Rockville, Maryland, June 8, 1979. PB-300 300. Report no. FDA/BVM-79/139. U.S. Dept. of Commerce, National Technical Information Service, Springfield, Virginia.
435. Mirocha, C.J., and C.M. Christensen. 1974. Oestrogenic mycotoxins synthesized by *Fusarium*, p. 129–148. *In* I.F.H. Purchase (ed.), Mycotoxins, Elsevier, Amsterdam.
436. Mirocha, C.J., and S. Pathre. 1973. Identification of the toxic principle in a sample of poaefusarin. Appl. Microbiol. 26: 719–724.
437. Mirocha, C.J., and S.V. Pathre. 1979. Mycotoxins—Their biosynthesis in fungi: zearalenone biosynthesis. J. Food Prot. 42: 821–824.
438. Mirocha, C.J., C.M. Christensen, and G.H. Nelson. 1967a. Estrogenic metabolite produced by *Fusarium graminearum* in stored corn. Appl. Microbiol. 15: 497–503.
439. Mirocha, C.J., C.M. Christensen, and G.H. Nelson. 1967b. An estrogenic metabolite produced by *Fusarium graminearum* in stored corn, p. 119–130. *In* R.I. Mateles and G.N. Wogan (ed.), Biochemistry of some foodborne toxins, MIT Press, Cambridge, Massachusetts.
440. Mirocha, C.J., C.M. Christensen, and G.H. Nelson. 1968a. Toxic metabolites produced by fungi implicated in mycotoxicoses. Biotech. Bioeng. 10: 469–482.
441. Mirocha, C.J., C.M. Christensen, and G.H. Nelson. 1968b. Physiologic activity of some fungal estrogens produced by *Fusarium*. Cancer. Res. 28: 2319–2322.
442. Mirocha, C.J., J. Harrison, A.A. Nichols, and M. McClintock. 1968c. Detection of a fungal estrogen (F-2) in hay associated with infertility in dairy cattle. Appl. Microbiol. 16: 797–798.
443. Mirocha, C.J., C.M. Christensen, and G.H. Nelson. 1969. Biosynthesis of the fungal estrogen F-2 and a naturally occurring derivative (F-3) by *Fusarium moniliforme*. Appl. Microbiol. 17: 482–483.
444. Mirocha, C.J., C.M. Christensen, and G.H. Nelson. 1971. F-2 (zearalenone) estrogenic mycotoxin from *Fusarium*, p. 107–138. *In* S. Kadis, A. Ciegler, and S.J. Ajl (ed.), Microbial toxins, Vol. VII., Academic Press, New York.
445. Mirocha, C.J., B. Schauerhamer, and S.V. Pathre. 1974. Isolation, detection, and quantitation of zearalenone in maize and barley. J. Assoc. Off. Anal. Chem. 57: 1104–1110.
446. Mirocha, C.J., S.V. Pathre, B. Schauerhamer, and C.M. Christensen. 1976. Natural occurrence of *Fusarium* toxins in feedstuff. Appl. Environ. Microbiol. 32: 553–556.
447. Mirocha, C.J., S.V. Pathre, and C.M. Christensen. 1977a. Zearalenone, p. 345–364. *In* J.V. Rodricks, C.W. Hesseltine, and M.A. Mehlman (ed.), Mycotoxins in human and animal health, Pathotox Publishers, Park Forest South, Illinois.
448. Mirocha, C.J., S.V. Pathre, and C.M. Christensen. 1977b. Chemistry of *Fusarium* and *Stachybotrys* mycotoxins, p. 365–420. *In* T.D. Wyllie and L.G. Morehouse (ed.), Mycotoxic fungi, mycotoxins, mycotoxicoses. An encyclopedic handbook, Vol. 1, Marcel Dekker, New York.
449. Mirocha, C.J., S.V. Pathre, J. Behrens, and B. Schauerhamer. 1978. Uterotropic activity of *cis* and *trans* isomers of zearalenone and zearalenol. Appl. Environ. Microbiol. 35: 986–987.
450. Mirocha, C.J., B. Schauerhamer, C.M. Christensen, and T. Kommedahl. 1979a. Zearalenone, deoxynivalenol, and T-2 toxin associated with stalk rot in corn. Appl. Environ. Microbiol. 38: 557–558.

451. Mirocha, C.J., B. Schauerhamer, C.M. Christensen, M.L. Niku-Paavola, and M. Nummi. 1979b. Incidence of zearalenol (*Fusarium* mycotoxin) in animal feed. Appl. Environ. Microbiol. 38: 749-750.
452. Mirocha, C.J., S.V. Pathre, and T.S. Robison. 1981. Comparative metabolism of zearalenone and transmission into bovine milk. Food Cosmet. Toxicol. 19: 25-30.
453. Mitchell, H.H., and J.R. Beadles. 1940. The impairment in nutritive value of corn grain damaged by specific fungi. J. Agric. Res. 61: 135-142.
454. Mitton, A., J.C. Collet, J. Szymanski, and R. Gousse. 1975. Avortements dans un élevage ovin et presence de zearalenone dans l'alimentation. Rev. Med. Vet. 126: 813-820.
455. Möller, J.M., A. Thalmann, and M. Hausmann. 1978. Uber das Vorkommen von Fusarien und Zearalenon in Futtermitteln. Landwirtsch. Forsch. 31: 38-44.
456. Monina, M.I., E.A. Moscotena, J. Ruager, J.R. Idiart, E.H. Reinoso, A. Muro, E.O. Nosetto, and E.R. Pons. 1981. Leucoencefalomalacia equina. Casos registrados en el pais. Rev. Milit. Vet. 28: 13-17.
457. Moreau, C. 1972. Mycotoxicose chez des vaches laitières liée au développement du *Fusarium roseum* var. *culmorum* sur l'herbe d'une prairie. Comp. Rend. Acad. Agric. France 58: 383-387.
458. Moreau, C. 1974. Trois cas de paralysie chez des porcs et des volailles vraisemblablement liés à une action toxique du *Fusarium moniliforme* Sheld. Bull. Soc. Mycol. France 90: 201-208.
459. Moreau, C. 1979. Troubles nerveux et digestifs liés à la consommation, par les animaux, d'aliments contaminés par des *Aspergillus, Penicillium* et *Fusarium*. Rev. Mycol. 43: 227-238.
460. Morooka, N., and T. Tatsuno. 1970. Toxic substances (fusarenon and nivalenol) produced by *Fusarium nivale*, p. 114-119. In M. Herzberg (ed.), Toxic micro-organisms, UJNR Joint Panels on toxic micro-organisms and the U.S. Department of Interior, Washington, D.C.
461. Morooka, N., N. Nakano, S. Nakazawa, and H. Tsunoda. 1971. On the chemical properties of fusarenon and the related compound obtained from toxic metabolites of *Fusarium nivale*. Nippon Nogeikagaku Kaishi 45: 151-155 (In Japanese).
462. Morooka, N., N. Uratsuji, T. Yoshizawa, and H. Yamamoto. 1972. Studies on the toxic substances in barley infected with *Fusarium* spp. J. Food Hyg. Soc. Japan 13: 368-375 (In Japanese).
463. Morooka, N., M. Ichinoe, H. Oku, M. Mayama, and K. Kitani. 1980. Occurrence of *Fusarium* disease and detection of mycotoxins in grains of wheat and barley in Kyushu and Shikoku area in 1978. Proc. Japan. Assoc. Mycotoxicol. No. 10: 19-26 (In Japanese).
464. Mower, R.L., W.C. Snyder, and J.G. Hancock. 1975. Biological control of ergot by *Fusarium*. Phytopathology 65: 5-10.
465. Müller, H.-M., and M. Thaler. 1981. Propionsäurekonservierung von Körnermais nach Zuimpfung von Schimmelpilzen und Hefen. Arch. Tierernährung 31: 789-799.
466. Mundkur, B.B. 1934. Some preliminary feeding experiments with scabby barley. Phytopathology 24: 1237-1243.
467. Mutert, W.-U., H. Lütfring, W. Barz, and D. Strack. 1981. Formation of fusaric acid by fungi of the genus *Fusarium*. Z. Naturforsch. Sect. C. Biol. Sci. 36: 338-339.
468. Nagao, M., M. Honda, T. Hamasaki, S. Natori, Y. Uneo, M. Yamasaki, Y. Seino, T. Yahagi, and T. Sugimura. 1976. Mutagenicities of mycotoxins on *Salmonella*. Proc. Japan. Assoc. Mycotoxicol. 3/4: 41-43 (In Japanese).
469. Naik, D.M., L.V. Busch, and G.L. Barron. 1978. Influence of temperature on the strain of *Fusarium graminearum* Schwabe in zearalenone production. Can. J. Plant Sci. 58: 1095-1097.
470. Neish, G.A., and H. Cohen. 1981. Vomitoxin and zearalenone production by *Fusarium graminearum* from winter wheat and barley in Ontario. Can. J. Plant Sci. 61: 811-815.
471. Neish, G.A., and M. Leggett. 1981. *Fusarium moniliforme* var. *intermedium*, a new variety in the Liseola section. Can. J. Bot. 59: 288-291.

472. Nelson, G.H., C.M. Christensen, and C.J. Mirocha. 1973. *Fusarium* and estrogenism in swine. J. Amer. Vet. Med. Assoc. 163: 1276–1277.
473. Nelson, R.R., C.J. Mirocha, D. Huisingh, and A. Tigerina-Menchaca. 1968. Effects of F-2, an estrogenic metabolite from *Fusarium,* on sexual reproduction of certain Ascomycetes. Phytopathology 58: 1061–1062.
474. Nelson, P.E., T.A. Toussoun, and W.F.O. Marasas. 1983. *Fusarium* species: an illustrated manual for identification. The Pennsylvania State University Press, University Park, Pennsylvania, 203 p.
475. Nesterov, A.I. 1964. The clinical course of Kashin-Beck disease. Arthritis and Rheumatism 7: 29–40.
476. Niku-Paavola, M.-L., and M. Nummi. 1977. Research into toxic metabolites of fungi. Kemia—Kemi 4: 151–153 (In Finnish).
477. Niku-Paavola, M.-L., T. Ilus, P.J. Ward, and M. Nummi. 1977. Thin layer analysis of *Fusarium* toxins in grain samples. Arch. Inst. Pasteur Tunis 3–4: 264–278.
478. Nirenberg, H. 1976. Untersuchungen über die morphologische und biologische Differenzierung in der *Fusarium* Section Liseola. Mitt. Biol. Bundesanst. Land-Forstwirtsch. Berlin-Dahlem 169: 1–117.
479. Noller, C.H., M. Stob, and J. Tuite. 1979. Effects of feeding *Gibberella zeae*-infected corn on feed intake, body weight gain, and milk production of dairy cows. J. Dairy Sci. 62: 1003–1006.
480. Norppa, H., M. Penttilä, M. Sorsa, E.-L. Hintikka, and T. Ilus. 1980. Mycotoxin T-2 of *Fusarium tricinctum* and chromosome changes in Chinese hamster bone marrow. Hereditas 93: 329–332.
481. Norton, J.B.S., and C.C. Chen. 1920. Another corn seed parasite. Science (New York) 52: 250–251.
482. Novakovskaya, Y.S. 1960. Changes in the nitrogenous compounds in a nutrient medium, in proteins of rye and wheat under the influence of some strains of the fungus *Fusarium sporotrichiella* (in connection with the determination of the etiology of Kashin-Beck disease), p. 37–50. In V.I. Bilai (ed.), Mycotoxicoses of man and agricultural animals, Kiev, 167 p. English translation, JPRS 7434, U.S. Joint Publications Research Service, U.S. Dept. of Commerce, Washington D.C.
483. Nummi, M. 1977. Recent studies on *Fusarium* toxins. Zeszyty Problemowe Postepow Nauk Rolniczych 189: 189–192.
484. Nummi, M., M.-L. Niku-Paavola, and T.-M. Enari. 1975a. Der Einfluss eines *Fusarium*-Toxins auf die Gersten-Vermälzung. Brauwissenschaft 28: 130–133.
485. Nummi, M., M.-L. Niku-Paavola, and T.-M. Enari. 1975b. Effect of a *Fusarium*-toxin on the development of α-amylase activity in germinating barley. BIOS 7–8: 266–269.
486. Nusrath, M. 1979. Behaviour of toxic substances produced by two isolates of *Fusarium oxysporum* Schlecht. from chick pea and pigeon pea. Curr. Trends Life Sci. (New Delhi) 7: 283–295.
487. Ohta, M., K. Ishii, and Y. Ueno. 1977. Metabolism of trichothecene mycotoxins. I. Microsomal deacetylation of T-2 toxin in animal tissues. J. Biochem. 82: 1591–1598.
488. Ohta, M., H. Matsumoto, K. Ishii, and Y. Ueno. 1978. Metabolism of trichothecene mycotoxins. II. Substrate specificity of microsomal deacetylation of trichothecenes. J. Biochem. 84: 697–706.
489. Ohtsubo, K., and M. Saito. 1977. Chronic effects of trichothecenes, p. 255–262. In J.V. Rodricks, C.W. Hesseltine, and M.A. Mehlman (ed.), Mycotoxins in human and animal health, Pathotox Publishers, Park Forest South, Illinois.
490. Olifson, L.E. 1960. The chemical activity of some species of fungi which affect the grain of cereals, p. 58–66. In V.I. Bilai (ed.), Mycotoxicoses of man and agricultural animals, Kiev, 167 p. English translation, JPRS 7434, U.S. Joint Publications Research Service, U.S. Dept of Commerce, Washington, D.C.
491. Olifson, L.E. 1965. Chemical and biological characteristics of toxic materials derived from

grain infected with the fungus *Fusarium sporotrichiella*. Dissertation, Doctor of Biological Sciences, Technological Institute of Food Industry, Moscow (In Russian).
492. Olifson, L.E., S.M. Kenina, and V.L. Kartashova. 1972. Chromatographic method to identify toxicity of grain of grass cultures (wheat, rye, millet and others) affected by toxigen strain of fungi *Fusarium sporotrichiella*. Orenburg Regional Administration of D.I. Mendelieev, All-Union Chem. Assoc., p. 3–8 (In Russian).
493. Olifson, L.E., S.M. Kenina, V.L. Kartashova, and K.G. Galhovich. 1975. Chromatographic determination of toxicity of the grain affected by *Fusarium sporotrichiella*. Vopr. Pit. 1975: 83–86 (In Russian).
494. Olifson, L.E., S.M. Kenina, and V.L. Kartashova. 1978. Examination of the fractional composition of the lipid complex of cereals infected by the microscopic fungus *Fusarium sporotrichiella* Bilai. Prikl. Biokh. Mikrobiol. 14: 630–634 (In Russian). English translation in Appl. Biochem. Microbiol. 14: 488–491, 1979.
495. Otokawa, M., Y. Shibahara, and Y. Egashira. 1979. The inhibitory effect of T-2 toxin on tolerance induction of delayed-type hypersensitivity in mice. Japan. J. Med. Sci. Biod. 32: 37–45.
496. Ozegovic, L. 1970. Mould-maize poisoning in swine (F-2 zearalenone-fusarium toxicosis). Veterinaria 19: 525–531 (In Yugoslavian).
497. Palmisano, F., A. Visconti, A. Bottalico, P. Lerario, and P.G. Zambonin. 1981. Differential-pulse polarography of trichothecene toxins: detection of deoxynivalenol in corn. Analyst 106: 992–998.
498. Palti, J. 1978. Toxigenic Fusaria, their distribution and significance as causes of disease in animals and man. Acta Phytomedica 6. Paul Parey, Berlin. 110 p.
499. Palyusik, M. 1973. Experimental vulvooedema of swine (fusariotoxicosis caused by *Fusarium graminearum*). Mag. Allat. Lapja 1973, No. 6: 297–303 (In Hungarian).
500. Palyusik, M., and E. Koplik-Kovacs. 1975a. The effect of feed containing T_2 and F_2 toxins on laying geese. Mag. Allat. Lapja 1975, No. 12: 842–847 (In Hungarian).
501. Palyusik, M., and E. Koplik-Kovacs. 1975b. Effect on laying geese of feeds containing the fusariotoxins T-2 and F-2. Acta Vet. Acad. Sci. Hung. 25: 363–368.
502. Palyusik, M., I. Szep, and F. Szoke. 1968. Data on susceptibility to mycotoxins on day-old goslings. Acta Vet. Acad. Sci. Hung. 18: 363–372.
503. Palyusik, M., E. Koplik-Kovacs, and E. Guzsal. 1971. Effect of *Fusarium graminearum* on the semen production in ganders and turkeycocks. Mag. Allat. Lapja 1971, No. 6: 300–303 (In Hungarian).
504. Palyusik, M., G. Nagy, and L. Zöldag. 1974. The effect of different *Fusarium* species on the spermiogenesis in ganders. Mag. Allat. Lapja 1974, No. 8: 551–553 (In Hungarian).
505. Palyusik, M., W.M. Hagler, L. Horvath, and C.J. Mirocha. 1980. Biotransformation of zearalenone to zearalenol by *Candida tropicalis*. Acta. Vet. Acad. Sci. Hung. 28: 159–166.
506. Pan, Y.G. 1981. Report of an investigation on mouldy maize toxicosis in horses. Liaoning Xumu Shouyi 1981, No. 4: 10–12 (In Chinese).
507. Pathre, S.V., and C.J. Mirocha. 1977. Assay methods for trichothecenes and review of their natural occurrence, p. 229–253. In J.V. Rodricks, C.W. Hesseltine, and M.A. Mehlman (ed.), Mycotoxins in human and animal health, Pathotox Publishers, Park Forest South, Illinois.
508. Pathre, S.V., and C.J. Mirocha. 1978. Analysis of deoxynivalenol from cultures of *Fusarium* species. Appl. Environ. Microbiol. 35: 992–994.
509. Pathre, S.V., and C.J. Mirocha. 1979. Trichothecenes: natural occurrence and potential hazard. J. Amer. Oil Chem. Soc. 56: 820–823.
510. Pathre, S.V., and C.J. Mirocha. 1980. Mycotoxins as estrogens. Develop. Toxicol. Environ. Sci. 5: 265–278.
511. Pathre, S.V., C.J. Mirocha, C.M. Christensen, and J. Behrens. 1976. Monoacetoxyscirpenol. A new mycotoxin produced by *Fusarium roseum* Gibbosum. J. Agric. Food Chem. 24: 97–103.

512. Pathre, S.V., S.W. Fenton, and C.J. Mirocha. 1980. 3'-Hydroxyzearalenones, two new metabolites produced by *Fusarium roseum*. J. Agric. Food Chem. 28: 421–424.
513. Patterson, D.S.P., J.G. Matthews, B.J. Shreeve, B.A. Roberts, S.M. McDonald, and A.W. Hayes. 1979. The failure of trichothecene mycotoxins and whole cultures of *Fusarium tricinctum* to cause experimental haemorrhagic syndromes in calves and pigs. Vet. Rec. 105: 252–255.
514. Patterson, D.S.P., B.J. Shreeve, and B.A. Roberts. 1980. Mycotoxin residues in body fluids and tissues of food-producing animals. Zentralbl. Bakt. Parasit. Infek. Hyg. 1980: 321–328.
515. Payen, J., P. Lafont, R.M. Parache, and F. Boller. 1978. Detection des souches toxiques de *Fusarium*. Collection Med. Leg. Toxicol. Med. 107: 111–117.
516. Pearson, A.W. 1978. Biochemical changes produced by *Fusarium* T-2 toxin in the chicken. Res. Vet. Sci. 24: 92–97.
517. Peck, H.M., S.E. McKinney, A. Tytell, and B.B. Byham. 1957. Toxicologic evaluation of gibberellic acid. Science 126: 1064–1065.
518. Peckham, J.C., F.E. Mitchell, O.H. Jones, Jr, and B. Doupnik, Jr. 1972. Atypical interstitial pneumonia in cattle fed moldy sweet potatoes. J. Amer. Vet. Med. Assoc. 160: 169–172.
519. Perkel, N.V. 1957. Toxicity of certain forms of *Fusarium sporotrichiella* extracted from East Siberian grain. Vopr. Pit. 16: 64–69 (In Russian).
520. Perkel, N.V. 1960. Study of the toxicity of Transbaikal strains of *Fusarium sporotrichiella* Bilai in connection with the etiology of Urov disease (Kashin-Beck disease), p. 117–125. In V.I. Bilai (ed.), Mycotoxicoses of man and agricultural animals, Kiev, 167 p. English translation, JPRS 7434, U.S. Joint Publications Research Service, U.S. Dept of Commerce, Washington, D.C.
521. Peters, A.T. 1904. A fungus disease in corn. Agric. Exp. Stn. Nebraska, 17th Ann. Rep., p. 13–22.
522. Petrie, L., J. Robb, and A.F. Stewart. 1977. The identification of T-2 toxin and its association with a haemorrhagic syndrome in cattle. Vet. Rec. 101: 326.
523. Pienaar, J.G. 1977. Nuwere veterinêre neuropatologiese toestande in Suid-Afrika. J. S. Afr. Vet. Assoc. 48: 13–18.
524. Pienaar, J.G., T.S. Kellerman, and W.F.O Marasas. 1981. Field outbreaks of leukoencephalomalacia in horses consuming maize infected by *Fusarium verticillioides* (= *F. moniliforme*) in South Africa. J. S. Afr. Vet. Assoc. 52: 21–24.
525. Pier, A.C., S.J. Cysewski, J.L. Richard, A.L. Baetz, and L. Mitchell. 1976. Experimental mycotoxicosis in calves with aflatoxin, ochratoxin, rubratoxin, and T-2 toxin. Proc. 80th Ann. Meeting U.S. Anim. Health Assoc., p. 130–148.
526. Piglionica, V., A. Bottalico, and A. Siniscalco. 1976. Importance of *Fusarium moniliforme* Sheld. within kernels of maize (*Zea mays* L.), in Italy. Agric. Conspec. Sci. 39: 161–165.
527. Polzhofer, K., and M. Niehuss. 1980. Nachweis von Fusarientoxinen in Kartoffeln über den Hühnerembryonen Test. Z. Lebensm. Unters.-Forsch. 170: 124–128.
528. Pozuelo, J. 1976. Suppression of craving and withdrawal in humans addicted to narcotics or amphetamines by administration of alpha-methyl-para-tyrosine (AMPT) and 5-butylpicolinic acid (fusaric acid). Cleveland Clin. Quart. 43: 89–94.
529. Prentice, N., and A.D. Dickson. 1968. Emetic material associated with *Fusarium* species in cereal grains and artificial media. Biotechnol. Bioeng. 10: 413–427.
530. Prentice, N., and W. Sloey. 1960. Studies on barley microflora of possible importance to malting and brewing quality. I. The treatment of barley during malting with selected microorganisms. Proc. Amer. Soc. Brew. Chem. 1960: 28–33.
531. Prentice, N., A.D. Dickson, and J.G. Dickson. 1959. Production of emetic material by species of *Fusarium*. Nature (London) 184: 1319.
532. Puls, R., and J.A. Greenway. 1976. Fusariotoxicosis from barley in British Columbia. II. Analysis and toxicity of suspected barley. Can. J. Comp. Mcd. 40: 16–19.
533. Qin, Q., B.-S. Zhang, Z.-S. Wang, G. Wang, Z. Fu, and R. Wang. 1981. Studies on the "sore foot disease" of cattle. Acta Vet. Zootech. Sin. 12: 137–143 (In Chinese).

534. Rabie, C.J., S.J. van Rensburg, J.J. van der Watt, and A. Lübben. 1975. Onyalai—the possible involvement of a mycotoxin produced by *Phoma sorghina* in the aetiology. S. Afr. Med. J. 49: 1647–1650.
535. Rabie, C.J., A. Lübben, A.I. Louw, E.B. Rathbone, P.S. Steyn, and R. Vleggaar. 1978. Moniliformin, a mycotoxin from *Fusarium fusarioides*. J. Agr. Food Chem. 26: 375–379.
536. Renault, L., M. Goujet, A. Monin, G. Boutin, M. Palisse, and A. Alamagny. 1979. Suspicion de mycotoxicose provoquée par les trichothécènes chez les poulets de chair. Bull. Acad. Vét. France 52: 181–187.
537. Ribelin, W.E. 1978. Trichothecene toxicosis in cattle, p. 36–45. *In* T.D. Wyllie and L.G. Morehouse (ed.), Mycotoxic fungi, mycotoxins, mycotoxicoses. An encyclopedic handbook, Vol. 2, Marcel Dekker, New York.
538. Richard, J.L., L.H. Tiffany, and A.C. Pier. 1969. Toxigenic fungi associated with stored corn. Mycopath. Mycol. Appl. 38: 313–326.
539. Richard, J.L., S.J. Cysewski, A.C. Pier, and G.D. Booth. 1978. Comparison of effects of dietary T-2 toxin on growth, immunogenic organs, antibody formation, and pathologic changes in turkeys and chickens. Amer. J. Vet. Res. 39: 1674–1679.
540. Ripperger, H., K. Seifert, A. Römer, and J. Rullkötter. 1975. Isolierung von Diacetoxyscirpenol aus *Fusarium solani* var. *coeruleum*. Phytochemistry 14: 2298–2299.
541. Rodriquez, J.A. 1945. Differenciacion entre la enfermedad de los rastrojos y la meningo encefalomielitis infecciosa de los equinas. Anal. Soc. Rur. Argentina 69: 305–307.
542. Roine, K., E.-L. Korpinen, and K. Kallela. 1971. Mycotoxicosis as a probable cause of infertility in dairy cows. A case report. Nord. Vet. Med. 23: 628–633.
543. Rosenstein, Y., P. Lafont, and G. Frayssinet-Lafarge. 1978. Effets immunosuppresseurs des toxines de *Fusarium*. Collection Med. Leg. Toxicol. Med. 107: 51–57.
544. Rosenstein, Y., C. Lafarge-Frayssinet, G. Lespinats, F. Loisillier, P. Lafont, and C. Frayssinet. 1979. Immunosuppressive activity of *Fusarium* toxins. Effects on antibody synthesis and skin grafts of crude extracts, T-2 toxin and diacetoxyscirpenol. Immunology 36: 111–117.
545. Rosenstein, Y., R.R. Kretschmer, and C. Lafarge-Frayssinet. 1981. Effect of *Fusarium* toxins, T2-toxin and diacetoxyscirpenol, on murine T-independent immune responses. Immunology 44: 555–560.
546. Rubinstein, Yu.I., 1953. The etiology of Urov (Kashin-Beck) disease. Vopr. Pit. 12: 73–81 (In Russian).
547. Rubinstein, Yu.I. 1960a. The effect of the cultivation conditions on toxin formation in *Fusarium sporotrichiella* Bilai, p. 66–73. *In* V.I. Bilai (ed.), Mycotoxicoses of man and agricultural animals, Kiev, 167 p. English translation, JPRS 7434, U.S. Joint Publications Research Service, U.S. Dept of Commerce, Washington, D.C.
548. Rubinstein, Yu.I. 1960b. Food fusariotoxicoses, p. 89–110. *In* V.I. Bilai (ed.), Mycotoxicoses of man and agricultural animals, Kiev, 167 p. English translation, JPRS 7434, U.S. Joint Publications Research Service, U.S. Dept. of Commerce, Washington, D.C.
549. Rubinstein, Yu.I., and L.S. Lyass. 1948. On the etiology of alimentary toxic aleukia. Gigiena i Sanitariia 7: 33–38 (In Russian).
550. Rubinstein, Yu.I., Y.P. Kukel, and G.P. Kudinova. 1967. Papillomatosis with hyperkeratosis of the proventriculus in rats following their feeding on grain infested with the fungus *Fusarium sporotrichiella* Bilai var. *poae* (Pk.) Bilai. Vopr. Pit. 26: 57–61 (In Russian).
551. Ruddick, J.A., P.M. Scott, and J. Harwig. 1976. Teratological evaluation of zearalenone administered orally to the rat. Bull. Environ. Contam. Toxicol. 15: 678–681.
552. Ruhr, L.P., G.D. Osweiler, and C.W. Foley. 1978. The effect of zearalenone on fertility in the boar. Amer. Assoc. Vet. Lab. Diagnosticians, 21st. Ann. Proc. 1978: 127–134.
553. Rukmini, C., and R.V. Bhat. 1978. Occurrence of T-2 toxin in *Fusarium*-infected sorghum from India. J. Agric. Food Chem. 26: 647–649.
554. Rukmini, C., J.S. Prasad, and K. Rao. 1980. Effects of feeding T-2 toxin to rats and monkeys. Food Cosmet. Toxicol. 18: 267–269.
555. Ruzsas, C., B. Mess, M. Biro-Gosztonyi, and L. Wöller. 1978. Effect of pre- and perinatal

administration of the fungous F2 toxin on the reproduction of the albino rat, p. 57–60. *In* G. Dörner and M, Kawakami (ed.), Hormones and brain development, Elsevier, Amsterdam.
556. Ruzsas, C., J. Biro-Gosztonyi, L. Wöller, and B. Mess. 1979. Effect of the fungal toxin (zearalenone) on the reproductive system and fertility of male and female rats. Acta Biol. Acad. Sci. Hung. 30: 335–345.
557. Saccardo, P.A. 1881. Fungi Italici Autographice Delinati, Fig. 789. Patavii.
558. Saito, M., and K. Ohtsubo. 1974. Trichothecene toxins of *Fusarium* species, p. 263–281. *In* I.F.H. Purchase (ed.), Mycotoxins, Elsevier, Amsterdam.
559. Saito, M., and K. Okubo. 1970. Studies on the target injuries in experimental animals with mycotoxins of *Fusarium nivale,* p. 82–93. *In* M. Herzberg (ed.), Toxic micro-organisms, UJNR Joint panels on toxic micro-organisms and the U.S. Department of Interior, Washington, D.C.
560. Saito, M., and T. Tatsuno. 1971. Toxins of *Fusarium nivale,* p. 293–316. *In* S. Kadis, A. Ciegler, and S.J. Ajl (ed.), Microbial toxins, Vol. VII, Academic Press, New York.
561. Saito, M., M. Enomoto, and T. Tatsuno. 1969. Radiomimetic biological properties of the new scirpene metabolites of *Fusarium nivale.* GANN 60: 599–603.
562. Saito, M., T. Horiuchi, K. Ohtsubo, Y. Hatanaka, and Y. Ueno. 1980. Low tumor incidence in rats with long-term feeding of fusarenon-X, a cytotoxic trichothecene produced by *Fusarium nivale.* Japan. J. Exp. Med. 50: 293–302.
563. Salazar, S., H. Fromentin, and F. Mariat. 1980. Action du diacetoxyscirpenol sur la candidose experimentale de la souris. Compt. Rend. Acad. Sci. Paris 290: 877–878.
564. Sandor, G., A. Vanyi, and A. Petrie. 1980. Effect of irradiation on the viability and toxin production of different fungus species. Acta Vet. Acad. Sci. Hung. 28: 361–369.
565. Sarkisov, A.K. 1960. General characterization of the alimentary mycotoxicoses of agricultural animals, p. 155–163. *In* V.I. Bilai (ed.), Mycotoxicoses of man and agricultural animals, Kiev, 167 p. English translation, JPRS 7434, U.S. Joint Publications Research Service, U.S. Dept of Commerce, Washington, D.C.
566. Sato, N., and R. Amano. 1976. Herstellung von T-2 toxin unter Einsatz von *Fusarium solani* in ruhender kultur. Die Fleischwirtschaft 56: 1354–1355.
567. Sato, N., and Y. Ueno. 1977. Comparative toxicities of trichothecenes, p. 295–321. *In* J.V. Rodricks, C.W. Hesseltine, and M.A. Mehlman (ed.), Mycotoxins in human and animal health, Pathotox Publishers, Park Forest South, Illinois.
568. Sato, N., Y. Ueno, and M. Enomoto. 1975. Toxicological approaches to the toxic metabolites of *Fusaria.* VIII. Acute and subacute toxicities of T-2 toxin in cats. Japan. J. Pharmacol. 25: 263–270.
569. Sato, N., T. Ito, H. Kumada, Y. Ueno, K. Asano, M. Saito, K. Ohtsubo, I. Ueno, and Y. Hatanaka. 1978. Toxicological approaches to the metabolites of Fusaria. XIII. Hematological changes in mice by a single and repeated administrations of trichothecenes. J. Toxicol. Sci. 3: 335–356.
570. Scharf, H.D., and H. Frauenrath. 1980. The mycotoxin "moniliformin" and related substances, p. 101–119. *In* Conference: Oxocarbons, Academic Press, New York.
571. Schmidt, R., E. Ziegenhagen, and K. Dose. 1981a. High-performance liquid chromatography of trichothecenes. I. Detection of T-2 toxin and HT-2 toxin. J. Chromatogr. 212: 370–373.
572. Schmidt, R., A. Bieger, E. Ziegenhagen, and K. Dose. 1981b. Bestimmung von T-2-Toxin in pflanzlichen Nahrungsmitteln. I. T-2-Toxin in verschimmeltem Reis und Mais. Fresenius Z. Anal. Chem. 308: 133–136.
573. Schoental, R. 1974. Role of podophyllotoxin in the bedding and dietary zearalenone on incidence of spontaneous tumors in laboratory animals. Cancer Res. 34: 2419–2420.
574. Schoental, R. 1975. Mycotoxicoses "by proxy." Intern. J. Environ. Studies 8: 171–175.
575. Schoental, R. 1977a. Health hazards due to T-2 toxins. Vet. Rec. 101: 473–474.
576. Schoental, R. 1977b. The role of nicotinamide and of certain other modifying factors in diethylnitrosamine carcinogenesis. Fusaria mycotoxins and "spontaneous" tumors in animals and man. Cancer 40: 1833–1840.

577. Schoental, R. 1978. Mycotoxins in food and the variations in tumor incidence among laboratory rodents. Nutr. Cancer 1: 13–14.
578. Schoental, R. 1979a. *Fusarium* mycotoxins in the aetiology of neonatal abnormalities, cardiovascular and sexual disorders and tumours in man and animals. Toxicon 17: 164.
579. Schoental, R. 1979b. The role of *Fusarium* mycotoxins in the aetiology of tumours of the digestive tract and of certain other organs in man and animals. Front. Gastrointest. Res. 4: 17–24.
580. Schoental, R. 1980a. Cancer of the digestive tract and mycotoxins of *Fusarium* species. The Lancet, 13 September 1980: 593.
581. Schoental, R. 1980b. Relationships of *Fusarium* mycotoxins to disorders and tumors associated with alcoholic drinks. Nutr. Cancer 2: 88–92.
582. Schoental, R. 1980c. Mouldy grain and the aetiology of pellagra: the role of toxic metabolites of *Fusarium*. Biochem. Soc. Trans. 8: 147–150.
583. Schoental, R. 1981a. Relationships of *Fusarium* toxins to tumours and other disorders in livestock. J. Vet. Pharmacol. Therap. 4: 1–6.
584. Schoental, R. 1981b. Mycotoxins and fetal abnormalities. Intern. J. Environ. Studies 17: 25–29.
585. Schoental, R. 1981c. *Fusarium* mycotoxins and the effects of high-fat diets. Nutr. Cancer 3: 57–62.
586. Schoental, R. 1981d. Carcinogenic mycotoxins, p. 109–148. *In* R. Schoental and T.A. Connors (ed.), Dietary influences on cancer: traditional and modern. CRC Press, Boca Raton, Florida.
587. Schoental, R. 1981e. Carcinogenic contaminants of food, p. 149–173. *In* R. Schoental and T.A. Connors (ed.), Dietary influences on cancer: tradition and modern. CRC Press, Boca Raton, Florida.
588. Schoental, R., and A.Z. Joffe. 1974. Lesions induced in rodents by extracts from cultures of *Fusarium poae* and *F. sporotrichioides*. J. Pathol. 112: 37–42.
589. Schoental, R., A.Z. Joffe, and B. Yagen. 1976. Chronic lesions in rats treated with crude extracts of *Fusarium poae* and *F. sporotrichioides*. The role of mouldy food in the incidence of oesophageal, mammary and certain other abnormalities and tumours in livestock and man. Br. J. Cancer 34: 310 (Abstr.).
590. Schoental, R., A.Z. Joffe, and B. Yagen. 1978a. The induction of tumours of the digestive tract and of certain other organs in rats given T-2 toxin, a secondary metabolite of *Fusarium sporotrichioides*. Br. J. Cancer 38: 171 (Abstr.).
591. Schoental, R., A.Z. Joffe, and B. Yagen. 1978b. Irreversible depigmentation of dark mouse hair by T-2 toxin (a metabolite of *Fusarium sporotrichioides*) and by calcium pantothenate. Experientia 34: 763–764.
592. Schoental, R., A.Z. Joffe, and B. Yagen. 1979. Cardiovascular lesions and various tumors found in rats given T-2 toxin, a trichothecene metabolite of *Fusarium*. Cancer Res. 39: 2179–2189.
593. Schroeder, H.W., and H. Hein, Jr. 1975. A note on zearalenone in grain sorghum. Cereal Chem. 52: 751–752.
594. Schwarte, L.H., H.E. Biester, and C. Murray. 1937. A disease of horses caused by feeding moldy corn. J. Amer. Vet. Med. Assoc. 90: 76–85.
595. Scott, D. 1965. Toxigenic fungi isolated from cereal and legume products. Mycopathol. Mycol. Appl. 25: 213–222.
596. Scott, P.M. 1978. Mycotoxins in feeds and ingredients and their origin. J. Food Prot. 41: 385–398.
597. Scott, P.M., J.W. Lawrence, and W. van Walbeek. 1970. Detection of mycotoxins by thin-layer chromatography: application to screening of fungal extracts. Appl. Microbiol. 20: 839–842.
598. Scott, P.M., W. van Walbeek, B. Kennedy, and D. Anyeti. 1972. Mycotoxins (ochratoxin A,

citrinin, and sterigmatocystin) and toxigenic fungi in grains and other agricultural products. J. Agric. Food Chem. 20: 1103–1109.
599. Scott, P.M., T. Panalaks, S. Kanhere, and W.F. Miles. 1978. Determination of zearalenone in cornflakes and other corn-based foods by thin layer chromatography, high pressure liquid chromatography, and gas-liquid chromatography/high resolution mass spectrometry. J. Assoc. Off. Anal. Chem. 61: 593–600.
600. Scott, P.M., J. Harwig, and B.J. Blanchfield. 1980. Screening *Fusarium* strains isolated from overwintered Canadian grains for trichothecenes. Mycopathologia 72: 175–180.
601. Scott, P.M., P.-Y. Lau, and S.R. Kanhere. 1981. Gas chromatography with electron capture and mass spectrometric detection of deoxynivalenol in wheat and other grains. J. Assoc. Off. Anal. Chem. 64: 1364–1371.
602. Seagrave, S. 1981. Yellow rain: a journey through the terror of chemical warfare. M. Evans and Company, New York. 316 p.
603. Seemüller, E. 1968. Untersuchungen über die morphologische und biologische Differenzierung in der *Fusarium* Sektion Sporotrichiella. Mitt. Biol. Bundesanst. Land.-Forstwirtsch. Berlin-Dahlem 127: 1–93.
604. Shands, R.G. 1937. Longevity of *Gibberella saubinetii* and other fungi in barley kernels and its relation to the emetic effect. Phytopathology 27: 749–762.
605. Shannon, G.M., O.L. Shotwell, A.L. Lyons, D.G. White, and G. Garcia-Aguirre. 1980. Laboratory screening for zearalenone formation in corn hybrids and inbreds. J. Assoc. Off. Anal. Chem. 63: 1275–1277.
606. Shapovalov, M. 1917. Intoxicating bread. Phytopathology 7: 384–386.
607. Sharby, T.F., G.E. Templeton, J.N. Beasley, and E.L. Stephenson. 1973. Toxicity resulting from feeding experimentally molded corn to broiler chicks. Poult. Sci. 52: 1007–1014.
608. Sharda, D.P., R.F. Wilson, L.E. Williams, L.A. Swiger, and R.F. Cross. 1971. Mold toxicity in swine and laboratory animals: effect of feeding corn inoculated with pure cultures of *Fusarium roseum* Ohio isolate C. J. Anim. Sci. 32: 1169–1173.
609. Sharma, V.D., R.F. Wilson, and L.E. Williams. 1974. Reproductive performance of female swine fed corn naturally molded or inoculated with *Fusarium roseum,* Ohio isolates B and C. J. Anim. Sci. 38: 598–602.
610. Sheldon, J.L. 1904. A corn mold (*Fusarium moniliforme* n. sp.). Agric. Exp. Stn. Nebraska, 17th Ann. Rep., p. 23–32.
611. Sheridan, J.J. 1980. Some observations on selected mycoses and mycotoxicoses affecting animals in Ireland. Irish Vet. J. 34: 148–154.
612. Sherwood, R.F., and J.F. Peberdy. 1972. Factors affecting the production of zearalenone by *Fusarium graminearum* in grain. J. Stored Prod. Res. 8: 71–75.
613. Sherwood, R.F., and J.F. Peberdy. 1974a. Production of the mycotoxin, zearalenone, by *Fusarium graminearum* growing on stored grain. I. Grain storage at reduced temperatures. J. Sci. Food Agric. 25: 1081–1087.
614. Sherwood, R.F., and J.F. Peberdy. 1974b. Production of the mycotoxin, zearalenone, by *Fusarium graminearum* growing on stored grain. II. Treatment of wheat grain with organic acids. J. Sci. Food Agric. 25: 1089–1093.
615. Shimizu, T., N. Nakano, T. Matsui, and K. Aibara. 1979. Hypoglycemia in mice administered with fusarenon-X. Japan. J. Med. Sci. Biol. 32: 189–198.
616. Shotwell, O.L. 1977. Assay methods for zearalenone and its natural occurrence, p. 403–413. In R.V. Rodricks, C.W. Hesseltine, and M.A. Mehlman (ed.), Mycotoxins in human and animal health, Pathotox Publishers, Park Forest South, Illinois.
617. Shotwell, O.L., C.W. Hesseltine, M.L. Goulden, and E.E. Vandegraft. 1970. Survey of corn for aflatoxin, zearalenone and ochratoxin. Cereal Chem. 47: 700–707.
618. Shotwell, O.L., C.W. Hesseltine, E.E. Vandegraft, and M.L. Goulden. 1971. Survey of corn from different regions for aflatoxin, ochratoxin, and zearalenone. Cereal Sci. Today 16: 266–273.

619. Shotwell, O.L., M.L. Goulden, G.A. Bennet, R.D. Plattner, and C.W. Hesseltine. 1977. Survey of 1975 wheat and soybeans for aflatoxin, zearalenone, and ochratoxin. J. Assoc. Off. Anal. Chem. 60: 778–783.
620. Shreeve, B.J., D.S.P. Patterson, and B.A. Roberts. 1975. Investigation of suspected cases of mycotoxicosis in farm animals in Britain. Vet. Rec. 97: 275–278.
621. Shreeve, B.J., D.S.P. Patterson, B.A. Roberts, and A.E. Wrathall. 1978. Effect of mouldy feed containing zearalenone on pregnant sows. Br. Vet. J. 134: 421–427.
622. Siegfried, R. 1977a. *Fusarium*-Toxine. Naturwissenschaften 64: 274.
623. Siegfried, R. 1977b. Fusariumtoxine (Trichothecentoxine) in Futtermais. Landwirtsch. Forsch. Sonderheft 34: 37–43.
624. Siegfried, R., and E. Langerfeld. 1978. Vorläufige Untersuchungen über die Produktion von Toxinen durch Fäuleerreger bei Kartoffeln. Potato Res. 21: 335–339.
625. Sigg, H.P., R. Mauli, E. Flury, and D. Hauser. 1965. Die Konstitution von Diacetoxyscirpenol. Helv. Chim. Acta 48: 962–988.
626. Siriwardana, T.M.G., and P. Lafont. 1978. New sensitive biological assay for 12,13-epoxy-trichothecenes. Appl. Environ. Microbiol. 35: 206–207.
627. Sloey, W., and N. Prentice. 1962. Effects of *Fusarium* isolates applied during malting on properties of malt. Proc. Amer. Soc. Brew. Chem. 1962: 24–29.
628. Smalley, E.B. 1973. T-2 toxin. J. Amer. Vet. Med. Assoc. 163: 1278–1281.
629. Smalley, E.B., and F.M. Strong. 1974. Toxic trichothecenes, p. 199–228. *In* I.F.H. Purchase (ed.), Mycotoxins, Elsevier, Amsterdam.
630. Smalley, E.B., W.F.O. Marasas, F.M. Strong, J.R. Bamburg, R.E. Nichols, and N.R. Kosuri. 1970. Mycotoxicoses associated with moldy corn, p. 163–173. *In* M. Herzberg (ed.), Toxic micro-organisms, UJNR Joint panels on toxic micro-organisms and the U.S. Department of the Interior, Washington, D.C.
631. Smalley, E., A. Joffe, M. Palyusik, H. Kurata, and W. Marasas. 1977. Panel on trichothecene toxins, p. 337–340. *In* J.V. Rodricks, C.W. Hesseltine, and M.A. Mehlman (ed.), Mycotoxins in human and animal health, Pathotox Publishers, Park Forest South, Illinois.
632. Smirnoff, W.A. 1970. Fungus diseases affecting *Adelges piceae* in the fir forest of the Gaspé Peninsula, Quebec. Can. Ent. 102: 799–805.
633. Smith, T.K. 1980. Influence of dietary fiber, protein and zeolite on zearalenone toxicosis in rats and swine. J. Anim. Sci. 50: 278–285.
634. Smith, T.K. 1981. Effect of dietary calcium, phosphorus and vitamin D on zearalenone toxicosis in rats. Can. J. Anim. Sci. 61: 191–197.
635. Smith, A.J., and J.W. Wells. 1978. The source of androgenic activity in the African wood *Funtumia latifolia:* a steroid hormone formed by the action of *Fusarium solani*. J. Sci. Food Agric. 29: 783–787.
636. Smith, R.H., R. Palmer, and A.E. Reade. 1975. A chemical and biological assessment of *Aspergillus oryzae* and other filamentous fungi as protein sources for simple stomached animals. J. Sci. Food Agric. 26: 785–795.
637. Snyder, W.C., and H.N. Hansen. 1945. The species concept in *Fusarium* with reference to Discolor and other sections. Amer. J. Bot. 32: 657–666.
638. Snyder, W.C., H.N. Hansen, and J.W. Oswald. 1957. Cultivars of the fungus, *Fusarium*. J. Madras Univ. B, 27: 185–192.
639. Sorsa, M., K. Linnainmaa, M. Penttilä, and T. Ilus. 1980. Evaluation of the mutagenicity of epoxytrichothecene mycotoxins in *Drosophila melanogaster*. Hereditas 92: 163–165.
640. Speers, G.M., R.A. Meronuck, D.M. Barnes, and C.J. Mirocha. 1971. Effect of feeding *Fusarium roseum* f. sp. *graminearum* contaminated corn and the mycotoxin F-2 on the growing chick and laying hen. Poult. Sci. 50: 627–633.
641. Speers, G.M., C.J. Mirocha, and C.M. Christensen. 1972. Effects of *Fusarium* mycotoxins on laying hens. Poult. Sci. 51: 1872 (Abstr.).
642. Speers, G.M., C.J. Mirocha, C.M. Christensen, and J.C. Behrens. 1977. Effects on laying

hens of feeding corn invaded by two species of *Fusarium* and pure T-2 mycotoxin. Poult. Sci. 56: 98–102.
643. Springer, J.P., J. Clardy, R.J. Cole, J.W. Kirksey, R.K. Hill, R.M. Carlson, and J.L. Isidor. 1974. Structure and synthesis of moniliformin, a novel cyclobutane microbial toxin. J. Amer. Chem. Soc. 96: 2267–2268.
644. Stähelin, H., M.E. Kalberer-Rüsch, E. Signer, and S. Lazary. 1968. Uber enige biologische Wirkungen des Cytostaticum Diacetoxyscirpenol. Arzneimittelforschung 18: 989–994.
645. Steele, J.A., J.R. Lieberman, and C.J. Mirocha. 1974. Biogenesis of zearalenone (F-2) by *Fusarium roseum* 'Graminearum.' Can. J. Microbiol. 20: 531–534.
646. Steele, J.A., C.J. Mirocha, and S.V. Pathre. 1976. Metabolism of zearalenone by *Fusarium roseum* Graminearum. J. Agric. Food Chem. 24: 89–97.
647. Steele, J.A., C.J. Mirocha, and S.V. Pathre. 1977a. Metabolism of zearalenone by *Fusarium roseum*. J. Toxicol. Environ. Health 3: 35–42.
648. Steele, J.A., C.J. Mirocha, and S.V. Pathre. 1977b. Metabolism of zearalenone by *Fusarium roseum*, p. 393–401. *In* J.V. Rodricks, C.W. Hesseltine, and M.A. Mehlman (ed.), Mycotoxins in human and animal health, Pathotox Publishers, Park Forest South, Illinois.
649. Steyn, D.G. 1933. Fungi in relation to health in man and animals. Onderstepoort J. Vet. Sci. Anim. Ind. 1: 183–212.
650. Steyn, D.G. 1950. Recent investigations into the toxicity of known and unknown poisonous plants in the Union of South Africa. XVI. Onderstepoort J. Vet. Sci. Anim. Ind. 24: 53–56.
651. Steyn, M., P.G. Thiel, and G.C. Van Schalkwyk. 1978. Isolation and purification of moniliformin. J. Assoc. Off. Anal. Chem. 61: 578–580.
652. Steyn, P.S., R. Vleggaar, C.J. Rabie, N.P.J. Kriek, and J.S. Harington. 1978. Trichothecene mycotoxins from *Fusarium sulphureum*. Phytochemistry 17: 949–951.
653. Steyn, P.S., P.L. Wessels, and W.F.O. Marasas. 1979. Pigments from *Fusarium moniliforme* Sheldon. Structure and ^{13}C nuclear magnetic resonance assignments of an azaanthraquinone and three naphthoquinones. Tetrahedron 35: 1551–1555.
654. Stipanovic, R.D., and H.W. Schroeder. 1975. Zearalenone and 8'-hydroxyzearalenone from *Fusarium roseum*. Mycopathologia 57: 77–78.
655. Stob, J., R.S. Baldwin, J. Tuite, F.N. Andrews, and K.G. Gillette. 1962. Isolation of an anabolic, uterotrophic compound from corn infected with *Gibberella zeae*. Nature 196: 1318.
656. Stodola, F.H., K.B. Raper, D.I. Fennell, H.F. Conway, V.E. Sohns, C.T. Langford, and R.W. Jackson. 1955. The microbiological production of gibberellins A and X. Arch. Biochem. Biophys. 54: 240–245.
657. Stoll, C. 1954. Uber Stoffwechsel und biologisch wirksame Stoffe von *Gibberella fujikuroi* (Saw.) Woll., dem Erreger der Bakanaëkrankheit. Phytopathol. Z. 22: 233–274.
658. Stoll, C., and J. Renz. 1957. Uber den Fusarinsäure- und Dehydrofusarinsäurestoffwechsel von *Gibberella fujikuroi* (Saw.) Woll. Phytopathol. Z. 29: 380–387.
659. Stoloff, L., S. Henry, and O.J. Francis. 1976. Survey for aflatoxins and zearalenone in 1973 crop corn stored on farms and in country elevators. J. Assoc. Off. Annal. Chem. 59: 118–121.
660. Stuart, B.P., R.J. Cole, E.F. Waller, and R.F. Vesonder. 1981. Proventricular hyperplasia (malabsorption syndrome) in broilers: involvement of mycotoxins and other dietary factors. 118th Ann. Meeting, Amer. Vet. Med. Assoc., p. 107 (Abstr.).
661. Surico, G., and A. Graniti. 1977. Produzione di tossine da *Fusarium oxysporum* Schl. f. sp. *albedinis*. Phytopath. Medit. 16: 30–33.
662. Sutton, J.C., W. Baliko, and H.S. Funnell. 1976. Evidence for translocation of zearalenone in corn plants colonized by *Fusarium graminearum*. Can. J. Plant Sci. 56: 7–12.
663. Sutton, J.C., W. Baliko, and H.S. Funnell. 1980a. Relation of weather variables to incidence of zearalenone in corn in southern Ontario. Can. J. Plant Sci. 60: 149–155.
664. Sutton, J.C., W. Baliko, and H.J. Liu. 1980b. Fungal colonization and zearalenone accumulation in maize ears injured by birds. Can. J. Plant Sci. 60: 453–461.

665. Suzuki, T., Y. Hoshino, M. Kurisu, N. Nose, and A. Watanabe. 1978. Gas chromatographic determination of zearalenone in the culture of genus *Fusarium* and cereal. J. Food Hyg. Soc. Japan 19: 201–207 (In Japanese).
666. Suzuki, T., M. Kurisu, Y. Hoshino, M. Ichinoe, N. Nose, Y. Tokumaru, and A. Watanabe. 1980. Production of trichothecene mycotoxins of *Fusarium* species in wheat and barley harvested in Saitama Prefecture. J. Food Hyg. Soc. Japan 21: 43–49 (In Japanese).
667. Suzuki, T., M. Kurisu, N. Nose, and A. Watanabe. 1981a. Determination of butenolide (*Fusarium* toxin) by gas chromatography with an electron-capture detector. J. Food Hyg. Soc. Japan 22: 197–202 (In Japanese).
668. Suzuki, T., M. Kurisu, and M. Ichinoe. 1981b. Trichothecenes-producing fungi of *Fusarium* species. Proc. Jap. Assoc. Mycotoxicol. 13: 34–36 (In Japanese).
669. Szathmary, Cs.I., C.J. Mirocha, M. Palyusik, and S.V. Pathre. 1976. Identification of mycotoxins produced by species of *Fusarium* and *Stachybotrys* obtained from eastern Europe. Appl. Environ. Microbiol. 32: 579–584.
670. Szathmary, Cs., C.J. Mirocha, M. Palyusik, and S.V. Pathre. 1977. Chemical and biologic studies on the toxins of various *Fusarium* and *Stachybotrys* strains. Mag. Allat. Lapja. 1977, No. 7: 455–459 (In Hungarian).
671. Szathmary, Cs., J. Galacz, L. Vida, and G. Alexander. 1980. Capillary gas chromatographic-mass spectrometric determination of some mycotoxins causing fusariotoxicosis in animals. J. Chrom. 191: 327–331.
672. Takatori, K., K. Sakamoto, K. Ohkubo, T. Konishi, and K. Yumino. 1980. Distribution of *Fusarium* among the feeds and the producibility of *Fusarium* toxins. Japan. J. Zootech. Sci. 51: 325–330 (In Japanese).
673. Tanaka, K., M. Minamisawa, M., Manabe, and S. Matuura. 1975. Biological test using the brine shrimp (Part 1). The influence of mycotoxins on the brine shrimp. Rep. Nat. Food Res. Inst. (Tokyo) 30: 43–48 (In Japanese).
674. Tatsuno, T. 1968. Toxicologic research on substances from *Fusarium nivale*. Cancer Res. 28: 2393–2396.
675. Tatsuno, T. 1976. Recherches chimiques et biologiques sur les mycotoxins des *Fusariums*. Ann. Pharm. France 34: 25–29.
676. Tatsuno, T., M. Saito, M. Enomoto, and H. Tsunoda. 1968. Nivalenol, a toxic principle of *Fusarium nivale*. Chem. Pharm. Bull. 16: 2519–2520.
677. Tatsuno, T., Y. Fujimoto, and Y. Morita. 1969. Toxicological research on substances from *Fusarium nivale* III. The structure of nivalenol and its monoacetate. Tetrahedron Lett. 33: 2823–2826.
678. Tatsuno, T., Y. Morita, H. Tsunoda, and M. Umeda. 1970. Recherches toxicologiques des substances métaboliques du *Fusarium nivale* VII. La troisième substance metabolique de *F. nivale*, le diacetate de nivalenol. Chem. Pharm. Bull. 18: 1485–1487.
679. Tatsuno, T., K. Ohtsubo, and M. Saito. 1973. Chemical and biological detection of 12,13-epoxytrichothecenes isolated from *Fusarium* species. Pure Appl. Chem. 35: 309–313.
680. Thaler, M., U. July, H. Kuppinger, H.-M. Müller, and R. Scherer. 1979. Die Belüftungstrockung von Körnermais unter mykotoxikologischem Aspekt. Getreide, Mehl und Brot 33: 8–11.
681. Thiel, P.G. 1978. A molecular mechanism for the toxic action of moniliformin, a mycotoxin produced by *Fusarium moniliforme*. Biochem. Pharmacol. 27: 483–486.
682. Tidd, B.K. 1967. Phytotoxic compounds produced by *Fusarium equiseti*. Part III. Nuclear magnetic resonance spectra. J. Chem. Soc. (C) 1967: 218–220.
683. Tiraboschi, C. 1905. Sopra alcuni ifomiceti del mais guasto di regioni pellagrose. Ann. di Bot. 2: 137–168.
684. Tookey, H.L., S.G. Yates, J.J. Ellis, M.D. Grove, and R.E. Nichols. 1972. Toxic effects of a butenolide mycotoxin and of *Fusarium tricinctum* cultures in cattle. J. Amer. Vet. Med. Assoc. 160: 1522–1526.
685. Trenholm, H.L., W.P. Cochrane, H. Cohen, J.I. Elliot, E.R. Farnworth, D.W. Friend, R.M.G.

Hamilton, G.A. Neish, and J.F. Standish. 1981. Survey of vomitoxin contamination of the 1980 white winter wheat crop in Ontario, Canada. J. Assoc. Off. Anal. Chem. 58: 992A–994A.
686. Tsunoda, H. 1970. Micro-organisms which deteriorate stored cereals and grains, p. 143–162. In M. Herzberg (ed.), Toxic micro-organisms, UJNR Joint panels on toxic micro-organisms and the U.S. Department of the Interior, Washington, D.C.
687. Tuite, J., G. Shaner, G. Rambo, J. Foster, and R.W. Caldwell. 1974. The *Gibberella* ear rot epidemics of corn in Indiana in 1965 and 1972. Cereal. Sci. Today 19: 238–241.
688. Turner, W.B. 1971. Fungal metabolites. Academic Press, New York, 446 p.
689. Tuttobello, L., C.O. Zavattiero, and A. Macri. 1974. Ricerca di sostanze ad attivita uterotropa con colture diverse di *Fusarium*. Atti. Soc. Ital. Sci. Vet. 28: 624–629.
690. Udagawa, S., M. Ichinoe, and H. Kurata. 1970. Occurrence and distribution of mycotoxin producers in Japanese foods, p. 174–184. In M. Herzberg (ed.), Toxic micro-organisms, UJNR Joint panels on toxic micro-organisms and the U.S. Department of Interior, Washington, D.C.
691. Ueda, A., T. Ono, and S. Yamagiwa. 1967. Neurohistopathological studies on bean-hulls poisoning of horses. Res. Bull. Obihiro Univ. 5: 149–154 (In Japanese).
692. Ueno, Y. 1971. Toxicological and biological properties of fusarenon-X, a cytotoxic mycotoxin of *Fusarium nivale* Fn-2B, p. 163–178. In I.F.H. Purchase (ed.), Mycotoxins in human health, Macmillan, London.
693. Ueno, Y. 1973a. Akakabi toxins (*Fusarium* toxins) Part 1. J. Food Hyg. Soc. Japan 14: 403–414 (In Japanese).
694. Ueno, Y., 1973b. Akakabi toxins (*Fusarium* toxins) Part 2. J. Food Hyg. Soc. Japan 14: 501–510 (In Japanese).
695. Ueno, Y. 1977a. Trichothecenes: overview address, p. 189–207. In J.V. Rodricks, C.W. Hesseltine, and M.A. Mehlman (ed.), Mycotoxins in human and animal health, Pathotox Publishers, Park Forest South, Illinois.
696. Ueno, Y., 1977b. Mode of action of trichothecenes. Pure Appl. Chem. 49: 1737–1745.
697. Ueno, Y. 1980. Trichothecene mycotoxins, mycology, chemistry, and toxicology, p. 301–353. In H.H. Draper (ed.), Advances in nutritional research, Vol. 3, Plenum Publishing Corporation, New York.
698. Ueno, Y., and K. Fukushima. 1968. Inhibition of protein and DNA synthesis in Ehrlich ascites tumour by nivalenol, a toxic principle of *Fusarium nivale*–growing rice. Experientia 24: 1032–1033.
699. Ueno, Y., and K. Kubota. 1976. DNA-attacking ability of carcinogenic mycotoxins in recombination deficient mutant cells of *Bacillus subtilis*. Cancer Res. 36: 445–451.
700. Ueno, Y., and N. Shimada. 1974. Reconfirmation of the specific nature of reticulocytes bioassay system to trichothec mycotoxins of *Fusarium* spp. Chem. Pharm. Bull. 22: 2744–2746.
701. Ueno, Y., and F. Tashiro. 1981. α-Zearalenol, a major hepatic metabolite in rats of zearalenone, an estrogenic mycotoxin of *Fusarium* species. J. Biochem. 89: 563–571.
702. Ueno, Y., I. Ueno, T. Tatsuno, K. Ohokubo, and H. Tsunoda. 1969a. Fusarenon-X, a toxic principle of *Fusarium nivale*–culture filtrate. Experientia 25: 1062.
703. Ueno, Y., M. Hosoya, and Y. Ishikawa. 1969b. Inhibitory effects of mycotoxins on the protein synthesis in rabbit reticulocytes. J. Biochem. 66: 419–422.
704. Ueno, Y., Y. Ishikawa, K. Saito-Amakai, and H. Tsunoda. 1970a. Environmental factors influencing the production of fusarenon-X, a cytotoxic mycotoxin of *Fusarium nivale* Fn 2B. Chem. Pharm. Bull. 18: 304–312.
705. Ueno, Y., Y. Ishikawa, K. Amakai, M. Nakajima, M. Saito, M. Enomoto, and K. Ohtsubo. 1970b. Comparative study of skin-necrotizing effect of scirpene metabolites of Fusaria. Japan. J. Exp. Med. 40: 33–38.
706. Ueno, Y., K. Saito, and H. Tsunoda. 1970. Isolation of toxic principles from the culture filtrate of *Fusarium nivale*, p. 120. In M. Herzberg (ed.), Toxic micro-organisms, UJNR

Joint panels on toxic micro-organisms and the U.S. Department of Interior, Washington, D.C.

707. Ueno, Y., Y. Ishikawa, M. Nakajima, K. Sakai, K. Ishii, H. Tsunoda, M. Saito, M. Enomoto, K. Ohtsubo, and M. Umeda. 1971a. Toxicological approaches to the metabolites of *Fusaria*. I. Screening of toxic strains. Japan. J. Exp. Med. 41: 257–272.

708. Ueno, Y., I. Ueno, K. Amakai, Y. Ishikawa, H. Tsunoda, K. Okubo, M. Saito, and M. Enomoto. 1971b. Toxicological approaches to the metabolites of *Fusaria*. II. Isolation of fusarenon-X from the culture filtrate of *Fusarium nivale* Fn 2B. Japan. J. Exp. Med. 41: 507–519.

709. Ueno, Y., I. Ueno, Y. Iitoi, H. Tsunoda, M. Enomoto, and K. Ohtsubo. 1971c. Toxicological approaches to the metabolites of *Fusaria*. III. Acute toxicity of fusarenon-X. Japan. J. Exp. Med. 41: 521–539.

710. Ueno, Y., K. Ishii, K. Sakai, S. Kanaeda, H. Tsunoda, T. Tanaka, and M. Enomoto. 1972a. Toxicological approaches to the metabolites of *Fusaria*. IV. Microbial survey on "Bean-hulls poisoning of horses" with the isolation of toxic trichothecenes, neosolaniol and T-2 toxin, from *Fusarium solani* M-1-1. Japan. J. Exp. Med. 42: 187–203.

711. Ueno, Y., N. Sato, K. Ishii, K. Sakai, and M. Enomoto. 1972b. Toxicological approaches to the metabolites of *Fusaria*. V. Neosolaniol, T-2 toxin and butenolide, toxic metabolites of *Fusarium sporotrichioides* NRRL 3510 and *Fusarium poae* 3287. Japan. J. Exp. Med. 42: 461–472.

712. Ueno, Y., N. Sato, K. Ishii, K. Sakai, H. Tsunoda, and M. Enomoto. 1973a. Biological and chemical detection of trichothecene mycotoxins of *Fusarium* species. Appl. Microbiol. 25: 699–704.

713. Ueno, Y., M. Nakajima, K. Sakai, K. Ishii, N. Sato, and M. Shimada. 1973b. Comparative toxicology of trichothec mycotoxins: inhibition of protein synthesis in animal cells. J. Biochem. 74: 285–296.

714. Ueno, Y., K. Ishii, N. Sato, and K. Ohtsubo. 1974a. Toxicological approaches to the metabolites of *Fusaria*. VI. Vomiting factor from moldy corn infected with *Fusarium spp.* Japan. J. Exp. Med. 44: 123–127.

715. Ueno, Y., N. Shimada, S. Yagasaki, and M. Enomoto. 1974b. Toxicological approaches to the metabolites of *Fusaria*. VII. Effects of zearalenone on the uteri of mice and rats. Chem. Pharm. Bull. 22: 2830–2835.

716. Ueno, Y., M. Sawano, and K. Ishii. 1975. Production of trichothecene mycotoxins by *Fusarium* species in shake culture. Appl. Microbiol. 30: 4–9.

717. Ueno, Y., S. Ayaki, N. Sato, and T. Ito. 1977a. Fate and mode of action of zearalenone. Ann. Nutr. Alim. 31: 935–948.

718. Ueno, Y., K. Ishii, M. Sawano, K. Ohtsubo, Y. Matsuda, T. Tanaka, H. Kurata, and M. Ichinoe. 1977b. Toxicological approaches to the metabolites of *Fusaria*. XI. Trichothecenes and zearalenone from *Fusarium* species isolated from river sediments. Japan. J. Exp. Med. 43: 177–184.

719. Ueno, Y., K. Kubota, T. Ito, and Y. Nakamura. 1978. Mutagenicity of carcinogenic mycotoxins in *Salmonella typhimurium*. Cancer Res. 38: 536–542.

720. Ullstrup, A.J. 1936. The occurrence of *Gibberella fujikuroi* var. *subglutinans* in the United States. Phytopathology 26: 685–693.

721. Uraguchi, K. 1971. Pharmacology of mycotoxins, p. 143–299. *In* G. Peters (ed.), International encyclopedia of pharmacology and therapeutics, Vol. II, Section 71, Pergamon Press, Oxford.

722. Urry, W.H., H.L. Wehrmeister, E.B. Hodge, and P.H. Hidy. 1966. The structure of zearalenone. Tetrahedron Lett. No. 27: 3109–3114.

723. Van der Walt, S.J., and D.G. Steyn. 1943. Recent investigations into the toxicity of plants, etc., XIII. Onderstepoort J. Vet. Sci. and Anim. Ind. 18: 207–224.

724. Van Rensburg, S.J., I.F.H. Purchase, and J.J. van der Watt. 1971. Hepatic and renal pathology induced in mice by feeding fungal cultures, p. 153–161. *In* I.F.H. Purchase (ed.), Mycotoxins in human health. Macmillan, London.

725. Vanyi, A., and A. Szeky. 1980. Fusariotoxicoses. VII. Disturbed spermatogenesis caused by zearalenone (F-2 Fusariotoxin) and by imperfect illumination in guinea-cocks. Mag. Allat. Lapja 35: 247-252 (In Hungarian).
726. Vanyi, A., A. Szeky, and E. Romvaryne-Szailer. 1974. Fusariotoxicoses. V. The effect of F2 toxin on the sexual activity of female swine. Mag. Allat. Lapja 1974, No. 11: 723-730 (In Hungarian).
727. Verdeal, K., and D.S. Ryan. 1979. Naturally-occurring estrogens in plant foodstuffs—a review. J. Food Prot. 42: 577-583.
728. Vesela, D., D. Vesely, and A. Adamkova. 1981. The occurrence of zearalenone and zearalenone-producing Fusaria in feeds. Vet. Med. (Praha), 26: 737-741 (In Czechoslovakian).
729. Vesonder, R.F., and A. Ciegler. 1979. Natural occurrence of vomitoxin in Austrian and Canadian corn. European J. Appl. Microbiol. Biotechnol. 8: 237-240.
730. Vesonder, R.F., and C.W. Hesseltine. 1981. Metabolites of *Fusarium*, p. 350-364. In P.E. Nelson, T.A. Toussoun, and R.J. Cook (ed.), *Fusarium*. Diseases, biology and taxonomy. The Pennsylvania State University Press, University Park, Pennsylvania.
731. Vesonder, R.F., A. Ciegler, and A.H. Jensen. 1973. Isolation of the emetic principle from *Fusarium*-infected corn. Appl. Microbiol. 26: 1008-1010.
732. Vesonder, R.F., A. Ciegler, A.H. Jensen, W.H. Rohwedder, and D. Weisleder. 1976. Co-identity of the refusal and emetic principle from *Fusarium*-infected corn. Appl. Environ. Microbiol. 31: 280-285.
733. Vesonder, R.F., L.W. Tjarks, A. Ciegler, G.F. Spencer, and L.L. Wallen. 1977a. 4-Acetamido-2-butenoic acid from *Fusarium graminearum*. Phytochemistry 16: 1296-1297.
734. Vesonder, R.F., A. Ciegler, and A.H. Jensen, 1977b. Production of refusal factors by *Fusarium* strains on grains. Appl. Environ. Microbiol. 34: 105-106.
735. Vesonder, R.F., A. Ciegler, R.F. Rogers, K.A. Burbridge, R.J. Bothast, and A.H. Jensen. 1978. Survey of 1977 crop year harvest corn for vomitoxin. Appl. Environ. Microbiol. 36: 885-888.
736. Vesonder, R.F., L.W. Tjarks, W.K. Rohwedder, H.R. Burmeister, and J.A. Laugal. 1979a. Equisetin, an antibiotic from *Fusarium equiseti* NRRL 5537, identified as a derivative of N-methyl-2,4-pyrollidone. J. Antibiotics 32: 759-761.
737. Vesonder, R.F., A. Ciegler, H.R. Burmeister, and A.H. Jensen. 1979b. Acceptance by swine and rats of corn amended with trichothecenes. Appl. Environ. Microbiol. 38: 344-346.
738. Vesonder, R.F., A. Ciegler, W.K. Rohwedder, and R. Eppley. 1979c. Re-examination of 1972 midwest corn for vomitoxin. Toxicon 17: 658-660.
739. Vesonder, R.F., J.J. Ellis, and W.K. Rohwedder. 1981a. Swine refusal factors elaborated by *Fusarium* strains and identified as trichothecenes. Appl. Environ. Microbiol. 41: 323-324.
740. Vesonder, R.F., J.J. Ellis, and W.K. Rohwedder. 1981b. Elaboration of vomitoxin and zearalenone by *Fusarium* isolates and the biological activity of *Fusarium*-produced toxins. Appl. Environ. Microbiol. 42: 1132-1134.
741. Von Deckenbach, C. 1907. Zur Frage über die Aetiologie der Pellagra. Centralbl. Bakt., Abt. 1, 45: 507-512.
742. Walser, M.M., N.K. Allen, and C.J. Mirocha. 1980. Tibial dyschondroplasia induced by toxins of *Fusarium roseum*. Poult. Sci. 59: 1669-1670 (Abstr.).
743. Ware, G.M., and C.W. Thorpe. 1978. Determination of zearalenone in corn by high pressure liquid chromatography and fluorescence detection. J. Assoc. Off. Anal. Chem. 61: 1058-1062.
744. Weaver, G.A., H.J. Kurtz, and C.J. Mirocha. 1977. The effect of *Fusarium* toxins on food-producing animals. Proc. Ann. Meeting. U.S. Anim. Health Assoc. 81: 215-218.
745. Weaver, G.A., H.J. Kurtz, F.Y. Bates, M.S. Chi, C.J. Mirocha, J.C. Behrens, and T.S. Robison. 1978a. Acute and chronic toxicity of T-2 mycotoxin in swine. Vet. Rec. 103: 531-535.
746. Weaver, G.A., H.J. Kurtz, C.J. Mirocha, F.Y. Bates, J.C. Behrens, T.S. Robison, and W.F. Gipp. 1978b. Mycotoxin-induced abortions in swine. Can. Vet. J. 19: 72-74.

747. Weaver, G.A., H.J. Kurtz, C.J. Mirocha, F.Y. Bates, and J.C. Behrens. 1978c. Acute toxicity of the mycotoxin diacetoxyscirpenol in swine. Can. Vet. J. 19: 267–271.
748. Weaver, G.A., H.J. Kurtz, C.J. Mirocha, F.Y. Bates, J.C. Behrens, and T.S. Robison. 1978d. Effect of T-2 toxin on porcine reproduction. Can. Vet. J. 19: 310–314.
749. Weaver, G.A., H.J. Kurtz, C.J. Mirocha, F.Y. Bates, J.C. Behrens, T.S. Robison, and S.P. Swanson. 1980. The failure of purified T-2 mycotoxin to produce hemorrhaging in dairy cattle. Can. Vet. J. 21: 210–213.
750. Weaver, G.A., H.J. Kurtz, F.Y. Bates, C.J. Mirocha, J.C. Behrens, and W.M. Hagler. 1981. Diacetoxyscirpenol toxicity in pigs. Res. Vet. Sci. 31: 131–135.
751. Wehner, F.C., W.F.O. Marasas, and P.G. Thiel. 1978. Lack of mutagenicity to *Salmonella typhimurium* of some *Fusarium* mycotoxins. Appl. Environ. Microbiol. 35: 659–662.
752. Wei, R.D., F.M. Strong, E.B. Smalley, and H.K. Schnoes. 1971. Chemical interconversion of T-2 and HT-2 toxins and related compounds. Biochem. Biophys. Res. Commun. 45: 396–401.
753. Wei, R.D., E.B. Smalley, and F.M. Strong. 1972. Improved skin test for detection of T-2 toxin. Appl. Microbiol. 23: 1029–1030.
754. White, E.P. 1967. Isolation of (±)-2-acetamido-2,5-dihydro-5-oxofuran from *Fusarium equiseti*. J. Chem. Soc. (C) 1967: 346–347.
755. Wiebe, L.A., and L.F. Bjeldanes. 1981. Fusarin C, a mutagen from *Fusarium moniliforme* grown on corn. J. Food Sci. 46: 1424–1426.
756. Willemart, J.P., and E. Schricke. 1981. Un cas d'infécondité chez les faisans, description-reproduction-etiologie. Avian Pathol. 10: 489–498.
757. Wilson, B.J. 1971. Recently discovered metabolites with unusual toxic manifestations, p. 223–229. In I.F.H. Purchase (ed.), Mycotoxins in human health, Macmillan, London.
758. Wilson, B.J. 1973. Toxicity of mold-damaged sweetpotatoes. Nutr. Rev. 31: 73–78.
759. Wilson, B.J., and M.R. Boyd. 1974. Toxins produced by sweet potato roots infected with *Ceratocystis fimbriata* and *Fusarium solani*, p. 327–344. In I.F.H. Purchase (ed.). Mycotoxins, Elsevier, Amsterdam.
760. Wilson, B.J., and L.T. Burka. 1979. Toxicity of novel sesquiterpenoids from the stressed sweet potato (*Ipomoea batatas*). Food Cosmet. Toxicol. 17: 353–355.
761. Wilson, B.J., and R.R. Maronpot. 1971. Causative fungus agent of leucoencephalomalacia in equine animals. Vet. Rec. 88: 484–486.
762. Wilson, R.F., V.D. Sharma, L.E. Williams, D.P. Sharda, and H.S. Teague. 1967. Effects of feeding *Fusarium roseum* mold to rats and hogs. J. Anim. Sci. 26: 1479–1480 (Abstr.).
763. Wilson, B.J., D.T.C. Yang, and M.R. Boyd. 1970. Toxicity of mould-damaged sweet potatoes (*Ipomoea batatas*). Nature 227: 521–522.
764. Wilson, B.J., M.R. Boyd, T.M. Harris, and D.T.C. Yang. 1971. A lung oedema factor from mouldy sweet potatoes (*Ipomoea batatas*). Nature 231: 52–53.
765. Wilson, B.J., R.R. Maronpot, and P.K. Hildebrandt. 1973. Equine leukoencephalomalacia. J. Amer. Vet. Med. Assoc. 163: 1293–1295.
766. Wineland, G.O. 1924. An ascigerous stage and synonymy for *Fusarium moniliforme*. J. Agric. Res. 28: 909–922.
767. Witlock, D.R., R.D. Wyatt, and M.D. Ruff. 1977. Morphological changes in the avian intestine induced by citrinin and lack of effect of aflatoxin and T-2 toxin as seen with scanning electron microscopy. Toxicon 15: 41–44.
768. Wolf, J.C., and C.J. Mirocha. 1973. Regulation of sexual reproduction in *Gibberella zeae* (*Fusarium roseum* "Graminearum") by F-2 (zearalenone). Can. J. Microbiol. 19: 725–734.
769. Wolf, J.C., and C.J. Mirocha. 1977. Control of sexual reproduction in *Gibberella zeae* (*Fusarium roseum* "Graminearum"). Appl. Environ. Microbiol. 33: 546–550.
770. Wolf, J.C., J.R. Lieberman, and C.J. Mirocha. 1972. Inhibition of F-2 (zearalenone) biosynthesis and perithecium production in *Fusarium roseum* "Graminearum." Phytopathology 62: 937–939.
771. Wollenweber, H.W., and O.A. Reinking. 1935. Die Fusarien. Paul Parey, Berlin. 355 p.

772. Woronin, M. 1891. Uber das "Taumelgetreide" in Süd-Ussurien. Bot. Z. 49: 81–93.
773. Wray, B.B., and K.G. O'Steen. 1975. Mycotoxin-producing fungi from house associated with leukemia. Arch. Environ. Health 30: 571–573.
774. Wray, B.B., E.J. Rushings, R.C. Boyd, and A.M. Schindel. 1979. Suppression of phytohemagglutinin response by fungi from a "leukemia" house. Arch. Environ. Health 34: 350–353.
775. Wright, D.E. 1968. Toxins produced by fungi. Ann. Rev. Microbiol. 22: 269–282.
776. Wu, J., B. Sun, Y. Wang, H. Kong, and J. Chiao. 1981. Studies on the toxins of *Gibberella zeae* (Schw.) Petch. II. Toxigenic studies of *Fusarium graminearum* FG-1 in solid and liquid cultures. Acta Microbiol. Sin. 21: 344–349 (In Chinese).
777. Wyatt, R.D., B.A. Weeks, P.B. Hamilton, and H.R. Burmeister. 1972a. Severe oral lesions in chickens caused by ingestion of dietary fusariotoxin T-2. Appl. Microbiol. 24: 251–257.
778. Wyatt, R.D., J.R. Harris, P.B. Hamilton, and H.R. Burmeister. 1972b. Possible outbreaks of fusariotoxicosis in avians. Avian Dis. 16: 1123–1130.
779. Wyatt, R.D., W.M. Colwell, P.B. Hamilton, and H.R. Burmeister. 1973a. Neural disturbances in chickens caused by dietary T-2 toxin. Appl. Microbiol. 26: 757–761.
780. Wyatt, R.D., P.B. Hamilton, and H.R. Burmeister. 1973b. The effects of T-2 toxin in broiler chickens. Poult. Sci. 52: 1853–1859.
781. Wyatt, R.D., J.A. Doerr, P.B. Hamilton, and H.R. Burmeister. 1975a. Egg production, shell thickness, and other physiological parameters of laying hens affected by T-2 toxin. Appl. Microbiol. 29: 641–645.
782. Wyatt, R.D., P.B. Hamilton, and H.R. Burmeister. 1975b. Altered feathering of chicks caused by T-2 toxin. Poult. Sci. 54: 1042–1045.
783. Yabuta, T., and T. Hayashi. 1939. Biochemical studies of bakanäe fungus on rice. II. Isolation of gibberellin, the active principle, which produces slender rice seedlings. J. Agric. Chem. Soc. Japan 15: 257–266 (In Japanese).
784. Yabuta, T., K. Kambe, and T. Hayashi. 1934. Biochemical studies of bakanäe fungus on rice. I. Fusarinic acid, a new product of the bakanäe fungus. J. Agric. Chem. Soc. Japan 10: 1059–1068 (In Japanese).
785. Yagen, B., and A.Z. Joffe. 1976. Screening of toxic isolates of *Fusarium poae* and *Fusarium sporotrichioides* involved in causing alimentary toxic aleukia. Appl. Environ. Microbiol. 32: 423–427.
786. Yagen, B., A.Z. Joffe, P. Horn, N. Mor, and I.I. Lutsky. 1977. Toxins from a strain involved in A.T.A., p. 329–336. *In* J.V. Rodricks, C.W. Hesseltine, and M.A. Mehlman (ed.), Mycotoxins in human and animal health, Pathotox Publishers, Park Forest South, Illinois.
787. Yagen, B., P. Horn, A.Z. Joffe, and R.H. Cox. 1980. Isolation and structural elucidation of a novel sterol metabolite of *Fusarium sporotrichioides* 921. J. Chem. Soc. Perkin I, 1980: 2914–2917.
788. Yang, C.S. 1980. Research on esophageal cancer in China: a review. Cancer Res. 40: 2633–2644.
789. Yang, D.T.C., B.J. Wilson, and T.M. Harris. 1971. The structure of ipomeamaronol: a new toxic furanosesquiterpene from moldy sweet potatoes. Phytochemistry 10: 1653–1654.
790. Yap, H.-Y., W.K. Murphy, A. DiStefano, G.R. Blumenschein, and G.P. Bodey. 1979. Phase II study of anquidine in advanced breast cancer. Cancer Treat. Rep. 63: 789–791.
791. Yates, S.G. 1962. Toxicity of tall fescue forage: a review. Econ. Bot. 16: 295–303.
792. Yates, S.G. 1971. Toxin-producing fungi from fescue pasture, p. 191–206. *In* S. Kadis, A. Ciegler, and S.J. Ajl (ed.), Microbial toxins, Vol. VII, Academic Press, New York.
793. Yates, S.G., H.L. Tookey, J.J. Ellis, and H.J. Burkhardt. 1967. Toxic butenolide produced by *Fusarium nivale* (Fries) Cesati isolated from tall fescue (*Festuca arundinacea* Schreb.). Tetrahedron Lett. 7: 621–625.
794. Yates, S.G., H.L. Tookey, J.J. Ellis, and H.J. Burkhardt. 1968. Mycotoxins produced by *Fusarium nivale* isolated from tall fescue (*Festuca arundinacea* Schreb.). Phytochemistry 7: 139–146.

795. Yates, S.G., H.L. Tookey, J.J. Ellis, W.H. Tallent, and I.A. Wolff. 1969. Mycotoxins as a possible cause of fescue toxicity. J. Agric. Food Chem. 17: 437-442.
796. Yates, S.G., H.L. Tookey, and J.J. Ellis. 1970. Survey of tall-fescue pasture: correlation of toxicity of *Fusarium* isolates to known toxins. Appl. Microbiol. 19: 103-105.
797. Ylimäki, A. 1981. The mycoflora of cereal seeds and some feedstuffs. Ann. Agric. Fenn. 20: 74-88.
798. Ylimäki, A., H. Koponen, E.-L. Hintikka, M. Nummi, M.-L. Niku-Paavola, T. Ilus, and T.-M. Enari. 1979. Mycoflora and occurrence of *Fusarium* toxins in Finnish grain. Tech. Res. Centre Finland, Materials and Processing Technology Publ. 21. Espoo, Finland. 28 p.
799. Yoshizawa, T., and N. Morooka. 1973. Deoxynivalenol and its monoacetate: New mycotoxins from *Fusarium roseum* and moldy barley. Agric. Biol. Chem. 37: 2933-2934.
800. Yoshizawa, T., and N. Morooka. 1974. Studies on the toxic substances in the infected cereals. III. Acute toxicities of new trichothecene mycotoxins: deoxynivalenol and its monoacetate. J. Food Hyg. Soc. Japan 15: 261-269 (In Japanese).
801. Yoshizawa, T., and N. Morooka. 1975a. Biological modification of trichothecene mycotoxins: acetylation and deacetylation of deoxynivalenols by *Fusarium* spp. Appl. Microbiol. 29: 54-58.
802. Yoshizawa, T., and N. Morooka. 1975b. Comparative studies on microbial and chemical modifications of trichothecene mycotoxins. Appl. Microbiol. 30: 38-43.
803. Yoshizawa, T., and N. Morooka. 1977. Trichothecenes from mold-infested cereals in Japan, p. 309-321. *In* J.V. Rodricks, C.W. Hesseltine, and M.A. Mehlman (ed.), Mycotoxins in human and animal health, Pathotox Publishers, Park Forest South, Illinois.
804. Yoshizawa, T., Y. Tsuchiya, M. Teraura, and N. Morooka. 1976. Studies on the toxic substances in the infected cereals. IV. Trichothecene from *Fusarium*-infected wheat in the crop field of Kagawa in 1975. Proc. Jap. Assoc. Mycotoxicol. 2: 30-32 (In Japanese).
805. Yoshizawa, T., T. Shirota, and N. Morooka. 1978. Deoxynivalenol and its acetate as feed refusal principles in rice cultures of *Fusarium roseum* No. 117 (ATCC 28114). J. Food Hyg. Soc. Japan 19: 178-184.
806. Yoshizawa, T., Y. Matsuura, Y. Tsuchiya, N. Morooka, K. Kitani, M. Ichinoe, and H. Kurata. 1979. On the toxigenic Fusaria invading barley and wheat in the southern Japan. J. Food Hyg. Soc. Japan 20: 21-26.
807. Yoshizawa, T., C. Onomoto, and N. Morooka. 1980. Microbial acetyl conjugation of T-2 toxin and its derivatives. Appl. Environ. Microbiol. 39: 962-966.
808. Young, L.G., R.F. Vesonder, H.S. Funnell, I. Simons, and B. Wilcock. 1981. Moldy corn in diets of swine. J. Anim. Sci. 52: 1312-1318.

Fusarium Strains Cross-Reference Index

The index for this book is divided into two parts. The first part provides a listing of all the culture numbers and strain numbers used in the text. The second part is a standard index.

In this part the reader will find all the culture numbers or designations used in the text in the left hand column, the correct species identification in the center column, and the strain number of each culture in our collections in the right hand column. Examples are given below:

Culture No.	*Fusarium* species	Strain
Fn-2B, as *F. nivale*	sporotrichioides	4.9

In this case the culture is listed in the literature as *F. nivale* Fn-2B, the correct species identification is *F. sporotrichioides,* and additional information will be found in the standard index on the page numbers listed under strain number 4.9.

Culture No.	*Fusarium* species	Strain
M-1	moniliforme	see index

312 Index

In this case the culture is listed in the literature as *F. moniliforme* M-1, the identification is *F. moniliforme,* and since we have not seen the culture and do not have it in our collections, the notation, see index, occurs in the strain number column. This means that all information on this isolate will appear on the page numbers listed under "*F. moniliforme,* M-1" in the standard index.

Culture No.	*Fusarium* species	Strain
1	equiseti	see index
1	graminearum	8.21
1	lateritium	see index
1, as *F. roseum*	graminearum	8.21
1	semitectum	see index
1/81	equiseti	7.1
2	equiseti	see index
2	lateritium	see index
2, as *F. roseum*	sporotrichioides	4.26
3	equiseti	see index
3-8-66, No. 10, as *F. roseum*	semitectum	6.4
3D-10	equiseti	see index
4	equiseti	see index
4	semitectum	see index
5, as *F. equiseti*	semitectum	6.2
5	graminearum	8.20
5	semitectum	6.2
5D-14	equiseti	see index
6	equiseti	see index
6W-9	equiseti	see index
7	moniliforme	see index
8	graminearum	see index
9	roseum 'Equiseti'	see index
9	graminearum	see index
10	graminearum	8.21
10D-18	equiseti	see index
10W-9	equiseti	see index
11W-8	equiseti	see index
13	graminearum	8.45
13	moniliforme	see index
15, as *F. tricinctum*	tricinctum sensu Snyd. & Hans.	see index
16	equiseti	see index
16, as *F. tricinctum*	sporotrichioides	4.5
17	lateritium	see index
18	equiseti	see index
18-3	equiseti	see index
19	moniliforme	10.14
20, as *F. equiseti*	semitectum	6.1
20	poae	see index
20	semitectum	6.1
21	equiseti	see index
22	equiseti	see index
25	lateritium	see index
26	larvarum	see index
26	lateritium	see index
27	larvarum	3.2
28	lateritium	see index
34	culmorum	8.18
45	moniliforme	see index
53	culmorum	see index
55-1	oxysporum	see index
57	moniliforme	see index
60/9	poae	see index
60/10	sporotrichioides	see index
63	sporotrichioides	see index
70-k-11, as *F. roseum* 'Culmorum'	graminearum	8.40
80	avenaceum	5.1
111, as *F. moniliforme*	sporotrichioides	4.24
112/2	solani	see index
117	graminearum	8.22
117, as *F. roseum*	graminearum	8.22
126	graminearum	see index
161	moniliforme	see index
161	tricinctum sensu Snyd. & Hans.	see index
162	roseum sensu Snyd. & Hans.	see index
182	culmorum	see index
203, as *F. tricinctum*	sporotrichioides	4.3
209, as *F. tricinctum*	sporotrichioides	4.29
214	sporotrichioides	see index
215, as *F. tricinctum*	poae	4.47
223, as *F. tricinctum*	sportrichioides	4.4
238, as *F. tricinctum*	sporotrichiodes	4.27
334, as *F. tricinctum*	poae	4.49
335, as *F. tricinctum*	poae	4.50
347, as *F. tricinctum*	sporotrichioides	4.28
348, as *F. tricinctum*	sporotrichioides	4.30
350, as *F. tricinctum*	sporotrichioides	4.34
351	sporotrichioides	see index
355, as *F. tricinctum*	poae	4.48
374	moniliforme	10.32
396	poae	see index
562/2	solani	see index
577	equiseti	see index
587, as *F. moniliforme*	proliferatum	10.33
597	oxysporum	11.1
722, as *F. sulphureum*	sambucinum	8.3
738	sporotrichioides	see index
792	poae	see index
844	lateritium	see index
845	lateritium	see index
877	poae	see index
881	moniliforme	10.28
883	moniliforme	10.30
903, as *F. coeruleum*	solani	12.2
917	moniliforme	see index
921	sporotrichioides	see index
953	monilifome var. subglutinans	see index
958	poae	see index
965	solani	see index
973	oxysporum	see index
975	moniliforme	10.31
989	moniliforme	see index

Index

Culture No.	*Fusarium* species	Strain
990	solani	see index
995	moniliforme	see index
999	proliferatum	see index
1000	moniliforme	10.29
1140/26	poae	see index
1172	poae	see index
1179	poae	see index
1182	sporotrichioides	see index
1183	sporotrichioides	see index
1238	poae	see index
1421	lateritium	see index
1536	poae	see index
1664	solani	see index
1671	poae	see index
1677	poae	see index
1857	moniliforme	see index
1862	poae	see index
1863	moniliforme	see index
1953	poae	see index
1973-33	sporotrichioides	4.20
1973-76	tricinctum sensu stricto	see index
2061-C, as *F. tricinctum*	sporotrichioides	4.6
2131	oxysporum	see index
2269	moniliforme	see index
2317	redolens	see index
2341	oxysporum	see index
2381	lateritium	see index
2518, as *F. poae*	sporotrichioides	4.25
3125	nivale	3.1
3163	oxysporum	see index
3737	culmorum	8.15
4150	moniliforme	10.24
5003	moniliforme	10.15
5008	equiseti	7.9
5013, as *F. lateritium*	oxysporum	11.6
5026	oxysporum	11.5
5027	solani	12.3
5029, as *Fusarium* sp.	oxysporum	11.9
5034	solani	see index
5036, as *F. lateritium*	oxysporum	11.7
5253	poae	see index
5328a	sporotrichioides	see index
5627	poae	see index
7085 II-19	moniliforme	10.14
7137	graminearum	8.45
7289	culmorum	8.14
7452, as *F. poae*	sporotrichioides	4.7
8549, as *F. moniliforme* var. *subglutinans*	subglutinans	10.35
8549, as *F. sacchari* var. *elongatum*	subglutinans	10.35
8998, as *F. moniliforme* var. *anthophilum*	anthophilum	10.38
62239	larvarum	3.2
62286 (Berlin)	oxysporum	11.2
62409 (Berlin)	solani	12.1
62448	tricinctum sensu stricto	4.59
63312	equiseti	see index
63313	equiseti	see index
72183	tricinctum sensu stricto	4.58
72187-10k, as *F. tricinctum*	sporotrichioides	4.23
72187-12	sporotrichioides	4.23
72188	poae	4.53
72188	tricinctum sensu stricto	4.57
72196	culmorum	8.12
72313	avenaceum	5.5
72313-20	culmorum	8.13
72322	graminearum	8.44
136418	semitectum	see index
A-1	sporotrichiodes	4.43
A-130	moniliforme	see index
A-B-2, as *F. tricinctum*	graminearum	8.33
Abashiri-2, as *F. tricinctum*	graminearum	8.33
Abashiri-3	oxysporum	see index
Alaska 2-2, as *F. roseum*	equiseti	7.6
A-R-5, as *F. tricinctum*	graminearum	8.32
ATCC 11852	sambucinum	8.5
ATCC 12616	moniliforme	see index
ATCC 12763, as *Gibberella fujikuroi*	moniliforme	10.24
ATCC 14164	moniliforme	10.27
ATCC 15620	culmorum	see index
ATCC 20227	lateritium	see index
ATCC 20272, as *Gibberella zeae*	graminearum	8.19
ATCC 24043, as *F. tricinctum*	sporotrichioides	4.3
ATCC 24044, as *F. tricinctum*	sporotrichioides	4.4
ATCC 24045, as *F. tricinctum*	sporotrichioides	4.5
ATCC 24088	moniliforme	10.10
ATCC 24089	moniliforme	10.11
ATCC 24529, as *F. tricinctum*	sporotrichioides	4.8
ATCC 24631, as *F. tricinctum*	sporotrichioides	4.3
ATCC 24688, as *Gibberella zeae*	graminearum	8.21
ATCC 24689, as *Gibberella zeae*	graminearum	8.21
ATCC 26263	moniliforme	10.23
ATCC 26531, as *F. episphaeria*	sporotrichioides	4.10
ATCC 26532, as *F. nivale*	sporotrichioides	4.9
ATCC 26533	sporotrichioides	4.12
ATCC 26553	sporotrichioides	4.12
ATCC 26557	graminearum	8.21
ATCC 26559, as *Micronectriella nivalis*	culmorum	8.18
ATCC 28112, as *F. roseum*	graminearum	8.30
ATCC 28114, as *F. roseum*	graminearum	8.22

Index

Culture No.	*Fusarium* species	Strain
M-1352	moniliforme	10.12
M-1353	moniliforme	10.9
M-1354, as *F. moniliforme* var. *subglutinans*	subglutinans	10.35
M-1355, as *F. moniliforme* var. *anthophilum*	anthophilum	10.38
Map. 10—Ascospore 55, as *Gibberella zeae*	graminearum	8.21
Mapleton 10, as *F. roseum* 'Graminearum'	graminearum	8.21
MC-103	moniliforme	10.19
MC-107	moniliforme	10.17
MC-121	moniliforme	10.18
MC-127	moniliforme	10.20
MC-130	moniliforme	10.21
MC-252	moniliforme	10.22
Melon-1, as *F. oxysporum* "niveum"	sporotrichioides	4.11
MHC 7452, as *F. poae*	sporotrichioides	4.7
MRC 35, as *F. fusarioides*	chlamydosporum	4.44
MRC 42	moniliforme	10.2
MRC 43	sporotrichioides	4.3
MRC 69	moniliforme	10.24
MRC 115, as *F. moniliforme* var. *subglutinans*	subglutinans	10.36
MRC 117	chlamydosporum	4.45
MRC 120	graminearum	8.37
MRC 121	graminearum	8.42
MRC 134, as *F. moniliforme* var. *subglutinans*	subglutinans	10.34
MRC 137	moniliforme	10.12
MRC 460	graminearum	8.43
MRC 514, as *F. sulphureum*	sambucinum	8.1
MRC 515	moniliforme	10.26
MRC 548, as *F. verticillioides*	moniliforme	10.9
MRC 602, as *F. verticillioides*	moniliforme	10.5
MRC 756, as *F. moniliforme* var. *subglutinans*	subglutinans	10.37
MRC 826, as *F. verticillioides*	moniliforme	10.4
MRC 846, as *F. sulphureum*	sambucinum	8.2
MRC 929	moniliforme	10.6
MRC 930	moniliforme	10.3
MRC 935	sporotrichioides	4.12
MRC 939	culmorum	8.8
MRC 940, as *F. moniliforme* var. *subglutinans*	subglutinans	10.35
MRC 941, as *F. moniliforme* var. *anthophilum*	anthophilum	10.38

Culture No.	*Fusarium* species	Strain
MRC 1369	culmorum	8.12
MRC 1371	culmorum	8.13
MRC 1374	avenaceum	5.5
MRC 1391	graminearum	8.44
MRC 1393	graminearum	8.45
MRC 1398	poae	4.53
MRC 1399	tricinctum sensu stricto	4.57
MRC 1400	tricinctum sensu stricto	4.58
MRC 1411	moniliforme	10.23
MRC 1412	moniliforme	10.24
MRC 1413	avenaceum	5.2
MRC 1414	oxysporum	11.3
MRC 1439	moniliforme	10.13
MRC 1445	semitectum	6.5
MRC 1574	tricinctum sensu stricto	4.60
MRC 1575	tricinctum sensu stricto	4.61
MRC 1576	tricinctum sensu stricto	4.62
MRC 1605	poae	4.49
MRC 1606	sporotrichiodes	4.35
MRC 1607	poae	4.48
MRC 1620	sporotrichioides	4.6
MRC 1621	graminearum	8.20
MRC 1623	culmorum	8.16
MRC 1642	semitectum	6.4
MRC 1646	graminearum	8.38
MRC 1657	sporotrichioides	4.27
MRC 1658	poae	4.47
MRC 1659	sporotrichioides	4.28
MRC 1660	sporotrichioides	4.29
MRC 1661	sporotrichioides	4.4
MRC 1662	poae	4.50
MRC 1663	sporotrichioides	4.30
MRC 1692	sporotrichioides	4.2
MRC 1693	graminearum	8.24
MRC 1694	oxysporum	11.4
MRC 1704	sporotrichioides	4.1
MRC 1705	sporotrichioides	4.8
MRC 1708	sporotrichioides	4.34
MRC 1709	sporotrichioides	4.33
MRC 1767	sporotrichioides	4.9
MRC 1768	sporotrichioides	4.3
MRC 1781	graminearum	8.41
MRC 1782	graminearum	8.19
MRC 1783	acuminatum	7.16
MRC 1784	moniliforme	10.25
MRC 1785	graminearum	8.23
MRC 1786	sporotrichioides	4.25
MRC 1787	culmorum	8.15
MRC 1799	tricinctum sensu stricto	4.59
MRC 1802	larvarum	3.2
MRC 1806	graminearum	8.21
MRC 1888	avenaceum	5.3
MRC 1889	culmorum	8.10
MRC 1890	culmorum	8.11
MRC 1891	equiseti	7.7
MRC 1892	equiseti	7.8
MRC 1893	poae	4.52

INDEX 317

Culture No.	*Fusarium* species	Strain	Culture No.	*Fusarium* species	Strain
MRC 1894	tricinctum sensu stricto	4.55	MRC 2389	moniliforme	10.31
			MRC 2390	moniliforme	10.29
MRC 1895	tricinctum sensu stricto	4.56	MRC 2393	graminearum	8.52
			MRC 2394	sambucinum	8.6
MRC 1902	sporotrichioides	4.26	MRC 2433	equiseti	7.14
MRC 1903	sambucinum	8.7	MRC 2434	culmorum	8.18
MRC 1904	graminearum	8.21	MRC 2435	equiseti	7.11
MRC 1905	culmorum	8.14	MRC 2436	graminearum	8.26
MRC 1946	graminearum	8.22	MRC 2476	sporotrichioides	4.9
MRC 1959	acuminatum	7.19	MRC 2478	sporotrichioides	4.12
MRC 1960	graminearum	8.34	MRC 2483	equiseti	7.2
MRC 1961	equiseti	7.15	MRC 2486	chlamydosporum	4.46
MRC 1962	sporotrichioides	4.9	MRC 2487	sporotrichioides	4.43
MRC 1963	graminearum	8.35	MRC 2488	sportrichioides	4.37
MRC 1976	graminearum	8.36	MRC 2489	sporotrichioides	4.36
MRC 2017	acuminatum	7.18	MRC 2523	sporotrichioides	4.31
MRC 2031	sporotrichioides	4.21	MRC 2531	sporotrichioides	4.10
MRC 2032	sporotrichioides	4.20	MRC 2532	avenaceum	5.4
MRC 2033	sporotrichioides	4.22	MRC 2533	sporotrichioides	4.9
MRC 2034	sporotrichioides	4.12	MRC 2534	sporotrichioides	4.11
MRC 2064	culmorum	8.17	MRC 2535	moniliforme	10.15
MRC 2065	graminearum	8.49	MRC 2536	oxysporum	11.9
MRC 2066	oxysporum	11.8	MRC 2557	sporotrichioides	4.12
MRC 2079	moniliforme	10.1	MRC 2558	equiseti	7.3
MRC 2181	poae	4.51	MRC 2559	equiseti	7.4
MRC 2182	sporotrichioides	4.13	MRC 2560	graminearum	8.40
MRC 2183	sporotrichioides	4.14	MRC 2561	equiseti	7.5
MRC 2184	sporotrichioides	4.15	MRC 2563	graminearum	8.33
MRC 2186	sporotrichioides	4.16	MRC 2564	oxysporum	11.5
MRC 2187	sambucinum	8.4	MRC 2565	solani	12.3
MRC 2188	sporotrichioides	4.17	MRC 2566	oxysporum	11.6
MRC 2189	sporotrichioides	4.18	MRC 2567	oxysporum	11.7
MRC 2190	acuminatum	7.17	MRC 2568	equiseti	7.9
MRC 2191	semitectum	6.3	MRC 2580	graminearum	8.29
MRC 2192	sporotrichioides	4.19	MRC 2581	graminearum	8.28
MRC 2193	sambucinum	8.3	MRC 2583	graminearum	8.30
MRC 2194	solani	12.2	MRC 2605	graminearum	8.32
MRC 2195	avenaceum	5.1	MRC 2606	graminearum	8.31
MRC 2196	merismoides	1.1	MRC 2609	equiseti	7.1
MRC 2197	culmorum	8.9	MRC 2610	semitectum	6.1
MRC 2198	solani	12.1	MRC 2611	semitectum	6.2
MRC 2199	oxysporum	11.2	MRC 2625	sporotrichioides	4.23
MRC 2228	moniliforme	10.1	MRC 2627	moniliforme	10.19
MRC 2229	nivale	3.1	MRC 2628	moniliforme	10.17
MRC 2231	equiseti	7.12	MRC 2629	moniliforme	10.18
MRC 2232	equiseti	7.13	MRC 2630	moniliforme	10.20
MRC 2233	graminearum	8.25	MRC 2631	moniliforme	10.21
MRC 2234	graminearum	8.27	MRC 2632	moniliforme	10.22
MRC 2316	moniliforme	10.10	MRC 2633	moniliforme	10.16
MRC 2317	moniliforme	10.11	MRC 2634	solani	12.4
MRC 2319	sporotrichioides	4.12	MRC 2635	solani	12.5
MRC 2320	sambucinum	8.5	MRC 2636	semitectum	6.6
MRC 2322	moniliforme	10.27	MRC 2637	equiseti	7.10
MRC 2323	sporotrichioides	4.24	MRC 2638	graminearum	8.46
MRC 2324	proliferatum	10.33	MRC 2639	graminearum	8.47
MRC 2325	oxysporum	11.1	MRC 2640	graminearum	8.48
MRC 2326	moniliforme	10.8	MRC 2677	moniliforme	10.14
MRC 2327	moniliforme	10.7	MRC 2728	sporotrichioides	4.7
MRC 2329	graminearum	8.21	MRC 2803	solani	12.6
MRC 2330	equiseti	7.6	MRC 2804	semitectum	6.7
MRC 2386	moniliforme	10.32	MRC 2805	solani	12.7
MRC 2387	moniliforme	10.28	MRC 2806	semitectum	6.8
MRC 2388	moniliforme	10.30	MRC 2863	sporotrichioides	4.5

318 Index

Culture No.	Fusarium species	Strain
MRC 2908	poae	4.54
MRC 2909	sporotrichioides	4.41
MRC 2910	sporotrichioides	4.40
MRC 2911	sporotrichioides	4.42
MRC 2912	sporotrichioides	4.38
MRC 2913	sporotrichioides	4.32
MRC 2914	sporotrichioides	4.39
MRC 2940	graminearum	8.39
MRC 3081	graminearum	8.50
MRC 3082	graminearum	8.51
N 5	moniliforme	see index
N-40	larvarum	3.2
N-41	nivale	3.1
N 136429	moniliforme	see index
NHL-F-111, as F. solani	sporotrichioides	4.12
NHL-5-1025	acuminatum	7.19
NHL-F-1104	graminearum	8.35
NHL-F-1112	graminearum	8.36
NHL-F-1118	graminearum	8.34
NHL-F-1121	equiseti	7.15
NRRL 1943	oxysporum	11.4
NRRL 2284	moniliforme	10.27
NRRL 2380, see NRRL 2830		
NRRL 2830, as F. moniliforme	graminearum	8.19
NRRL 3020, as F. anguioides	equiseti	7.12
NRRL 3197, as F. moniliforme	sporotrichioides	4.24
NRRL 3211, as F. diversisporum	equiseti	7.14
NRRL 3212, as F. scirpi	equiseti	7.11
NRRI 3213	equiseti	see index
NRRL 3214, as F. concolor	equiseti	7.13
NRRL 3249, as F. nivale or F. tricinctum	sporotrichioides	4.8
NRRL 3287, as F. poae or F. tricinctum	sporotrichioides	4.25
NRRL 3288	culmorum	8.15
NRRL 3289	nivale	3.1
NRRL 3299, as F. tricinctum	sporotrichioides	4.3
NRRL 3509, as F. nivale	sporotrichioides	4.9
NRRL 3510, as F. tricinctum	sporotrichioides	4.1
NRRL 3511, as F. tricinctum	sporotrichioides	4.2
NRRL 5537, as F. roseum 'Gibbosum'	equiseti	7.2
NRRL 5860	moniliforme	10.23
NRRL 5864	graminearum	8.24
NRRL 5883	graminearum	8.23
NRRL 5908, as F. tricinctum	graminearum	8.41
NRRL 6022, as Gibberella fujikuroi	moniliforme	10.24
NRRL 6151, as F. tricinctum	sporotrichioides	4.12
NRRL 6202	graminearum	see index
NRRL 6206	graminearum	see index
NRRL 6207	graminearum	see index

Culture No.	Fusarium species	Strain
NRRL 6227, as F. roseum 'Acuminatum'	acuminatum	7.16
NRRL 6251, as F. tricinctum	sporotrichioides	4.12
NRRL 6322	moniliforme	10.25
NRRL 6325, as F. tricinctum	chlamydosporum	4.46
NRRL 6442	moniliforme	10.1
NRRL 6450	graminearum	8.25
NRRL 6451	graminearum	8.26
NRRL 6452	graminearum	8.27
NRRL 6478	roseum 'Sambucinum'	see index
NRRL 6490, as F. tricinctum	sporotrichioides	4.9
NRRL 6491, as F. tricinctum	sporotrichioides	4.10
NRRL 13318, as F. nivale	sporotrichioides	4.8
NRRL A-15666	roseum	see index
NRRL A-23377, as F. tricinctum	chlamydosporum	4.46
O-916	oxysporum	11.3
O-1011	oxysporum	11.4
O-1042	oxysporum	11.8
O-1055	oxysporum	11.1
O-1071	oxysporum	11.2
O-1166	oxysporum	11.9
O-1173	oxysporum	11.5
O-1174	oxysporum	11.6
O-1175	oxysporum	11.7
Ohio Isolate C	roseum	see index
Ohoita-II, as G. zeae	graminearum	see index
OP-6	moniliforme	10.6
OP-32	moniliforme	10.2
OP-32B	moniliforme	10.2
OP-124	moniliforme	10.3
oxyrose, as F. roseum oxyrose	semitectum	6.4
PE-8-17	moniliforme	see index
PL W-76	solani	12.4
PL W-168	solani	12.5
PL W-517	semitectum	6.6
PL W-604	moniliforme	10.16
PL W-741	equiseti	7.10
PL W-771	solani	12.7
PL W-865	semitectum	6.8
PL Z-65	solani	12.6
PL Z-81	semitectum	6.7
PP96	solani	see index
R-13c	roseum 'Sambucinum'	see index
R-4053	graminearum	8.43
R-4057	graminearum	8.42
R-4482	equiseti	7.7
R-4485	semitectum	6.4
R-4606	avenaceum	5.2
R-4607	culmorum	8.10
R-4608	avenaceum	5.3
R-4610	graminearum	8.50
R-4611	graminearum	8.51
R-4613	graminearum	8.39
R-4783	equiseti	7.8
R-4838	graminearum	8.38
R-5145	culmorum	8.12

INDEX 319

Culture No.	Fusarium species	Strain
R-5146	culmorum	8.13
R-5147	avenaceum	5.5
R-5169	semitectum	6.5
R-5176	graminearum	8.44
R-5250	graminearum	8.20
R-5251	culmorum	8.16
R-5316	graminearum	8.24
R-5317	graminearum	8.41
R-5318	graminearum	8.19
R-5319	acuminatum	7.16
R-5320	graminearum	8.23
R-5321	culmorum	8.15
R-5389	sambucinum	8.1
R-5390	sambucinum	8.2
R-5391	culmorum	8.11
R-5452	culmorum	8.8
R-5453	graminearum	8.21
R-5454	graminearum	8.21
R-5455	sambucinum	8.7
R-5456	culmorum	8.14
R-5466	acuminatum	7.19
R-5467	graminearum	8.34
R-5468	equiseti	7.15
R-5469	graminearum	8.35
R-5470	graminearum	8.36
R-5701	acuminatum	7.18
R-5796	graminearum	8.22
R-5797	culmorum	8.17
R-5798	graminearum	8.49
R-5953	graminearum	8.45
R-6053	equiseti	7.12
R-6054	equiseti	7.13
R-6055	graminearum	8.25
R-6056	graminearum	8.27
R-6112	sambucinum	8.5
R-6136	graminearum	8.21
R-6137	equiseti	7.6
R-6182	graminearum	8.52
R-6239	sambucinum	8.6
R-6324	equiseti	7.14
R-6325	equiseti	7.2
R-6336	equiseti	7.11
R-6337	graminearum	8.26
R-6353	culmorum	8.18
R-6354	sambucinum	8.4
R-6355	semitectum	6.3
R-6379	acuminatum	7.17
R-6380	sambucinum	8.3
R-6381	avenaceum	5.1
R-6382	culmorum	8.9
R-6520	semitectum	6.6
R-6521	equiseti	7.10
R-6574	graminearum	8.46
R-6575	graminearum	8.47
R-6576	graminearum	8.48
R-6605	equiseti	7.1
R-6606	semitectum	6.1
R-6607	semitectum	6.2
R-6750	avenaceum	5.4
R-6765	equiseti	7.3
R-6766	equiseti	7.4
R-6767	graminearum	8.40
R-6768	equiseti	7.5
R-6777	graminearum	8.33
R-6784	equiseti	7.9
R-6795	graminearum	8.29
R-6796	graminearum	8.28
R-6798	graminearum	8.30
R-6800	graminearum	8.37
R-6801	graminearum	8.32
R-6802	graminearum	8.31
R-6983	semitectum	6.7
R-6984	semitectum	6.8
R-P-1	tricinctum sensu stricto	4.60
S-74-1c	roseum	see index
S-679	solani	12.1
S-682	solani	12.2
S-693	solani	12.4
S-694	solani	12.5
S-735	solani	12.3
S-787	solani	12.6
S-788	solani	12.7
Sp. 803	avenaceum	5.3
Sp. 879	tricinctum sensu stricto	4.55
Sp. 880	tricinctum sensu stricto	4.56
Sp. 884	poae	4.52
Sp. 885	culmorum	8.10
Sp. 887	culmorum	8.11
Sp. 889	avenaceum	5.2
Sp. 890	equiseti	7.7
Sp. 897, as F. tricinctum	sporotrichioides	4.3
Sp. 898	graminearum	8.50
Sp. 899	graminearum	8.51
Sp. 900, as F. solani	sporotrichioides	4.12
Sp. 914	sporotrichioides	4.21
Sp. 934, as F. solani	sporotrichioides	4.12
Sp. 950	sporotrichioides	4.20
Sp. 954	sporotrichioides	4.22
Sp. 980, as F. solani	sporotrichioides	4.12
Sp. 1028	oxysporum	11.3
Sp. 1090	equiseti	7.8
SUF 374	moniliforme	10.32
SUF 881	moniliforme	10.28
SUF 883	moniliforme	10.30
SUF 975	moniliforme	10.31
SUF 1000	moniliforme	10.29
T-2, as F. tricinctum	sporotrichioides	4.3
T-227, as F. fusarioides	chlamydosporum	4.45
T-245	sporotrichioides	4.21
T-246	tricinctum sensu stricto	4.56
T-247	poae	4.52
T-248	sporotrichioides	4.22
T-249	sporotrichioides	4.20
T-280	poae	4.54
T-285	sporotrichioides	4.33
T-286	sporotrichioides	4.41
T-288	sporotrichioides	4.34
T-290	sporotrichioides	4.40
T-306	sporotrichioides	4.42
T-336	sporotrichioides	4.38
T-338	tricinctum sensu stricto	4.57

Index

Culture No.	*Fusarium* species	Strain
T-339	poae	4.53
T-340	sporotrichioides	4.32
T-341	sporotrichioides	4.39
T-344	sporotrichioides	4.8
T-345	sporotrichioides	4.1
T-346	sprootrichioides	4.9
T-347	sporotrichioides	4.2
T-348	sporotrichioides	4.3
T-399	tricinctum sensu stricto	4.60
T-402	tricinctum sensu stricto	4.62
T-403	poae	4.49
T-405	sporotrichioides	4.6
T-406	sporotrichioides	4.27
T-407	sporotrichioides	4.28
T-408	sporotrichioides	4.29
T-409	sporotrichioides	4.4
T-410	poae	4.50
T-411	sporotrichioides	4.30
T-412	poae	4.47
T-422	sporotrichioides	4.12
T-423	sporotrichioides	4.25
T-424	sporotrichioides	4.3
T-428, as *F. fusarioides*	chlamydosporum	4.44
T-429	tricinctum sensu stricto	4.55
T-431	tricinctum sensu stricto	4.59
T-432	sporotrichioides	4.26
T-436	sporotrichioides	4.9
T-490	sporotrichioides	4.12
T-492	sporotrichioides	4.12
T-494	sporotrichioides	4.24
T-502	chlamydosporum	4.46
T-503	poae	4.51
T-504	sporotrichioides	4.13
T-505	sporotrichioides	4.14
T-506	sporotrichioides	4.15
T-507	sporotrichioides	4.16
T-508	sporotrichioides	4.17
T-509	sporotrichioides	4.18
T-510	sporotrichioides	4.19
T-520	sporotrichioides	4.43
T-521	sporotrichioides	4.37
T-522	sporotrichioides	4.36
T-544	sporotrichioides	4.23
T-563	sporotrichioides	4.9
T-564	sporotrichioides	4.9
T-565	sporotrichioides	4.10
T-566	sporotrichioides	4.11
T-567	sporotrichioides	4.12
T-568	sporotrichioides	4.12
T-570	sporotrichioides	4.31
T-592	sporotrichioides	4.7
T-602	sporotrichioides	4.5
T-624	tricinctum sensu stricto	4.58
T-625	sporotrichioides	4.35
T-626	poae	4.48
T-701	tricinctum sensu stricto	4.61
T-M-1, as *F. oxysporum*	sporotrichioides	4.11
V-18, as *F. roseum*	sambucinum	8.7
W-8, as *Gibberella zeae*	graminearum	8.48
W-P-1	tricinctum sensu stricto	4.61
Washington	culmorum	8.16
YN-13, as *F. tricinctum*	sporotrichioides	4.26

Index

This index is divided into four parts: Animals Affected, *Fusarium* Species and Their Perfect States, Mycotoxicoses, and Mycotoxins. In all sections, page numbers printed in boldface type indicate the principal discussion of a subject. In the section on *Fusarium* species, when isolates of a species are identified only by a number, those numbers are printed in italics so that the user may more readily distinguish them.

Animals Affected

baboons, 221
brine shrimp (*Artemia salina*)
    F. acuminatum, 127
    F. avenaceum, 93
    F. coeruleum, 265
    F. culmorum, 144
    F. equiseti, 114
    F. heterosporum, 174
    F. lateritium, 213
    F. merismoides, 2
    F. oxysporum, 254
    F. poae, 76
    F. proliferatum, 247
    F. sambucinum, 135

    F. semitectum, 99
    F. solani, 265
    F. sporotrichioides, 20
    F. sulphureum, 135
    T-2 toxin, 20
    zearalenone, 135
buffaloes
    degnala disease, 99, 100, 114
    F. equiseti, 99, 100, 114
    F. semitectum, 99, 100
cats
    alimentary toxic aleukia (ATA), 13, 14, 20, 21, 74, 76
    fusarenon-X, 21

    F. sporotrichioides, 20
    hemorrhagic syndrome, 15
    T-2 toxin, 14, 21
cattle, 6, 13, 74
    butenolide, 22
    deoxynivalenol, 168
    fescue foot, 17–18
    furanoterpenoids, 266
    F. equiseti, 99, 100, 114
    F. graminearum, 167
    F. javanicum, 265
    F. poae, 77
    F. semitectum, 99, 100, 114
    F. solani, 265, 266

F. sporotrichioides, 21, 22
hemorrhagic syndrome, 15, 22
hyperestrogenism, 167
4-ipomeanol, 266
sore foot disease, 18, 99, 100, 114
T-2 toxin, 21, 22
zearalenol, 168
zearalenone, 167
chickens
  acetyldeoxynivalenol, 144
  androgenic syndrome, 266
  deoxynivalenol, 144, 168, 169
  diacetoxyscirpenol, 23
  diamines, 23
  fusarins, 222
  F. acuminatum, 127, 134
  F. avenaceum, 93
  F. chlamydosporum, 70
  F. culmorum, 144
  F. equiseti, 114, 115
  F. graminearum, 168
  F. heterosporum, 127
  F. lateritium, 213
  F. moniliforme, 221, 222
  F. oxysporum, 254, 255
  F. proliferatum, 221, 247
  F. semitectum, 100, 247
  F. solani, 266
  F. sporotrichioides, 23
  F. tricinctum, 87
  hemorrhagic syndrome, 15, 169
  histamines, 23
  malabsorption syndrome, 23
  monoacetoxyscirpenol, 100
  moniliformin, 93, 222, 254
  T-2 toxin, 22, 23, 77
  tibial dyschondroplasia, 115
  zearalenone, 100, 144, 168, 169
dogs, 75, 77, 78
donkeys, 222, 223
ducklings
  deoxynivalenol, 169
  diacetoxyscirpenol, 24, 136
  fusarenon-X, 24
  F. chlamydosporum, 20
  F. equiseti, 115
  F. graminearum, 169
  F. moniliforme, 223, 255
  F. oxysporum, 255
  F. sambucinum, 135-36
  F. sporotrichioides, 24
  F. subglutinans, 248
  F. verticillioides, 223
  moniliformin, 70, 223, 248
  neosolaniol, 24
  T-2 toxin, 24, 78
  zearalenone, 169
geese
  F. graminearum, 144-45, 169, 170
  F. moniliforme, 223
  F. poae, 78
  F. sporotrichioides, 24
  HT-2 toxin, 24
  neosolaniol, 24
  T-2 tetraol, 24

T-2 toxin, 24
zearalenone, 169, 170
goats, 24
guinea pigs
  diacetoxyscirpenol, 25, 213
  diacetylnivalenol, 25
  feed refusal, 170
  fusarenon-X, 25
  F. equiseti, 115
  F. graminearum, 170
  F. lateritium, 213
  F. poae, 74, 78
  F. sambucinum, 136
  F. semitectum, 100, 101
  F. sporotrichiella var. *sporotrichioides,* 25
  F. sporotrichioides, 6, 13, 25, 55
  F. tricinctum, 87
  nivalenol, 25
  T-2 toxin, 25, 101
hamsters, 170
horses
  fusarenon-X, 20, 25
  F. moniliforme, 223, 224
  F. poae, 74, 78
  F. sporotrichioides, 13, 20, 25
  leukoencephalomalacia (LEM), 223, 224
  T-2 toxin, 20, 25
insects
  anhydrofusarubin, 266
  blowfly (*Calliphora erythrocephala*), 10, 11, 266
  confused flour beetle (*Tribolium confusum*), 170
  diacetoxyscirpenol, 213
  dihydroisocoumarin derivatives, 10
  fusaric acid, 266
  F. acuminatum, 127
  F. equiseti, 115
  F. graminearum, 170
  F. larvarum, 10
  F. nivale, 6
  F. solani, 266
  F. sporotrichioides, 25
  fusarubin, 266
  javanicin, 266
  lesser meal worm (*Alphitobius diaperinus*), 170
  Lucilia sericata, 213
  T-2 toxin, 26
  yellow meal worm (*Tenebrio molitor*), 6, 115, 127
  zearalenone, 170
mice
  butenolide, 101, 115, 171, 213
  deoxynivalenol, 171
  diacetoxyscirpenol, 26, 27, 79, 116, 171, 225, 255
  diacetylnivalenol, 27, 255
  7,8-dihydroxydiacetoxyscirpenol, 255
  fescue foot, 101, 115, 128, 213
  furanoterpenoids, 267
  fusarenon-X, 27, 171
  fusaric acid, 257

fusariocins, 225
F. acuminatum, 128
F. avenaceum, 93, 94
F. culmorum, 145
F. equiseti, 116
F. graminearum, 170-71
F. javanicum, 266
F. larvarum, 10
F. lateritium, 213-14, 255
F. moniliforme, 224, 225, 267
F. oxysporum, 214, 255
F. sambucinum, 136
F. semitectum, 101
F. solani, 266, 267
F. sporotrichiella var. *sporotrichioides,* 26, 27
F. sporotrichioides, 6, 13, 17, 26, 27, 74
Gibberella zeae, 224
HT-2 toxin, 26, 27, 79
7-hydroxydiacetoxyscirpenol, 255
hyperestrogenism, 170, 225
neosolaniol, 26, 27, 79, 116, 171, 255
nivalenol, 27, 171
T-2 toxin, 26, 27, 79, 115, 116, 127, 171, 213, 225
zearalenone, 171, 224, 225, 255
monkeys
  alimentary toxic aleukia (ATA), 13, 14, 27, 28
  diacetoxyscirpenol, 14, 28
  F. sporotrichioides, 14, 28
  hemorrhagic syndrome, 15
  T-2 toxin, 14, 28
pigeons
  deoxynivalenol, 6, 28, 145
  diacetoxyscirpenol, 225, 255-56
  F. culmorum, 145
  F. graminearum, 171-72
  F. moniliforme, 225
  F. nivale, 6
  F. oxysporum f. sp. *carthami,* 255-56
  F. poae, 79
  F. sporotrichiella var. *sporotrichioides,* 28
  F. sporotrichioides, 13, 28
  T-2 toxin, 28, 225, 255-56
  zearalenone, 116, 172, 225
pigs
  deoxynivalenol, 6, 29, 145, 172, 173, 174, 175
  diacetoxyscirpenol, 16, 29, 74
  emetic syndrome, 172, 174, 175
  estrogenic syndrome, 172-75
  feed refusal syndrome, 29, 172, 174, 175
  F. equiseti, 116
  F. graminearum, 172-75
  F. moniliforme 225, 226
  F. nivale, 6
  F. poae, 79
  F. semitectum, 101
  F. sporotrichioides, 13, 28, 29

F. verticillioides, 226
Gibberella zeae, 173
hemorrhagic syndrome, 15, 16, 28, 74
T-2 toxin, 16, 29
zearalenone, 116, 145, 172, 173, 174, 175
rabbits
  butenolide, 30
  diacetoxyscirpenol, 30, 128, 175
  diacetylnivalenol, 30, 117
  fescue foot, 30
  fusarenon-X, 30
  F. acuminatum, 128
  F. anthophilum, 252
  F. avenaceum, 94
  F. chlamydosporum, 71
  F. culmorum, 145, 146
  F. equiseti, 116, 117
  F. graminearum, 175
  F. lateritium, 214
  F. moniliforme, 226
  F. nivale, 6
  F. oxysporum, 256
  F. poae, 74, 79–80
  F. redolens, 256
  F. sambucinum, 136
  F. semitectum, 101
  F. solani, 267, 268
F. sporotrichioides, 13, 21, 29, 30
F. subglutinans, 249
F. tricinctum, 6, 87–88
mycotic keratitis, 268
neosolaniol, 175
nivalenol, 30
T-2 toxin, 29, 30, 80, 175
rats
  acetyldeoxynivalenol, 147, 178
  2-acetyl-quinazolin-4(3H)-one, 147
  butenolide, 31
  Calonectria nivalis, 147
  calonectrin, 147
  culmorin, 147
  15-deacetylcalonectrin, 147
  deoxynivalenol, 176–79
  diacetoxyscirpenol, 16, 31, 32, 82, 137, 228, 256
  diacetylnivalenol, 31
  fusarenon-X, 31, 32
  F. chlamydosporum, 71
  F. culmorum, 146
  F. equiseti, 117
  F. graminearum, 147, 176–79
  F. heterosporum, 134
  F. moniliforme, 226–28
  F. oxysporum, 256
  F. oxysporum f. sp. *carthami,* 256
  F. poae, 74, 75, 76, 80–82
F. sambucinum, 136, 137
F. semitectum, 101, 102
F. sporotrichioides, 7, 13, 31, 32
F. subglutinans, 249
F. verticillioides, 227
HT-2 toxin, 31, 82
moniliformin, 249
monoacetoxyscirpenol, 102
neosolaniol, 31, 82
nitrosamines, 228
nivalenol, 31, 32
T-2 toxin, 16, 31, 32, 82, 102, 228, 256
zearalenol, 102, 137, 146, 227, 256
zearalenone 117, 146, 176, 177, 178, 179, 227, 228
sheep
  F. moniliforme, 228
  F. sporotrichioides, 13, 15, 16, 32
  F. verticillioides, 228
  T-2 toxin, 32
turkeys
  F. graminearum, 179, 180
  F. semitectum, 107, 117, 256
  F. sporotrichioides, 32
  monoacetoxyscirpenol, 102
  T-2 toxin, 33
  zearalenone, 102, 179, 180

Fusarium Species and Their Perfect States

Calonectria
  nivalis, 147
  rigidiuscula, 3
F. acuminatum 122, **127–32**, 138, 142
  strain 7.16, 122, 128, **130**
  strain 7.17, 122, 127, 129, 130, **131**, 135, 138, 142
  strain 7.18, 122, 127, 129, 130, **131**, 132, 134, 142; new trichothecenes, 127, 129, 130, 131
  strain 7.19, 122, 128, 129, 130, **132**
F. anguioides, 98, 177
F. anthophilum, 216, **252**
  strain 10.38, 236, **252**
F. aquaeductuum var. *medium,* 57
F. avenaceum, **92–97**
  strain 5.1, 93, **95–96**
  strain 5.2, 93, 94, **96–97**
  strain 5.3, 93, 94, 96, **97**
  strain 5.4, 94, 96, **97**, 218, 225, 229, 233
  strain 5.5, 96, **97**
F. chlamydosporum, **70–72**, 87
  strain 4.44, 46, 70, 71, **72**
  strain 4.45, 46, **72–73**
  strain 4.46, 46, 70, 71, **73**
F. coeruleum, 265
F. concolor, 98, 117
F. culmorum, 133, 140, **143–55**
  53, 147, 148
  182, 149
ATCC 15620, 149
F-104, 151
F-247, 146, 149, 150
ISS Roma, 146, 151
strain 8.8, 140, 143, 146, **152**
strain 8.9, 140, 144, **152**
strain 8.10, 140, 144, 147, 148, 149, 151, **152**
strain 8.11, 140, 144, 147, 148, 149, 151, **152**
strain 8.12, 140, 145, 150, 151, **152–53**
strain 8.13, 140, 146, 150, 151, **153**
strain 8.14, 140, 148, **153**
strain 8.15, 140, 145, 148, 149, 151, **153–54**
strain 8.16, 140, 148, 149, 150, 151, **154**
strain 8.17, 140, 148, **154**
strain 8.18, 7, 9, 140, 147, 148, **154–55**
F. decemcellulare, **3**, 4
  M-10-1, 3
  M-13-3, 3
F. diversisporum, 98, 117, 118
F. episphaeria sensu Snyd. & Hans., 1
F. equiseti, 18, 93, 96, 98, 99, 100, **111–26**, 145
  1, 120, 123
  2, 120, 123
  3, 120, 123
3D-10, 118, 126
4, 120, 123
5, 120, 123
5D-14, 103
6, 120, 123
6W-9, 126
10D-18, 126
10W-9, 126
11W-8, 126
16, 117, 120, 123
18, 117, 120, 123
18-3, 119, 126
21, 117, 120, 123
22, 117, 120, 123
577, 120
63312, 117
63313, 17
CMI 35100, 118
DE 7255-56-57, 115
DE 7255-59-60, 115
NRRL 3213, 118
strain 7.1, 100, 106, 107, 112, 114, 116, 118, **122–23**
strain 7.2, 113, 115, 117, 119, 120, 122, **123**
strain 7.3, 113, 116, 118, 122, **123–24**
strain 7.4, 113, 116, 118, 122, **124**, 142, 145, 148
strain 7.5, 93, 95, 96, 113, 116, 122, **124**

strain 7.6, 113, 115, 122, **124–25**
strain 7.7, 114, 118, 120, 121, 122, **125**
strain 7.8, 114, 118, 120, 121, 122, **125**
strain 7.9, 116, 119, 120, 122, **125**
strain 7.10, 116, 122, **125–26**
strain 7.11, 117, 118, 122, **126**
strain 7.12, 106, 117, 118, 122, **126**
strain 7.13, 106, 117, 118, 122, **126**
strain 7.14, 106, 117, 118, 122, **126**
strain 7.15, 118, 119, 122, **126**
F. fujikuroi, 229–30
    ETH M82, 229–30
F. fusarioides. See *F. chlamydosporum*
F. graminearum, 3, 4, 6, 18, 30, 34, 40, 44, 47, 109, 121, 123, 133, 140, 141, 145, **155–211**
    8, 172
    9, 179
    126, 183
    DAOM 177406, 183
    DAOM 177407, 183
    DAOM 177408, 183
    DAOM 177409, 183
    DAOM 177408, 183
    DAOM 177409, 183
    DAOM 177410, 183
    F-135, 187
    F-184a, 195
    FG-1, 172
    NRRL 6202, 184
    NRRL 6206, 185
    NRRL 6207, 184, 185
    Ohoita-II, 165, 171, 185, 186, 188
    strain 8.19, 140, 157, 170, 171, 173, 178, 182, 185, 186, 188, 192, **197–198**, 234
    strain 8.20, 140, 146, 149, 157, 173, 176, 188, **198**
    strain 8.21, 140, 157, 169, 173, 178, 179, 180, 188, 190, 196, **198–201**
    strain 8.22, 140, 162, 165, 166, 171, 178, 180, 181, 182, 185, **201–2**
    strain 8.23, 140, 163, 174, 180, 182, 185, 188, 192, **202–3**
    strain 8.24, 140, 163, 182, 192, **203**
    strain 8.25, 140, 163, 183, **203**
    strain 8.26, 140, 163, 183, **203**
    strain 8.27, 140, 163, 183, **204**
    strain 8.28, 140, 166, 186, 193, **204**
    strain 8.29, 3, 4, 140, 166, 171, 184, 186, 187, **204**
    strain 8.30, 140, 145, 150, 166, 171, 190, 193, **204–5**
    strain 8.31, 121, 123, 140, 166, 193, **205**
    strain 8.32, 140, 166, 186, 187, **205**
    strain 8.33, 44, 47, 140, 166, 171, 184, 186, 187, 193, **205**
    strain 8.34, 140, 165, 181, 194, **205–6**
    strain 8.35, 140, 165, 185, 187, **206**
    strain 8.36, 140, 165, 185, 187, 194, **206**
    strain 8.37, 140, 169, 178, 196, **206**
    strain 8.38, 140, 169, 173, 176, 177, 183, 184, 186, 187, 188, 189, 195, **206–7**
    strain 8.39, 140, 145, 146, 149, 151, 169, 176, 177, 186, 188, 189, 195, 196, **207**
    strain 8.40, 140, 145, 148, 149, 184, 186, 187, **207**
    strain 8.41, 30, 40, 47, 140, 175, 187, **208**
    strain 8.42, 140, 177, 178, 183, 186, 188, 189, 196, **208**
    strain 8.43, 140, 177, 178, 183, 186, 188, 189, 196, **208**
    strain 8.44, 141, 178, 179, **208–9**
    strain 8.45, 141, 177, 178, 183, 188, 189, 195, **209**
    strain 8.46, 141, 184, **209**
    strain 8.47, 141, 184, **209**
    strain 8.48, 141, 184, **210**
    strain 8.49, 141, 183, 194, **210**
    strain 8.50, 141, 195, **210**
    strain 8.51, 141, 195, **210–11**
    strain 8.52, 141, 193, **211**
F. heterosporum, 127, 129, 130, **133–34**
F. incarnatum, 101, 102, 104, 105, 109
F. javanicum, 264, 265, 267
F. larvarum, 9, **10–11**
    26, 10
    strain 3.2, 9, **11**
F. lateritium, **212–15**
    1, 215
    2, 215
    17, 213, 214, 215
    25, 213, 214
    26, 213, 214
    28, 213, 214
    844, 213
    845, 213
    1421, 214
    2381, 214
    ATCC 20227, 213, 214
    AUA 553, 213
    BRL 886, 214
    IMI 140879, 213, 214
F. merismoides, **1–2**
    strain 1.1, **2**
F. moniliforme, 20, **216–46**
    7, 227, 234
    13, 222
    45, 226
    57, 224
    161, 221
    917, 231
    989, 226
    995, 226
    1857, 226
    1863, 226
    2269, 226
    A-130, 225, 230
    ATCC 12616, 231
    AUA 573, 221
    AUA 893, 221
    AUA 1093, 221
    CBS 183.29, 231
    CMI IMI 204057, 225, 227, 228, 232, 233
    IMI 58290, 231
    IMI 173211, 221
    ISS Roma, 227, 234
    M-1, 222, 230
    M-702, 224, 227
    N 5, 234
    N 136429, 234
    PE-8-17, 224
    strain 10.1, 218, 222, 232, **235**, 236, **237**
    strain 10.2, 218, 223, 236, **237–38**
    strain 10.3, 218, 223, 236, **238**
    strain 10.4, 218, 220, 223, 226, 227, 228, 232, 236, **238–39**
    strain 10.5, 218, 220, 221, 223, 227, 232, 236, **239**
    strain 10.6, 218, 223, 236, **240**
    strain 10.7, 220, 228, 236, **240**
    strain 10.8, 220, 236, **240–41**, 264, 266, 268
    strain 10.9, 223, 227, 232, 236, **241**
    strain 10.10, 219, 222, 236, **241**
    strain 10.11, 219, 222, 236, **241**
    strain 10.12, 223, 227, 236, **241**
    strain 10.13, 223, 236, **242**
    strain 10.14, 224, 236, **242**
    strain 10.15, 225, 229, 236, **242**
    strain 10.16, 221, 236, **242**
    strain 10.17, 223, 236, **242**
    strain 10.18, 223, 236, **242–43**
    strain 10.19, 224, 236, **243**
    strain 10.20, 224, 236, **243**
    strain 10.21, 224, 236, **243**
    strain 10.22, 224, 236, **243**
    strain 10.23, 222, 231, 236, **243–44**
    strain 10.24, 231, 236, 244
    strain 10.25, 222, 232, 236, **244–45**
    strain 10.26, 232, 236, **245**
    strain 10.27, 229, 230, 231, 236, **245**
    strain 10.28, 229, 230, 231, 236, **245–46**
    strain 10.29, 229, 230, 231, 236, **246**
    strain 10.30, 229, 230, 231, 236, **246**
    strain 10.31, 229, 230, 231, 236, **246**
    strain 10.32, 229, 230, 231, 236, **246**
F. moniliforme var. *anthophilum*, 252
F. moniliforme var. *subglutinans*
    953, 249, 250
    CMI IMI 225231, 249, 250
    IMI 131473, 249, 250
F. nivale, **5–10**, 54
    DE 7247, 6
    DE 7249, 6
    strain 3.1, 6, **8**
F. oxysporum, 214, 215, **253–62**
    55-1, 257
    973, 256
    2131, 256
    2341, 255

INDEX 325

3163, 256
Abashiri-3, 255
AUA 1078, 254
Gaüman 1536/9, 256
IMI 166917, 255, 256, 258
IMI 186539, 255, 256, 257, 258
strain 11.1, 254, **259–60**
strain 11.2, 254, **260**
strain 11.3, 258, **260**
strain 11.4, 255, 257, 259, **260–61**
strain 11.5, 255, 257, 259, 260, **261**
strain 11.6, 214, 215, 255, 259, 260, **261**
strain 11.7, 214, 215, 255, 256, 257, 258, 260, **261–62**
strain 11.8, 258, 260, **262**
strain 11.9, 259, 260, **262**
F. poae, 14, **73–86**, 87
20, 78
60/9, 78, 80, 83
396, 80, 83
792, 80
877, 75, 81
958, 77, 79, 80, 82, 83
1140/26, 81
1172, 78
1179, 75, 81
1238, 75, 77, 80
1536, 80
1671, 78
1677, 78
1862, 81
1953, 81
5253, 81, 82, 83, 84
5627, 14, 81
HPB 071178-11, 76, 82
strain 4.47, 47, 74, 75, 82, 83, 84, **85**
strain 4.48, 46, 83, 84, **85**
strain 4.49, 46, 74, 75, 83, 84, **85**
strain 4.50, 46, 74, 75, 84, **85**
strain 4.51, 46, 74, 75, 76, 82, 83, 84, **86**
strain 4.52, 46, 74, 77, **86**
strain 4.53, 46, 74, 82, **86**
strain 4.54, 46, 83, **86**
F. proliferatum, 216, 221, **247**
999, 247
strain 10.33, 221, 236, **247**
F. redolens, 256
2317, 256
F. rigidiusculum, 3
F. roseum sensu Snyd. & Hans., 92, 93, 94, 95, 98, 100, 104, 109, 111, 117, 120, 121, 127, 133, 136, 143, 170, 172, 179, 183, 184, 186, 189, 190
9, 116, 117
162, 171
DE 7246, 127
FR 22, 149
M-7-2, 95
M-11-1, 93, 94, 95
NRRL 6478, 136
NRRL A-15666, 161, 171
Ohio Isolate C, 170, 175–76

R-13c, 136
S-74-1c, 104, 105, 120, 121, 190
F. sacchari, 248
var. *subglutinans*, 248
var. *elongatum*, 251
F. sambucinum, 43, 133, **134–43**
CBS 185.29, 138
CBS 229.31, 138
CBS 260.53, 138
strain 8.1, 135, 136, 137, 138, **139**, 140
strain 8.2, 135, 136, 137, 140, **141**
strain 8.3, 135, 140, **141**
strain 8.4, 43, 135, 139, 140, **141–42**
strain 8.5, 137, 138, 140, **142**
strain 8.6, 138, 140, **142**
strain 8.7, 137, 138, 140, **142–43**
var. *coeruleum*, 139
F. scirpi, 117, 118, 119, 120, 124, 128
CMI 45490, 118, 119, 120
var. *acuminatum*, 128
var. *compactum*, 131
var. *filiferum*, 124
F. semitectum, 18, **98–110**, 112, 113, 115, 117, 118, 119, 121, 123
1, 103, 104
4, 103, 104
136418, 102
IMI 135410, 101, 102
strain 6.1, 99, 100, **106**, 107, 112, 114, 123
strain 6.2, 99, 100, 106, **107**, 112, 113, 117, 123
strain 6.3, 99, 106, **107**
strain 6.4, 100, 101, 102, 103, 104, 105, 106, **107–9**, 115, 118, 119, 120, 121, 123, 168, 255, 256, 259, 261
strain 6.5, 101, 102, 104, 105, 106, **109–10**
strain 6.6, 101, 106, **110**
strain 6.7, 101, 106, **110**
strain 6.8, 101, 106, **110**
F. solani, **263–72**
112/2, 267
562/2, 267
965, 267
990, 267
1664, 267
5034, 267
AUA 625, 266
C166, 269
CMI 172215, 266
CMI 197459, 266, 269
IMI 144799, 265, 266
IMI 187207, 265, 266
PP96, 269
strain 12.1, 265, **270**, 271
strain 12.2, 265, **270**, 271
strain 12.3, 269, **270–71**
strain 12.4, 267, **271**
strain 12.5, 267, **271**
strain 12.6, 267, **271**
strain 12.7, 267, 271, **272**
F. sporotrichiella sensu Bilai, 20, 21, 24, 27, 28, 74

var. *poae*, 75, 81
var. *poae* f. *osteodystrophica*, 75, 77, 80
var. *sporotrichioides*, 20, 21, 24, 27, 28, 74
F. sporotrichioides, **13–69**, 71, 74, 79, 80, 82, 83, 84, 87, 88, 89, 158
60/10, 29, 40
63, 27, 40
214, 43
347, 24, 29, 40
351, 29, 40
738, 26, 29, 33, 34, 36, 37, 40
921, 14, 21, 22, 29, 36, 37, 40, 42
1182, 29, 40
1183, 30, 40
5328a, 14, 37, 40, 43, 44
strain 4.1, 13, 14, 26, 33, 34, 35, 37, 39, 40, **44**, 45
strain 4.2, 13, 26, 36, **44**, 45, **46**, 74
strain 4.3, 15, 22, 25, 26, 28, 30, 31, 33, 34, 35, 36, 38, 39, 40, 42, 45, **46–50**, 80, 84
strain 4.4, 15, 23, 31, 32, 39, 45, **50**
strain 4.5, 15, 23, 32, 45, **50**
strain 4.6, 15, 23, 30, 32, 40, 43, 45, **50–51**
strain 4.7, 15, 25, 45, **51**
strain 4.8, 5, 6, 7, 9, 17, 18, 21, 26, 30, 33, 39, 40, 45, **51–53**, 117
strain 4.9, 6, 7, 8, 9, 16, 18, 19, 21, 24, 25, 26, 27, 30, 31, 34, 35, 37, 38, 40, 45, **53–57**, 88, 89, 166
strain 4.10, 1, 19, 26, 30, 35, 38, 45, 56, **57**
strain 4.11, 19, 26, 34, 35, 45, **57–58**, 255, 257, 259, 261
strain 4.12, 19, 20, 26, 34, 35, 38, 39, 45, **58–60**, 265, 267, 268, 269, 272
strain 4.13, 20, 41, 43, 45, **60**
strain 4.14, 20, 41, 43, 45, **60**
strain 4.15, 20, 36, 37, 41, 42, 43, 45, **60**
strain 4.16, 20, 36, 37, 41, 43, 45, **60–61**
strain 4.17, 20, 41, 43, 45, **61**
strain 4.18, 20, 37, 41, 43, 45, **61**
strain 4.19, 20, 37, 41, 43, 45, **61**
strain 4.20, 24, 36, 37, 41, 42, 43, 44, 45, **61–62**
strain 4.21, 36, 37, 41, 42, 45, **62**
strain 4.22, 36, 37, 41, 45, **62–63**
strain 4.23, 27, 36, 37, 38, 41, 45, **63**
strain 4.24, 28, 29, 34, 42, 45, **63–64**, 225, 228, 229, 233
strain 4.25, 26, 28, 29, 30, 33, 37, 40, 45, **64–65**, 79, 80, 82, 83, 84, 176
strain 4.26, 30, 40, 42, 43, 45, **65–66**
strain 4.27, 31, 33, 39, 45, **66**
strain 4.28, 31, 39, 45, **66**
strain 4.29, 31, 33, 39, 45, **66–67**
strain 4.30, 31, 39, 45, **67**
strain 4.31, 41, 45, **67**

strain 4.32, 41, 42, 45, **67**, 88, 89
strain 4.33, 41, 42, 45, **67-68**, 88, 89
strain 4.34, 41, 45, **68**, 88, 89
strain 4.35, 41, 43, 44, 45, **68**, 88, 89
strain 4.36, 41, 43, 45, **68**, 88, 89
strain 4.37, 41, 43, 44, 45, **68**, 88, 89
strain 4.38, 41, 43, 44, 45, **68-69**, 88, 89
strain 4.39, 41, 42, 43, 44, 45, **69**
strain 4.40, 41, 45, **69**
strain 4.41, 41, 46, **69**
strain 4.42, 41, 46, **69**
strain 4.43, 41, 46, **69**
var. *chlamydosporum*, 71
var. *tricinctum*, 88
F. *subglutinans*, 216, **248-51**
  CMI IMI 204057, 228, 229, 232, 234, 249
  strain 10.34, 236, **251**
  strain 10.35, 236, 249, **251**
  strain 10.36, 232, 236, 248, **251**
  strain 10.37, 232, 236, 250, **251**
F. *sulphureum*, 127, 129, 130, 133, 134, 135, 137

BBA 11125, 135
BBA 11126, 135
F. *tricinctum* sensu Snyd. & Hans., 12, 34, 42, 44, 87
  15, 32
  161, 43
  FT 2, 42, 44
  FT 3, 42, 44
  FT 12, 42, 44
F. *tricinctum* sensu stricto, 12, **86-91**
  1973-76, 88, 89
  M-320, 88
  strain 4.55, 46, 88, **89**
  strain 4.56, 46, 88, **89**
  strain 4.57, 46, 88, **89-90**
  strain 4.58, 46, 88, 89, **90**
  strain 4.59, 46, 88, 89, **90**
  strain 4.60, 46, 87, 88, 89, **90**
  strain 4.61, 46, 87, 88, 89, **91**
  strain 4.62, 46, 87, 88, 89, **91**
F. *verticillioides*, 217, 220, 221, 223, 226, 227, 232
Gerlachia nivalis. See *F. nivale*
Gibberella, 86, 93, 111, 127, 133, 134, 155, 157, 165, 171, 172, 173, 183, 184, 185, 186, 187, 188, 191, 192, 193, 212, 214, 216, 248
  acuminata, 127
  avenacea, 93
  baccata, 212
  fujikuroi, 216
  gordonii, 133
  intricans, 111, 118, 119
  pulicaris, 134
  roseum 'Graminearum,' 172
  subglutinans, 248
  tricincta, 86
  zeae, 155, 157, 165, 171, 173, 183, 184, 185, 186, 187, 188, 191, 192, 193
Micronectriella nivalis, 147
Monographella nivalis, 5
Nectria
  aurantiicola, 10
  haematococca, 263
Oospora verticillioides. See *F. verticillioides*

Mycotoxicoses

abnormal bone development, 219
akakabi-byo
  butenolide, 166
  deoxynivalenol, 165-67
  diacetoxyscirpenol, 163-64, 165-66
  diacetylnivalenol, 166
  fusarenon-X, 166
  F. *graminearum*, **164-67**
  F. *sporotrichioides*, 6, **18-19**, 166
  neosolaniol, 165-66
  nivalenol, 165-66
  T-2 toxin, 166
  zearalenone, 166-67
alimentary toxic aleukia (ATA)
  diacetoxyscirpenol, 14, 17, 25, 28, 34, 128
  F. *acuminatum*, 128
  F. *avenaceum*, 94
  F. *equiseti*, 115, 116
  F. *lateritium*, 212-13, 214
  F. *moniliforme*, 226
  F. *nivale*, 7
  F. *poae*, **74**, 76-81 passim, 82, 93
  F. *sambucinum*, 136
  F. *semitectum*, 100, 101
  F. *sporotrichioides*, **13-15**, 20-33 passim, 34, 35, 37, 39, 40
  F. *tricinctum*, 87
  monkeys, 13, 14, 27, 28
  poaefusarin, 14, 74
  sporofusarin, 14
  T-2 toxin, 14, 17, 22, 25, 28, 29, 31, 39, 40
androgenic syndrome, 265
bean hulls poisoning
  diacetoxyscirpenol, 19, 113, 118

equine leukoencephalomalacia (LEM), 20, 218
fusarenon-X, 20
F. *avenaceum*, 95
F. *decemcellulare*, 3
F. *equiseti*, **113**, 118
F. *moniliforme*, 218
F. *solani*, 265
F. *sporotrichioides*, **19-20**, 265
horses, 19, 20, 113, 218
HT-2 toxin, 19
neosolaniol, 19
NT-1, 19
NT-2, 19
T-2 toxin, 19, 95
degnala disease
  buffaloes, 17, 99, 100, 112
  butenolide, 112, 117
  cattle, 17, 99, 100, 112
  fescue foot, 17-18, 112, 117
  F. *equiseti*, 18, 99, 100, **112-13**, 117, 120, 122-23
  F. *semitectum*, 18, **99**, 100
  sore foot disease, 18, 99, 100, 112
endemic panmyelotoxicosis. See alimentary toxic aleukia
equine leukoencephalomalacia (LEM)
  bean hulls poisoning, 20, 218
  donkeys, 217-18, 222
  F. *moniliforme*, 20, **217-19**
  horses, 20, 217-18, 223
  moniliformin, 218, 232
  mules, 217
estrogenic syndrome
  cattle, 157, 158
  deoxynivalenol, 158, 159
  F. *culmorum*, **143-44**, 150, 151, 157, 155, 157, 165, 171, 172, 173, 183, 184, 185, 186, 187, 188, 191, 192, 193, 212, 214, 216, 248
  F. *graminearum*, 143, **156-61**, 169, 170, 172-74
  nivalenol, 159
  pigs, 143, 150, 156, 157, 158, 159, 172
  T-2 toxin, 158
  zearalenol, 159, 160
  zearalenone, 143-44, 149-51, 156-61, 169, 188-90
feed refusal and emetic syndromes
  akakabi-byo, 161
  butenolide, 164
  deoxynivalenol, 29, 144, 148, 162, 163, 164
  diacetoxyscirpenol, 29, 163, 164
  F. *culmorum*, **144**, 148
  F. *graminearum*, 144, **161-64**, 172-73
  F. *sporotrichioides*, 19, 29
  nivalenol, 164
  pigs, 29, 144, 145, 161-64, 172-73
  Rd toxin, 162
  T-2 toxin, 29, 163, 164
  vomitoxin, 162
  zearalenone, 163, 164
fescue foot
  butenolide, 17, 33, 117
  cattle, 5, 17, 21-22, 30, 99
  degnala disease, 99, 112-13
  F. *semitectum*, 99, 101
  F. *sporotrichioides*, 5, **17-18**, 21-22, 30
  T-2 toxin, 39, 120
hemorrhagic syndrome
  ATA, 16-17
  cattle, 15, 22
  diacetoxyscirpenol, 17, 74-75

F. poae, 74–75
F. sporotrichioides, 15–17, 28
pigs, 15, 16, 28, 74
T-2 toxin, 17, 22
human esophageal cancer
  4-acetoxyscirpenediol, 135, 137
  diacetoxyscirpenol, 135, 137–38
  F. moniliforme, 220–21
  F. poae, 76, 82
  F. sambucinum, 135, 136
  F. semitectum, 99, 101
  F. subglutinans, 248
  F. sulphureum, 135

monoacetoxyscirpenol, 135
mutagens, 221
nitrosamines, 220–21
triacetoxyscirpenol, 135
hyperestrogenism. See estrogenic syndrome
Kashin-Beck disease. See Urov disease
leukemia, 113–14
moldy corn toxicosis, 143. See also hemorrhagic syndrome
moldy sweet potato toxicosis
  furanoterpenoids, 220, 254, 264, 268
  F. moniliforme, 220, 264

F. oxysporum, 254
F. solani, 220, 254, 264–65
impomeamarone, 254
onyali disease, 70, 71, 245
scabby grain intoxication. See akakabi-byo
septic angina. See alimentary toxic aleukia
tribial dyschondroplasia, 113
Urov disease, 75–76, 77–78, 80–81
vulvo vaginitis. See estrogenic syndrome

Mycotoxins and other chemicals

4-acetamido-2-butenoic acid, 180
4-acetoxyscirpenediol
  F. equiseti, 117
  F. sambucinum, 135, 137
  F. semitectum, 103
3-acetyldeoxynivalenol
  Calonectria nivalis, 147
  feed refusal, 181
  F. acuminatum, 128
  F. culmorum, 147
  F. graminearum, 180–81
8-acetylneosolaniol
  F. chlamydosporum, 70, 71
  F. sambucinum, 137
4- or 15-acetylscirpentriol. See 4-acetoxyscirpenediol
acetyl T-2 toxin, 9, 33, 82
2-acetylquinazolin-4(3H)-one, 147
anguidin. See diacetoxyscirpenol
butenolide
  degnala disease, 115, 117
  fescue foot, 17–18, 21, 33, 115, 117, 214
  F. acuminatum, 128
  F. avenaceum, 94
  F. equiseti, 18, 117
  F. graminearum, 181
  F. lateritium, 214
  F. nivale, 7
  F. poae, 82
  F. sambucinum, 137
  F. semitectum, 101, 103
  F. sporotrichioides, 21–22, 30–31, 33–34, 117
  F. sulphureum, 137
calonectrin, 7, 147
culmorin, 7, 147
cyclic peptide "swelling factor," 128, 136
15-deacetylcalonectrin, 7, 147
7-0-demethylmonocerin, 10, 11
deoxynivalenol
  feed refusal, 29, 144, 162–64, 182–83
  F. acuminatum, 128
  F. culmorum, 144, 148
  F. graminearum, 148, 159, 165–67, 181–84
  F. moniliforme, 228

F. nivale, 6, 7, 9
F. sporotrichioides, 18, 28, 29, 34
Gibberella zeae, 184
pigs, 29, 148, 182, 183
deoxynivalenol diacetate. See diacetyl-deoxynivalenol
deoxynivalenol monoacetate. See 3-acetyldeoxynivalenol
diacetoxyscirpendiol. See 7α-hydroxydiacetoxyscirpenol
diacetoxyscirpenol
  alimentary toxic aleukia (ATA), 14, 17, 28, 34, 82
  bean hulls poisoning, 19, 113, 116, 118
  feed refusal, 163–64
  F. acuminatum, 128–29
  F. avenaceum, 94, 95
  F. culmorum, 148
  F. equiseti, 113, 116, 118, 119
  F. graminearum, 4, 165–66, 184
  F. lateritium, 214
  F. moniliforme, 228–29
  F. moniliforme var. subglutinans, 228
  F. nivale, 6, 9
  F. oxysporum, 256
  F. oxysporum f. sp. carthami, 256–57
  F. poae, 74–75, 79, 82–83
  F. sambucinum, 135, 137–38
  F. semitectum, 103, 104
  F. solani var. coeruleum, 268
  F. sporotrichioides, 16, 19, 26–28, 31, 32, 34
  F. subglutinans, 228, 250
  F. sulphureum, 135, 137, 138
  Gibberella intricans, 118
  hemorrhagic syndrome, 15–17, 74–75
  Mangifera indica, 228
diacetoxyscirpentriol. See 7α,8α-dihydroxydiacetoxyscirpenol
diacetyldeoxynivalenol, 185
diacetylnivalenol
  F. culmorum, 148
  F. equiseti, 118, 119
  F. lateritium, 214, 257
  F. oxysporum, 257
  F. semitectum, 101, 104

F. sporotrichioides, 7, 18, 19, 27, 30, 31, 34, 257
dihydroisocoumarin derivatives, 10, 11
7α,8α-dihydroxydiacetoxyscirpenol, 214, 257
3α,4β-dihydroxy-15-acetoxy-8α-(3-hydroxy-3-methylbutyryloxy)-12,13-epoxytrichothec-9-ene, 129
3,15-dihydroxy-12,13-epoxytrichothec-9-ene-8-one, 185
equisetin, 119
furanoterpenoids
  F. moniliforme, 264, 268
  F. oxysporum, 254, 268
  F. solani, 220, 254, 264–65, 268
fusarenon. See fusarenon-X
fusarenon-X
  akakabi-byo, 18–19
  F. equiseti, 118, 119
  F. graminearum, 185
  F. nivale, 6, 7, 9
  F. oxysporum, 257
  F. sambucinum, 138
  F. semitectum, 103
  F. sporotrichioides, 18–19, 27, 30, 31, 35, 257
  F. sulphureum, 138
  F. proliferatum, 247
  F. solani, 269
  F. subglutinans, 249
  hemorrhagic syndrome, 16, 32
  Mangifera indica, 249
fusarinic acid. See fusaric acid
fusarins, 230
fusariocins, 230
fusarentin 6,7-dimethyl ether, 10, 11
fusarentin 6,8-dimethyl ether, 10, 11
fusarentin 6-methyl ether, 10, 11
fusaric acid
  abnormal bone development, 219, 229, 257
  F. fujikuroi, 229–30
  F. heterosporum, 229
  F. moniliforme, 229–30, 250
  F. moniliforme var. subglutinans, 249
  F. oxysporum, 257–58
F-2. See zearalenone
gibberellins

F. anthophilum, 252
F. fujikuroi, 231, 250
F. moniliforme, **230–31**, 250, 252
F. moniliforme var. *anthophilum*, 252
F. moniliforme var. *subglutinans*, 250
F. proliferatum, 247
F. subglutinans, **250**
HT-2 toxin
  alimentary toxic aleukia (ATA), 35
  F. acuminatum, **129**, 138
  F. graminearum, **186**
  F. nivale, 9
  F. poae, 79, 82, 83
  F. solani, 269
  F. sporotrichioides, 19, 24, 26, 27, 31, **35–36**, 269
7-hydroxydiacetoxyscirpenol, 215, 258
8-hydroxydiacetoxyscirpenol. *See* neosolaniol
impomeamarone. *See* furanoterpenoids
1,4-ipomeadiol. *See* furanoterpenoids
ipomeamaronol. *See* furanoterpenoids
ipomeanine. *See* furanoterpenoids
1-ipomeanol. *See* furanoterpenoids
4-ipomeanol. *See* furanoterpenoids
(+)-mellein, 10, 11
moniliformin
  donkeys, 232
  F. avenaceum, **94**
  F. chlamydosporum, **70**, 71
  F. fusarioides, 70, 71
  F. moniliforme, **231–32**
  F. moniliforme var. *subglutinans*, 250
  F. oxysporum, 258
  F. subglutinans, 232, 248, **250**
  F. verticillioides, 232
  Gibberella fujikuroi, 231
  horses, 232
  leukoenchephalomalacia (LEM), 218, 232
  onyalai disease, 70, 71
monoacetoxyscirpenol
  F. graminearum, 186
  F. nivale, 6, 9
  F. sambucinum, 135, 138
  F. semitectum, 100, 102, **103**, 119
  F. sulphureum, 135, **138**, 139
monoacetylnivalenol. *See* fusarenon-X
monocerin, 10, 11
monodeactylanguidin. *See* 4-acetoxyscirpenediol
neosolaniol
  alimentary toxic aleukia (ATA), 37
  bean hulls poisoning, 19, 94, 116
  F. acuminatum, **129**, 138
  F. avenaceum, 94–95
  F. culmorum, 149
  F. equiseti, 116, 118, **119**

F. graminearum, 4, 149, **186**
F. lateritium, 215, 258
F. oxysporum, 215, 258
F. poae, 79, 82, 83
F. sambucinum, **138**
F. semitectum, 103–4
F. solani, 269
F. sporotrichioides 19, 20, 26, 27, 31, **36–37**, 269
neosolaniolacetate. *See* 8-acetylneosolaniol
neosolaniol monoacetate. *See* 8-acetylneosolaniol
nivalenol
  F. equiseti, 118, **119**
  F. graminearum, 159, **186–87**
  F. nivale, 6, 7, 9
  F. sambucinum, 138
  F. semitectum, 103–4
  F. sporotrichioides, 7, 18–19, 27, 30, 31, 32, **37–38**
  F. sulphureum, 138
nivalenol diacetate. *See* diacetylnivalenol
nivalenol monoacetate. *See* fusarenon-X
NT-1 toxin, 19, 27, 38
NT-2 toxin, 19, 38
pigments, 269
poaefusarin, 14, 74
poaefusariogenin, 14, 74
poin, 83
Rd toxin. *See* deoxynivalenol
scirpentriol, 104, 119
sesquiterpene. *See* culmorin
solaniol. *See* neosolaniol
sporofusarin, 14
sporofusariogenin, 14
steroidal hormones, 269
T-1 toxin. *See* NT-1 toxin
T-2 toxin
  alimentary toxic aleukia (ATA), 14, 17, 28, 31, 39, 40
  bean hulls poisoning, 19, 20, 39, 95, 116
  buffaloes, 120
  cattle, 22, 120
  degnala disease, 115, 120
  feed refusal, 29
  fescue foot, 39, 101, 115, 120
  F. acuminatum, **130**
  F. avenaceum, 94, 95
  F. chlamydosporum, 72
  F. culmorum, 149
  F. equiseti, 115, 116, 119, **120**
  F. fusarioides, 72
  F. graminearum, 4, 30, 40, 149, 158, **187–88**

F. lateritium, 215
F. moniliforme, 232–33
F. nivale, 6, 7, 9
F. oxysporum, 258–59
F. poae, 77, 79, 80, 82, **83–84**
F. sambucinum, 138
F. semitectum, 101, 102, **104**
F. solani, 269
F. sporotrichioides, 7, 15–34 passim, **38–42**, 187, 269
F. subglutinans, 233
F. sulphureum, 138
F. tricinctum, 88
T-2 tetraol
  F. acuminatum, **130**
  F. heterosporum, 130
  F. poae, 82, 84
  F. sporotrichioides, **42**
triacetoxyscirpendiol
  F. equiseti, 118
  F. sambucinum, 135
  F. scirpi, 118, 119, 120
  F. sulphureum, 135
triacetoxyscirpenol, 138
$3\alpha,4\beta,8\alpha$-trihydroxy-15-acetoxy-12,13-epoxytrichothec-9-ene, 129–30
$3\alpha,4\beta,15$-trihydroxy-8α-(3-hydroxy-3-methylbutyryloxy-12,13-epoxytrichothec-9-ene, 130
vomitoxin. *See* deoxynivalenol
zearalenol, 102, **104**, 105, 120, 121, 139, 146, 151, 159, 160, 168, 196
zearalenone
  estrogenic effect, 156–61, 188, 190
  feed refusal, 163–64
  F. avenaceum, **95**
  F. chlamydosporum, 72
  F. culmorum, **149–51**, 194, 196
  F. equiseti, 116, 117, **120–21**
  F. fusarioides, 72
  F. graminearum, 121, 149, 150, 151, 156–61, **188–96**
  F. lateritium, 215, 259
  F. moniliforme, **233–35**
  F. nivale, 8
  F. oxysporum, 215, **259**
  F. poae, 84
  F. sambucinum, **138–39**
  F. semitectum, 100, 102, **105**, 121
  F. solani, **269–70**
  F. sporotrichioides, **42–44**
  F. subglutinans, 234
  F. tricinctum, 88–89
  pigs, 116, 120, 188, 190, 195
  porcine hyperestrogenism, 150, 188